Atmospheric Chemistry

Atmospheric Chemistry

Edited by Mary D'souza

SYRAWOOD
PUBLISHING HOUSE

New York

Published by Syrawood Publishing House,
750 Third Avenue, 9ᵗʰ Floor,
New York, NY 10017, USA
www.syrawoodpublishinghouse.com

Atmospheric Chemistry
Edited by Mary D'souza

International Standard Book Number: 978-1-68286-450-0 (Hardback)

Cataloging-in-publication Data

Atmospheric chemistry / edited by Mary D'souza.
    p. cm.
Includes bibliographical references and index.
ISBN 978-1-68286-450-0
1. Atmospheric chemistry. 2. Meteorology. 3. Air--Pollution--Environmental aspects. I. D'souza, Mary.
QC879.6 .A86 2017
551.511--dc23

Printed in the United States of America.

TABLE OF CONTENTS

PREFACE

This book unfolds the innovative aspects of atmospheric chemistry which will be crucial for the progress of this field in the future. It strives to provide a fair idea about this discipline and to help develop a better understanding of the latest advances within this field. A sub-discipline of atmospheric science, atmospheric chemistry combines principles of environmental chemistry, computational modeling, meteorology, etc to study and analyze the chemical processes within the Earth's atmosphere. This book will serve as a reference to a broad spectrum of readers including climatologists, meteorologists, students, researchers and professionals engaged in the field of atmospheric science at various levels. It will help the readers in keeping pace with the rapid changes in this field.

The researches compiled throughout the book are authentic and of high quality, combining several disciplines and from very diverse regions from around the world. Drawing on the contributions of many researchers from diverse countries, the book's objective is to provide the readers with the latest achievements in the area of research. This book will surely be a source of knowledge to all interested and researching the field.

In the end, I would like to express my deep sense of gratitude to all the authors for meeting the set deadlines in completing and submitting their research chapters. I would also like to thank the publisher for the support offered to us throughout the course of the book. Finally, I extend my sincere thanks to my family for being a constant source of inspiration and encouragement.

Editor

Exposure to Ozone Modulates Human Airway Protease/Antiprotease Balance Contributing to Increased Influenza A Infection

Matthew J. Kesic[1]*, **Megan Meyer**[1,4], **Rebecca Bauer**[1,2], **Ilona Jaspers**[1,2,3,4]

1 Center for Environmental Medicine, Asthma, and Lung Biology, University of North Carolina Chapel Hill, North Carolina, United States of America, 2 Curriculum in Toxicology, University of North Carolina Chapel Hill, North Carolina, United States of America, 3 Department of Pediatrics University of North Carolina Chapel Hill, North Carolina, United States of America, 4 Department of Microbiology and Immunology University of North Carolina Chapel Hill, North Carolina, United States of America

Abstract

Exposure to oxidant air pollution is associated with increased respiratory morbidities and susceptibility to infections. Ozone is a commonly encountered oxidant air pollutant, yet its effects on influenza infections in humans are not known. The greater Mexico City area was the primary site for the spring 2009 influenza A H1N1 pandemic, which also coincided with high levels of environmental ozone. Proteolytic cleavage of the viral membrane protein hemagglutinin (HA) is essential for influenza virus infectivity. Recent studies suggest that HA cleavage might be cell-associated and facilitated by the type II transmembrane serine proteases (TTSPs) human airway trypsin-like protease (HAT) and transmembrane protease, serine 2 (TMPRSS2), whose activities are regulated by antiproteases, such as secretory leukocyte protease inhibitor (SLPI). Based on these observations, we sought to determine how acute exposure to ozone may modulate cellular protease/antiprotease expression and function, and to define their roles in a viral infection. We utilized our *in vitro* model of differentiated human nasal epithelial cells (NECs) to determine the effects of ozone on influenza cleavage, entry, and replication. We show that ozone exposure disrupts the protease/antiprotease balance within the airway liquid. We also determined that functional forms of HAT, TMPRSS2, and SLPI are secreted from human airway epithelium, and acute exposure to ozone inversely alters their expression levels. We also show that addition of antioxidants significantly reduces virus replication through the induction of SLPI. In addition, we determined that ozone-induced cleavage of the viral HA protein is not cell-associated and that secreted endogenous proteases are sufficient to activate HA leading to a significant increase in viral replication. Our data indicate that pre-exposure to ozone disrupts the protease/antiprotease balance found in the human airway, leading to increased influenza susceptibility.

Editor: Andrew Pekosz, Johns Hopkins University - Bloomberg School of Public Health, United States of America

Funding: The project described was in part supported by grant number HL095163 from the National Heart, Lung, and Blood Institute (NHLBI), National Institutes of Health, a grant from the Flight Attendant Medical Research Institute (FAMRI), as well as a grant from the Environmental Protection Agency (CR829522) (all to IJ). This research was also supported by grant number T32 ES007126-26 National Institute of Environmental Health Sciences Curriculum of Toxicology Training Grant. The funders had no role in study design, data collection and analysis, decision to publish, or preparation of the manuscript.

Competing Interests: The authors have declared that no competing interests exist.

* E-mail: kesic@email.unc.edu

Introduction

Influenza A virus is responsible for the seasonal epidemics and reoccurring pandemics, which represents a worldwide threat to global public health [1]. It is a major cause of morbidity and mortality worldwide, as during the recent H1N1 pandemic. In the U.S. alone, over 100,000 individuals are hospitalized and over 20,000 people die every year due to influenza virus infection and related diseases [2,3]. Despite large-scale efforts in vaccination and antiviral therapies, the morbidity and mortality rates associated with influenza infections have not significantly changed in recent years [4,5]. In the context of potentially pandemic respiratory viral infections, it is important to identify molecular targets/pathways for therapeutic intervention to protect susceptible sub-populations. Epidemiologic studies show that exposure to inhaled oxidants such as cigarette smoke, diesel exhaust, and ozone can modulate immune function and increase susceptibility to respiratory viral infections [6,7,8,9,10,11]. Despite the extensive study into different factors influencing susceptibility to infection, the mechanism(s) by which inhaled oxidants modify viral pathogenesis are very complex and have yet to be fully elucidated.

The ability of oxidants to cause lung injury/dysfunction is dependent, in part, on the delicate equilibrium that exists between oxidants and antioxidants. Oxidative stress is caused by an imbalance between the production of reactive oxygen species (ROS) and the body's ability to readily detoxify reactive intermediates. Ozone is one of the most abundant components of air pollution in urban areas, and has been shown to be a potent inducer of oxidative stress causing airway inflammation and increased respiratory morbidities [8,9,12]. The Mexico City Metropolitan Area (MCMA), one of the most densely populated areas in the world, experiences high levels of air pollutants such as environmental ozone and particulate matter (PM) [13,14,15]. The MCMA is located 2240 m above sea level and is surrounded by mountains. Due to this geographical location, there is less available oxygen, making combustion less efficient, which produces more polycyclic aromatic hydrocarbon (PAH) pollutants. For these reasons, Mexico City receives higher levels of environmental ozone and various other types of photochemical smog [13]. It was here that the first influenza pandemic of the 21st century emerged in March 2009. It was responsible for an estimated ~258,698 laboratory confirmed cases and roughly ~1,370 deaths by December 2009 [1,16,17]. The normal "Flu Season" occurs during the colder half of the year in each hemisphere [18]. Interestingly, this outbreak emerged during the dry and warmer months when ozone levels were significantly higher [19,20].

Possible mechanisms by which oxidative stress alters the airway environment leading to broadened cellular tropism and/or susceptibility to viral infections have been proposed. [21,22,23] The relationship between oxidative stress and the cellular protease/antiprotease balance has been considered to be a major contributor in the development of numerous airway pathologies [24]. Early studies demonstrated that the cleavage of influenza HA is essential for viral penetration into host cells and that this mechanism was mediated by cellular trypsin-like serine proteases. These proteases, in turn, are regulated by mucus antiproteases, such as SLPI and α_1-antitrypsin (A1AT) [25,26,27,28,29]. Recent studies have identified specific cellular proteases and antiproteases that may be involved in influenza infection, which include transmembrane protease serine 2 (TMPRSS2), human airway trypsin-like protease (HAT), and secretory leukocyte proteinase inhibitor (SLPI) [30,31,32,33,34]. The expression of these proteases in the lung are necessary for cleavage of the viral HA surface protein, thus allowing viral fusion and entry into the host cell. Studies have shown a correlation between inflammation and oxidative stress which alters expression of these proteases and antiproteases [29,35]. Specifically, HAT has been shown to be released into the airway fluids under inflammatory conditions, particularly in asthmatics [31,36]. In addition, a murine study showed SLPI expression was suppressed and protease expression was increased in Nrf2-deficient mice which led to increased inflammation, further demonstrating a relationship between oxidative stress and protease expression [37]. Oxidative stress derived from cigarette smoke or addition of reactive oxygen intermediates has been shown to decrease antiprotease activity [35,38] and a murine study demonstrated that decreased antiprotease expression increased influenza infectivity [39]. The imbalance of protease/antiprotease expression caused by oxidative stress has been attributed to both an increase in inflammatory cells, which release proteolytic enzymes capable of destroying lung tissue, and a functional deficit of antiproteases due to oxidation of their active site [35,40]. In addition to influenza, studies have shown that regulated proteolysis is required for the spread/propagation of many human viruses, including Human immunodeficiency virus (HIV), Nipah, Ebola, severe acute respiratory syndrome coronavirus (SARS-CoV), and metapneumoviruses [41,42,43,44,45].

Although it has been shown that oxidative stress increases severity of viral infections, the exact mechanism as to how and why this happens and the role of ozone exposure in these responses are not fully understood. While the deleterious health effects of both ozone and influenza have been well documented, few studies have looked at the effects of ozone in the context of an influenza infection. Herein, we used our established cell culture model of differentiated human nasal epithelial cells (NECs) [46], exposed them to 0.4ppm of ozone, and determined the effects on susceptibility to an influenza A infection. We show that there is a delicate balance of proteases and antiproteases present in the human airway surface liquid. Additionally, we found exposure to ozone disrupts this equilibrium by enhancing protease secretion

while inhibiting antiprotease expression leading to increased susceptibility to viral infection. We directly show that ozone increases soluble protease expression and that they are functional for cleaving the viral HA protein and activation of the intact influenza virion.

Materials and Methods

Ethics Statement

Primary human nasal epithelial cells (NECs) were obtained from healthy adult volunteers. This protocol was approved by the University of North Carolina School of Medicine Institutional Review Board for Biomedical Research. In addition, written informed consent was provided by each study participants and/or their legal guardians.

Nasal epithelial cell cultures, cell lines, and growth conditions

Primary human nasal epithelial cells (NECs) were obtained as previously described [6]. Briefly, NECs were obtained from healthy adult volunteers by gently stroking the inferior surface of the turbinate several times with a Rhino-Probe curette (Arlington Scientific, Arlington, TX), which was inserted through a nasoscope. This protocol was approved by the University of North Carolina School of Medicine Institutional Review Board for Biomedical Research. NEC were expanded to passage 2 in bronchial epithelial growth medium (BEGM, Cambrex Bioscience Walkersville, Inc., Walkersville, MD) and then plated on collagen-coated filter supports with a 0.4 μM pore size (Trans-CLR; Costar, Cambridge, MA) and cultured in a 1:1 mixture of bronchial epithelial cell basic medium (BEBM) and DMEM-H with SingleQuot supplements (Cambrex), bovine pituitary extracts (13 mg/ml), bovine serum albumin (BSA, 1.5 μg/ml), and nystatin (20 units). Upon confluency, all-trans retinoic acid was added to the basolateral medium to establish air liquid interface (ALI) culture conditions (removal of the apical medium) to promote differentiation. Mucociliary differentiation was achieved after 18–21 days post-ALI. We obtained the Madin Darby canine kidney (MDCK) cell line from the American Type Culture Collection (ATCC, Manassas, VA). The BEAS-2B cell line was derived by transforming human bronchial cells with an adenovirus 12-simian virus 40 construct [47]. We obtained our BEAS-2B cells from the American Type Culture Collection (ATCC, Manassas, VA). BEAS-2B cells were grown in keratinocyte basal medium (KBM) supplemented with 30 μg/ml bovine pituitary extract, 5 ng/ml human epidermal growth factor, 500 ng/ml hydrocortisone, 0.1 mM ethanolamine, 0.1 mM phosphoethanolamine, and 5 ng/ml insulin.

The primary human NECs were exposed under ALI conditions to filtered air or 0.4 ppm ozone for 4 h in the exposure chambers operated by the U.S. Environmental Protection Agency, Environmental Public Health Division. In noted experiments, 10 mM of a cell-permeable form of reduced glutathione, glutathione-ethylester (GSH-ET) (Sigma-Aldrich St. Louis. Mo) or 1 μM of EGCG (Sigma-Aldrich St. Louis. Mo) was added to the basolateral medium 30 min prior to ozone exposure, similar to our previous studies [6,46].

Virus-like particle (VLP) entry assay

Construction of VLP expression plasmids. To generate the β-lactamase-M1 fusion expression plasmid (pCAGGS-β-lacM1 PR8) the influenza A/Puerto Rico/8/34/Mount Sinai (H1N1) M1 sequence was PCR amplified from the M1 expression vector pDZ-M (which was kindly provided by Dr. Adolfo Garcia-

Sastre, Mount Sinai School of Medicine) and inserted into the pCAGGS vector [48,49,50]. The β-lactamase gene was PCR amplified from pcDNA3.1 and fused N-terminally to M1 within pCAGGS to create a modified β-lactamase-M1 fusion. In the modified β-lactamase, the first 24 amino acids were excluded to remove a secretion signal and His 24 was substituted with Asp to create an optimal Kozak consensus sequence. The pcDNA3.1-β-lactamase construct has been described previously [51]. The influenza A/Puerto Rico/8/34/Mount Sinai (H1N1) HA (pCAGGS-HA) and NA (pDZ-NA) over-expression vectors were generously provided by Dr. Aldolfo Garcia-Sastre and have been described previously [48,49,50,52].

Production of VLPs. To generate influenza A/PR/8/H1N1 β-lactamaseM1VLPs (PR8 β-lacM1 VLPs), 293T cells were co-transfected with 3 μg each of pCAGGS-HA, pDZ-NA, and pCAGGS-β-lacM1 PR8 using FuGENE® HD (Roche Applied Science, Indianapolis, IN) according to manufacturer's instructions. The supernatant containing the VLPs were collect 48 h post-transfection and clarified of floating cell debris by centrifugation at 3,000 rpm for 10 min. The VLPs were concentrated once by low-speed centrifugation through an Amicon Ultra 100 kD centrifuge filter unit (Millipore; Billerica, MA), and the retentates were aliquoted and stored at −80°C.

VLP entry assay. The VLP entry assay was performed as previously described [6] with modifications. Briefly, target cells were exposed to filtered air or 0.4 ppm ozone for 4 h. 24 h post-exposure, VLPs were added to the apical surface in a total volume of 100 μl and incubated at 37°C for 3 h. The cells were washed twice with HBSS to remove unbound virus and infected cells were detected by using GeneBLAzer^{TM} FRET *in vivo* cell-based assay system substrate CCF2-AM according to manufacturer's recommendations (Invitrogen). Samples were lysed in HBSS by freeze-thaw method repeatedly treated for 3 cycles (frozen in liquid nitrogen for 3 min and thawed in a 65°C water bath for 3 min). Viral entry was quantified by using the POLARstar OPTIMA plate reader (BMG LABTECH, Inc.). All experiments were performed in triplicate in three independent experiments.

Infection with Influenza A

We used influenza A/Bangkok/1/79 H3N2 serotype (which was kindly provided by Dr. Melinda Beck, University of North Carolina). We also obtained influenza A/Malaya/302/1954 H1N1 serotype from the American Type Culture Collection (ATCC, Manassas, VA) for Western blot analysis of the cleavage products of the viral HA protein. Virus was propagated in 10-day-old embryonated hen's eggs. The virus was collected in the allantoic fluid and titered by 50% tissue culture infectious dose in Madin-Darby canine kidney cells and hemagglutination as described before [53]. Stock virus was aliquoted and stored at −80°C until use. Unless otherwise indicated, for infection about 5×10^5 cells were infected with approximately 128 hemagglutination units (HAU) of influenza A, which resulted in approximately 10% of the cells being infected with influenza 24 h post-infection. Total RNA, total protein, basolateral supernatants, and apical washes were collected 24 h post-infection.

Immunoblotting

Apical surface liquid was collected at 24 h post-exposure by washing the apical surface of the cells with HBSS and total protein concentrations were determined by Bradford protein assay (Bio-Rad). Cell lysates were prepared at 24 h post-infection in Passive Lysis Buffer (Promega, Madison, WI) with a protease inhibitor mixture (Cocktail Set III; Calbiochem, San Diego, CA) as well as phosphatase inhibitors (0.5 mM NaVO4, 1 mM β-glyceropho-

phate) on ice for 30 min. After centrifugation, total protein concentrations were determined by Bradford protein assay (Bio-Rad). Both the apical supernatants and cellular lysates were subjected to 12% sodium dodecyl sulfate-polyacrylamide gel electrophoresis (SDS-PAGE) and transferred to nitrocellulose (Schleicher & Schuell Biosciences, Keene, NH). Proteins were detected using specific antibodies to HAT, TMPRSS2, and SPLI (1:1,000; Santa Cruz Biotechnology, Santa Cruz, CA) or Influenza A virus hemagglutinin H1 antibody (1:1,000; Abcam, Cambridge, MA). β-actin (1:2,000; US Biological, Swampscott, MA), which was used as a loading control. Antigen-antibody complexes were stained with anti-rabbit or anti-mouse, horseradish peroxidase (HRP)-conjugated antibody (1:2000, Santa Cruz Biotechnology) and detected with SuperSignal West Pico Chemiluminescent Substrate (Pierce, Rockford, IL). Densitometry was performed using a Fujifilm LAS-3000 imager (Fuji Film Global Tokyo, Japan).

Influenza HA cleavage

Apical surface liquid was collected by washing the apical surface of the cells with HBSS and total protein concentrations were determined by Bradford protein assay (Bio-Rad). 50 μg of total protein were immunoprecipated overnight at 4°C with 200 ng of anti-TMPRSS2, anti-HAT, or IgG isotype as a control, followed by 1 h incubation with 50 μl of protein G-agrose beads (Pierce, Rockford, IL). Cleared supernatants were incubated with 50 ul of Influenza A/Malaya/302/1954 H1N1 for 30 min at 32°C. Virus was subjected to 12% sodium dodecyl sulfate-polyacrylamide gel electrophoresis (SDS-PAGE) and transferred to nitrocellulose (Schleicher & Schuell Biosciences, Keene, NH). Proteins were detected using specific antibodies to HAT, TMPRSS2, and SPLI (1:1,000; Santa Cruz Biotechnology, Santa Cruz, CA) or Influenza A virus hemagglutinin H1 antibody (1:1,000; Abcam, Cambridge, MA).

Influenza virus titer

Influenza virus titers in apical washes were determined by 50% tissue-culture infectious dose (TCID50) in Madin Darby canine kidney (MDCK) cells and evaluated using agglutination of red blood cells as an indicator according to a modified protocol described before [54]. Briefly, MDCK cells grown in round-bottom 96-well plates were inoculated with virus-containing samples diluted in serum-free DMEM containing 20 μg/ml trypsin using log10 dilutions. After 3 days incubation, a suspension of human erythrocytes (0.5%) was added to each well. Wells exhibiting hemagglutination were considered positive and virus titers were expressed as log $TCID_{50}$.

Protease activity assay

A modification to the above protocol was performed to test secreted protease activity. In these experiments, NECs were exposed to either 0.4ppm of ozone or air for 4 h. 24 h post-exposure, apical supernatants were collected and protein concentrations were determined by Bradford protein assay. 100 μg of protein within the apical supernatant, which contained the endogenous secreted proteases, was incubated with 80 μl of virus for 30 min at room temperature. MDCK cells grown in round-bottom 96-well plates were inoculated with the above samples diluted in serum-free DMEM or serum-free DMEM containing 20 μg/ml trypsin, as a positive control, using 10-fold dilutions. After 3 days incubation, a suspension of human erythrocytes (0.5%) was added to each well. Wells exhibiting hemagglutination were considered positive and virus titers were expressed as log $TCID_{50}$. 10 mM Phenylmethyl sulfonyl floride (PMSF) (Sigma-

Aldrich St. Louis Mo) was used to inhibit trypsin protease activity in control samples.

Protease inhibition assay

Recombinant human SLPI (rhSLPI) (R&D Systems Minneapolis MN) was added to serum-free DMEM containing 20 μg/ml trypsin and incubated for 60 min at room temperature prior to addition of 80 μl of virus. MDCK cells grown in round-bottom 96-well plates were inoculated with the above samples diluted in serum-free DMEM using 10-fold dilutions. After 3 days incubation, a suspension of human erythrocytes (0.5%) was added to each well. Wells exhibiting hemagglutination were considered positive and virus titers were expressed as log $TCID_{50}$. All experiments were performed in triplicate in three independent experiments.

RT-PCR

Total RNA was extracted using TRizol (Invitrogen) according to the supplier's instruction. First-strand cDNA synthesis and real-time RT-PCR was performed as described previously [46,53] using commercially available primers and probes for HA (inventoried Taqman Gene Expression Assays) purchased from Applied Biosystems (Foster City, CA).

Measurement of Lactate Dehydrogenase (LDH) and Interleukin-6 (IL-6) levels

NECs were exposed to filtered air or 0.4 ppm ozone for 4 h. 24 h post exposure, basolateral supernatants were collected for the measurement of LDH concentration according to the manufacturer's recommendations (Takara Bio Inc. Madison, WI). Released LDH concentration was expressed as optical density. Measurements of IL-6 concentration were determined according to the manufacturer's recommendations (BD Biosciences San Diego, CA). Released IL-6 concentration was expressed as pg/ml.

SLPI trans-activation reporter assay

To measure trans-activation of the SLPI promoter, 1.2×10^5 BEAS-2B were cotransfected with either 0.1 μg of SLPI-luciferase reporter plasmid (which was kindly provided by Dr. Rosalia Simmen, Arkansas Children's Hospital Research Institute) along with 0.02 μg of Thymidine kinase-*Renilla* luciferase by using Fugene 6 (Roche) according to the manufacturer's recommendations. After 24 h, cells were treated with 1 μM EGCG or DMSO. 8 h post-treatment cells were pelleted and lysed in passive lysis buffer (Promega, Madison, WI). The trans-activation activity was measured as luciferase light units as described previously [55]. All transfection experiments were performed in triplicate and normalized for transfection efficiency by using Renilla Luciferase.

Statistical Analysis

Unpaired Student's t-test and One-way ANOVA were used for determination of statistically significant differences. The use of the term significant in text refers to a comparison of values for which $p < 0.05$.

Results

Ozone exposure enhances viral replication

Our group has recently demonstrated that oxidative stress increases susceptibility to influenza virus and that these responses may be mediated via increased viral entry [6,46]. Therefore, we wanted to evaluate how exposure to ozone affects viral entry and replication. We chose to examine influenza infection in human NECs exposed to 0.4ppm ozone for 4 h, based on our group's recent *in vivo* [12,56] and *in vitro* [9] studies. To assess the effects of ozone on NECs, we determined LDH and IL-6 levels in basolateral supernatants as markers of cytotoxicity and inflammation, respectively (Figure 1A and 1B). As expected, exposure to 0.4ppm of ozone for 4 h causes a significant increase in both inflammation and cytotoxicity [57]. Subsequently, 24 h post-exposure, cells were infected with influenza A/Bangkok/1/79 or a mock control. 24 h post-infection total RNA was subjected to real time RT-PCR to quantitate the influenza viral hemagglutinin transcript (HA) (Figure 1C). As depicted, we saw a significant increase in the amount of viral *HA* mRNA produced in the cells previously exposed to ozone as compared to the filtered air control. Similarly, by analyzing the apical washes for influenza viral titers 24 h post-infection, we saw a significant increase in viral titers in ozone exposed cells as compared to the control (Figure 1D). These results demonstrate that pre-exposure to ozone significantly increased viral replication in NECs.

Ozone exposure does not affect cellular antiviral responses

To determine the potential mechanism(s) mediating the enhancement of viral replication following acute ozone exposure, we assessed whether innate antiviral immune response mediators were modulated. Specifically, we analyzed the expression of interferon-β (IFN-β), interferon-α (IFN-α), retinoic acid inducible gene I (RIG-I), and Toll-Like recptor-3 (TLR-3). Human NECs were exposed to ozone and 24 h post-exposure were infected with influenza A. 24 h post-infection total RNA was isolated and RT-PCR was performed to quantitate the amount of each cellular transcript. Ozone alone had no significant effect on mRNA expression in any of the four antiviral genes as compared to the air control (Figure 2A–D). As expected, we found substantial inductions of these four antiviral mediators 24 h post-infection with influenza A, but without significant differences in ozone exposed compared to control cells. These data demonstrate that exposure to ozone does not alter baseline expression of host antiviral immune response genes, and does not interfere with their induction in response to influenza infection.

Antioxidant supplementation suppresses viral replication

Since we have shown that ozone exposure did not abrogate antiviral responses, we wanted to determine if the increase in influenza virus infection is due to the oxidative stress pathway/mechanism induced by ozone exposure. To test this hypothesis, we treated our NECs with either a potent phase-II antioxidant inducer or a direct antioxidant. We have previously shown that oxidative stress increases susceptibility to influenza infection [6]. Previous reports have shown that *Nrf2* gene expression and protein expression can be induced via antioxidant supplementation [58,59,60], more specifically by the addition of the polyphenolic catechin, epigallocatechin-3-gallate (EGCG). This compound has the ability to induce *Nrf2* activation which in turn up-regulates the expression of multiple phase-II antioxidants [6,61,62]. Based on our previous published data [6,46], we determined that 1 μM of EGCG or 10 mM of GSH-ET is sufficient to detect an increase in Nrf2 expression or counteract oxidative stress induced by diesel exhaust exposure in NECs. These concentrations also correlate with levels of EGCG that are achievable *in vivo* [63,64,65]. NECs were treated with either 1 μM EGCG, 10 mM GSH-ET, or DMSO as a vehicle control for 30 min prior to a 4 h 0.4ppm ozone exposure. 24 h post-ozone exposure cells were infected with influenza A virus and total RNA was isolated 24 h post-infection to determine the level of influenza HA transcripts by RT-PCR. As shown in Figure 3A, exposure to ozone significantly increases viral

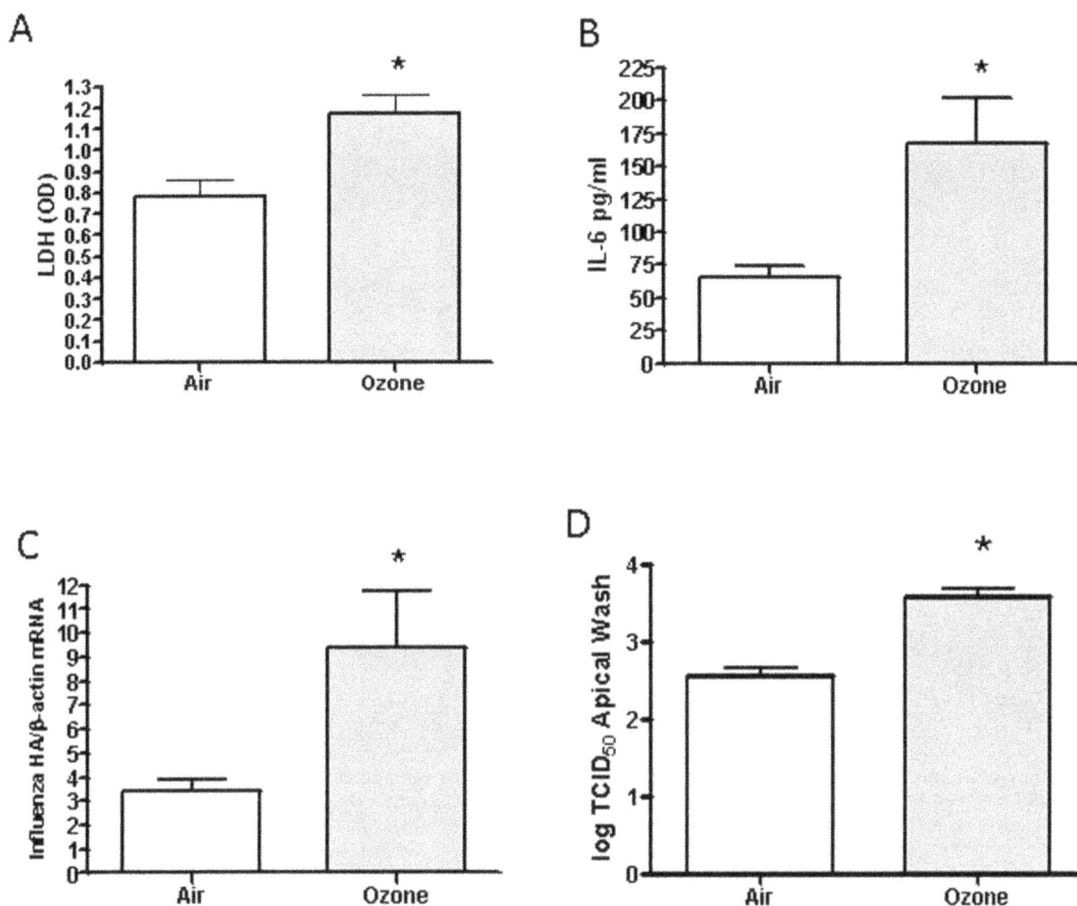

Figure 1. Exposure to ozone enhances influenza infection. Cultures of differentiated human epithelial cells were exposed to either 0.4ppm of ozone or air for 4 h. **A**) Basolateral supernatants collected 24 h post-exposure were analyzed for LDH levels. **B**) Basolateral supernatants collected 24 h post-exposure were analyzed for IL-6 levels. **C**) 24 h post-exposure, cells were infected with influenza A/Bangkok/1/79. Total RNA was isolated and subjected to RT-PCR to quantify Influenza *HA* transcripts and normalized to the expression of *β-actin*. **D**) 24 h post-exposure, cells were infected with influenza A/Bangkok/1/79. Apical supernatants collected 24 h post-infection were analyzed for determination of the viral titer indicated in log $TCID_{50}$. NECs were obtained from six healthy volunteers (n = 6), and each experiment was performed in triplicate. Asterisk indicates statistical significance between test sample and control, * $p < 0.05$.

HA expression. Interestingly, the addition of either EGCG or GSH-ET significantly inhibited viral HA transcription as compared to the vehicle control. We again collected apical washes to perform viral titer assays. Our results, shown on a log scale, indicate that both EGCG and GSH-ET significantly reduced viral replication (Figure 3B). Taken together, these results demonstrate that antioxidants, either through Nrf2 activation (EGCG) or by direct addition of an antioxidant (GSH-ET), counteract the oxidative stress induced by ozone exposure and can suppress influenza virus replication in NECs.

Ozone exposure increases viral entry

Ozone not only triggers intracellular oxidative stress, but has also been shown to directly affect and modify the cellular lipid membrane [66,67] and transmembrane molecules [68]. Since the influenza virion must first bind and enter the target cell prior to replication, we hypothesize that the increased susceptibility to influenza after ozone exposure most likely happens early in the virus lifecycle, upstream of viral replication. To determine which step(s) in the virus life cycle are affected by ozone exposure, we employed an enzymatic virus-like particle (VLP) assay. This assay quantitatively measures the amount of virus that enters the cells.

Similar to our previous work [6], cells were infected with a VLP that only express the hemagglutinin (HA), neuraminidase (NA), and matrix (M) proteins along with a functional β-lactamase reporter fusion (PR8 β-lacM1 VLP) prior to loading with the fluorogenic substrate CCF2-AM . In cells in which the PR8 β-lacM1 VLP has entered, the CCF2-AM substrate will be cleaved, disrupting FRET of the substrate, and resulting in increased CCF2 emission at 447 nm. To determine the effects of ozone exposure on virus entry, NECs and the cell line MDCK as a control were exposed to air or ozone and 24 h post-exposure, cells were infected with PR8 β-lacM1 VLP. In this experiment, MDCK cells are used as a control cell line which does not produce proteases required to activate the viral HA protein. Figure 4 shows that exposure to ozone results in a significant increase in viral entry in the NECs alone. In summary, these data demonstrate that acute exposure to ozone enhances viral entry, consistent with the increased viral replication shown in Figure 1C.

Ozone exposure disrupts host protease/antiproteases balance

As stated earlier, recent reports have identified two endogenous human trypsin-like serine proteases (TMPRSS2 and HAT) that

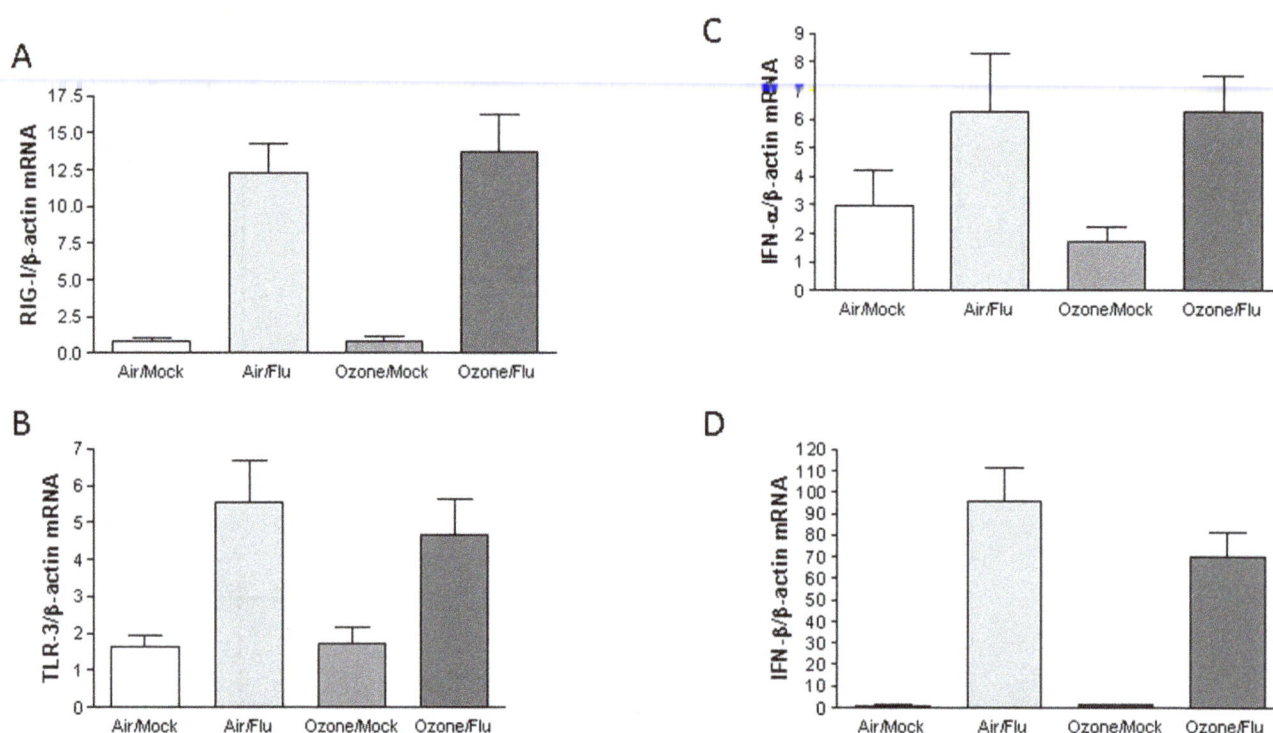

Figure 2. Exposure to ozone does not affect cellular antiviral responses. Cultures of differentiated human epithelial cells were exposed to either 0.4ppm of ozone or air for 4 h. 24 h post-exposure, cells were infected with influenza A/Bangkok/1/79. 24 h post-infection **A)** Total RNA was isolated and subjected to RT-PCR to quantify RIG-I transcripts and normalized to the expression of β-actin. **B)** TLR-3 transcripts and normalized to the expression of β-actin. **C)** IFN-α transcripts and normalized to the expression of β-actin. **D)** IFN-β transcripts and normalized to the expression of β-actin. NECs were obtained from six healthy volunteers (n=6), and each experiment was performed in triplicate. Asterisk indicates statistical significance between test sample and control, * $p < 0.05$.

possess the ability to cleave influenza virus *in vitro* [30,32,33,69]. Viral HA proteolytic cleavage is required for viral fusion and entry into the host cell. Proteases are regulated by antiproteases, such as SLPI and to a lesser extent A1AT. Studies have shown that a disruption of the protease/antiprotease balance is a hallmark of numerous lung diseases and pathologies including COPD, emphysema, asthma, and cancer [40,70,71,72]. Hennet et al, demonstrated a link between oxidative stress and protease expression which led to increased influenza infection in mice [39]. Linking this work with our current hypothesis, we investigated the protease/antiprotease balance on the epithelial surface in the context of an acute ozone exposure. For these experiments we exposed our NECs to either air or ozone for 4 h. 24 h post-exposure, both apical surface liquid and cell lysates were collected to characterize secreted/soluble and membrane-bound intracellular levels of HAT, TMPRSS2, and SLPI. Western blots revealed that intracellular levels of SLPI decreased slightly with ozone exposure, while HAT and TMPRSS2 expression remained relatively unchanged (Figure 5A). Although it is well known that SLPI is secreted and is present in normal human airway surface liquid, only recently was the transmembrane protease, HAT, found to be secreted in an *in vitro* overexpression system [33]. We found the levels of the secreted forms of HAT, TMPRSS2, and SLPI to be significantly changed by ozone exposure (Figure 5B). Under normal conditions, SLPI is in greater abundance than the proteases, yet 24 h post ozone exposure, we found a decrease in SLPI but a significant increase in both HAT and TMPRSS2. It is worth noting that the time at which these samples were taken corresponds to the exact time post ozone exposure when the cells were infected with influenza in the data shown in Figures 1, 2 and

3, thus representing the levels of protease and antiprotease present at the time of infection. To determine if disruption in the protease/ antiprotease equilibrium is due to oxidative stress imposed by the ozone exposure, we treated our NECs with the potent Nrf2 inducer EGCG prior to ozone challenge. Figure 5C shows that EGCG did increase levels of SLPI which corresponds with a decrease in protease expression. Finally, Figure 5D demonstrate that the addition of 1 μM EGCG significantly increases the trans-activation of the SLPI promoter which correlates with the increase in SLPI protein expression displayed in Figure 5C. Taken together, these results indicate that ozone-induced oxidative stress disrupts the protease/antiprotease balance, in favor of protease activity. Furthermore, we demonstrate that the equilibrium can be restored with the induction of Nrf2 via antioxidant supplementa-tion resulting in an increase in the antiprotease SLPI.

Secreted proteases are functional for hemagglutinin cleavage

We characterized the cleavage products of the viral HA protein to determine if the proteases produced by acute ozone exposure can cleave an intact influenza virion. We first investigated where HA cleavage takes place. NECs were exposed to air or ozone and 24 h post-exposure, cells were infected with Influenza A/Malaya/302/1954 H1N1. The cell lysates were analyzed by Western blotting for detection of the HA cleavage products. Figure 6A shows that the majority of the viral HA protein within the cell is cleaved and there is not a significant change with ozone exposure. In addition, Western blot analysis of apical washings incubated with virus showed that HA0 was cleaved and that there was an

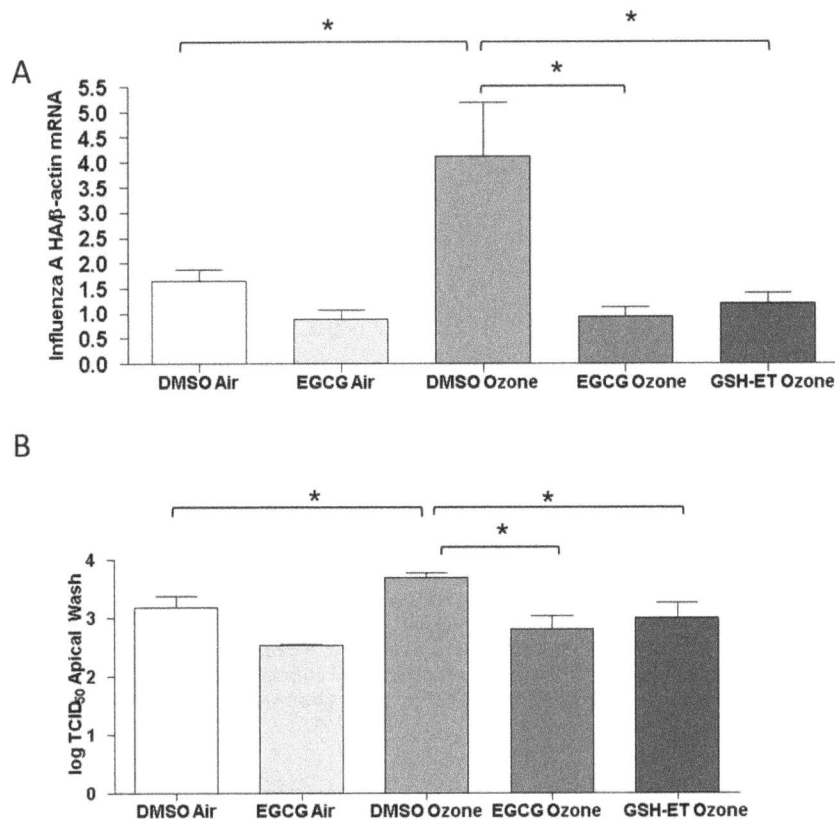

Figure 3. Antioxidant supplementation suppresses viral replication. Cultures of differentiated human epithelial cells were treated with EGCG, GSH-ET, or DMSO as a vehicle control for 30 min before exposure to either 0.4ppm of ozone or air for 4 h. **A)** 24 h post-exposure, cells were infected with influenza A/Bangkok/1/79. Total RNA was isolated and subjected to RT-PCR to quantify Influenza *HA* transcripts and normalized to the expression of *β-actin*. **B)** 24 h post-exposure, cells were infected with influenza A/Bangkok/1/79. Apical supernatants collected 24 h post-infection were analyzed for determination of the viral titer indicated in log $TCID_{50}$. NECs were obtained from 6 healthy volunteers (n = 6), and each experiment was performed in triplicate. Asterisk indicates statistical significance between test sample and control, * $p < 0.05$.

increase in the cleavage of HA from the ozone exposed cells as measured by densitometry. Quantification of the H2 cleavage products demonstrated that ozone exposure leads to enhanced

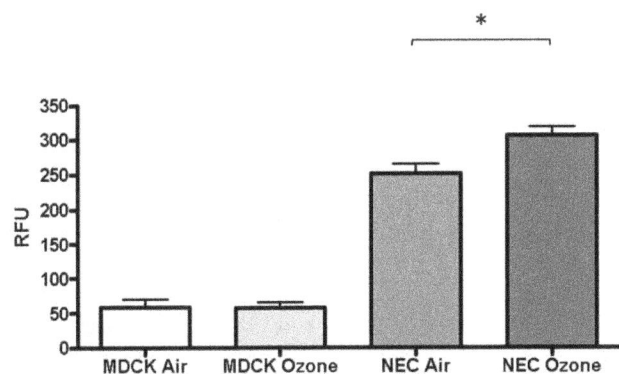

Figure 4. Ozone exposure increases viral entry. Cultures of differentiated human epithelial cells and MDCK cells used as controls, were exposed to either 0.4ppm of ozone or air for 4 h. 24 h post ozone exposure cells were infected with PR8 β-lacM1 VLP. At 4 h post-infection, cells were loaded with CCF2-AM substrate and assayed for viral entry. Data from a representative experiment performed in triplicate. Asterisk indicates statistical significance between test sample and control, * $p < 0.05$.

proteolytic activation of the virion. This data demonstrate that HA cleavage can take place in both the intra- and extracellular environment and that ozone exposure increases the proteolytic activation of the virus in the airway (Figure 6B). To characterize the function of TMPRSS2 and HAT in HA cleavage, we depleted the proteases via immunoprecipitation prior to addition of the virus. Figure 6C shows the apical washes from NECs that were immunoprecipitated with anti-HAT, anti-TMPRSS2, and IgG as a control. The cleared supernatants were incubated with Influenza A/Malaya/302/1954 H1N1. As shown in Figure 6D, there is little difference in the cleavage of the HA protein when the protease are depleted separately, but we demonstrate a significant decrease in the cleavage of HA when both proteases were removed. This data show that the protease(s) found in the apical surface liquid are able to cleave an intact influenza A virion. In addition we show that there are additional proteases secreted into the airway liquid space that are able to activate the virus.

Secreted proteases are functional and can activate Influenza A virions to propagate a viral infection

Although we show that SLPI, HAT, and TMPRSS2 are secreted by NECs, the question remains as to whether they are functional. To test functionality and the role they play in a viral infection, we employed a modified infectivity viral titer assay. These experiments are very similar to the viral titer assays mentioned earlier, but with a few modifications. Similar to the

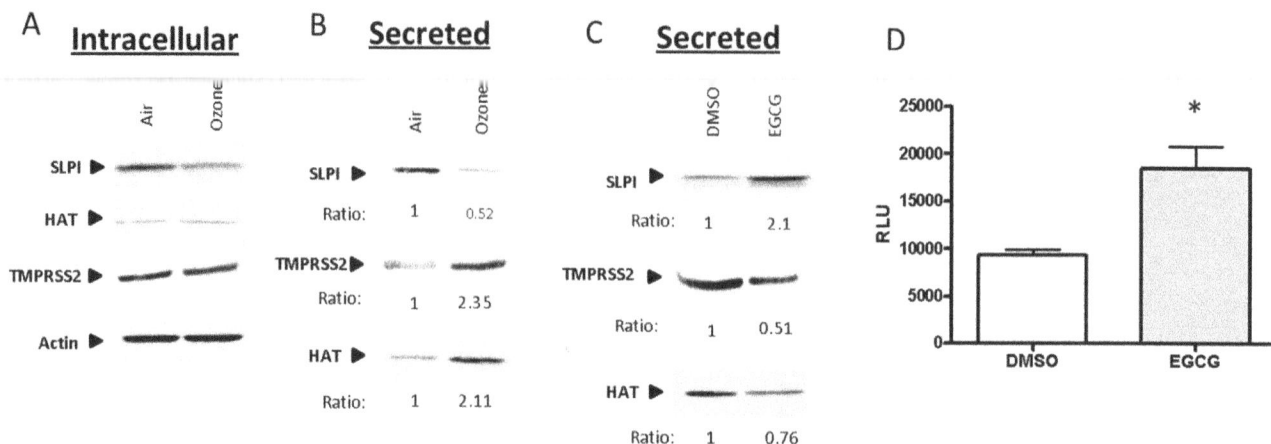

Figure 5. Ozone exposure modulates cellular protease and anti-protease levels. Cultures of differentiated human epithelial cells were exposed to either 0.4ppm of ozone or air for 4 h. 24 h post-exposure, samples were collected and subjected to Western blot analysis to detect SLPI, HAT, TMPRSS2, and β-actin. **A)** Total cellular lysates were analyzed for intracellular SLPI, TMPRSS2, and HAT protein expression by Western blot. Membrane was stripped and analyzed for β-actin as a loading control. **B)** 24 h post-exposure, apical supernatants were analyzed for secreted SLPI, TMPRSS2, and HAT protein levels. **C)** Cultures of differentiated human epithelial cells were treated with EGCG or DMSO as a vehicle control for 30 min before exposure to either 0.4ppm of ozone or air for 4 h. 24 h post-exposure, apical supernatants were analyzed for secreted SLPI, TMPRSS2, and HAT protein levels. **D)** Function activity of the SLPI promoter was determined using a dual-luciferase reporter assay. BEAS-2B cells were co-transfected with SLPI-Luc and TK-Renilla. Cells were treated with EGCG or DMSO as a vehicle control. Cells were harvested 8 h post-treatment and analyzed for luciferase expression. The values represent luciferase production from a representative experiment performed in triplicate. NECs were obtained from four healthy volunteers (n = 4). Densitometry was used to quantitate the amounts of protein, and the numbers below the gel indicate the ozone/air or EGCG/DMSO sample ratio.

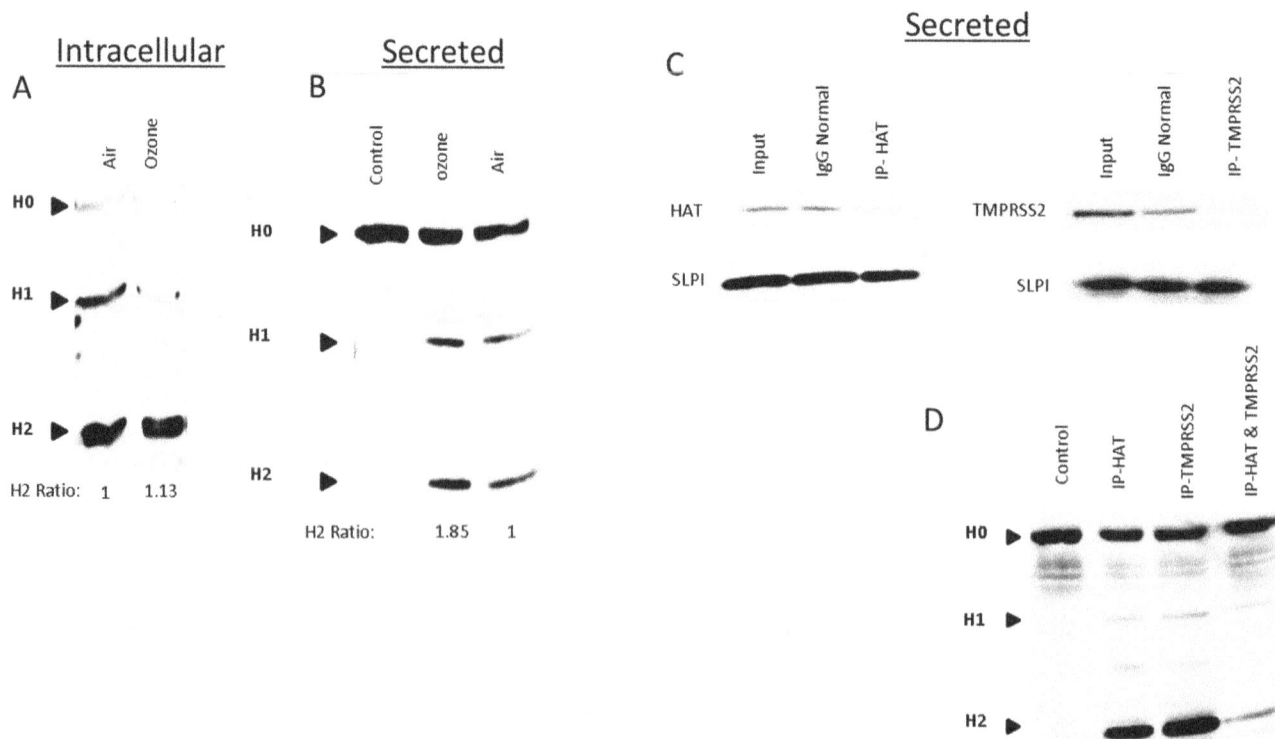

Figure 6. Secreted proteases are functional for Hemagglutinin cleavage. Cultures of differentiated human epithelial cells were exposed to either 0.4ppm of ozone or air for 4 h. A) 24 h post-exposure, cells were infected with influenza A/Malaya/302/1954. 24 h post-infection, total cellular lysates were analyzed for intracellular cleavage of the viral HA protein by Western blot. B) 24 h post-exposure, apical supernatants were collected and incubated with influenza A/Malaya/302/1954. Cleavage of the viral HA protein was analyzed by Western blot. C) Cultures of differentiated human epithelial cells were exposed to either 0.4ppm of ozone or air for 4 h. 24 h post-exposure, apical supernatants were immunoprecipitated (IP) with anti- HAT, anti- TMPRSS2, or an isotype control (IgG). Protein levels were analyzed by Western blot. D) Apical supernatants of differentiated human epithelial cells, were immunoprecipitated (IP) with anti- HAT, anti- TMPRSS2, or an isotype control (IgG) followed by incubation with influenza A/Malaya/302/1954. Cleavage of the viral HA protein was analyzed by Western blot. NECs were obtained from four healthy volunteers (n = 4). Densitometry was used to quantitate the amounts of cleaved H2 protein, and the numbers below the gel indicate the ozone/air ratio.

assays before, we utilized the well-characterized MDCK cell line which does not produce the proper proteases to activate the influenza virion, which can be overcome by the addition of exogenous trypsin. In contrast to the previous titer experiments above, we did not add exogenous trypsin to enable multicycle replication of influenza virus in these cells. These experiments allowed us to test if secreted proteases present in the apical surface liquid from NECs exposed to air or ozone were able to facilitate multiple rounds of viral replication in MDCKs. We show that the secreted proteases present in the apical supernatants from NECs are functional and can propagate viral entry and replication. We also demonstrate that the ozone-exposed NEC supernatants displayed significantly greater infection kinetics than the air control (Figure 7A). This correlates with the increase in protease expression post ozone exposure (Figure 5B), viral entry (Figure 4), and viral replication (Figure 1C and 1D). In addition we show that this is a protease-mediated event, since the addition of a known serine protease inhibitor, PMSF, significantly inhibits viral replication.

To demonstrate that SLPI has the ability to protect airway epithelium from influenza infection by inhibiting proteolytic cleavage, we incubated various amounts of recombinant human SLPI (rhSLPI) to our traditional viral titer assay. Briefly, we incubated our viral titer media, serum-free DMEM containing 20 µg/ml of exogenous trypsin, with varying amounts of rhSLPI prior to addition of influenza. As shown in Figure 7B, addition of rhSLPI significantly inhibits multicycle viral replication. We conclude that the addition of rhSLPI directly inhibits trypsin-mediated cleavage of viral HA leading to decreased infection and replication. Taken together, these data demonstrate that exposure to ozone disrupts the protease/antiprotease balance resulting in greater secretion of endogenous proteases leading to increased viral cleavage/activation culminating in enhanced viral entry and replication.

Discussion

Numerous epidemiological studies link increased ambient ozone levels with respiratory viral infections [1,73,74]. However, the role of ozone exposure in viral replication and pathogenesis remains to be fully defined. Because ambient ozone is a major source of oxidative stress in the airway epithelium and this is the primary site for influenza infections, it is important to determine the role ozone exposure plays in viral infections. Although it has been shown that ozone exposure can modulate many aspects of the immune response [8], how these alterations influence a viral infection have yet to be elucidated. Recent studies have demonstrated the importance of the protease/antiprotease balance in the context of normal lung homeostasis, and have shown that oxidative stress can disrupt this delicate equilibrium [40,75]. It has been well documented that many lung pathologies are dependent on the regulatory interplay between oxidative stress and protease expression [40,70,71]. Although there is speculation that Mexico City's increased influenza-related morbidity and mortality rates were due to exposures to higher levels of ozone, little research has been done looking at the effects of pre-exposure to ozone in the context of an influenza infection in humans. Herein, we demonstrate that acute ozone exposure disrupts the protease/antiprotease balance resulting in increased secreted protease expression in primary human epithelial cells, resulting in increased

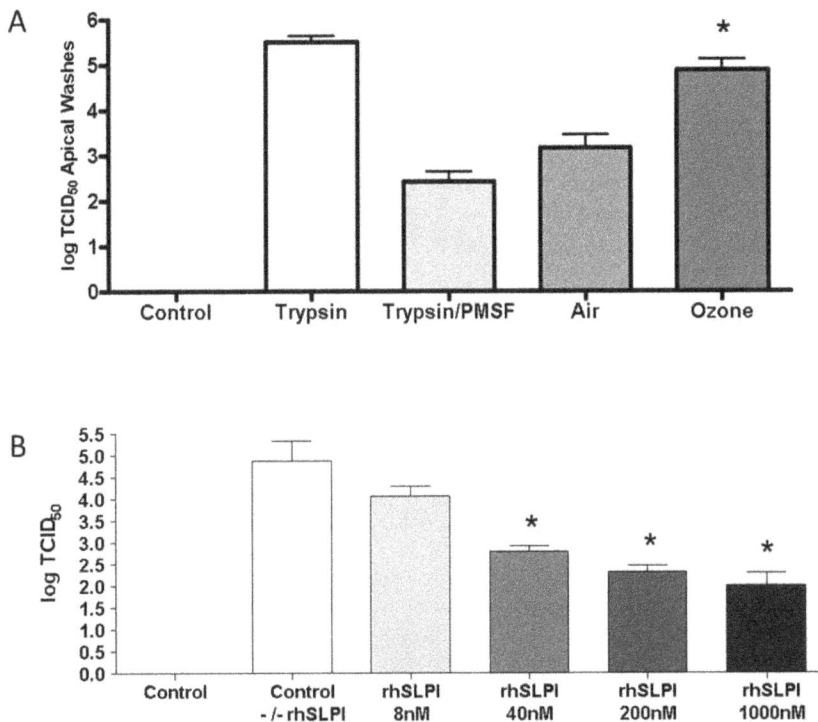

Figure 7. Secreted proteases are functional and can activate virions to propagate a viral infection. A) Cultures of differentiated human epithelial cells were exposed to either 0.4ppm of ozone or air for 4 h. 24 h post-exposure apical supernatants were incubated with influenza A/Bangkok/1/79 for 30 min, and then analyzed for determination of the viral titer indicated in log $TCID_{50}$. **B)** Increasing amounts of rhSLPI (8 nM, 40 nM, 200 nM, 1000 nM) were incubated with media containing 20 µg/ml trypsin for 60 min, then analyzed for determination of the viral titer indicated in log $TCID_{50}$. NECs were obtained from four healthy volunteers (n = 4), and each experiment was performed in triplicate. Data from a representative experiment performed in triplicate. Asterisk indicates statistical significance between test sample and control, * $p < 0.05$.

HA cleavage and activation of the influenza virion, which correlated with enhanced viral entry and replication.

Numerous host cell-dependent factors can affect and control influenza virus attachment and uptake by: (i) proteolytic cleavage of viral HA by host cell-derived serine proteases [30,33], (ii) host cell derived innate immune defense molecules aimed at inhibiting the infectious virions [25,76], and (iii) antiviral mediators limiting viral replication and shedding of virus particles [6,52]. We first focused on potential effects of ozone on host antiviral defense responses. Our data showed that exposure to ozone elicited the typical pro-inflammatory and cellular damage responses as seen with increased release of IL-6 and LDH. We found that the same exposure regimen also resulted in significant increase in viral replication. We initially hypothesized that this could be due to a disruption in one of the cellular antiviral response mechanism/pathways, and assessed the transcription levels of 4 classical antiviral mediators; RIG-I, IFN-α, IFN-β, and TLR-3. Similar to our previous studies using exposure to diesel exhaust [46], we found that ozone did not decrease expression of these mediators or their response when challenged with influenza. In our previous study [46], we found that exposure to an oxidant pollutant increased influenza virus attachment and/or entry, but did not distinguish between these two steps in the viral infection cycle. Therefore, we focused our attention on steps upstream of viral replication to identify mechanisms by which oxidants, such as ozone, affect influenza infectivity.

To dissect specific points in the virus life cycle upstream of viral replication that could determine the role ozone exposure plays in viral susceptibility, which ultimately dictates viral pathogenesis and outcome, we utilized our previously described enzymatic virus-like particle (VLP) assay [6]. Our data demonstrated that ozone exposure significantly increased influenza virus entry. This is consistent with our previous work showing that exposure to other oxidant pollutants and suppression of Nrf2 increases influenza virus entry [6,46]. Similar to these previous studies, data shown here demonstrated that the mechanism(s) through which ozone exposure alters influenza virus entry is mediated by oxidative stress. Since ozone directly interacts with the apical surface of the respiratory epithelium we hypothesize that ozone either directly, or via the induction of biologically active ozone reaction products, modifies either the expression or function of surface proteins, as previously been demonstrated for human surfactant protein A (SP-A) [77]. Among the most prominent proteins on the apical surface potentially regulating infection are the proteases. To date, at least five different proteases have been identified in the airways of animals and humans [78]. Despite many years of effort, the exact protease(s) that activates influenza has yet to be identified. Numerous groups have reported that the family of type II transmembrane serine proteases and trypsin-like proteases are responsible for viral cleavage [26,30,33,79]. Previous studies have examined these proteases in the context of influenza infections, specifically TMPRSS2 and HAT. Although these studies demonstrated that these proteases are capable of cleavage, the experiments were performed in cell lines such as MDCKs [33] or the colorectal adenocarcinoma cell line Caco-2 [32], neither of which are natural target cells for influenza. Since the cellular protease determines the tropism as well as efficiency of viral replication, we utilized fully differentiated human primary nasal epithelial cells (NECs) in this study, a natural host cell for influenza virus. We show that HAT and TMPRSS2 are not only present, but secreted from NECs (Figure 5B), and that ozone exposure enhances the release of the proteases into the apical compartment. This is consistent with reports showing that oxidative stress increases cellular protease activity [38]. Since proteases in the

upper airway are inhibited by SLPI [78], we focused on this antiprotease and its expression in relation to HAT and TMPRSS2. Similar to previous reports, we report that SLPI is detected in the secreted airway and oxidative stress reduces the expression and function of the protein (Figure 5B) [35,80]. These data are consistent with our hypothesis that influenza virions can become activated in the airway surface liquid prior to binding to the epithelial surface. Virion activation in the airway prior to binding would eliminate the need for a membrane-bound protease thus broadening viral tropism from airway epithelium, that contain the necessary proteases, to any mammalian cell, since 2,6-linked sialic acid is expressed by most mammalian cells ranging from lymphoid to glial cells [81,82].This could explain the increased morbidity and mortality associated with the Mexico City 2009 H1N1 pandemic which resulted in increased infection of multiple cell types including distal airway epithelium and immune cells including alveolar macrophages [83,84] as compared to previous seasonal outbreaks.

EGCG has been shown to have potent antioxidant capabilities; in part by inducing the expression of a number of antioxidant enzymes [85]. In vitro and in vivo studies have shown that this supplement induces the expression of phase II antioxidant genes such as GSTM1, which was associated with Nrf2-ARE signaling [58,86]. Whether and how activation of antioxidants expression is involved in the potential antiviral effects of EGCG is not known. Previous studies have indicated that it directly binds influenza virus and therefore prevents attachment and entry into host cells [63,87]. However, these studies were conducted in MDCK cells, which are not a natural host cell for influenza and require addition of exogenous proteases to achieve viral entry [63]. Iizuka et al, demonstrated that Nrf2-knockout mice display protease/antiprotease imbalance leading to increased susceptibility to cigarette smoke-induced emphysema [37]. In addition, the same group showed that Nrf2 expression exerts a protective effect through the transcriptional activation of antiproteases [88]. Our data demonstrate that supplementation of differentiated nasal epithelial cells with EGCG from the basolateral side (to eliminate direct interaction with the virus during infection) significantly increases SLPI production (Figure 5C). It is worthy to note that we show a direct inverse relationship between the levels of secreted SLPI and secreted protease expression, suggesting a potential novel antiviral mechanism by which EGCG could be protective against influenza infections.

We next determined that the secreted proteases are functional. We demonstrate that proteases present in the apical surface liquid were able to activate the virus (Figure 6B) and produce multiple rounds of viral replication (Figure 7A). In addition, apical supernatants from ozone-exposed epithelial cells had significantly higher viral titers, which correlate with the increased protease expression displayed in Figure 5B. In the final experiment we utilized rhSLPI and show a dose dependent protective response (Figure 7B). This demonstrates that that the antiprotease SLPI has a protective affect against viral activation and replication.

In conclusion, this is the first study to demonstrate that secreted proteases from primary human respiratory nasal epithelium proteolytically activate influenza virions and that exposure to a common ambient air oxidant pollutant increases these effects. We speculate that individuals with increased airway oxidative stress, whether due to an underlying medical condition or exposure to high levels of air pollutants, would develop disruption of the epithelial protease/antiprotease balance and become more susceptible to infections with viruses like influenza, SARS-CoV, and metapneumovirus. Disruption in the protease/antiprotease balance has been described in several respiratory diseases including COPD, emphysema, and asthma [70,71,89]. This mechanism could be exploited as a novel anti-viral therapy in combination with conventional antiviral treatments. Antiproteases, such as SLPI appear to protect the respiratory epithelium from influenza virus infection and nutritional supplements may increase the antiprotease:protease ratio. Taken together, the data shown here provide further support to the concept that targeting HA-activating proteases and decreasing oxidative stress to arrest viral infection and spread in conjunction with antiviral agents is a promising potential therapeutic intervention approach against influenza.

References

Acknowledgments

We thank Ms. Luisa E. Brighton for her expert technical assistance and Dr. Philip Bromberg, M.D., for his critical review of the manuscript.

Author Contributions

Conceived and designed the experiments: MK MM IJ. Performed the experiments: MK MM RB. Analyzed the data: MK MM RB IJ. Wrote the paper: MK.

1. Chowell G, Echevarria-Zuno S, Viboud C, Simonsen L, Tamerius J, et al. (2011) Characterizing the epidemiology of the 2009 influenza A/H1N1 pandemic in Mexico. PLoS medicine 8: e1000436.
2. Monto AS (2004) Occurrence of respiratory virus: time, place and person. The Pediatric infectious disease journal 23: S58–64.
3. Thompson WW, Shay DK, Weintraub E, Brammer L, Cox N, et al. (2003) Mortality associated with influenza and respiratory syncytial virus in the United States. JAMA : the journal of the American Medical Association 289: 179–186.
4. Lambert LC, Fauci AS (2010) Influenza vaccines for the future. The New England journal of medicine 363: 2036–2044.
5. Thompson WW, Shay DK, Weintraub E, Brammer L, Bridges CB, et al. (2004) Influenza-associated hospitalizations in the United States. JAMA : the journal of the American Medical Association 292: 1333–1340.
6. Kesic MJ, Simmons SO, Bauer R, Jaspers I (2011) Nrf2 expression modifies influenza A entry and replication in nasal epithelial cells. Free radical biology & medicine 51: 444–453.
7. Cho HY, Imani F, Miller-DeGraff L, Walters D, Melendi GA, et al. (2009) Antiviral activity of Nrf2 in a murine model of respiratory syncytial virus disease. American journal of respiratory and critical care medicine 179: 138–150.
8. Jakab GJ, Spannhake EW, Canning BJ, Kleeberger SR, Gilmour MI (1995) The effects of ozone on immune function. Environmental health perspectives 103 Suppl 2: 77–89.
9. Wu W, Doreswamy V, Diaz-Sanchez D, Samet JM, Kesic M, et al. (2011) GSTM1 modulation of IL-8 expression in human bronchial epithelial cells exposed to ozone. Free radical biology & medicine 51: 522–529.
10. Jakab GJ, Bassett DJ (1990) Influenza virus infection, ozone exposure, and fibrogenesis. The American review of respiratory disease 141: 1307–1315.
11. Razani-Boroujerdi S, Singh SP, Knall C, Hahn FF, Pena-Philippides JC, et al. (2004) Chronic nicotine inhibits inflammation and promotes influenza infection. Cellular immunology 230: 1–9.
12. Hernandez ML, Lay JC, Harris B, Esther CR, Jr., Brickey WJ, et al. (2010) Atopic asthmatic subjects but not atopic subjects without asthma have enhanced inflammatory response to ozone. The Journal of allergy and clinical immunology 126: 537–544 e531.
13. Garcia-Suastegui WA, Huerta-Chagoya A, Carrasco-Colin KL, Pratt MM, John K, et al. (2011) Seasonal variations in the levels of PAH-DNA adducts in young adults living in Mexico City. Mutagenesis 26: 385–391.
14. O'Neill MS, Loomis D, Borja Aburto VH, Gold D, Hertz-Picciotto I, et al. (2004) Do associations between airborne particles and daily mortality in Mexico City differ by measurement method, region, or modeling strategy? Journal of exposure analysis and environmental epidemiology 14: 429–439.
15. Borja-Aburto VH, Castillejos M, Gold DR, Bierzwinski S, Loomis D (1998) Mortality and ambient fine particles in southwest Mexico City, 1993–1995. Environmental health perspectives 106: 849–855.
16. Dawood FS, Jain S, Finelli L, Shaw MW, Lindstrom S, et al. (2009) Emergence of a novel swine-origin influenza A (H1N1) virus in humans. The New England journal of medicine 360: 2605–2615.
17. Fraser C, Donnelly CA, Cauchemez S, Hanage WP, Van Kerkhove MD, et al. (2009) Pandemic potential of a strain of influenza A (H1N1): early findings. Science 324: 1557–1561.
18. Cannell JJ, Zasloff M, Garland CF, Scragg R, Giovannucci E (2008) On the epidemiology of influenza. Virology journal 5: 29.
19. Gold DR, Damokosh AI, Pope CA, III, Dockery DW, McDonnell WF, et al. (1999) Particulate and ozone pollutant effects on the respiratory function of children in southwest Mexico City. Epidemiology 10: 8–16.
20. Torres-Jardon R, Garcia-Reynoso JA, Jazcilevich A, Ruiz-Suarez LG, Keener TC (2009) Assessment of the ozone-nitrogen oxide-volatile organic compound sensitivity of Mexico City through an indicator-based approach: measurements and numerical simulations comparison. Journal of the Air & Waste Management Association 59: 1155–1172.
21. Ciriolo MR, Palamara AT, Incerpi S, Lafavia E, Bue MC, et al. (1997) Loss of GSH, oxidative stress, and decrease of intracellular pH as sequential steps in viral infection. The Journal of biological chemistry 272: 2700–2708.
22. Schwarz KB (1996) Oxidative stress during viral infection: a review. Free radical biology & medicine 21: 641–649.
23. Cho HY, Kleeberger SR (2010) Nrf2 protects against airway disorders. Toxicology and applied pharmacology 244: 43–56.
24. Keller JN (2006) The many nuances of oxidative stress and proteolysis. Antioxidants & redox signaling 8: 119–120.
25. Kido H, Beppu Y, Imamura Y, Chen Y, Murakami M, et al. (1999) The human mucus protease inhibitor and its mutants are novel defensive compounds against infection with influenza A and Sendai viruses. Biopolymers 51: 79–86.
26. Kido H, Yokogoshi Y, Sakai K, Tashiro M, Kishino Y, et al. (1992) Isolation and characterization of a novel trypsin-like protease found in rat bronchiolar epithelial Clara cells. A possible activator of the viral fusion glycoprotein. The Journal of biological chemistry 267: 13573–13579.
27. Sakai K, Kawaguchi Y, Kishino Y, Kido H (1993) Electron immunohisto-chemical localization in rat bronchiolar epithelial cells of tryptase Clara, which determines the pneumotropism and pathogenicity of Sendai virus and influenza virus. The journal of histochemistry and cytochemistry : official journal of the Histochemistry Society 41: 89–93.
28. Ying QL, Simon SR (2000) DNA from bronchial secretions modulates elastase inhibition by alpha(1)-proteinase inhibitor and oxidized secretory leukoprotease inhibitor. American journal of respiratory cell and molecular biology 23: 506–513.
29. Cavarra E, Bartalesi B, Lucattelli M, Fineschi S, Lunghi B, et al. (2001) Effects of cigarette smoke in mice with different levels of alpha(1)-proteinase inhibitor and sensitivity to oxidants. American journal of respiratory and critical care medicine 164: 886–890.
30. Bottcher E, Matrosovich T, Beyerle M, Klenk HD, Garten W, et al. (2006) Proteolytic activation of influenza viruses by serine proteases TMPRSS2 and HAT from human airway epithelium. Journal of virology 80: 9896–9898.
31. Yasuoka S, Ohnishi T, Kawano S, Tsuchihashi S, Ogawara M, et al. (1997) Purification, characterization, and localization of a novel trypsin-like protease found in the human airway. American journal of respiratory cell and molecular biology 16: 300–308.
32. Bertram S, Glowacka I, Blazejewska P, Soilleux E, Allen P, et al. (2010) TMPRSS2 and TMPRSS4 facilitate trypsin-independent spread of influenza virus in Caco-2 cells. Journal of virology 84: 10016–10025.
33. Bottcher-Friebertshauser E, Freuer C, Sielaff F, Schmidt S, Eickmann M, et al. (2010) Cleavage of influenza virus hemagglutinin by airway proteases TMPRSS2 and HAT differs in subcellular localization and susceptibility to protease inhibitors. Journal of virology 84: 5605–5614.
34. Bottcher-Friebertshauser E, Stein DA, Klenk HD, Garten W (2011) Inhibition of influenza virus infection in human airway cell cultures by an antisense peptide-conjugated morpholino oligomer targeting the hemagglutinin-activating protease TMPRSS2. Journal of virology 85: 1554–1562.
35. Cavarra E, Lucattelli M, Gambelli F, Bartalesi B, Fineschi S, et al. (2001) Human SLPI inactivation after cigarette smoke exposure in a new in vivo model of pulmonary oxidative stress. American journal of physiology Lung cellular and molecular physiology 281: L412–417.
36. Szabo R, Wu Q, Dickson RB, Netzel-Arnett S, Antalis TM, et al. (2003) Type II transmembrane serine proteases. Thrombosis and haemostasis 90: 185–193.
37. Iizuka T, Ishii Y, Itoh K, Kiwamoto T, Kimura T, et al. (2005) Nrf2-deficient mice are highly susceptible to cigarette smoke-induced emphysema. Genes to cells : devoted to molecular & cellular mechanisms 10: 1113–1125.
38. Vogelmeier C, Biedermann T, Maier K, Mazur G, Behr J, et al. (1997) Comparative loss of activity of recombinant secretory leukoprotease inhibitor and alpha 1-protease inhibitor caused by different forms of oxidative stress. The European respiratory journal : official journal of the European Society for Clinical Respiratory Physiology 10: 2114–2119.
39. Hennet T, Peterhans E, Stocker R (1992) Alterations in antioxidant defences in lung and liver of mice infected with influenza A virus. The Journal of general virology 73(Pt 1): 39–46.
40. Skrzydlewska E, Farbiszewski R, Gacko M (1997) [The effect of protein oxidation modification on protease-antiprotease balance and intracellular proteolysis]. Postepy higieny i medycyny doswiadczalnej 51: 443–456.
41. Shulla A, Heald-Sargent T, Subramanya G, Zhao J, Perlman S, et al. (2011) A transmembrane serine protease is linked to the severe acute respiratory syndrome coronavirus receptor and activates virus entry. Journal of virology 85: 873–882.

42. Matsuyama S, Nagata N, Shirato K, Kawase M, Takeda M, et al. (2010) Efficient activation of the severe acute respiratory syndrome coronavirus spike protein by the transmembrane protease TMPRSS2. Journal of virology 84: 12658–12664.

43. Chandran K, Sullivan NJ, Felbor U, Whelan SP, Cunningham JM (2005) Endosomal proteolysis of the Ebola virus glycoprotein is necessary for infection. Science 308: 1643–1645.

44. Matsuyama S, Taguchi F (2009) Two-step conformational changes in a coronavirus envelope glycoprotein mediated by receptor binding and proteolysis. Journal of virology 83: 11133–11141.

45. Simmons G, Reeves JD, Rennekamp AJ, Amberg SM, Piefer AJ, et al. (2004) Characterization of severe acute respiratory syndrome-associated coronavirus (SARS-CoV) spike glycoprotein-mediated viral entry. Proceedings of the National Academy of Sciences of the United States of America 101: 4240–4245.

46. Jaspers I, Ciencewicki JM, Zhang W, Brighton LE, Carson JL, et al. (2005) Diesel exhaust enhances influenza virus infections in respiratory epithelial cells. Toxicological sciences : an official journal of the Society of Toxicology 85: 990–1002.

47. Reddel RR, Ke Y, Gerwin BI, McMenamin MG, Lechner JF, et al. (1988) Transformation of human bronchial epithelial cells by infection with SV40 or adenovirus-12 SV40 hybrid virus, or transfection via strontium phosphate coprecipitation with a plasmid containing SV40 early region genes. Cancer research 48: 1904–1909.

48. Niwa H, Yamamura K, Miyazaki J (1991) Efficient selection for high-expression transfectants with a novel eukaryotic vector. Gene 108: 193–199.

49. Quinlivan M, Zamarin D, Garcia-Sastre A, Cullinane A, Chambers T, et al. (2005) Attenuation of equine influenza viruses through truncations of the NS1 protein. Journal of virology 79: 8431–8439.

50. Neumann G, Watanabe T, Ito H, Watanabe S, Goto H, et al. (1999) Generation of influenza A viruses entirely from cloned cDNAs. Proceedings of the National Academy of Sciences of the United States of America 96: 9345–9350.

51. Manicassamy B, Rong L (2009) Expression of Ebolavirus glycoprotein on the target cells enhances viral entry. Virology journal 6: 75.

52. Marsh GA, Hatami R, Palese P (2007) Specific residues of the influenza A virus hemagglutinin viral RNA are important for efficient packaging into budding virions. Journal of virology 81: 9727–9736.

53. Jaspers I, Zhang W, Brighton LE, Carson JL, Styblo M, et al. (2007) Selenium deficiency alters epithelial cell morphology and responses to influenza. Free radical biology & medicine 42: 1826–1837.

54. Farag-Mahmod FI, Wyde PR, Rosborough JP, Six HR (1988) Immunogenicity and efficacy of orally administered inactivated influenza virus vaccine in mice. Vaccine 6: 262–268.

55. Reed KL, Badinga L, Davis DL, Chung TE, Simmen RC (1996) Porcine endometrial glandular epithelial cells in vitro: transcriptional activities of the pregnancy-associated genes encoding antileukoproteinase and uteroferrin. Biology of reproduction 55: 469–477.

56. Hernandez ML, Harris B, Lay JC, Bromberg PA, Diaz-Sanchez D, et al. (2010) Comparative airway inflammatory response of normal volunteers to ozone and lipopolysaccharide challenge. Inhalation toxicology 22: 648–656.

57. Gabrielson EW, Yu XY, Spannhake EW (1994) Comparison of the toxic effects of hydrogen peroxide and ozone on cultured human bronchial epithelial cells. Environmental health perspectives 102: 972–974.

58. Nair S, Barve A, Khor TO, Shen GX, Lin W, et al. (2010) Regulation of Nrf2-and AP-1-mediated gene expression by epigallocatechin-3-gallate and sulforaphane in prostate of Nrf2-knockout or C57BL/6J mice and PC-3 AP-1 human prostate cancer cells. Acta pharmacologica Sinica 31: 1223–1240.

59. Shinkai Y, Sumi D, Fukami I, Ishii T, Kumagai Y (2006) Sulforaphane, an activator of Nrf2, suppresses cellular accumulation of arsenic and its cytotoxicity in primary mouse hepatocytes. FEBS letters 580: 1771–1774.

60. Reddy NM, Kleeberger SR, Cho HY, Yamamoto M, Kensler TW, et al. (2007) Deficiency in Nrf2-GSH signaling impairs type II cell growth and enhances sensitivity to oxidants. American journal of respiratory cell and molecular biology 37: 3–8.

61. Na HK, Surh YJ (2008) Modulation of Nrf2-mediated antioxidant and detoxifying enzyme induction by the green tea polyphenol EGCG. Food and chemical toxicology : an international journal published for the British Industrial Biological Research Association 46: 1271–1278.

62. Zhu H, Jia Z, Zhang L, Yamamoto M, Misra HP, et al. (2008) Antioxidants and phase 2 enzymes in macrophages: regulation by Nrf2 signaling and protection against oxidative and electrophilic stress. Experimental biology and medicine 233: 463–474.

63. Wu CC, Hsu MC, Hsieh CW, Lin JB, Lai PH, et al. (2006) Upregulation of heme oxygenase-1 by Epigallocatechin-3-gallate via the phosphatidylinositol 3-kinase/Akt and ERK pathways. Life sciences 78: 2889–2897.

64. Van Amelsvoort JM, Van Hof KH, Mathot JN, Mulder TP, Wiersma A, et al. (2001) Plasma concentrations of individual tea catechins after a single oral dose in humans. Xenobiotica; the fate of foreign compounds in biological systems 31: 891–901.

65. Ye L, Dinkova-Kostova AT, Wade KL, Zhang Y, Shapiro TA, et al. (2002) Quantitative determination of dithiocarbamates in human plasma, serum, erythrocytes and urine: pharmacokinetics of broccoli sprout isothiocyanates in humans. Clinica chimica acta; international journal of clinical chemistry 316: 43–53.

66. Friedman M, Madden MC, Samet JM, Koren HS (1992) Effects of ozone exposure on lipid metabolism in human alveolar macrophages. Environmental health perspectives 97: 95–101.

67. Pryor WA, Squadrito GL, Friedman M (1995) The cascade mechanism to explain ozone toxicity: the role of lipid ozonation products. Free radical biology & medicine 19: 935–941.

68. Qu F, Qin XQ, Cui YR, Xiang Y, Tan YR, et al. (2009) Ozone stress down-regulates the expression of cystic fibrosis transmembrane conductance regulator in human bronchial epithelial cells. Chemico-biological interactions 179: 219–226.

69. Bertram S, Glowacka I, Muller MA, Lavender H, Gnirss K, et al. (2011) Cleavage and activation of the severe acute respiratory syndrome coronavirus spike protein by human airway trypsin-like protease. Journal of virology 85: 13363–13372.

70. Abboud RT, Vimalanathan S (2008) Pathogenesis of COPD. Part I. The role of protease-antiprotease imbalance in emphysema. The international journal of tuberculosis and lung disease : the official journal of the International Union against Tuberculosis and Lung Disease 12: 361–367.

71. Cox LA, Jr. (2009) A mathematical model of protease-antiprotease homeostasis failure in chronic obstructive pulmonary disease (COPD). Risk analysis : an official publication of the Society for Risk Analysis 29: 576–586.

72. Dillard GH, Chanutin A (1949) The protease and antiprotease of plasmas of patients with cancer and other diseases. Cancer research 9: 665–668.

73. Becker S, Soukup JM, Reed W, Carson J, Devlin RB, et al. (1998) Effect of ozone on susceptibility to respiratory viral infection and virus-induced cytokine secretion. Environmental toxicology and pharmacology 6: 257–265.

74. Purvis MR, Miller S, Ehrlich R (1961) Effect of atmospheric pollutants on susceptibility to respiratory infection. I. Effect of ozone. The Journal of infectious diseases 109: 238–242.

75. Bogadel'nikov IV, Kubyshkin AV, Gavalova NG (1990) Changes in protease inhibitor balance of the bronchoalveolar secretion in children with inflammatory lung diseases. Pediatriia. pp 17–20.

76. Garcia-Sastre A (2011) Induction and evasion of type I interferon responses by influenza viruses. Virus research 162: 12–18.

77. Mikerov AN, Umstead TM, Gan X, Huang W, Guo X, et al. (2008) Impact of ozone exposure on the phagocytic activity of human surfactant protein A (SP-A) and SP-A variants. American journal of physiology Lung cellular and molecular physiology 294: L121–130.

78. Kido H, Okumura Y, Yamada H, Le TQ, Yano M (2007) Proteases essential for human influenza virus entry into cells and their inhibitors as potential therapeutic agents. Current pharmaceutical design 13: 405–414.

79. Hooper JD, Clements JA, Quigley JP, Antalis TM (2001) Type II transmembrane serine proteases. Insights into an emerging class of cell surface proteolytic enzymes. The Journal of biological chemistry 276: 857–860.

80. Fischer BM, Pavlisko E, Voynow JA (2011) Pathogenic triad in COPD: oxidative stress, protease-antiprotease imbalance, and inflammation. International journal of chronic obstructive pulmonary disease 6: 413–421.

81. Liu CK, Wei G, Atwood WJ (1998) Infection of glial cells by the human polyomavirus JC is mediated by an N-linked glycoprotein containing terminal alpha(2–6)-linked sialic acids. Journal of virology 72: 4643–4649.

82. Gagneux P, Cheriyan M, Hurtado-Ziola N, van der Linden EC, Anderson D, et al. (2003) Human-specific regulation of alpha 2–6-linked sialic acids. The Journal of biological chemistry 278: 48245–48250.

83. Mukhopadhyay S, Philip AT, Stoppacher R (2010) Pathologic findings in novel influenza A (H1N1) virus ("Swine Flu") infection: contrasting clinical manifestations and lung pathology in two fatal cases. American journal of clinical pathology 133: 380–387.

84. Mauad T, Hajjar LA, Callegari GD, da Silva LF, Schout D, et al. (2010) Lung pathology in fatal novel human influenza A (H1N1) infection. American journal of respiratory and critical care medicine 181: 72–79.

85. Cho HY, Reddy SP, Debiase A, Yamamoto M, Kleeberger SR (2005) Gene expression profiling of NRF2-mediated protection against oxidative injury. Free radical biology & medicine 38: 325–343.

86. Zhu M, Chen Y, Li RC (2000) Oral absorption and bioavailability of tea catechins. Planta medica 66: 444–447.

87. Nakayama M, Suzuki K, Toda M, Okubo S, Hara Y, et al. (1993) Inhibition of the infectivity of influenza virus by tea polyphenols. Antiviral research 21: 289–299.

88. Ishii Y, Itoh K, Morishima Y, Kimura T, Kiwamoto T, et al. (2005) Transcription factor Nrf2 plays a pivotal role in protection against elastase-induced pulmonary inflammation and emphysema. Journal of immunology 175: 6968–6975.

89. Venkatasamy R, Spina D (2007) Protease inhibitors in respiratory disease: focus on asthma and chronic obstructive pulmonary disease. Expert review of clinical immunology 3: 365–381.

2

UV Irradiance and Albedo at Union Glacier Camp (Antarctica): A Case Study

Raul R. Cordero[1]*, Alessandro Damiani[1], Jorge Ferrer[1], Jose Jorquera[1,2], Mario Tobar[1], Fernando Labbe[3], Jorge Carrasco[4], David Laroze[5]

1 Departamento de Física, Universidad de Santiago de Chile, Santiago, Chile, 2 Escuela Superior Politécnica del Litoral, Guayaquil, Ecuador, 3 Departamento de Ingeniería Mecánica, Universidad Técnica Federico Santa María, Valparaíso, Chile, 4 Dirección Meteorológica de Chile, Santiago, Chile, 5 Instituto de Alta Investigación, Universidad de Tarapacá, Arica, Chile

Abstract

We report on the first spectral measurements of ultraviolet (UV) irradiance and the albedo at a Camp located in the southern Ellsworth Mountains on the broad expanse of Union Glacier (700 m altitude, 79° 46′ S; 82° 52′W); about 1,000 km from the South Pole. The measurements were carried out by using a double monochromator-based spectroradiometer during a campaign (in December 2012) meant to weight up the effect of the local albedo on the UV irradiance. We found that the albedo measured at noon was about 0.95 in the UV and the visible part of the spectrum. This high surface reflectivity led to enhancements in the UV index under cloudless conditions of about 50% in comparison with snow free surfaces. Spectral measurements carried out elsewhere as well as estimates retrieved from the Ozone Monitoring Instrument (OMI) were used for further comparisons.

Editor: Juan A. Añel, University of Oxford, United Kingdom

Funding: Although the support of the Escuela Superior Politécnica del Litoral (Guayaquil, ECUADOR), CONICYT-REDES (Preis 130047), CONICYT-BMBF, FONDEF (IT13I10034), FONDECYT (Preis 1120639, 1140239 and Preis 1120764), Millennium Scientific Initiative (P10-061-F), CEDENNA, UTA-Project 8750-12, USACH-DICYT ASOCIATIVO, and UTFSM-DGIP, is gratefully acknowledged, the funders had no role in study design, data collection and analysis, decision to publish, or preparation of the manuscript.

Competing Interests: The authors have declared that no competing interests exist.

* E-mail: raul.cordero@usach.cl

Introduction

In general, surface UV climate is determined by the total ozone column, cloudiness, ground reflectivity (i.e. the albedo), and local aerosols. Although the latter may significantly modulate the surface UV, in Antarctica the load of aerosols is extremely low [1]. Moreover, while heavily overcast conditions can reduce surface UV irradiance up to 90% in the Antarctic Peninsula [2,3], the role of clouds is less important on the Antarctic plateau. Therefore, surface UV in Antarctica is driven by ozone and by albedo. The UV climatology at several Antarctic sites has been already studied [2–6].

On the Antarctic plateau, large seasonal ozone losses (which occur every year in the late August to early October period [7]) lead to a significant increase in surface UV radiation. Ground-based measurements have shown that the average spring erythemal irradiance for 1990–2006 is up to 85% greater than the modeled irradiance for 1963–1980 [8].

The high albedo of a snow-covered surface has a large effect on the global UV radiation, due to multiple reflections between the ground and the scattering atmosphere [9]. Model calculations of UV irradiance for cloudless sky show enhancements in irradiance levels of nearly 50% at 320 nm for a snow-covered surface, in comparison with snow-free surfaces [10]. Snow albedo depends on snow characteristics (grain size, age of snow, snow height, and soot on the snow) and on other factors such as the solar zenith angle (SZA), atmospheric parameters, geometric pattern of the snow surface, and morphology of the area surrounding the measurement site [11].

So far, the ozone depletion in Antarctica has been the dominant factor in the alterations observed in the UV irradiance. However, as a consequence of climate change, the albedo (that is the other major parameter determining the surface UV) is likely to change in its absolute amount, in its qualitative structure or in its temporal pattern.

The consequences of changes in the Antarctic albedo are beyond local variations in the UV. The surface energy budget of the Antarctic continent is significantly dependent on the surface albedo. Therefore a change in prevailing climatic conditions (triggered for example by a change in temperature) can enhance feedback mechanisms. The so-called albedo feedback is a mechanism that enhances climate change, particularly in regions with snow and ice cover [12]. As the Earth warms, the surface reflects less shortwave radiation to space due to changes in the coverage and reflectivity of the surface ice cover, which leads to additional warming [13]. If temperature increases in Polar Regions we can expect both a decrease of the surface albedo in some areas (due to ice melting) and an enhancement in other areas (due to the freshly fallen snow caused by increased precipitations) [14]. More snow and ice melting, more water vapor and in turn more clouds may lead to changes in the Earth's albedo and alterations in the global energy budget. Indeed, general circulation models (GCMs) [15–17] have shown a high sensitivity to changes in the Earth's albedo [16].

Since albedo significantly affects the UV climatology in Polar Regions (and also plays an important role in the Earth's climate warming pattern), its characterization is required. However, quality-controlled measurements of the albedo in Polar Regions are sparse and few spectral measurements have been reported [9,18–23].

Prior measurements at Amundsen-Scott (South Pole) Station and at Vostok Station found that the albedo was nearly independent of snow grain size and ranged from 0.96 to 0.98 in the UV and the visible part of the spectrum [18]; it has been also observed that the albedo slightly increased with the SZA (which agrees with prior results [24]).

Measurements in Sodankylä (67°22′N, 26°39′E, 179 m altitude) [21], and at Neumayer station (70°39′S, 8°15′W) [9], detected that the albedo decreased (up to 10%) as the SZA decreased through the day. The reductions range from 0.77 to 0.67 in the Arctic, and from 0.96 to 0.86 in the Antarctica [25]. The changes in the Antarctic albedo essentially agree with measurements carried out by using broadband instruments at Hells Gate Station (74°51′S, 163°48′E, 20 m altitude), at Neumayer Station, and at Dome Concordia Station (75°09′S, 123°06′E, 3232 m altitude) [11]. Snow metamorphism, sublimation during the day, and refreezing and/or crystal formation/precipitation during the night can explain the observed trends [11]. The spectral albedo of melting snow has been also measured near Barrow (Alaska) [19], near Davis Station (68°34′S, 77°58′E) [20] and in Sodankylä [22].

In this paper, we report on the first quality-controlled spectral measurements of the UV irradiance and the albedo at Union Glacier Camp, located in the southern Ellsworth Mountains on the broad expanse of Union Glacier; 3,030 km from the southern tip of Chile and about 1,140 km from the South Pole (700 m altitude, 79° 46′ S; 82° 52′W). The mean ice thickness of Union Glacier is 1450 m (with a deep subglacial topography; about 900 m below sea level [23]). The measurements were carried out by using the so-called USACH spectroradiometer from the Universidad de Santiago de Chile (USACH, Chile) during a campaign (in December 2012) meant to weight up the effect of the local albedo on the UV irradiance.

Materials and Methods

1. Ground-based Measurements

Quality-controlled measurements of the surface UV require double monochromator-based spectroradiometers. Instruments developed according to the specifications defined by the World Meteorological Organization (WMO) [26] and the Network for the Detection of Atmospheric Composition Change (NDACC) [27] can produce a radiometric stability better than 1% [28].

The accuracy of NDACC-certified instruments has been tested by intercomparison Campaigns [29,30,31]. The differences between NDACC-certified instruments are normally within the bounds defined by the involved uncertainties; up to 4% for UVA wavelengths and up to 10% for UVB wavelengths [32]. By comparison, for broadband instruments (calibrated by using spectroradiometers) uncertainties in the range 7%–16% have been reported [33,34].

The UV Index (UVI) can be retrieved from spectral measurements by integrating the UV spectra (weighted by the Erythema action spectrum [35]). The uncertainty of UVI values computed from measurements by NDACC-certified instruments has been estimated to be about 5% [36]. Uncertainties of erythemal daily doses (computed also from spectral measurements) are expected to be similar.

Quality-controlled UV time series are limited. In Antarctica, the spectral UV irradiance has been monitored for almost two decades by the U.S. National Science Foundation [8]. Based on those time series, the UV climatology at several Antarctic locations (Palmer Station, McMurdo station and at South Pole station) has been reported [2,4,5].

Spectral measurements reported below were carried out by using the USACH spectroradiometer, which is based on a double monochromator Bentham DMc150F-U, 150 mm focal length, and 1800 lines/mm gratings, fitted with a photomultiplier (PMT) as detector. The Full With at Half Maximum (FWHM) of the USACH spectroradiometer is 2 nm. The system is operated within a temperature-controlled weatherproof box. An input optics for global irradiance fitted with a flat diffuser was used. Although it is not a NDACC-certified instrument, our spectroradiometer complies with NDACC specifications [27] and the WMO recommendations [26]. The USACH spectroradiometer sampled the irradiance every 1 nm (in the range 280–400 nm); scans were carried at a 30 min interval. The absolute calibration of the system was achieved by using a field calibrator fitted with a baffled 100 W quartz halogen lamp. In addition to in situ calibration, quality assurance included corrections for dark signal and for cosine error of the diffuser (see [26] for details). Based on the certificate of the lamp and the transfer of calibrations, we estimated the uncertainty involved in the absolute calibration to be up to 4% for UVA wavelengths and up to 10% for UVB wavelengths (see [32] for details).

2. Satellite Estimates

Satellite-derived estimates of the total ozone column (TOC) data have been compared with ground-based measurements [37,38]. However, under specific conditions (e.g., high satellite SZA, high surface albedo) and/or for Polar Regions (e.g., at high latitudes), a poorer agreement has been reported [39].

Satellite-derived estimates of the surface UV are normally derived from other products (e.g., ozone, albedo, aerosols and cloud cover) by using radiative models and then uncertainties tend to be somewhat higher especially for partly cloudy and overcast conditions [40,41], and over snow-covered surfaces [42,43].

Over polluted areas, validations of OMI-derived data have generally reported that the UV estimates are biased high [44–52]. Among the factors that explain the overestimation have been pointed out the limited spatial resolution [46,51], the lack of sensitivity of OMI to the boundary layer [39,41,44,45,48,52], and the effect of high SZA on the nadir-viewing instruments [39]. However, over snow-covered surfaces the OMI-derived dose is generally lower than the ground-based measurement because the OMI algorithm uses climatological surface albedo that may then be lower than the actual effective surface albedo [42,43]. Part of the problem is that a portion of the observed reflectivity may be incorrectly interpreted as cloud cover, which reduces the estimated irradiance [3].

As shown below, in this study our ground-based measurements were compared with OMI data. The UV estimates were retrieved from OMUVB products (version 1.3). We selected overpass OMI data minimizing the distance between the ground station and the center of the satellite pixel and, due to the presence of the mountains, discarding pixels at altitudes greater than 1000 m asl. The ozone estimates shown below over the period 2004–2012 were retrieved from OMTO3 products.

3. Radiative Transfer Models

Surface UV can also be computed by using radiative transfer models from a set of parameters that include the albedo, the total ozone column, the SZA, as well as the characteristics of aerosols and clouds. Under cloudless conditions and for UVB wavelengths the uncertainties of model products can be and up to 20% for sites with very large aerosol load up, and up to 9% for unpolluted sites [53,54].

The reflectivity of the areas with spatial variation in the albedo [55,56] is often characterized by using the "effective" albedo (the albedo that when used as input into a model reproduces the measured spectrum) [57,58].

Comparisons with ground-based measurements have validated model products [59,60] under cloudless conditions. Although the characterization of the cloud effect is still difficult, models are useful for checking the consistency of surface measurements [2]. In this work, as radiative transfer model we used UVSPEC [61]. The selected radiative transfer solver was the DISORT solver [62]; the extraterrestrial spectrum was quoted from Gueymard [63].

4. Meteorology at Union Glacier Camp

In general, the meteorology at the Union Glacier Camp is defined by the high pressure centered in the interior of the continent, the orography of the region, and the fact that it is located a distance of ~750 km from the edge of Ronne Ice Shelf and the coast line of Ellsworth Land. An automatic Weather Station (AWS) deployed by the Antarctic Logistic & Expeditions (ALE) Company, nearby the Union Glacier runway (79°46 S, 83°16 W) in early February 2010, provides meteorological data to characterize the area. According with the almost 3 years of data, the near-surface air temperature shows a "winter coreless" behavior reaching around −26 to −28°C from April to September and around −6°C in January. The average relative humidity after temperature correction for freezing environment [64] is ~77%; while, the prevailing wind direction year-round is from the South-South west (190–200° magnetic direction) with an average wind speed of ~20 ms^{-1}, revealing the katabatic origin of the wind field [65] affecting the area. On the other hand, the annual cloud cover fraction, according with recent results of the joint CloudSat-CALIPSO satellite data [66] over the southern Ellsworth Mountains region, can range between 30–40% in summer and 60–70% in winter.

Results and Discussion

1. Irradiance Measurements

Despite the fact that no specific permissions are required for the locations/activities reported in our paper, the Chilean Antarctic Institute (INACH, www.inach.cl) issued a permission after our request for measurements that we carried out at an installation in Antarctica (at Union Glacier Camp) under their supervision. Moreover, we confirm that the field studies did not involve endangered or protected species.

Figure 1a shows the spectral measurements carried out on 04.12.2012 (day.month.year) under cloudless conditions. Figure 1b depicts the spectra measured on 11.12.2012, also under cloudless conditions at noon (SZA = 57°) and at midnight (SZA = 77°). Figure 1c shows the UVI computed from our spectral measurements for 4-13.12.2012. The predominant cloudless conditions led to the smooth curves shown in figure 1c.

Figures 1b and 1c also show some satellite estimates. Crosses in figure 1b stand for OMI readings at 305 nm, at 310 nm, at 324 nm and 380 nm. OMI-derived estimates of UVI are also depicted by crosses in figure 1c. OMI measures the solar reflected and backscattered radiation in the wavelength range from 270 nm to 500 nm, that are used to retrieve the TOC, aerosol and cloud cover characteristics, surface UV irradiance and gas traces. The

ground pixel size at nadir position is 13×24 km (along × across track) for total ozone column.

Despite the different spectral resolution (OMI resolution is 0.55 nm in the ultraviolet and 0.63 nm in the visible while the FWHM of our instrument was 2 nm), figure 1b shows a reasonable good agreement between our measurements and the OMI readings at 305 nm, at 310 nm, at 324 nm and 380 nm. Without correcting the effect of the resolution, the differences were within the range ±10% at 305 nm, at 324 nm and 380 nm; differences up to 30% were found at 310 nm.

In figure 1c, it can be observed that the UVI peaked at 3.8 on 13.12.2012; this value is about 10% lower than the corresponding UVI estimate retrieved from OMI. Also on 13.12.2012, the difference between the Erythemal Daily Dose (EDD) computed from our ground-based measurements (4113 J/m2), and the corresponding OMI-derived estimate (4448 J/m2), was about 8%. The OMI estimates of the EDD during the Campaign ranged from 4000 J/m2 on 11.12.2012 to 4448 J/m2 on 13.12.2012. When comparing ground-based estimates and satellite-derived data, we found during the campaign differences that ranged from 2% to 10% (in the case of the UVI) and from 0.1% to 8% (in the case of the EDD). These differences were anyway always within the expanded uncertainty bounds of our measurements [36].

Figure 2a shows the time series of the OMI-derived estimates of UVI at noon at Union Glacier Camp. Due to the high SZA, OMI readings are not available for the period May–August. As shown in figure 2a, annually in spring, the noon-time UVI is typically lower than 6 (although UVI estimates higher than 8 have been retrieved due to the seasonal ozone depletion). As shown in figure 2a, the average of UVI estimates computed for November 2008 was about 20% greater than the average for the same month over the 4 preceding years.

Figure 2b shows the ozone estimates over the period 2004–2012 retrieved from OMTO3 products. Large seasonal ozone losses are apparent in figure 2b and values close to 100 DU have been recorded every year in October [67]. By comparing figures 2a and 2b, becomes clear that the seasonal ozone depletion leads to significant increments in the UVI and in its variability. In this work, the monthly average and the corresponding variability were taken as average and the standard deviation, respectively, of the daily estimates. Table 1 shows the monthly average and the corresponding variability of OMI products. The average of the daily ozone estimates for September (158 DU) is typically about 40% lower than for January (275 DU). The variability of ozone data ranges between 5% in January and 26% in November (see Table 1). In the case of UV products, the variability of the "clear-sky UVI" computed for October (34%) is roughly 3 times greater than for January (11%).

We compared the ozone estimates retrieved from OMI, with the ozone computed from our measurements of the UV spectra carried out in December 2012. The ozone was retrieved from our measurements of the global irradiance by applying a method that implied comparing the ratio (between irradiances measured at different wavelengths) with a synthetic chart of this ratio computed for a variety of ozone values [68]. Figure 3 shows the total ozone column progression for 4-13.12.2012 computed from our ground-based measurements. As shown in the figure, our daily measurements agree with OMI-derived estimates of ozone (see crosses in Figure 3). Although the detected differences were slight (ranging from 0.1% to 3% (i.e. up to about 10 DU)). due to the high sensitivity of the UVI to the ozone, they can explain most of the differences in the UVI estimates shown in figure 1c. Estimations of the total ozone column retrieved from spectral measurements at

A

B

C

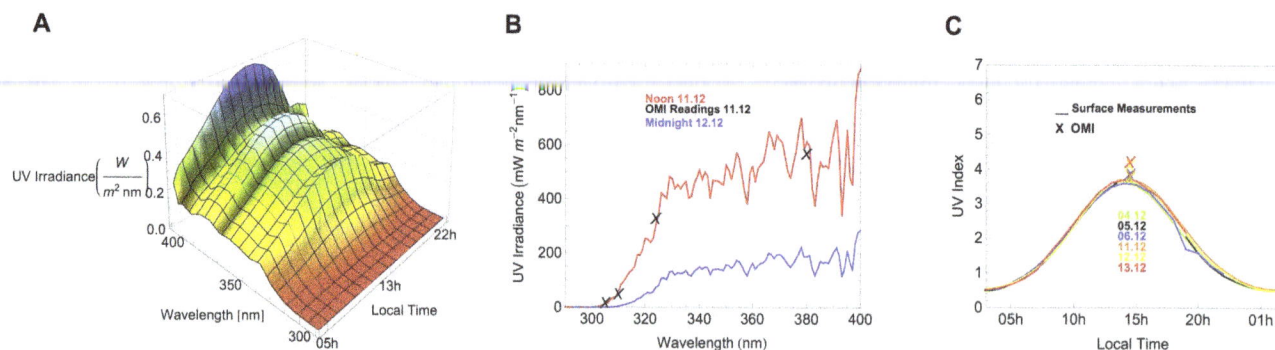

Figure 1. UV Irradiance at Union Glacier Camp. a) Ground-based measurements of the UV irradiance on 04.12.2012 (cloudless conditions). b) Spectra measured at noon on 11.12.2012 (cloudless conditions, SZA = 57°, red line) and at midnight on 12.12.2012 (cloudless conditions, SZA = 77°, blue line). Crosses stand for OMI readings at 305 nm, at 310 nm, at 324 nm and at 380 nm. c) Lines stand for UVI computed from ground-based spectral measurements on 04.12.2012 (green line), on 05.12.2012 (black line), on 06.12.2012 (blue line), on 11.12.2012 (orange line), on 12.12.12 (yellow line), on 13.12.2012 (red line). Crosses stand for OMI estimates of UVI.

other Antarctic locations have been also found to agree with TOMS data [2,5,69].

We further exploited our ground-based measurements studying the possibility of an unlikely aerosol modulation. The sky in the Antarctic interior is breathtakingly clear: there is no source of dust or other particulate matter but no measurements of the properties of aerosol loading at Union glacier Camp have been reported. We carried a limited number of spectral measurements of both the

global and the diffuse irradiance (by using a shadow ring) at Union Glacier Camp around noon on 09.12.2012. Both, the AOD and the single scattering albedo (SSA) were retrieved from these quality-controlled spectral measurements by applying methods based on the comparison of the measured spectral irradiance with UV spectra computed by using a radiative transfer model [70–72]. According to these methods, the retrieved values of the AOD and SSA are those leading to the best match between the measured

A

B

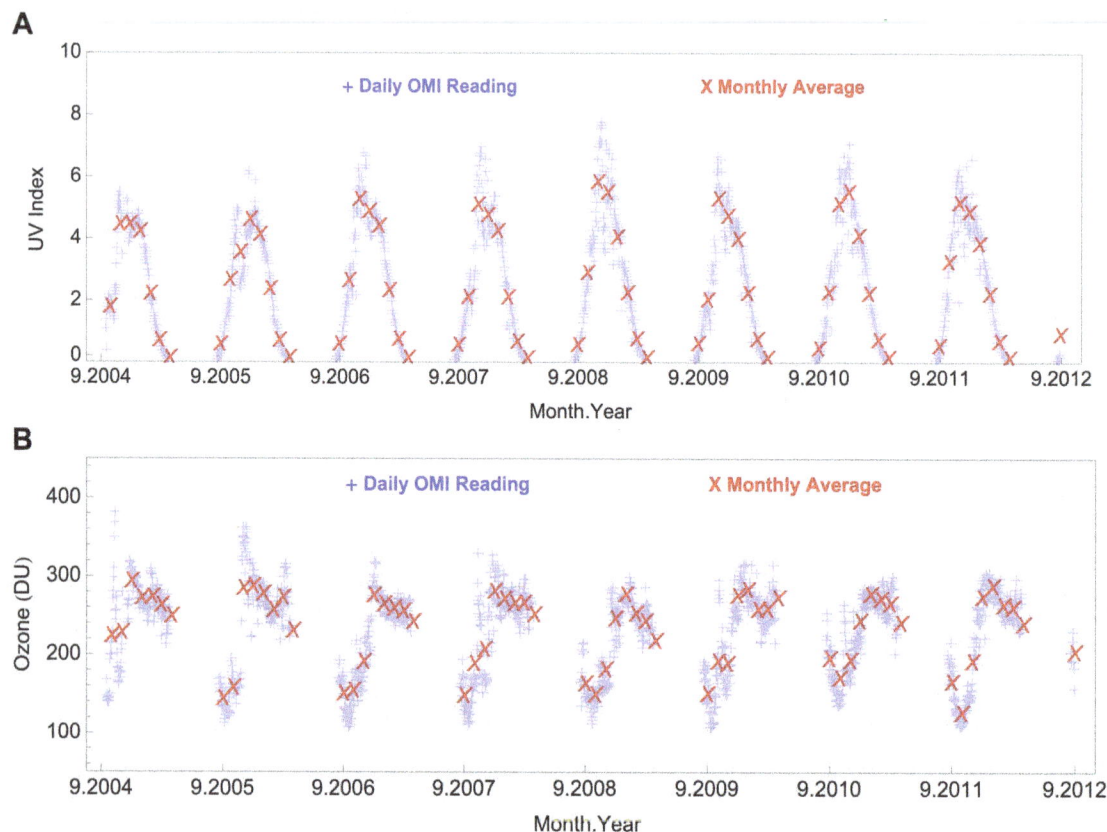

Figure 2. Time series of OMI-derived data at Union Glacier Camp. The number "9" in the abscise tick labels means "September". The blue crosses stand for the daily estimates; the red crosses stand for the monthly averages of the OMI-derived data. (a) Noontime UV index (UVI). (b) Total ozone column.

Table 1. UV Climatology at Union Glacier Camp computed by using the OMI-derived data over the period 2004–2012.

		Jan	Feb	March	April	Sept	Oct	Nov	Dec
Ozone [DU]	Average	275	261	259	241	158	161	203	267
	Variability	13	16	20	18	28	40	53	30
	Variability [%]	5	6	8	7	18	25	26	11
UVI (Clear-Sky)	Average	4.14	2.27	0.70	0.11	0.53	2.59	5.17	5.03
	Variability	0.46	0.61	0.34	0.03	0.40	0.89	1.17	0.74
	Variability [%]	11	27	49	30	77	34	22	14
UVI	Average	4.09	2.22	0.68	0.11	0.53	2.52	5.02	4.95
	Variability	0.53	0.64	0.35	0.03	0.40	0.90	1.22	0.78
	Variability [%]	13	29	51	30	77	36	24	16
Erythemal Daily Dose (J/m²)	Average	3825	1691	425	38	278	1767	4058	4849
	Variability	756	657	249	23	245	799	889	782
	Variability [%]	20	39	58	59	88	45	22	16
Surface Albedo	Average	0.89	0.87	0.83	0.78	0.76	0.85	0.88	0.89

The variability was taken as being equal to the standard deviation of the OMI-derived values.

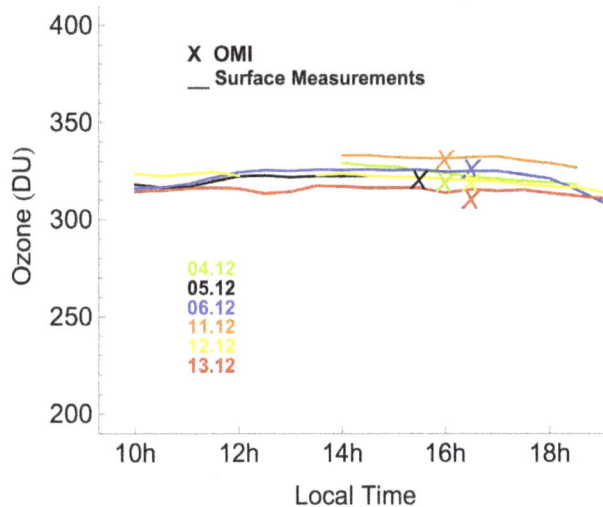

Figure 3. Ozone Column at Union Glacier Camp. Lines stand for total ozone column retrieved from ground-based measurements. Crosses stand for ozone estimates retrieved from OMI. Color indicates the dates: 04.12.2012 (Green); 05.12.2012 (black); 06.12.2012 (blue); 11.12.2012 (orange), 12.12.2012 (yellow), 13.12.2012 (red).

and the computed spectra [70–72]. By using the methods described above, we estimated that the AOD and the SSA at Union Glacier Camp are about 0.02 and 0.99, respectively. These values stand for an extremely low aerosol loading. Actually, due to the uncertainty associated with our aerosol estimates, assuming the total lack of aerosols in the area is also reasonable [9]. At Neumayer, extremely low aerosol loading has been recently measured [1].

Since our measurements were carried our under cloudless conditions, they did not allow us to weigh up the effect of local clouds on the UV irradiance. However, the effect of the high albedo of the surrounding snow-covered surface was assessed (see sections 4 and 5).

2. Albedo Measurements

The albedo is normally taken as being equal to the ratio between the up- and downwelling global radiation. Although measurements of the broadband albedo are normally carried out by using two radiometers [26], our spectral measurements were carried out (at 300–800 nm wavelength range) by using a single spectroradiometer and a rotating input optics that was set up at about 2.5 m above the snow surface. This means that the albedo assessments required measuring sequentially the up- and downwelling global radiation. Due to the time needed to accomplish a scan, the up- and downwelling spectra were separated in time about 5 min. Pairs of scans were carried out at 30 min intervals. A similar method based on the rotation of the input optics has been previously used in Antarctica [9]. Since the measurements of the up- and downwelling radiation were not carried out at the same time (i.e. at the same SZA), the spectra measured at 30 min intervals were linearly interpolated. Then, we estimated the up- and downwelling radiation for certain common SZAs.

Figure 4 shows the spectral albedo measured around noon. No significant differences were observed between measurements carried out on different days around noon under cloudless conditions. As shown in figure 4, the spectral albedo around noon was nearly constant in the UV and visible range but decreased at wavelengths longer than 600 nm; the lowest albedo (about 0.86) was measured at 800 nm. The diminishment of the albedo with

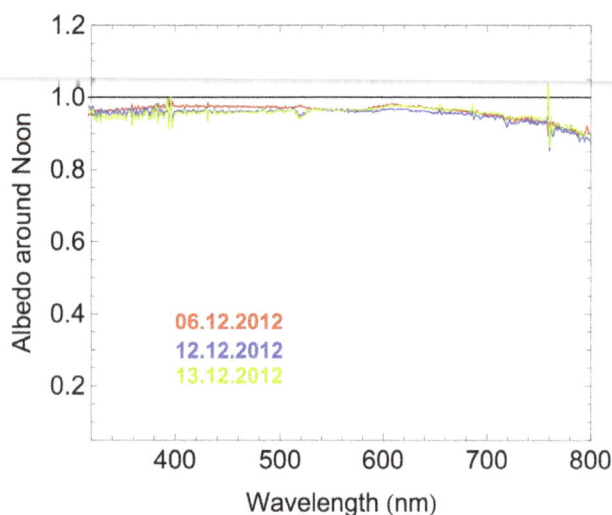

Figure 4. Spectral albedo measured under cloudless conditions around noon at Union Glacier Camp. Color indicates the dates: 06.12.2012 (Red); 12.12.2012 (Blue); 13.12.2012 (Green). Note that the narrow spectral variations in figure 4a have no physical meaning and are due to insignificant misalignments in the grating of our scanning double monochromator when measuring sequentially the up- and downwelling global radiation.

the wavelength in the near infrared was expected since it has been reported by all prior spectral measurements of the albedo over snow-covered surfaces [9,18–23].

Note that the albedo measurements shown in figure 4 are slightly lower than the values measured at Neumeyer station [9]. The difference maybe linked with characteristic of the snow, which is ultimately determined by the local meteorology. While at Neumeyer station the albedo of fresh snow was often measured, our measurements were carried out over a glacier covered by 2 m of old snow with grain radio of about 1 mm; old coarse-grained has lower albedo than fresh snow [20].

Measurements through the day have detected that the albedo decreased (up to 10%) as the SZA decreased through the day [9,21]. These diminishments range from 0.77 to 0.67 in the Arctic, and from 0.96 to 0.86 in the Antarctica [25]. The change in the albedo through the day has been attributed to changes in the snow characteristics [11]. Moreover, snow melting during daytime and refreezing during night may lead to albedo asymmetries (i.e. differences in the albedo measured at the same SZA, at different times of day) [25]. However, we did not detect significant chances in the albedo through the day at Union Glacier Camp. We attributed the apparent albedo stability to the relatively small variations through the day (from $-12°C$ to $-9°C$) in the air temperature logged by the ALE's AWS during our albedo measurements.

Note that our measurements were carried out about 3 km away from the closest peak (1450 m altitude). Therefore, the slopes of the surrounding mountains were in the field of view of the input optics. However, the complex topography surrounding Union Glacier Camp did not lead to any measurable effect on our ground-based measurements. Indeed, we checked the possibility of asymmetries (i.e. differences between spectral measurements carried out at the same SZA, at different times of day). We found that the differences were always within the uncertainty bounds of our measurements [32].

The surrounding topography did affect satellite data. Figure 5a depicts the surface reflectance map of Antarctica from OMI Lambertian equivalent reflectivity (LER) at 354 nm. LER is the required reflectance of an isotropic surface needed to match the observed top of the atmosphere (TOA) reflectance in a pure Rayleigh scattering atmosphere [73]. Therefore, it can be used to roughly estimate the actual surface albedo. Figure 5b shows the average of the surface reflectance data at 354 nm computed for December by using the data over the period 2005–2009 [73]. This is the albedo used by OMI algorithm to compute the UV surface irradiance (see Table 1). As expected the general pattern shows a very high surface albedo over almost all the Antarctic continent (with values roughly equal to or greater than 0.95). Lower reflectivities can be observed over the Transantarctic Mountains (close to the Ross Ice Shelf) and over the Ellsworth Mountains. Due to the katabatic winds the Antarctic mountain slopes are often snow-free and showing their underlying dark bedrock. Considering the relatively large OMI pixel (maps in figures 5a and 5b have a 0.5×0.5 degree resolution), the difference between ground-based measurement of albedo (about 0.95 in the UV and visible) and the lower reflectivity recorded by satellite (about 0.9 at 354 nm, see Table 1) may be partly explained by these small dark areas inside satellite bright pixels and by the Lambertian approximation.

Figure 5b shows the annual average of the reflectance data at 354 nm of snow-covered scenes (with relaxed cloud-screening criteria [74]). In contrast to figure 5a, clear-sky OMI scenes have been selected by using cloud and aerosol data from the MODIS/Aqua satellite instrument that flies 12 min ahead of OMI. The result is a reflectivity dataset that does not rely on statistical methods to eliminate cloud effect [74]. Despite some minor differences in absolute values (likely related to of the different periods of time considered in the datasets), the patterns are nearly the same.

3. Albedo Effect on UV

In order to weight up the effect of the albedo, we computed the UVI by using the UVSPEC model under the conditions observed at Union Glacier Camp at 11:40 LT on 06.12.2012, 2012 (SZA = 60°, AOD = 0.02; SSA = 0.99; altitude = 700 m; cloudless conditions) but assuming different ozone values as well as different albedo levels. As pointed out above, as radiative transfer solver was used the DISORT solver [69]; the extraterrestrial spectrum was quoted from Gueymard [68]. Figure 6 shows that despite of the ozone, a surface reflectivity greater than 0.9 in the UV (as that measured at Union Glacier Camp) led to enhancements in the UVI under cloudless conditions of about 50% in comparison with snow free surfaces (albedo = 0.1).

The effect of the albedo can be also weighted up by comparing our ground-based measurements at Union Glacier Camp with other spectra measured at snow free locations (albedo = 0.1). We used some of our measurements carried out in the austral summer at Paranal Observatory (2,635 m altitude, 24°37′S, 70°24′W) and in Santiago de Chile (500 m altitude, 33°27′ S, 70°41′ W).

Figure 7a shows the UV index computed from spectral measurements carried out at Union Glacier Camp on 05.12.2012, at Paranal Observatory on 09.01.2013, and in Santiago de Chile on 10.12.2011. Mostly due to the latitude (and in turn to the SZA), peak UVI values at Union Glacier Camp are significantly lower than those in Santiago and at Paranal Observatory. The figure changes when further comparisons are carried at common SZA.

Figure 7b shows the ratio between UV spectra at SZA = 60° measured at Union Glacier Camp (on 05.12.2012) and at Paranal Observatory (on 09.01.2013). Figure 7c shows the ratio between

A

B

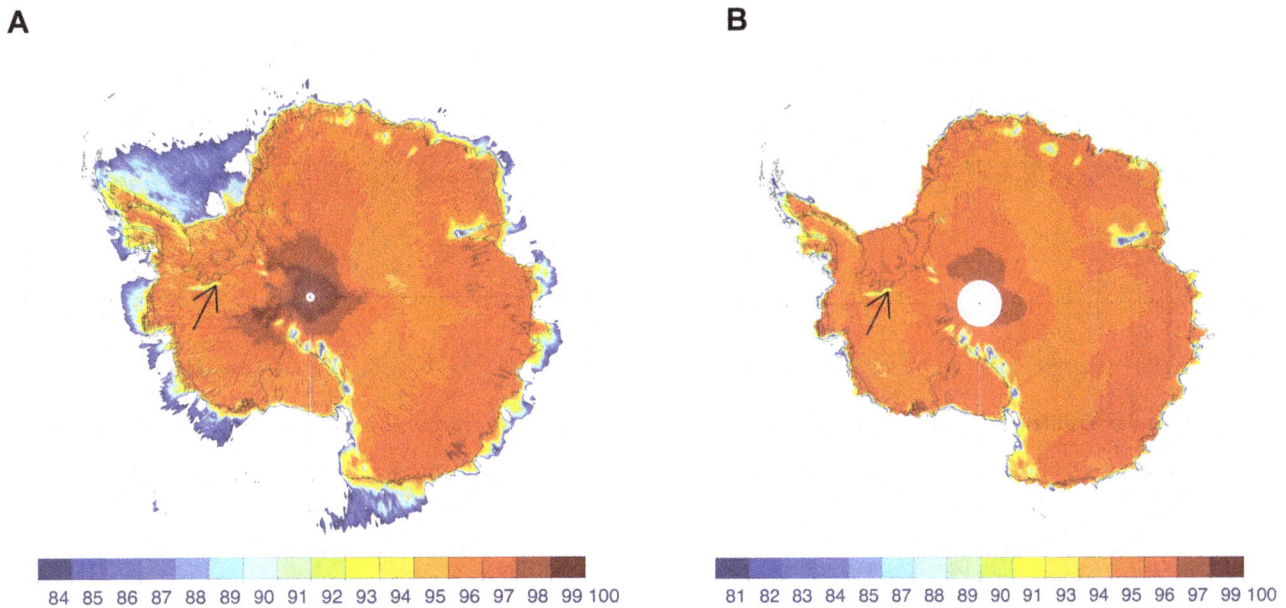

Figure 5. Satellite Albedo. a) Climatology albedo from OMI surface reflectance data (LER) at 354 nm for December. Arrow indicates the location of the Union Glacier Camp. b) Climatology albedo from OMI reflectance data (LER) at 354 nm for snow-covered scenes. Arrow indicates the location of the Union Glacier Camp. See the text for details.

UV spectra at SZA = 60° measured at Union Glacier Camp (on 05.12.2012) and at USACH station in Santiago de Chile (on 10.12.2011). In figures 7b and 7c we also show the ratio between modeled UV spectra computed by using the UVSPEC model under the conditions observed at the moment of the ground-based measurements (see details in the captions).

It can be seen in figures 7b that the UVA irradiance at Union Glacier Camp is significantly greater than at Paranal Observatory (the difference is 60% at 330 nm). Most of this difference is due to the albedo, and it would be even greater but the altitude of the observatory attenuates the difference. In figures 7c, it can be observed that the UVA irradiance at Union Glacier Camp is also significantly greater than in Santiago (the difference is 100% at 330 nm). Although again most of the difference is due to the albedo, the aerosol load in the case of Santiago boosts the

difference. As shown in figures 7b and 7c, the differences in the UVB range are counterbalanced by the ozone (about 30% lower in Santiago and at Paranal than at Union Glacier Camp, see captions for details). However, despite the relatively high ozone, the UVI at Union Glacier Camp on 05.12.2012 at SZA = 60° (330 DU), was greater than in Santiago on 10.12.2011 at SZA = 60° (265 DU) and than at Paranal observatory on 09.01.2013 at SZA = 60° (260 DU).

Summary and Conclusions

We report on the first quality-controlled spectral measurements of the albedo and the UV irradiance at Union Glacier Camp, located in the southern Ellsworth Mountains on the broad expanse of Union Glacier. The measurements were carried out in December 2012 by using a double monochromator-based spectro-radiometer during a campaign meant to weight up the effect of local albedo on the UV irradiance.

The UVI, the TOC, the AOD and the SSA, were computed from our spectra. We found that our surface measurements roughly agreed with OMI estimates of UVI and TOC; the detected differences ranged from 0.1% to 3% (i.e. up to about 10 DU) in the case of the TOC, and from 2% to 10% in the case of the UVI. Moreover, the AOD and the SSA retrieved from our surface measurements allowed us to conclude that the aerosol loading at Union Glacier Camp is extremely low (i.e. the aerosol influence in the area is negligible).

We also found that the spectral albedo measured at noon was about 0.95 in the UV and the visible part of the spectrum. We did not detect significant asymmetries (i.e. differences in the albedo measured at the same SZA, at different times of day). The differences between albedo values sampled at different times of day were always within the uncertainty bounds of our measurements.

In order to weight up the effect of the albedo, we compared our ground-based measurements at Union Glacier Camp with other spectra measured at snow-free locations. We used some of our measurements carried out in the austral summer at Paranal

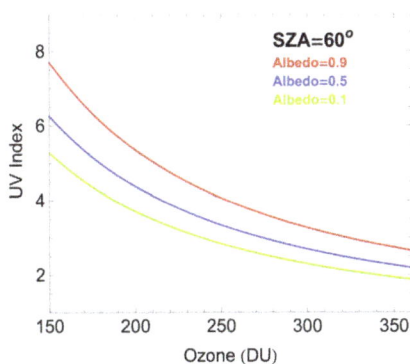

Figure 6. Change in the UVI with the albedo. The UVI was computed by using the UVSPEC model under the conditions observed at Union Glacier Camp at 11:40 LT on 06.12.2012 (SZA = 60°, AOD = 0.02; SSA = 0.99; altitude = 700 m) but assuming different ozone column values as well as different albedo levels (Red line: 0.9; Blue line: 0.5; Green line: 0.1).

A   **B**   **C**

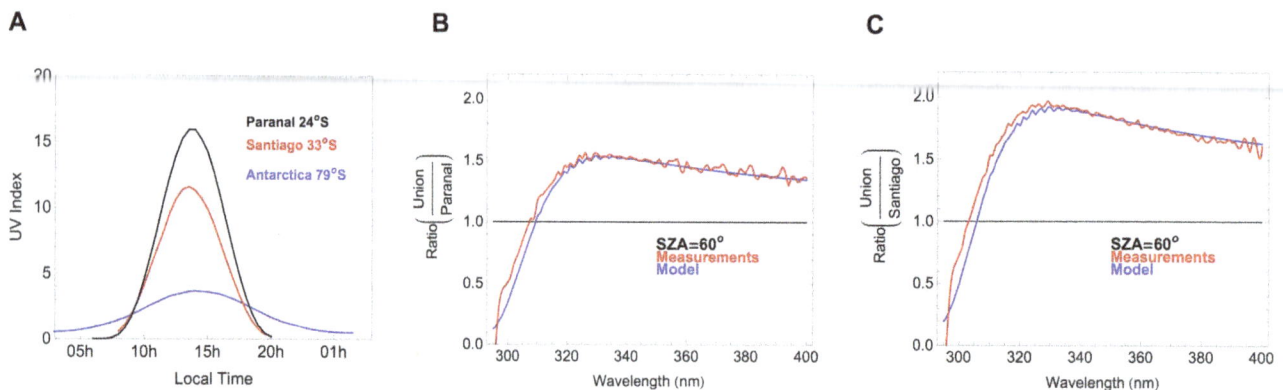

Figure 7. UV index at different locations. a) UV index computed from spectral measurements carried out by using the USACH spectroradiometer at Paranal Observatory on 09.01.2013 (Black line; minimum SZA = 2.6°), in Santiago de Chile on 10.12.2011 (Red line; minimum SZA = 11°), and at Union Glacier Camp on 05.12.2012 (Blue line; minimum SZA = 56.7°). b) Red line: ratio between UV spectra at SZA = 60° measured at Union Glacier Camp (on 05.12.2012) and at Paranal Observatory on (09.01.2013). Blue line: ratio between modeled UV spectra at Union Glacier Camp (albedo = 0.95; AOD = 0.02; SSA = 0.99; altitude = 700 m; ozone = 330 DU) and at Paranal Observatory (albedo = 0.1; AOD = 0.15; SSA = 0.7; altitude = 2,635 m; ozone = 260 DU). c) Red line: ratio between UV spectra at SZA = 60° measured at Union Glacier Camp (on 05.12.2012) and at USACH station in Santiago de Chile (on 10.12.2011). Blue line: ratio between modeled UV spectra at Union Glacier Camp (albedo = 0.95; AOD = 0.02; SSA = 0.99; altitude = 700 m; ozone = 330 DU) and in Santiago de Chile (albedo = 0.1; AOD = 0.3; SSA = 0.7; altitude = 500 m; ozone = 265 DU).

Observatory (2,635 m altitude, 24°37′S–70°24′W) and in Santiago de Chile (500 m altitude, 33°27′ S–70°41′ W). Further modeling allowed us to confirm that the surface reflectivity similar to that at Union Glacier Camp around noon (albedo = 0.95) led to enhancements in the UVI under cloudless conditions of about 50% in comparison with snow free surfaces (albedo = 0.1).

Author Contributions

Conceived and designed the experiments: RRC AD FL. Performed the experiments: JF JJ MT. Analyzed the data: RRC AD JC DL. Wrote the paper: RRC AD FL.

References

1. Weller R, Minikin A, Petzold A, Wagenbach D, König-Langlo G (2013) Characterization of long-term and seasonal variations of black carbon (BC) concentrations at Neumayer, Antarctica. Atmos Chem Phys 13: 1579–1590.
2. Bernhard G, Booth CR, Ehramjian JC (2005) UV climatology at Palmer Station, Antarctica. In: Ultraviolet Ground- and Space-based Measurements, Models, and Effects V, Bernhard G, Slusser JR, Herman JR, Gao W editors, Proceedings of SPIE, 5886, 51–62.
3. Cordero RR, Damiani A, Seckmeyer G, Riechelmann S, Laroze D, et al. (2013) Satellite-derived UV Climatology at Escudero Station (Antarctic Peninsula). Antarctic Science 25(6): 791–803.
4. Bernhard G, Booth CR, Ehramjian JC (2004) Version 2 data of the National Science Foundation's Ultraviolet Radiation Monitoring Network: South Pole. J Geophys Res doi: 10.1029/2004JD004937.
5. Bernhard G, Booth CR, Ehramjian JC, Nichol SE (2006) UV climatology at McMurdo Station, Antarctica, based on version 2 data of the National Science Foundation's Ultraviolet Radiation Monitoring Network. J Geophys Res doi: 10.1029/2005JD005857.
6. Bernhard G, Booth CR, Ehramjian JC, Stone R, Dutton EG (2007) Ultraviolet and visible radiation at Barrow, Alaska: Climatology and influencing factors on the basis of version 2 National Science Foundation network data. J Geophys Res doi:10.1029/2006JD007865.
7. Flemming J, Inness A, Jones L, Eskes HJ, Huijnen V, et al. (2011) Forecasts and assimilation experiments of the Antarctic ozone hole 2008. Atmos Chem Phys 11: 1961–1977.
8. Bernhard G, Booth CR, Ehramjian JC (2010) Climatology of Ultraviolet Radiation at High Latitudes Derived from Measurements of the National Science Foundation's Ultraviolet Spectral Irradiance Monitoring Network. In: UV Radiation in Global Climate Change: Measurements, Modeling and Effects on Ecosystems, Gao W, Schmoldt DL, and Slusser JR editors, Springer-Verlag and Tsinghua University Press, ISBN: 978-3-642-03312-4.
9. Wuttke S, Seckmeyer G, Koenig-Langlo G (2006) Measurements of spectral snow albedo at Neumayer, Antarctica. Annales Geophysicae 24: 7–21.
10. Lenoble J (1998) Modelling of the influence of snow reflectance on ultraviolet irradiance for cloudless sky. Appl Opt 37: 2441–2447.
11. Pirazzini R (2004) Surface albedo measurements over Antarctic sites in summer. J Geophys Res doi: 10.1029/2004JD004617.
12. Hall A (2004) The Role of Surface Albedo Feedback in Climate. J Climate 17: 1550–1568.
13. Winton M. (2006) Surface Albedo Feedback Estimates for the AR4 Climate Models. J Climate 19: 359–365.
14. Vaughan DG, Marshall GJ, Connolley WM, King JC, Mulvaney R (2001) Devil in the Detail. Science 293: 1777–1779.
15. Aoki T, Kuchiki K, Niwano M, Kodama Y, Hosaka M, et al. (2011) Physically based snow albedo model for calculating broadband albedos and the solar heating profile in snowpack for general circulation models. Journal of Geophysical Research doi: 10.1029/2010JD015507.
16. Pedersen CA, Roeckner E, Lüthje M, Winther JG (2009) A new sea ice albedo scheme including melt ponds for ECHAM5 general circulation model. Journal of Geophysical Research doi: 10.1029/2008JD010440.
17. Collins WD, Bitz CM, Blackmon ML, Bonan GB, Bretherton CS, et al. (2006) The community climate system model version 3 (CCSM3). Journal of Climate 19 (11): 2122–2143.
18. Grenfell TC, Warren SG, Mullen PC (1994) Reflection of solar radiation by the Antarctic snow surface at ultraviolet, visible, and near-infrared wavelengths. Journal of Geophysical Research doi: 10.1029/94JD01484.
19. Grenfell TC, Perovich DK (2004) Seasonal and spatial evolution of albedo in a snow-ice-land-ocean environment. Journal of Geophysical Research Oceans doi: 10.1029/2003JC001866.
20. Brandt RE, Warren SG, Worby AP, Grenfell TC (2005) Surface albedo of the Antarctic sea ice zone. Journal of Climate 18(17): 3606–3622.
21. Meinander O, Kontu A, Lakkala K, Heikkilä A, Ylianttila L, et al. (2008) Diurnal variations in the UV albedo of arctic snow. Atmos Chem Phys 8: 6551–6563.
22. Meinander O, Kazadzis S, Arola A, Riihelä A, Räisänen P, et al. (2013) Spectral albedo of seasonal snow during intensive melt period at Sodankylä, beyond the Arctic Circle. Atmos Chem Phys 13: 3793–3810.
23. Cordero RR, Damiani A, Ferrer J, Rayas J, Jorquera J, et al. (2013) Downwelling and Upwelling Radiance Distributions sampled under Cloudless Conditions in Antarctica, Applied Optics 52(25): 6287–94.
24. Li S, Zhou X (2003) Assessment of the Accuracy of Snow Surface Direct Beam Spectral Albedo under a Variety of Overcast Skies Derived by a Reciprocal Approach through Radiative Transfer Simulation. Appl Opt 42: 5427–5441.
25. Meinander O, Wuttke S, Seckmeyer G, Kazadzis S, Lindfors A, et al. (2009) Solar Zenith Angle Asymmetry Cases in Polar Snow UV Albedo. Geophysica 45(1–2): 183–198.
26. Seckmeyer G, Bais A, Bernhard G, Blumthaler M, Booth CR, et al. (2001) Part 1: Spectral instruments Instruments to Measure Solar Ultraviolet Radiation WMO-GAW No. 125 (Geneva, Switzerland: World Meteorological Organization).

27. Wuttke S, Seckmeyer G, Bernhard G, Ehramjian J, McKenzie R, et al. (2006) New spectroradiometers complying with the NDSC standards. J Atmos Ocean Technol 23(2): 241–251.

28. Cede A, Herman J, Richter A, Krotkov N, Burrows J (2006) Measurements of nitrogen dioxide total column amounts using a Brewer double spectrophotometer in direct sun mode. J Geophys Res doi: 10.1029/2005JD006585.

29. Gröbner J, Albold A, Blumthaler M, Cabot T, de la Casinière A, et al. (2000) The variability of spectral solar ultraviolet irradiance in an Alpine environment. J Geophys Res doi: 10.1029/2000JD900395.

30. Gröbner J, Blumthaler M, Kazadzis S, Bais A, Webb A, et al. (2006) Quality assurance of spectral solar UV measurements: result from 25 UV monitoring sites in Europe, 2002 to 2004. Metrologia 43: S66–S71.

31. Bais AF, Gardiner BG, Slaper H, Blumthaler M, Bernhard G, et al. (2001) The SUSPEN intercomparison of ultraviolet spectroradiometers. J Geophys Res 106: 12509–12525.

32. Cordero RR, Seckmeyer G, Pissulla D, DaSilva L, Labbe F (2008) Uncertainty Evaluation of Spectral UV Irradiance Measurements. Meas. Sci. Technol. 19: 1–15.

33. Gröbner J, Hülsen G, Vuilleumier L, Blumthaler M, Vilaplana JM, et al. (2007) Report of the PMOD/WRC-COST calibration and intercomparison of erythemal radiometers.

34. Antón M, Serrano A, Cancillo ML, Vilaplana JM (2011) Quality assurance of broadband erythemal radiometers at the Extremadura UV Monitoring Network (Southwestern Spain). Atmospheric Research 100: 83–92.

35. McKinlay AF, Diffey BL (1987) A reference action spectrum for ultraviolet induced erythema in human skin. Commission Internationale de l'Eclairage Journal 6: 17–22.

36. Cordero RR, Seckmeyer G, Pissulla D, Labbe F (2008) Uncertainty of experimental integrals: application to the UV index calculation. Metrologia, 45: 1–10.

37. Balis D, Kroon M, Koukouli ME, Brinksma EJ, Labow G, et al. (2007) Validation of Ozone Monitoring Instrument total ozone column measurements using Brewer and Dobson spectrophotometer ground-based observations. Journal of Geophysical Research doi: 10.1029/2007JD008796.

38. McPeters R, Kroon M, Labow G, Brinksma E, Balis D, et al. (2008) Validation of the Aura Ozone Monitoring Instrument total column ozone product. Journal of Geophysical Research doi: 10.1029/2007JD008802.

39. Damiani A, De Simone S, Rafanelli C, Cordero RR, Laurenza M (2012) Three years of ground-based total ozone measurements in the Arctic: comparison with OMI, GOME and SCIAMACHY satellite data. Remote Sensing of Environment 127: 162–180.

40. Antón M, Cachorro VE, Vilaplana JM, Toledano C, Krotkov NA, et al. (2010) Comparison of UV irradiances from Aura/Ozone Monitoring Instrument (OMI) with Brewer measurements at El Arenosillo (Spain) - Part 1: Analysis of parameter influence. Atmos Chem Phys 10: 5979–5989.

41. Damiani A, Cabrera S, Muñoz RC, Cordero RR, Labbe F (2013) Satellite-derived UV irradiance for a region with complex morphology and meteorology: comparison against ground measurements in Santiago de Chile. International Journal of Remote Sensing doi: 10.1080/01431161.2013.796101.

42. Douglass A, Fioletov V, Godin-Beekmann S, Müller R, Stolarski RS, et al. (2011) Stratospheric ozone and surface ultraviolet radiation 1–80; Global Ozone Research and Monitoring Project-Report No. 52, 516 pp., Geneva, Switzerland.

43. Tanskanen A, Manninen T (2007) Effective UV surface albedo of seasonally snow-covered lands. Atmos Chem Phys 7: 2759–2764.

44. Cabrera S, Ipiña A, Damiani A, Cordero RR, Piacentini RD (2012) UV index values and trends in Santiago, Chile (33.5°S) based on ground and satellite data. Journal of Photochemistry and Photobiology B: Biology 115: 73–84.

45. Kazadzis S, Bais A, Balis D, Kouremeti N, Zempila M, et al. (2009) Spatial and temporal UV irradiance and aerosol variability within the area of an OMI satellite pixel. Atmos Chem Phys 9: 4593–4601.

46. Kazadzis S, Bais A, Arola A, Krotkov N, Kouremeti N, et al. (2009) Ozone Monitoring Instrument spectral UV irradiance products: comparison with ground based measurements at an urban environment. Atmos Chem Phys 9: 585–594.

47. Ialongo I, Buchard V, Brogniez C, Casale GR, Siani AM (2009) Aerosol Single Scattering Albedo retrieval in the UV range: an application to OMI satellite validation. Atmos Chem Phys Discuss 9: 19009–19033.

48. Buchard V, Brogniez C, Auriol F, Bonnel B, Lenoble J, et al. (2008) Comparison of OMI ozone and UV irradiance data with ground-based measurements at two French sites. Atmos Chem Phys 8: 4517–4528.

49. Tanskanen A, Krotkov NA, Herman JR, Arola A (2006) Surface ultraviolet irradiance from OMI. IEEE transactions on Geoscience and Remote Sensing 44(5): 1267–1271.

50. Tanskanen A, Lindfors A, Määttä A, Krotkov N, Herman J, et al. (2007) Validation of daily erythemal doses from Ozone Monitoring Instrument with ground-based UV measurement data. J Geophys Res doi: 10.1029/2007JD008830.

51. Weihs P, Blumthaler M, Rieder HE, Kreuter A, Simic S, et al. (2008) Measurements of UV irradiance within the area of one satellite pixel. Atmos Chem Phys 8: 5615–5626.

52. Ialongo I, Casale GR, Siani AM (2008) Comparison of total ozone and erythemal UV data from OMI with ground-based measurements at Rome station. Atmos Chem Phys 8: 3283–3289.

53. Cordero RR, Seckmeyer G, Pissulla D, DaSilva L, Labbe F (2007) Uncertainty evaluation of the spectral UV irradiance evaluated by using the UVSPEC Radiative Transfer Model. Optics Communications 276: 44–53.

54. Cordero RR, Seckmeyer G, Damiani A, Labbe F, Laroze D (2013) Monte Carlo-based Uncertainties of Surface UV Estimates from Models and from Spectroradiometers. Metrologia 50: 1–5.

55. Cordero RR, Damiani A, Laroze D, Dasilva L, Labbe F (2013) Spectral UV radiance measured at a coastal site: a case study. Photochemical & Photobiological Sciences 12: 1193.

56. Blumthaler M, Kreuter A, Webb A, Bais A, Kift R, et al. (2009) Albedo Effect on UV Irradiance, MOCA Joint Assembly, Montreal, Quebec, July 2009, Available: http://www.moca-09.org.Accessed 1 October 2013.

57. Kylling A, Persen T, Mayer B, Svenøe T (2000) Determination of an effective spectral surface albedo from ground-based global and direct UV irradiance measurements. J Geophys Res 105(D4): 4949–4959.

58. Smolskaia I, Masserot D, Lenoble J, Brogniez C, de la Casinière A (2003) Retrieval of the Ultraviolet Effective Snow Albedo During 1998 Winter Campaign in the French Alps. Appl Opt 42(9): 1583.

59. Badosa J, McKenzie RL, Kotkamp M, Calbó J, González JA, et al. (2007) Towards closure between measured and modelled UV under clear skies at four diverse sites. Atmospheric Chemistry and Physics 7: 2817–2837.

60. Satheesh SK, Srinivasan J, Vinoj V, Chandra S (2006) New Directions: How representative are aerosol radiative impact assessments? Atmospheric Environment 40(16): 3008–3010.

61. Mayer B, Kylling A (2005) Technical note: The libRadtran software package for radiative transfer calculations - description and examples of use. Atmos Chem Phys 5: 1855–1877.

62. Dahlback A, Stamnes K (1991) A new spherical model for computing the radiation field available for photolysis and heating at twilight. Planet Space Sci 39: 671–683.

63. Gueymard CA (2004) The sun's total and spectral irradiance for solar energy applications and solar radiation models. Solar Energy 76: 423–453.

64. van den Broeke MR, Reijmer CH, van de Wal RSW (2004) A study of the surface mass balance in Dronning Maud Land, Antarctica, using automatic weather stations. Journal of Glaciology 50(171): 565–582.

65. Parish TR, Bromwich DH (1987) The surface windfield over the Antarctic Ice Sheets. Nature 328: 51–54.

66. Bromwich DH, Nicolas JP, Hines KM, Kay JE, Key E, et al. (2012) Tropospheric Clouds in Antarctica. Rev Geophys doi: 10.1029/2011RG000363.

67. Grooß JU, Brautzsch K, Pommrich R, Solomon S, Müller R (2011) Stratospheric ozone chemistry in the Antarctic: what determines the lowest ozone values reached and their recovery? Atmospheric Chemistry and Physics 11: 12217–12226.

68. Stamnes K, Slusser J, Bowen M (1991) Derivation of Total Ozone Abundance and Cloud Effects from Spectral Irradiance Measurements. Applied Optics 30: 4418–4426.

69. Bernhard G, Booth CR, McPeters RD (2003) Calculation of total column ozone from global UV spectra at high latitudes. J Geophys Res doi: 10.1029/2003JD003450.

70. Cordero RR, Seckmeyer G, Pissulla D, Labbe F (2009) Exploitation of Spectral Direct UV Irradiance Measurements. Metrologia 46: 19–25.

71. Bais AF, Kazantzidis A, Kazadzis S, Balis DS, Zerefos CS, et al. (2005) Deriving an effective aerosol single scattering albedo from spectral surface UV irradiance measurements. Atmos Environ 39(6): 1093–1102.

72. Buchard V, Brogniez C, Auriol F, Bonnel B (2011) Aerosol single scattering albedo retrieved from ground-based measurements in the UV and visible region. Atmospheric Measurement Techniques 4: 1–7.

73. Kleipool QL, Dobber MR, de Haan JF, Levelt PF (2008) Earth surface reflectance climatology from 3 years of OMI data. J Geophys Res doi: 10.1029/2008JD010290.

74. O'Byrne G, Martin RV, van Donkelaar A, Joiner J, Celarier EA (2010) Surface reflectivity from the Ozone Monitoring Instrument using the Moderate Resolution Imaging Spectroradiometer to eliminate clouds: Effects of snow on ultraviolet and visible trace gas retrievals. J Geophys Res doi: 10.1029/2009JD013079.

3

Role of Surface Chemistry in Protein Remodeling at the Cell-Material Interface

Virginia Llopis-Hernández[1]◑, Patricia Rico[1,2]◑, José Ballester-Beltrán[1], David Moratal[1], Manuel Salmerón-Sánchez[1,2,3]*

1 Center for Biomaterials and Tissue Engineering, Universidad Politécnica de Valencia, Valencia, Spain, **2** CIBER de Bioingeniería, Biomateriales y Nanomedicina (CIBER-BBN), Valencia, Spain, **3** Regenerative Medicine Unit, Centro de Investigación Príncipe Felipe, Valencia, Spain

Abstract

Background: The cell-material interaction is a complex bi-directional and dynamic process that mimics to a certain extent the natural interactions of cells with the extracellular matrix. Cells tend to adhere and rearrange adsorbed extracellular matrix (ECM) proteins on the material surface in a fibril-like pattern. Afterwards, the ECM undergoes proteolytic degradation, which is a mechanism for the removal of the excess ECM usually approximated with remodeling. ECM remodeling is a dynamic process that consists of two opposite events: assembly and degradation.

Methodology/Principal Findings: This work investigates matrix protein dynamics on mixed self-assembled monolayers (SAMs) of –OH and –CH$_3$ terminated alkanethiols. SAMs assembled on gold are highly ordered organic surfaces able to provide different chemical functionalities and well-controlled surface properties. Fibronectin (FN) was adsorbed on the different surfaces and quantified in terms of the adsorbed surface density, distribution and conformation. Initial cell adhesion and signaling on FN-coated SAMs were characterized via the formation of focal adhesions, integrin expression and phosphorylation of FAKs. Afterwards, the reorganization and secretion of FN was assessed. Finally, matrix degradation was followed via the expression of matrix metalloproteinases MMP2 and MMP9 and correlated with Runx2 levels. We show that matrix degradation at the cell material interface depends on surface chemistry in MMP-dependent way.

Conclusions/Significance: This work provides a broad overview of matrix remodeling at the cell-material interface, establishing correlations between surface chemistry, FN adsorption, cell adhesion and signaling, matrix reorganization and degradation. The reported findings improve our understanding of the role of surface chemistry as a key parameter in the design of new biomaterials. It demonstrates the ability of surface chemistry to direct proteolytic routes at the cell-material interface, which gains a distinct bioengineering interest as a new tool to trigger matrix degradation in different biomedical applications.

Editor: Wei-Chun Chin, University of California Merced, United States of America

Funding: The support of the Spanish Ministry of Science and Innovation through project MAT2009-14440-C02-01 is acknowledged. CIBER-BBN is an initiative funded by the VI National R&D&i Plan 2008–2011, Iniciativa Ingenio 2010, Consolider Program, CIBER Actions and financed by the Instituto de Salud Carlos III with assistance from the European Regional Development Fund. This work was supported by funds for research in the field of Regenerative Medicine through the collaboration agreement from the Conselleria de Sanidad (Generalitat Valenciana), and the Instituto de Salud Carlos III. The funders had no role in study design, data collection and analysis, decision to publish, or preparation of the manuscript.

Competing Interests: The authors have declared that no competing interests exist.

* E-mail: masalsan@fis.upv.es

◑ These authors contributed equally to this work.

Introduction

The interaction of cells with foreign materials takes place via the adsorbed layer of proteins such as fibronectin (FN), vitronectin, and fibrinogen, representing the soluble matrix proteins in the biological fluids [1]. Cells primarily interact with these proteins via integrins, a family of transmembrane cell adhesion receptors [2]. Integrin-mediated adhesion is a complex process that involves integrin association with the actin cytoskeleton and clustering into focal adhesions: supramolecular complexes that contain structural proteins (vinculin, talin, tensin, etc.) and signaling molecules (focal adhesion kinase – FAK, etc.) [2,3]. FAK is a nonreceptor protein-tyrosine kinase that becomes activated in response to cell-matrix adhesion. FAK is a key signaling protein contributing to integrin control of cell motility, invasion, survival, and proliferation [4].

The cell-material interaction is a complex bi-directional and dynamic process that mimics to a certain extent the natural interactions of cells with the extracellular matrix [5,6]. Cells in the tissues are constantly accepting information from their environment from cues in the extracellular matrix ECM [7] and, at the same time, cells are producing and frequently remodeling their matrix [1,2,8]. Therefore, it is not surprising that many cells cannot adapt and poorly survive *in vitro* and, conversely, when a foreign material is implanted in the body, the adjacent tissue cells do not interact properly because of lack of their ECM.

A line of previous investigations has shown that cells tend to rearrange adsorbed matrix proteins at the material interface, such as FN, fibrinogen and collagen [9–11], in a fibril-like pattern. Using model surfaces – mostly self-assembled monolayers (SAMs) – it has been shown that this cellular activity is abundantly dependent on the surface properties of materials, such as wettability [9], surface chemistry and charge [12]. This evidence raises the possibility that tissue compatibility of such materials may be connected with the allowance of cells to remodel surface associated proteins presumably as an attempt to form their own matrix. Much is known about the interactions between different ECM proteins, but surprisingly less is our knowledge about the ECM composition, organization, and stability at the materials interface.

ECM remodeling is a dynamic process which consists of two opposite events: assembly and degradation. These processes are mostly active during development and regeneration of tissues but, when miss-regulated, can contribute to diseases such as atherosclerosis, fibrosis, ischemic injury and cancer [13–16]. The proteolytic cleavage of ECM components represents a main mechanism for ECM degradation and removal [17,18]. The major enzymes that degrade ECM and cell surface associated proteins are matrix metalloproteinases (MMPs). MMPs are a family (24 members) of zinc dependent endopeptidases, which together with adamalysin-related membrane proteinases that contain disintegrin and metalloproteinase domains (ADAMs or MDCs), such as thrombin, tissue plasminogen activator (tPA), urokinase (uPA) and plasmin are involved in the degradation of ECM proteins. MMPs are either secreted or anchored to the cell membrane by a transmembrane domain or by their ability to bind directly uPA receptor (uPAR) and integrin $\alpha_v\beta_3$ [19].

The role of MMPs in both development and diseases has been recently extensively studied and reviewed [20] because it is tightly linked with the mechanisms for tumor invasion and metastasis [18]. Also, MMPs regulate cell behavior through finely tuned and tightly controlled proteolytic processing of a large variety of signaling molecules that can also trigger beneficial effects in disease resolution [21].

This work investigates matrix protein dynamics on FN-coated mixed self-assembled monolayers (SAMs) of –OH and –CH$_3$ terminated alkanethiols, which constitute an excellent model to vary surface wettability in a broad range while maintaining controlled and simple surface chemistry. SAMs are model organic surfaces that provide defined chemical functionalities and well-controlled surface properties [22,23]. FN adsorption was investigated (adsorbed surface density, distribution and conformation) and correlated to cell behavior. Cell adhesion and signaling on FN-coated SAMs were characterized via the formation of focal adhesions, integrin expression and phosphorylation of FAKs. The reorganization and secretion of FN was linked to the activity of FN after adsorption on the different chemistries. Finally, the expression (gene and protein) of MMP2 and MMP9 metalloproteinases was used to follow matrix degradation. This work provides a broad overview of matrix remodeling at the cell-material interface, establishing correlations between surface chemistry, FN adsorption, cell adhesion and signaling, matrix reorganization and degradation.

Results

Fibronectin adsorption

The SAMs prepared in this work have been extensively used and characterized in previous studies making use of XPS, FTIR and ellipsometry [24,25]. As a routine control, we have measured the water contact angle (WCA) to assess that is in accordance with published results. WCA decreases as the fraction of hydroxy groups increases from 115° on the methyl terminated SAM to 20° on the hydroxyl terminated one (Figure 1a).

The surface density of adsorbed FN was quantified by western blot analyzing the amount of protein remaining in the supernatant after adsorption on the material surface. A calibration curve was built loading gels with known amounts of FN and the resulting

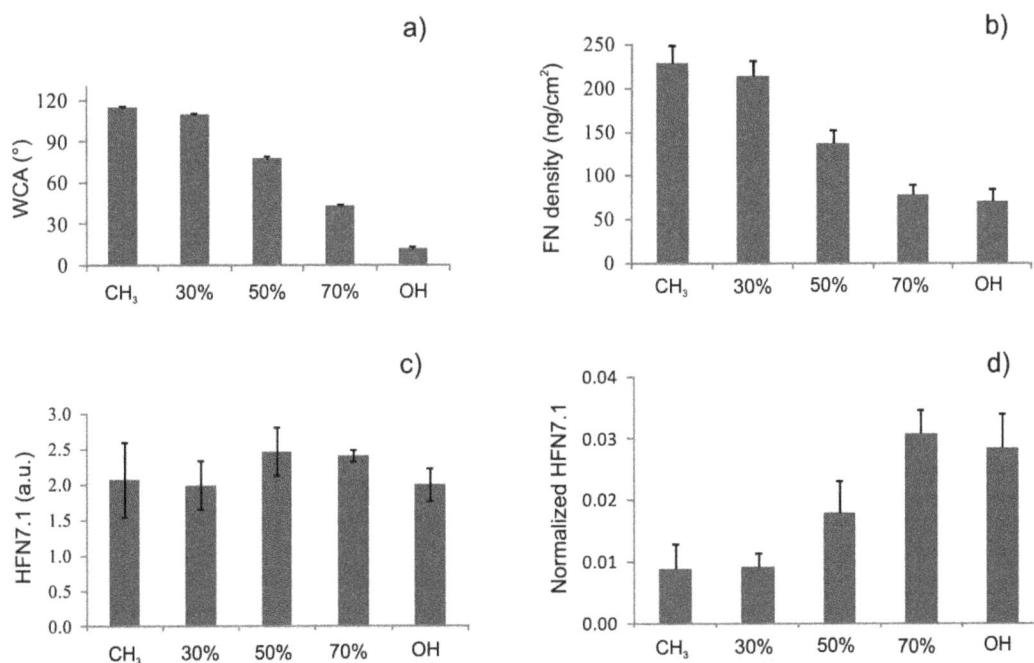

Figure 1. Surface wettability and FN adsorption on the CH₃/OH mixed SAMs. The horizontal axis displays the percentage of OH groups in SAMs. a) Water contact angle on the different SAMs. b) FN surface density after adsorption from a solution of concentration 20 µg/mL. c) Monoclonal antibody binding for HFN7.1 on the different SAMs after FN adsorption from a solution of concentration 20 µg/mL. d) Activity of the adsorbed FN on the different SAMs obtained by normalizing the monoclonal antibody binding for HFN7.1 relative to the FN surface density calculated in b).

bands were quantified by image analysis making use of the Otsu's algorithm to systematically identify the band borders [26]. Each experiment of FN adsorption on SAMs included the loading in the gel of two known amounts of FN (reference points) that correspond to points included in the calibration curve so that the position of the whole calibration curve could be verified for each adsorption experiment [26]. Figure 1b shows the surface density of FN on the different SAMs after adsorption from a solution of concentration 20 µg/mL. The amount of adsorbed protein diminishes monotonically as the –OH density increases from 225 ng/cm^2 on the methyl terminated SAM to 50 ng/cm^2 on the hydroxyl terminated one.

The availability of the cell adhesion domains in the adsorbed FN was evaluated by ELISA with monoclonal antibodies, which is a well established method to probe for structural or conformational changes in adsorbed proteins [27,28]. The antibody used (HFN7.1) was directed against the flexible linker between the 9th and 10th type III repeats of FN [29]. It has been previously demonstrated that HFN7.1 is a receptor-mimetic probe for integrin binding and cell adhesion [29]. HFN7.1 antibody binding is similar on the different SAMs regardless the composition of the surface after FN adsorption from a solution of concentration 20 µg/mL (Figure 1c). However, taking into account that the amount of adsorbed FN differs among SAMs, the availability of the HFN7.1 antibody was obtained by normalizing to the total amount of adsorbed FN on each surface (Figure 1d). This magnitude increases as the fraction of hydroxyl groups on the surface does.

The molecular distribution of FN upon adsorption on the different SAMs can be obtained by AFM. Figure 2 shows the organization of FN on three of the surfaces (CH$_3$, OH and the surface with 70% OH, that display qualitatively different WCA) after FN adsorption from solutions of different concentrations. FN fibrils are found on the methyl-terminated SAM after adsorption from a solution of 2 µg/mL (average thickness of the fiber is approximately 13±5 nm), less organized molecules are observed on the 70% OH surface that became isolated globular-like molecules on the hydroxyl terminated SAM (average size of the globular aggregates 20±4 nm). Increasing the concentration of the FN solution results in a dense network-like structure of FN on the methyl terminated surface and large molecular aggregates that cover the whole surface for the more hydrophilic surfaces (Figure 2). Figures S1, S2, S3 show AFM images for FN adsorption on the different substrates at different magnifications for the sake of completeness. The fibrillar nature of the adsorbed FN on the methyl-terminated SAM and the globular distribution on the other two surfaces is clearly grasped from this Figures S1, S2, S3.

Cell adhesion and signaling

The organization of proteins involved in the formation of focal adhesion complexes provides an opportunity to learn more about the effectiveness of cell-to-substrate interactions. Figure 3 shows the distribution of vinculin in cells adhering on the different model substrates. Well-defined focal adhesions were found only on the more hydrophilic substrates (OH- terminated and 70% OH). Even if vinculin is expressed also in cells on the more hydrophobic substrates, it is not afterwards organized into focal contacts but randomly distributed throughout the cell. Likewise, the formation of prominent F-actin fibers terminating in well-developed focal adhesion complexes occurs on the hydroxyl-terminated surfaces. More dispersed actin distribution (either lacking stress fiber formation or mostly peripheral staining) is observed as the fraction of OH groups on the surface diminishes (Figure 3).

Focal adhesion kinase (FAK) localizes to focal adhesions to activate multiple signaling pathways that regulate cell migration,

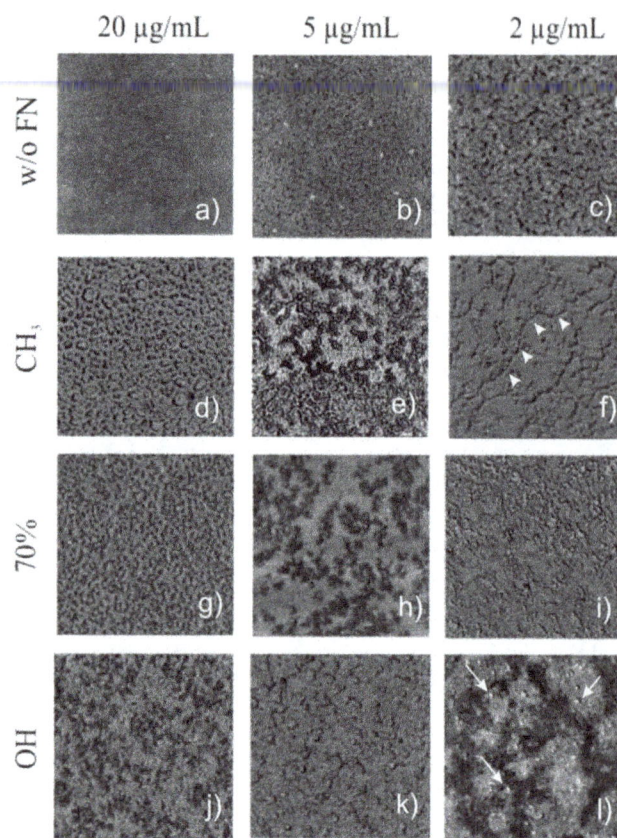

Figure 2. Fibronectin distribution on the different SAMs as observed by the phase magnitude in AFM. The protein was adsorbed for 10 min from different solutions of concentration 20 µg/ mL, 5 µg/mL and 2 µg/mL. The first row is the SAM surface without any FN at different magnifications: 5 µm (a), 2 µm (b) and 1 µm (c). Arrowheads in f) identify one of the FN fibers assembled on the material surface upon adsorption (fiber diameter 13±5 nm), arrows in l) identify globular aggregates of molecular size (diameter 20±4 nm). Images including FN are 1 µm side.

survival, proliferation, and differentiation [30–34]. We examined the phosphorylation of Y-397, the autophosphorylation site in FAK and a binding site for Src and PI-3 kinases [35,36]. According to Figure 4 the level of FAK remains constant (both as obtained by analysis of western-blot and PCR bands). By contrast, the ratio between phosphorylated and total FAKs on the different mixed SAMs decreases as the fraction of hydroxyl - terminated groups diminishes (Figure 4c). That is to say, the phosphorylation of specific sites in FAKs depends monotonically on the hydroxyl content of the surface. Likewise, gene expression for FAK as obtained by RT-PCR shows no difference among the different surfaces, while integrin (β$_1$) gene expression increases as the fraction of OH on the SAMs does (Figure S4).

Fibronectin reorganization and secretion

Figure 5 shows the cellular reorganization of adsorbed FN after 2.5 h of culture on the different SAMs. It is observed that cells are able to reorganize FN on the hydroxyl-terminated and the 70%-OH SAMs, as it is shown by movements of the adsorbed FN layer with dark zones in the pericellular area, mostly coincident with focal adhesion plaques. Late FN matrix formation was studied for longer times on the different SAMs (Figure S5). It is observed that matrix production increases as time goes by on every substrate.

Actin Vinculin Merge

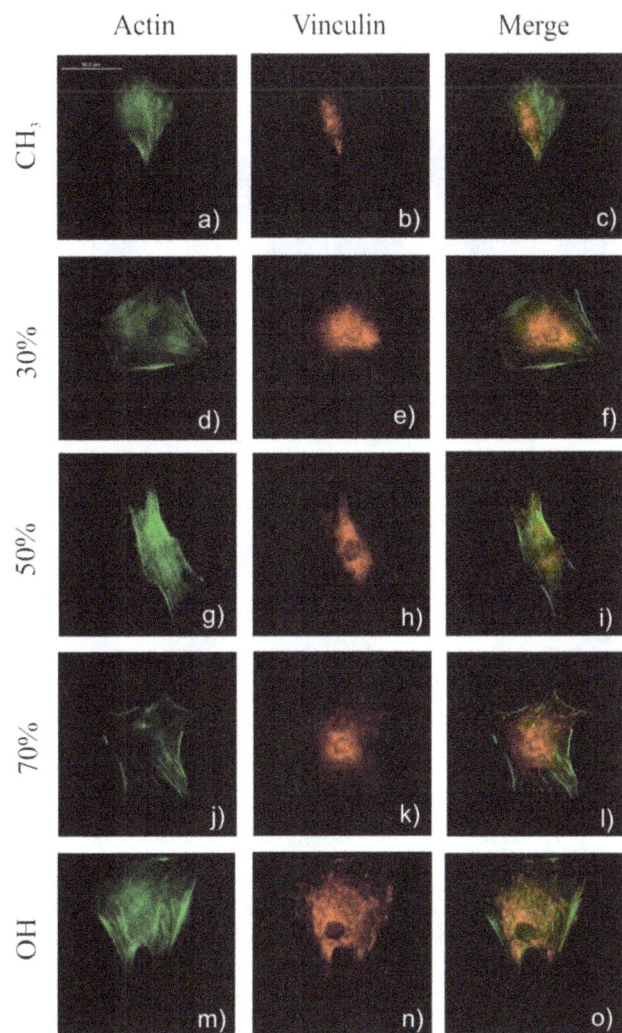

Figure 3. Adhesion of MC3T3-E1 cells after 3 hours on FN coated SAMs. To identify each SAM the percentage of OH groups has been used. First column shows F-actin cytoskeleton, second one the distribution of focal adhesion protein vinculin and its incorporation into focal contact plaques, which is enhanced as the fraction of OH groups increases (see e.g. peripheral organization of well-defined focal contacts in k) and n)). The third column is the superposition of the other two ones. The scale bar in a) is 50 μm.

However, cells are able to synthesize and deposit FN matrix more abundantly and better organized into fibrillar networks on the hydroxyl terminated and the 70%-OH SAMs surfaces.

Matrix degradation

The ability of cells to degrade ECM was investigated by characterizing the expression of two different matrix metalloproteinases (MMPs) and correlated with Runx2 expression. Figure 6 shows characteristic western blot bands for Runx2, MMP2 and MMP9 as well as their relative quantification after 1 day of culture. MMP9 and Runx2 expression increases as the fraction of hydroxyl terminated groups in the surface does. However, MMP2 remain constant regardless the hydroxyl/methyl composition of the material surface.

To gain further insights, we investigated gene expression by RT-PCR (Figure 7). Similar levels of MMP2 are found on the different surfaces. By contrast, MMP9 and Runx2 expressions are

highly dependent on surface chemistry and with enhanced level on the hydrophilic surfaces. Further, immunofluorescence was used to spatially locate MMP2 and MMP9 during cell culture (Figure S6).

Discussion

There is a lack of understanding of the cell-material interaction from an integrated point of view that includes the amount and state of the adsorbed layer of proteins on the material surface, cell adhesion - including integrin expression and focal adhesion formation - cell signaling, matrix reorganization, secretion and degradation, i.e. matrix protein dynamics at the cell-material interface. Some efforts have been devoted in the literature to correlate the material surface properties, especially surface chemistry, to protein adsorption and cell adhesion [37–40]. Here we present results that provide a link between surface chemistry and cell-mediated matrix protein remodeling (including reorganization, secretion and degradation) on a family of model surfaces (SAMs) with controlled ratio of methyl/hydroxyl groups. From a mechanistic point of view, it is known that the influence of surface chemistry on cell behavior is a consequence of the intermediate layer of proteins adsorbed on the material surface. That is to say, cells interact with synthetic material surfaces via the previously deposited layer of FN. The sequence of events would be the following: FN is a macromolecule that display a globular conformation in solution; upon adsorption on a particular surface chemistry, interactions between the chemical groups of the surface and the FN domains triggers changes in the conformation of the protein that might lead to complete unfolding and exposure of groups that were hidden in solution. Consequently, the effect of the material surface chemistry is indirectly received by cells via the adsorbed layer of FN.

The amount of adsorbed FN on the mixed CH_3/OH surfaces is lower as the fraction of hydroxyl terminated chains increases (Figure 1b). This is in agreement with results obtained on this family of SAMs by radiolabeling the protein [41]. That is to say, it is known that FN is adsorbed in higher amount on hydrophobic (CH_3) surfaces than hydrophilic ones (OH) [23]. Our results established the existence of a linear correlation between surface wettability (Figure 1a) and the density of adsorbed FN (Figure 1b) for this family of mixed SAMs. By contrast, the activity of FN after adsorption is higher as the fraction of OH groups on SAMs increased due to the better availability of cell adhesion domains of FN, as it is proved by the HFN7.1 antibody directed to the flexible linker between the 9th and 10th type III repeats of FN [29]. That the activity of FN upon adsorption on SAMs was greater on OH terminated SAMs than CH_3 terminated ones was previously assessed [23,42], and our results confirm the finely tuned chemistry-mediated conformation of FN that leads to a monotonically dependence of FN activity on surface composition, as the CH_3/OH balance on the surface is altered (Figure 1d). It is known that FN has a compact folded structure in physiological buffer that is stabilized through ionic interactions between arms [43]. FN interactions with chemical groups of the substrate (CH_3) give rise to conformational changes in the molecule that must lead to the occlusion of the cell binding domains (III_{9-10}). It is likely that FN orients at the CH_3 surface, so that its hydrophobic segments interact with the methyl groups in PEA, maybe throughout the heparin-binding fragment [44]. Different supramolecular organization of the protein at the material interface is also reflected in protein distribution on the material surface, as directly observed with AFM images in Figure 2 and Figures S1, S2, S3: globular aggregates on the hydrophilic surfaces and fibrillar-like structures on the methyl terminated SAMs.

a)

b)

c)

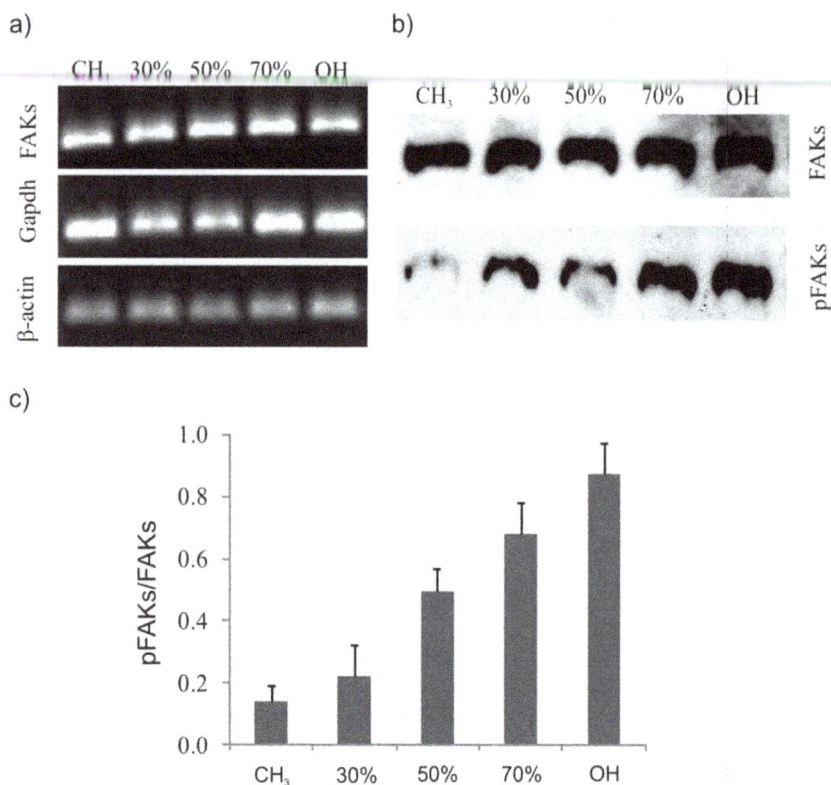

Figure 4. Total FAK expression (protein and gene) and phosphorylation of tyrosine Y-397, the autophosphorylation site in FAK, for MC3T3-E1 cells on FN coated surfaces. SAMs are identified by the percentage of OH groups. a) RT-PCR analysis of FAKs gene expression, β-actin and Gapdh are included as constitutive genes. b) Representative Western blot for total and phophorylated tyrosine residue Y-397 on FAK. c) Quantification of the fraction of phosphorylated FAKs relative to the total FAK expression by image analysis of the western blot bands in b). Error bars represent the standard deviation of three independent experiments; enhanced phosphorylation is obtained as the fraction of OH groups increases.

Differences in the availability of FN adhesion domains on the different SAMs influence the initial cell-material interaction, as determined by focal adhesion formation and F-actin cytoskeleton development (Figure 3). Gene expression of β_1 integrin subunit increases with the fraction of OH groups in the sample (Figure S4), which leads to the development of vinculin plaques and actin fibers only on those SAMs on which FN adsorption occurs with the most favorable conformation, i.e. on those chemistries with the highest fraction of OH groups (Figure 3). The influence of surface chemistry on FN conformation and cell adhesion has been established for SAMs based on different chemical groups. In particular, differences in integrin binding and focal adhesion assembly between OH and CH3 SAMs most likely resulted from surface chemistry dependent differences in the functional presentation of adsorbed FN, whose major integrin-binding RGD domain is particularly sensitive to the underlying chemistry [41,45]. Likewise, it was previously found that the number of cells on FBS-coated CH3/OH mixed SAMs increases as the fraction of OH groups does; up to 80% OH and then it remains constant [46].

Phosphorylation of FAK has been shown to be sensitive to surface chemistry [45]. In our case, increasing the fraction of

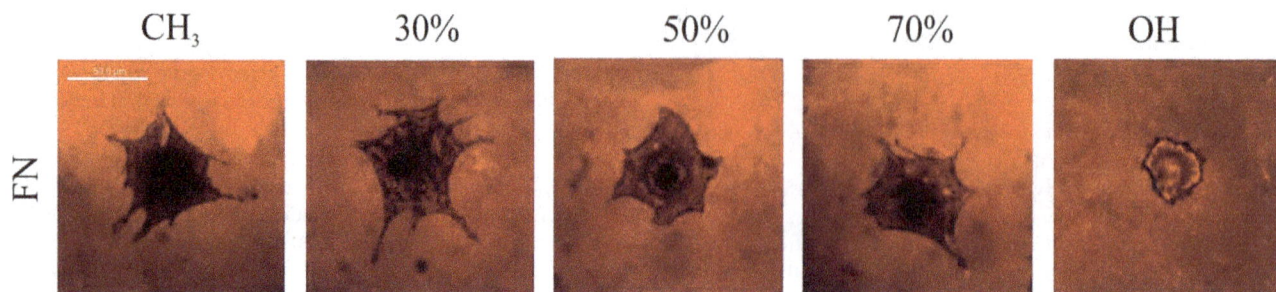

Figure 5. Cellular reorganization of adsorbed FN on the different SAMs after 2.5 h of culture as obtained by immunofluorecence of FN. The red bottom shows FN homogeneously distributed on the material surface. When reorganization of adsorbed FN occurs, black areas (related to the removal of substrate-bound FN) and fibrillar bright areas (as a result of enhanced fluorescence for the incorporation of removed FN into FN-fibrils) are observed. Only the cell shadow in observed for low OH contents (CH3 and 30%). The scale bar represents 50 μm.

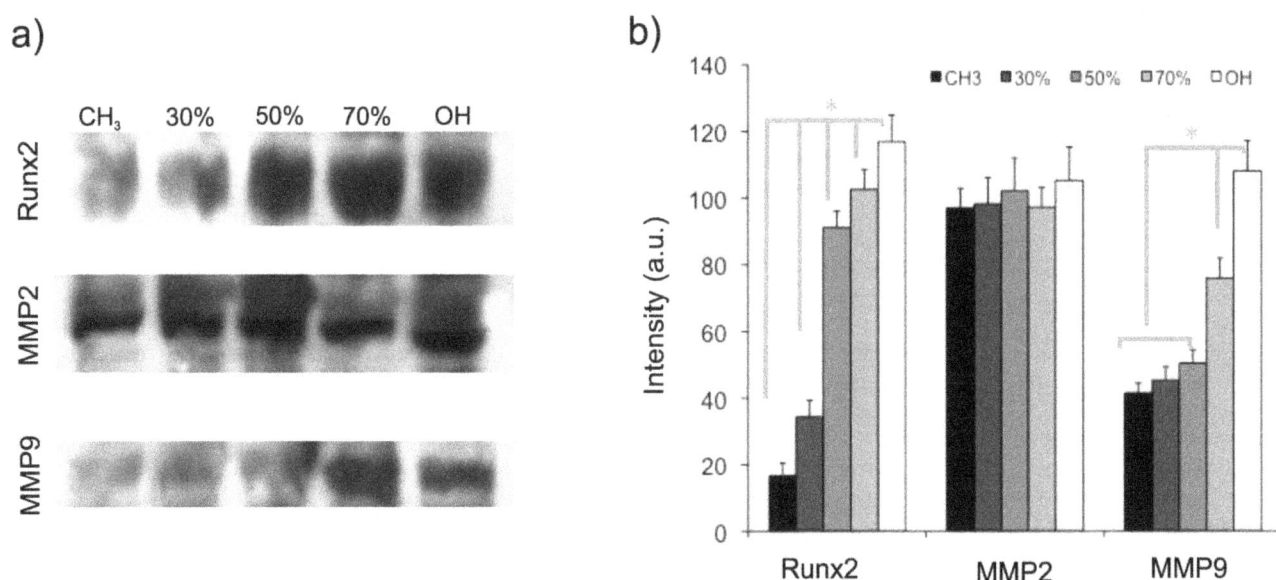

Figure 6. Matrix degradation on the different SAMs quantified by protein expression of matrix metalloproteinases (MMP2, MMP9) and the transcription factor Runx2, which is a target for MMP9. SAMs are identified by the percentage of OH groups. a) Representative Western blot for Runx2, MMP2 and MMP9. b) Quantification of the protein expression by image analysis of the western blot bands. Error bars represent the standard deviation of three independent experiments.

hydroxyl groups on the sample leads to similar FAK levels (both for gene and protein expression, Figure 4) but with higher and higher levels of phosphorylation of Y-397, the autophosphorylation site in FAK and a binding site for Src and PI-3 kinases [47], which suggests a stepwise activation of signaling cascades as a function of hydroxyl groups on the surface increases. That is to

say, activation of signaling pathways is directly related to integrin binding and focal adhesion formation, which are regulated by the availability of binding domains in FN upon adsorption on different chemistries (Figures 1, 2, 3). It has been demonstrated that FAK regulates cell adhesion strengthening via integrin activation and binding [48]. Moreover, our results are consistent with the role Y-

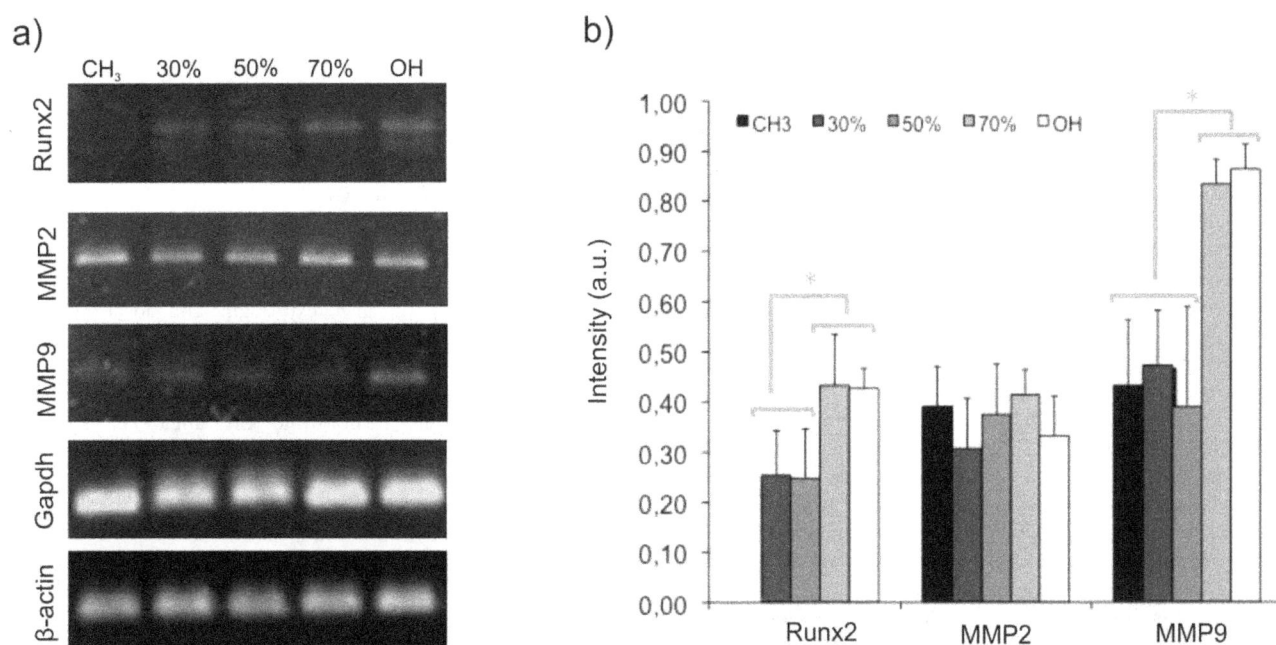

Figure 7. Matrix degradation on the different SAMs quantified by gene expression of matrix metalloproteinases (MMP2, MMP9) and the transcription factor Runx2, which is a target for MMP9. SAMs are identified by the percentage of OH groups. a) Representative RT-PCR bands for Runx2, MMP2 and MMP9; Gapdh and β-actin have been included as constitutive genes. b) Quantification of gene expression by image analysis of RT-PCR bands. The intensity of each band was referred to the level of Gapdh on the same sample. Error bars represent the standard deviation of three independent experiments.

397 autophosphorylation site plays in adhesion strengthening and integrin binding rate. Mutation or blocking of the Y-397 autophosphorylation site blocked FAK-mediated adhesive responses, cell migration and spreading [48–51].

After initial cell adhesion, cells tend to reorganize the adsorbed layer of proteins at the material interface before secreting their own matrix. In this way, FN synthesized by cells assembles into a network of fibrils. During this assembly, however, FN needs to undergo distinct conformational changes, which on adsorption to the substrate can be limited. This may explain why materials surfaces affect FN matrix formation [52,53]. After 2.5 h, cells are able to reorganize the adsorbed layer of FN on the most hydrophilic surfaces (Figure 5) and this ability decreases as the fraction of CH_3 groups on the surface increases. It has been suggested that the ability of cells to reorganize the adsorbed layer of proteins at the material interface must be a consequence of the strength of interaction between the ECM proteins and the material surface, e.g. materials that bind proteins loosely will support the organization of a provisional ECM [11,52–55]. However, additional reasons must be considered when seeking the molecular origin of this fact, which must also be a consequence of the following sequence of events: i) the availability of cell adhesion domains after FN adsorption on the SAM surface is higher in the samples with higher OH content (Figure 1); ii) integrin expression and focal adhesion formation is enhanced on the more hydrophilic surfaces (Figure 3, Figure S4); iii) phosphorylation of FAK is enhanced on the SAMs with higher OH contents (Figure 4). To reorganize the adsorbed layers of proteins, cells must develop mechanical forces on the substrate through a contractile mechanism. Contractility results from dynamic interactions between actin filaments and myosin, which are regulated via phosphorylation of myosin light chain (MLC). Rho GTPases control the formation of stress fibers and focal adhesion assembly by modulating MLC phosphorylation and generating actin-myosin contractility [56]. It is well known that inhibitors of contractility also down-regulated tyrosine phosphorylation of FAK [57–59]; more recently it has been shown that contractility-mediated cell forces also require FAK phosphorylation [60], a fact that supports our reorganization patterns in dependence of the fraction of OH groups: FN is better reorganized on those substrates on which FAK phosphorylation occurs more efficiently (Figures 4, 5).

The dynamics of FN secretion and formation of a fibrillar matrix (late matrix) occurs preferentially on the samples with the higher contents of OH groups (Figure S5); see e.g. the 70%-OH SAM in Figure S5, where the presence of defined FN fibrils of higher fluorescence intensity can be observed. SAMs that promote FN secretion are precisely the substrates on which FN reorganization takes place more intensively (Figure 5). These results support the hypothesis that late matrix formation is in need not only of cell adhesion on the substrate, but some cell movements, in the range of the size of the focal adhesion plaques, must take place so matrix deposition occurs normally [61]. Late matrix formation has been related to the ability of cells to rearrange the initially adsorbed protein layer, especially when comparing cell adhesion on hydrophilic and hydrophobic substrates [52–54].

Except organization, the ECM undergoes proteolytic degradation, which is a mechanism for the removal of the excess ECM usually approximated with remodeling. Matrix remodeling is a subject of an extensive biomedical research, but how it relates to the biocompatibility of materials remains unclear. The importance of the proteolytic activity of cells has been already considered in the design of biomaterials by incorporating MMP sensitive sequences, which have shown to be mandatory in tissue

regeneration in 3D, including cell proliferation, migration and angiogenesis [62–65]. Nevertheless, the effect of material chemistry on the proteolytic activity of cells has not been addressed so far.

Expressions of MMP2 and MMP9 have been observed in MC3T3-E1 cells cultured on tissue culture polystyrene dishes [66]. Our results show that the activation of proteolytic routes in these cells is an MMP-dependent phenomenon sensitive to surface chemistry. MMP2 has FN type II repeats inserted into the catalytic domain [67] and it has been found to cleavage FN and vitronectin into small fragments in vivo, which leads to increased cell adhesion and migration [67,68]. In this sense, MMP2 expression was constant on every FN-coated surface, regardless the underlying chemistry (Figure 6, 7). By contrast, MMP9 expression increases as the fraction of OH groups in the sample does (Figures 6, 7), which suggests a direct relationship between FN activity at the cell-material interface and MMP9 expression, as a consequence of a sequence of events that include integrin expression (Figure S4), focal adhesion formation (Figure 3), matrix reorganization (Figure 5) and FAK phosphorylation (Figure 4). While mechanical strain is known to be able to enhance MMP expression [69], only a few examples in the literature have related the use of synthetic materials on the transcription and activity of MMPs [70–73], which we make explicit here by using SAMs with controlled ratio of methyl/hydroxyl groups.

Runx2 is a key transcription factor in regulation of bone development and osteoblast differentiation. The consequence of interfering with endogenous Runx2 is a defect in normal osteoblast development or function [73]. It has been reported a direct relationship between MMP activity and osteblasts markers [74]. In this sense, MMP9 is a direct target of Runx2 in bone tissue, suggesting a regulatory link between Runx2, the expression of MMP9, and cell migration [75,76]. Figures 6 and 7 also suggest a correlation between Runx2 and MMP9 activation on every surface chemistry. That is to say, Figures 6 and 7 show that both protein and gene expression levels of Runx2 and MMP9 are directly correlated, with low values on the CH_3-rich SAMs, that increases as the OH content in the surface does. This result supports the idea that surface chemistry-mediated activation of MMP9 occurs in a physiological-like way, as its activation at the cell-material interface involves also the upregulation of its direct target Runx2, as occurs in vivo.

Overall, surface chemistry modulates FN dynamics at the cell-material interface. The ratio CH_3/OH in mixed SAMs modulates FN adsorption (in terms of the adsorbed density and conformation), cell adhesion (integrin expression and focal adhesion formation), matrix reorganization and secretion. Further, our results demonstrate that surface chemistry is an external parameter able to trigger proteolytic routes in cells in an MMP-dependent manner. Our results demonstrate the ability of synthetic biomaterials as new tools to direct matrix degradation, which must provide the field with new strategies to investigate fundamental aspects of the phenomenon, as well as the inclusion of parameters to take into account during the design of scaffolds for regenerative medicine, aiming at controlling matrix protein dynamics at the cell-material interface.

Materials and Methods

Preparation of SAMs

SAM surfaces were prepared and characterized as described elsewhere [23] from alkanethiols 1-dodecanethiol (HS-$(CH_2)_{11}$-CH_3), 11-mercapto-1-undecanol (HS-$(CH_2)_{11}$-OH) (Sigma). Au-coated glass coverslips (Fisher Scientific) were prepared by deposition of thin films of Ti (150 Å) followed by Au (150 Å) using a high vacuum evaporator (Polaron E6100) at a deposition rate of 2 Å/s and a chamber base-pressure of $2 \cdot 10^{-6}$ Torr. Glass coverslips were cleaned with 70% H_2SO_4 and 30% H_2O_2 at room temperature for 1 h, rinsed with deionized H_2O, rinsed with 95% ethanol, and dried under a stream of N_2 prior to metal deposition.

Freshly prepared Au-coated surfaces were immersed in alkanethiol solutions (1 mM in absolute ethanol) with different ratios (CH_3/OH), and SAMs were allowed to assemble overnight. SAMs were rinsed in 95% ethanol, dried under N_2 and allowed to equilibrate in DPBS prior to incubation in FN solutions. Surfaces were validated by water contact angle measurements (Dataphysics OCA).

Atomic force microscopy, AFM

AFM experiments were performed using a Multimode AFM equipped with NanoScope IIIa controller from Veeco (Manchester, UK) operating in tapping mode in air; the Nanoscope 5.30r2 software version was used. Si-cantilevers from Veeco (Manchester, UK) were used with force constant of 2.8 N/m and resonance frequency of 75 kHz. The phase signal was set to zero at a frequency 5–10% lower than the resonance one. Drive amplitude was 600 mV and the amplitude setpoint A_{sp} was 1.8 V. The ratio between the amplitude setpoint and the free amplitude A_{sp}/A_0 was kept equal to 0.8.

Protein adsorption

FN from human plasma (Sigma) was adsorbed from solutions of concentrations of 2, 5 and 20 µg/mL in PBS. After adsorption, samples were rinsed in PBS to eliminate the non-adsorbed protein. AFM was performed in the tapping mode immediately after sample preparation.

Separation of FN adsorbed on different samples was performed using 5%-SDS PAGE and denaturing standard conditions as described elsewhere [26]. Proteins were transferred to a PVDF membrane (GE Healthcare) using a semidry transfer cell system (Biorad), and blocked by immersion in 5% skimmed milk in PBS. The blot was incubated with rabbit anti-human FN polyclonal antibody (Sigma, 1:500) in PBS/0.1% Tween-20/2% skimmed milk for 1 h at room temperature and washed with PBS/0.1% Tween-20. The blot was subsequently incubated in HRP-conjugated secondary antibody (GE Healthcare) diluted 1:20000 in PBS/0.1% Tween-20/2% skimmed milk. The enhanced chemiluminescence detection system (GE Healthcare) was used prior to exposing the blot to X-ray. Image analysis of the western bands was done using in house software [26].

Antibody assay for FN conformation

After FN adsorption, surfaces were rinsed in PBS and blocked in 1% BSA/DPBS. Primary monoclonal antibody HFN7.1 (Developmental Hybridoma, Inc., Iowa City, IA) directed against the flexible linker between the 9th and 10th type III repeat was used. Substrates were incubated in primary antibody (1:4000) for 1 h at 37°C. After washing (0.5% Tween 20/DPBS), substrates were incubated in alkaline phosphatase conjugated secondary antibody (1:5000) for 1 h at 37°C and incubated in 4-methylumbelliferyl phosphate (4-MUP) (Sigma) for 45 min at 37°C. Reaction products were quantified using a fluorescence plate reader (Victor III, PerkinElmer) at 365 m /465 nm.

Cell culture

MC3T3-E1 cells were obtained from the RIKEN Cell Bank (Japan). Prior to seeding on FN-coated substrates, cells were maintained in DMEM medium supplemented with 10% foetal bovine serum and 1% penicillin-streptomycin and passaged twice a week using standard procedures. Sample disks placed in a 24-well tissue culture plate were coated with a solution of FN 20 µg/mL. Then, $3 \cdot 10^3$ cells per substrate were seeded and maintained at 37°C in a humidified atmosphere under 5% CO_2 for 3 h. Each experiment was performed in triplicate.

Immunofluorescence (FAKs, MMP, FN)

After 3 h of culture, MC3T3-E1 cells were washed in DPBS (Gibco) and fixed in 10% formalin solution (Sigma) at 4°C. Cells were incubated with permeabilizing buffer (103 g/L sucrose, 2.92 g/L NaCl, 0.6 g/L MgCl2, 4.76 g/L HEPES buffer, 5 mL/L Triton X-100, pH 7.2) for 5 min, blocked in 1% BSA/DPBS and incubated with primary antibody against vinculin (Sigma, 1:400), MMP2 (abcam, 2 µg/mL;) or MMP9 (abcam, 1:100). Samples were then rinsed in 0.5% Tween-20/DPBS. Cy3-conjugated secondary antibody in 1% BSA/DPBS (Invitrogen) and BODIPY FL phallacidin (Invitrogen) were used. Finally, samples were washed and mounted in Vectashield containing DAPI (Vector Laboratories). A Leica DM6000B fluorescent microscope was used for cellular imaging.

The ability of cells to reorganize adsorbed FN (i.e., early matrix) was monitored by coating all samples with 20 µg/mL solution prior seeding in serum containing medium. The evolution of FN in the ECM was followed by immunofluorescence after different culture times and following the same procedure as described before. Samples were incubated with anti-FN antibody (1:400, Sigma) and Cy3-conjugated secondary antibody before washed and mounted with Vectashield containing DAPI.

Protein expression analysis

Total protein extraction was performed lysing the cells with RIPA buffer (50 mM Tris-HCl pH 7.4, 1% nonidet p-40, 0.25% Na-deoxycholate, 150 mM NaCl and 1 mM EDTA) supplemented with protease inhibitor cocktail tablets (Roche). The lysates were concentrated with Microcon YM-30 Centrifugal Filters units (Millipore) and separated in 7%–10%-SDS PAGE under denaturing conditions. To analyze the different expression patterns of FAKs, p-FAKs, MMPs and Runx2 a conventional Western blot procedure was done as previously described. The blots were

Table 1. Primer sequences used in gene expression analysis.

Gen	Sequence (5′-3′)	References
β-actin F	TTCTACAATGAGCTGCGTGTG	M_007393.3
β-actin R	GGGGTGTTGAAGGTCTAAA	
Gapdh F	GTGTGAACGGATTTGGCCGT	NM_008084.2
Gadph R	TTGATGTTAGTGGGGTCTCG	
β integrin F	GGAGGAATGTAACACGACTG	[77]
β integrin R	TGCCCACTGCTGACTTAGGAATC	
FAK F	GGAGTTTTCAGGGTCCGACTG	[77]
FAK R	CATTTTCATATACCTTGTCATTGG	
Runx2 F	GTGCTCTAACCACAGTCCATGCAG	NM_001146038.1
Runx2 R	GTCGGTGCGGACCAGTTCGG	
MMP2 F	TGGTGTGGCACCACCGAGGA	NM_008610.2
MMP2 R	GCATCGGGGGAGGGCCCATA	
MMP9 F	AGCACGGCAACGGAGAAGGC	NM_013599.2
MMP9 R	AGCCCAGTGCATGGCCGAAC	

incubated separately with primary antibody against FAK (abcam, 400 ng/ml), pFAKs (abcam, 1 µg/mL), MMP2, MMP9 and Runx2 (abcam, 1 µg/mL). In all cases the secondary antibody was HRP linked and the dilutions used were: 1:50000 for FAKs, 1:10000 for p-FAKs and 1:20000 for MMP2, MMP9 and Runx2.

The Supersignal West Femto Maximum Sensitivity Substrate (Pierce) was used prior to exposing the blot to X-ray film.

Gene expression analysis

Gene expression (mRNA) of β_1 integrin, Runx2, FAKs, MMP2 and MMP9 was analyzed after 24 h of culture. Total RNA was extracted from cells using RNAeasy Mini Kit (Qiagen). The quantity and integrity of the RNA was measured with NanoDrop (ThermoScientific) and used 3 µg RNA as template for Super-Script III RT (Invitrogen) and oligo(dT)$_{12-18}$ (Invitrogen) as specific primer for amplification of mRNA. PCR reactions were performed with Ampli Taq Gold 360 DNA polymerase (Invitrogen). The oligonucleotides sequence used for PCR reactions are listed in Table 1. All reactions were done at least per triplicate and RNA template was obtained from independent experiments.

Statistical analysis

All experiments were performed at least three times in triplicate unless otherwise noted. Data are reported as mean ± standard error. Results were analyzed by one-way ANOVA using SYSTAT 8.0 (SPSS). If treatment level differences were determined to be significant, pair-wise comparisons were performed using a Tukey post hoc test. A 95% confidence level was considered significant.

Supporting Information

Figure S1 Fibronectin distribution on the different substrates as observed by the phase magnitude in AFM at different magnifications. The protein was adsorbed for 10 min from a solution of concentration 20 µg/mL.

Figure S2 Fibronectin distribution on the different substrates as observed by the phase magnitude in AFM at different magnifications. The protein was adsorbed for 10 min from a solution of concentration 5 µg/mL.

Figure S3 Fibronectin distribution on the different substrates as observed by the phase magnitude in AFM at different magnifications. The protein was adsorbed for 10 min from a solution of concentration 2 µg/mL.

Figure S4 β_1 integrin expression increases with the percentage of OH groups in SAMs. A) Representative bands for gene expression (RT-PCR) of integrin β_1. B) Image quantification of RT-PCR bands on the different surfaces.

Figure S5 Cellular reorganization of adsorbed FN and synthesized FN fibrils on the different surfaces after 2.5 h, 5 h, 1 d and 3 d of culture. The technique employed in these figures is immunofluorescence with anti-FN antibody. It is shown the adsorbed FN on the material surface (red bottom) and the way cells rearrange this layer of FN resulting in black-dark areas as well as enhanced intensity of the fluorescence as a consequence of the formation of FN fibrils by cells. It is shown a broad cell population (20–30 cells per image) after different culture times, so that not only FN reorganization is observed but also FN secretion can be accounted for. The adsorbed FN (red bottom) superimposed with cell-secreted FN fibrils on some SAMS (e.g. 70%).

Figure S6 Immunofluorescence for matrix metalloproteinases MMP2 and MMP9 after 1 day of culture on the FN-coated SAMs (identified by the percentage of OH groups). Fluorescence distribution and intensity is in agreement with protein expression displayed in Figure 6. The corresponding image for F-actin is also included for the sake of cell identification. The scale bar is 50 µm.

Acknowledgments

AFM was performed under the technical guidance of the Microscopy Service at the Universidad Politécnica de Valencia, whose advice is greatly appreciated.

Author Contributions

Conceived and designed the experiments: PR MS-S. Performed the experiments: VL-H PR JB-B MS-S. Analyzed the data: VL-H PR JB-B MS-S. Contributed reagents/materials/analysis tools: DM. Wrote the paper: VL-H PR MS-S.

References

1. Grinnell F (1986) Focal adhesion sites and the removal of substratum-bound fibronectin. J Cell Biol 103: 2697–2706.
2. Hynes RO (2002) Integrins: bidirectional, allosteric signaling machines. Cell 110: 673–687.
3. García AJ (2005) Get a grip: integrins in cell-biomaterial interactions. Biomaterials 26: 7525–7529.
4. Mitra S, Hanson D, Schlaepfer D (2005) Focal adhesion kinase: in command and control of cell motility. Nature Reviews Molecular Cell Biology 5: 56–68.
5. Spie J (2002) Tissue engineering and reparative medicine. Ann NY Acad Sci 961: 1–9.
6. Griffin L, Naughton G (2002) Tissue engineering – Current challenges and expanding opportunities. Science 259: 1009–1014.
7. Altankov G, Groth T (1994) Reorganization of substratum-bound fibronectin on hydrophilic and hydrophobic materials is related to biocompatibility. J Mater Sci Mater M 5: 732–737.
8. Avnur Z, Geiger B (1981) The removal of extracellular fibronectin from areas of cell-substrate contact. Cell 25: 121–132.
9. Altankov G, Groth T, Krasteva N, Albrecht W, Paul D (1997) Morphological evidence for different fibronectin receptor organization and function during fibroblast adhesion on hydrophilic and hydrophobic glass substrata. J Biomat Sci Polym E 8: 721–740.
10. Tzoneva R, Groth T, Altankov G, Paul D (2002) Remodeling of fibrinogen by endothelial cells in dependence of fibronectin matrix assembly. Effect of substratum wettability. J Mater Sci Mater M 13: 1235–1244.

11. Altankov G, Groth T (2006) Fibronectin matrix formation and the biocompatibility of materials. J Mater Sci Mater M 7: 425–429.
12. Pompe T, Keller K, Mitdank C, Werner C (2005) Fibronectin fibril pattern displays the force balance of cell-matrix adhesion. Eur Biophys J 34: 1049–1056.
13. Heyman S, Pauschinger M, De Plama A, Kollwellis-Opara A, Rutschow S, et al. (2006) Inhibition of urokinase type plasminogen activator or matrix metalloproteinases prevents cardiac injury and dysfunction during viral myocarditis. Circulation 114: 565–573.
14. Holmbeck K, Bianco P, Caterina S, et al. (1999) MT1 MMP deficient mice develop dwarfism, osteopenia, arthritis and connective tissue disease due to unadequate collagen turnover. Cell 99: 81–92.
15. Reisenawer A, Eickelberg O, Wille A, Heimburg A, Reinhold A, et al. (2007) Increased carcinogenic potential of myeloid tumor cells induced by aberrant TGF-beta-signaling of cathepsin B. Biol Chem 288: 639–650.
16. Carino AC, Engelholm LH, Yamada SS, Holmbeck K, Lund LR, et al. (2005) Intracellular collagen degradation mediated my uPAR/Endo 180 is a major pathway of extracellular matrix turnover during malignancy. J Cell Biol 169: 977–985.
17. Koblinski J, Ahram M, Sloane BF (2000) Unraveling the role of proteases in cancer. Clin Chem Acta 291: 113–135.
18. Mohamed M, Sloane BF (2006) Cysteine cathepsins: multifunctional enzymes in cancer. Nat Rev Cancer 6: 764–775.
19. Buck MR, Karustic DG, Day NA, Honn KV, Sloane BF (1992) Degradation of extracellular matrix proteins by human cathepsin B from normal and tumour tissues. Biochem J 282: 273–278.

20. Page-McCaw A, Ewald AJ, Werb Z (2007) Matrix metaloproteinases and the regulation of tissue remodeling. Nat Rev Mol Cell Biol 8: 221–233.

21. Rodríguez D, Morrison C, Over CM (2010) Matrix metalloproteinases: What do they not do? New substrates and biological roles identified by murine models and proteomics. Biochimica et Biophysica Acta 1803: 39–54.

22. Raynor JE, Capadona JR, Collard DM, Petrie TA, García AJ (2009) Polymer brushes and self-assembled monolayers: Versatile platforms to control cell adhesion to biomaterials. Biointerphases 4: FA3–16.

23. Keselowsky BG, Collard DM, Garcia AJ (2003) Surface chemistry modulates fibronectin conformation and directs integrin binding and specificity to control cell adhesion. J Biomed Mater Res 66: 247–259.

24. Martins MC, Ratner BD, Barbosa MA (2003) Protein adsorption on mixtures of hydroxyl- and methyl-terminated alkanethiols self-assembled monolayers. J Biomed Mater Res A 67: 158–71.

25. Rodrigues SN, Gonçalves IC, Martins MC, Barbosa MA, Ratner BD (2006) Fibrinogen adsorption, platelet adhesion and activation on mixed hydroxyl/methyl terminated self-assembled monolayers. Biomaterials 27: 5357–5367.

26. Rico P, Rodríguez Hernández JC, Moratal D, Altankov G, Monleón Pradas M, et al. (2009) Substrate-induced assembly of fibronectin into networks: influence of surface chemistry and effect on osteoblast adhesion. Tiss Eng Part A 15: 3271–3281.

27. Ugarova TP, Zamarron C, Veklich Y, Bowditch RD, Ginsberg MH, et al. (1995) Conformational Transitions in the Cell Binding Domain of Fibronectin. Biochemistry 34: 4457–4466.

28. McClary KB, Ugarova T, Grainger DW (2000) Modulating fibroblast adhesion, spreading, and proliferation using self-assembled monolayer films of alkylthiolates on gold. J Biomed Mater Res 50A: 428–439.

29. Schoen RC, Bentley KL, Klebe RJ (1982) Monoclonal antibody against human fibronectin which inhibits cell attachment. Hybridoma 1: 99–108.

30. Ilic D, Furuta Y, Kanazawa S, Takeda N, Sobue K, et al. (1995) Reduced cell motility and enhanced focal adhesion contact formation in cells from FAK-deficient mice. Nature 377: 539–544.

31. Cary LA, Chang JF, Guan JL (1996) Stimulation of cell migration by overexpression of focal adhesion kinase and its association with Src and Fyn. J Cell Sci 109: 1787–1794.

32. Frisch SM, Vuori K, Ruoslahti E, Chan-Hui PY (1996) Control of adhesion-dependent cell survival by focal adhesion kinase. J Cell Biol 134: 793–799.

33. Zhao JH, Reiske H, Guan JL (1998) Regulation of the cell cycle by focal adhesion kinase. J Cell Biol 143: 1997–2008.

34. Thannickal VJ, Lee DY, White ES, Cui Z, Larios JM, et al. (2003) Myofibroblast differentiation by transforming growth factor-beta1 is dependent on cell adhesion and integrin signaling via focal adhesion kinase. J Biol Chem 278: 12384–12389.

35. Schaller MD, Hildebrand JD, Shannon JD, Fox JW, Vines RR, et al. (1994) Autophosphorylation of the focal adhesion kinase, pp125FAK, directs SH2-dependent binding of pp60src. Mol Cell Biol 14: 1680–1688.

36. Reiske HR, Kao SC, Cary LA, Guan JL, Lai JF, et al. (1999) Requirement of phosphatidylinositol 3-kinase in focal adhesion kinase-promoted cell migration. J Biol Chem 274: 12361–12366.

37. Shin H (2007) Fabrication methods of an engineered microenvironment for analysis of cell-biomaterial interactions. Biomaterials 28: 126–133.

38. Monchaux E, Vermette P (2010) Effects of surface properties and bioactivation of biomaterials on endothelial cells. Front Biosci (Schol Ed) 2: 239–255.

39. Wilson CJ, Clegg RE, Leavesley DI, Pearcy MJ (2005) Mediation of biomaterial-cell interactions by adsorbed proteins: a review. Tissue Eng 11(1–2): 1–18.

40. Palacio ML, Schricker SR, Bhushan B (2011) Bioadhesion of various proteins on random, diblock and triblock copolymer surfaces and the effect of pH conditions. J R Soc Interface 8: 630–640.

41. Barrias CC, Martins MCL, Almeida-Porada G, Barbosa M, Granja PL (2009) The correlation between the adsorption of adhesive proteins and cell behaviour on hidroxi-methyl mixed self-assembled monolayer. Biomaterials 30: 307–316.

42. Michael KE, Vernekar VN, Keselowsky BG, Meredith JC, Latour RA, et al. (2003) Adsorption-Induced Conformational Changes in Fibronectin Due to Interactions with Well-Defined Surface Chemistries. Langmuir 19: 8033–8040.

43. Aota S, Nomizu M, Yamada KM (1994) The short amino acid sequence Pro-His-Ser-Arg-Asn in human fibronectin enhances cell-adhesive function. J Biol Chem2 69: 24756–24761.

44. Gugutkov D, Hernandez JCR, González-García C, Altankov G, Salmerón-Sánchez M (2009) Biological Activity of the Substrate-Induced Fibronectin Network: Insight into the Third Dimension through Electrospun Fibers. Langmuir 25(18): 10893–10900.

45. Keselowsky BG, Collard DM, García AJ (2004) Surface chemistry modulates focal adhesion composition and signaling through changes in integrin. Biomaterials 25: 5947–5954.

46. Arima Y, Iwata H (2007) Effect of wettability and surface functional groups on protein adsorption and cell adhesion using well-defined mixed self-assembled monolayers. Biomaterials 28: 3074–3082.

47. Schaller MD, Hildebrand JD, Shannon JD, Fox JW, Vines RR, et al. (1994) Autophosphorylation of the focal adhesion kinase, pp125FAK, directs SH2-dependent binding of pp60src. Mol Cell Biol 14: 1680–1688.

48. Michael KE, Dumbauld DW, Burns KL, Hanks SK, García AJ (2009) FAK modulates cell adhesion strengthening via integrin activation. Mol Biol Cell 20: 2508–2519.

49. Sieg DJ, Hauck CR, Ilic D, Klingbeil CK, Schaefer E, et al. (2000) FAK integrates growth-factor and integrin signals to promote cell migration. Nat Cell Biol 2: 249–256.

50. Wang HB, Dembo M, Hanks SK, Wang YY (2001) Focal adhesion kinase is involved in mechanosensing during fibroblast migration. Proc Natl Acad Sci USA 98: 11295–11300.

51. Webb DJ, Donais K, Whitmore LA, Thomas SM, Turner CE, et al. (2004) FAK-Src signalling through paxillin, ERK and MLCK regulates adhesion disassembly. Nat Cell Biol 6: 154–161.

52. Altankov G, Groth T (1994) Reorganization of substratum-bound fibronectin on hydrophilic and hydrophobic materials is related to biocompatibility. J Mater Sci Mater Med 5: 732–737.

53. Altankov G, Groth T (1996) Fibronectin matrix formation and the biocompatibility of materials. J Mater Sci Mater Med 7: 425–429.

54. Altankov G, Groth T, Krasteva N, Albrecht W, Paul D (1997) Morphological evidence for different fibronectin receptor organization and function during fibroblast adhesion on hydrophilic and hydrophobic glass substrata. J Biomat Sci Polym E 8: 721–740.

55. Tzoneva R, Groth T, Altankov G, Paul D (2002) Remodeling of fibrinogen by endothelial cells in dependence of fibronectin matrix assembly. Effect of substratum wettability. J Mater Sci Mater M 13: 1235–1244.

56. Kaibuchi K, Kuroda S, Amano M (1999) Regulation of the cytoskeleton and cell adhesion by the Rho family GTPases in mammalian cells. Annu Rev Biochem 68: 459–486.

57. Chrzanowska-Wodnicka M, Burridge K (1996) Rho-stimulated contractility drives the formation of stress fibers and focal adhesions. J Cell Biol 133: 1403–1415.

58. Gallagher PJ, Herring BP, Stull JT (1997) Myosin light chain kinases. J Muscle Res Cell Motil 18: 1–16.

59. Wozniak MA, Desai R, Solski PA, Der CJ, Keely PJ (2003) ROCK-generated contractility regulates breast epithelial cell differentiation in response to the physical properties of a three-dimensional collagen matrix. J Cell Biol 163: 583–595.

60. Dumbauld DW, Shin H, Gallant N, Michael K, Radhakrishna H, et al. (2010) Contractility Modulates Cell Adhesion Strengthening Through Focal Adhesion Kinase and Assembly of Vinculin-Containing Focal Adhesions. J Cell Physiol 223: 746–756.

61. González-García C, Sousa S, Moratal D, Rico P, Salmerón-Sánchez M (2010) Effect of nanoscale topography on fibronectin adsorption, focal adhesion size and matrix organisation. Col Surf B 77: 181–190.

62. Bott K, Upton Z, Schrobback K, Ehrbar M, Hubbell JA, et al. (2010) The effect of matrix characteristics on fibroblasts proliferation in 3D gels. Biomaterials 31: 8454–8464.

63. Phelps EA, Landázuri N, Thulé PM, Taylor WR, García AJ (2010) Bioartificial matrices for therapeutic vascularization. Proc Nat Acad Sci 107: 3323–3328.

64. Lutolf MP, Lauer-Fields JL, Schmoekel HG, Metters AT, Weber FE, et al. (2003) Synthetic matrix metalloproteinase-sensitive hydrogels for the conduction of tissue regeneration: engineering cell-invasion characteristics. Proc Nat Acad Sci 100: 5413–5418.

65. Schneider R, Puellen A, Kramann R, Raupach K, Bornemann I, et al. (2010) The osteogenic differentiation of adult bone marrow and perinatal umbilical mesenchymal stem cells and matrix remodelling in three-dimensional collagen scaffolds. Biomaterials 31: 467–480.

66. Uchida M, Shima M, Shimoaka T, Fujieda A, Obara K, et al. (2000) Regulation of matrix metalloproteinases (MMPs) and tissue inhibitors of metalloproteinases (TIMPs) by bone resorptive factors in osteoblastic cells. J Cell Physiol 185: 207–214.

67. Page-MacCaw A, Ewald AJ, Werb Z (2007) Matriz metaloproteinases and the regulation of tissue remodelling. Nat Mol Cell Biol 8: 221–233.

68. Kenny HA, Kaur S, Coussens LM, Lengyel E (2008) The inicial stops of ovarian cancer cell metastasis are mediated by MMP-2 cleavage of vitronectin and fibronectin. J Clin Invest 118: 1367–1379.

69. Yang CM, Chien CS, Yao CC, Hsiao LD, Huang YC, et al. (2004) Mechanical strain induces collagenase-3 (MMP-13) expresión in MC3T3-E1 osteoblastic cells. J Biol Chem 279: 22158–22165.

70. Wan R, Mo Y, Zhang X, Chien S, Tollerud DJ, et al. (2009) Matrix metalloproteinase-2 and -9 are induced differently by metal nanoparticles in human monocytes: The role of oxidative stress and protein tyrosine kinase activation. Toxicology and Applied Pharmacology 233: 276–285.

71. Zambuzzi W, Paiva KB, Menezes R, Oliveira RC, Taga R, et al. (2009) MMP-9 and CD68+ cells are required for tissue remodeling in response to natural hydroxyapatite. J Mol Hist 40: 301–309.

72. Chung AS, Waldeck H, Schmidt DR, Kao WJ (2009) Monocyte inflammatory and matrix remodeling response modulated by grafted ECM-derived ligand concentration. J Biomed Mater Res 91A: 742–752.

73. Ducy P, Zhang R, Geoffroy V, Ridall AL, Karsenty G (1997) Osf/Cbfa1: a transcripcional activator of osteoblast differentiation. Cell 89: 747–754.

74. Hayami T, Kapila YL, Kapila S (2008) MMP-1 (collagenase-1) and MMP-13 (collagenase-3) differentially regulate markers of osteoblastic differentiation in osteogenic cells. Matrix Biol 27: 682–692.

75. Pratap J, Javed A, Languino LR, van Wijnen AJ, Stein JL, et al. (2005) The Runx2 osteogenic transcription factor regulates matrix metalloproteinase 9 in bone metastático cancer cells and controls cell invasión. Mol Cell Biol 25: 0501–0591

76. Hess J, Porte D, Munz C, Angel P (2001) AP-1 and Cbfa/Runt physically interact and regulate PTH-dependent MMP13 expression in osteoblasts through a new OSE2/AP-1 composite element. J Biol Chem 276: 20029–20038.

77. Rouahi M, Champion E, Hardouin P, Anselme K (2006) Quantitative kinetic analysis of gene expression during human osteoblastic adhesion on orthopaedic materials. Biomaterials 27: 2829–2844.

4

Effects of Soil Warming and Nitrogen Addition on Soil Respiration in a New Zealand Tussock Grassland

Scott L. Graham[1,2]*, John E. Hunt[2], Peter Millard[2], Tony McSeveny[2], Jason M. Tylianakis[1,3], David Whitehead[2]

1 School of Biological Sciences, University of Canterbury, Christchurch, New Zealand, **2** Landcare Research, Lincoln, New Zealand, **3** Department of Life Sciences, Imperial College London, Silwood Park Campus, Ascot, Berkshire, United Kingdom

Abstract

Soil respiration (R_S) represents a large terrestrial source of CO_2 to the atmosphere. Global change drivers such as climate warming and nitrogen deposition are expected to alter the terrestrial carbon cycle with likely consequences for R_S and its components, autotrophic (R_A) and heterotrophic respiration (R_H). Here we investigate the impacts of a 3°C soil warming treatment and a 50 kg ha^{-1} y^{-1} nitrogen addition treatment on R_S, R_H and their respective seasonal temperature responses in an experimental tussock grassland. Average respiration in untreated soils was 0.96 ± 0.09 μmol m^{-2} s^{-1} over the course of the experiment. Soil warming and nitrogen addition increased R_S by 41% and 12% respectively. These treatment effects were additive under combined warming and nitrogen addition. Warming increased R_H by 37% while nitrogen addition had no effect. Warming and nitrogen addition affected the seasonal temperature response of R_S by increasing the basal rate of respiration (R_{10}) by 14% and 20% respectively. There was no significant interaction between treatments for R_{10}. The treatments had no impact on activation energy (E_0). The seasonal temperature response of R_H was not affected by either warming or nitrogen addition. These results suggest that the additional CO_2 emissions from New Zealand tussock grassland soils as a result of warming-enhanced R_S constitute a potential positive feedback to rising atmospheric CO_2 concentration.

Editor: Xiujun Wang, University of Maryland, United States of America

Funding: SLG was supported by a fellowship from the Miss E.L. Hellaby Indigenous Grassland Research Trust. JEH, PM, TM, and DW received funding from the Ministry for Business, Innovation and Employment. JMT was funded by a Rutherford Discovery Fellowship administered by the Royal Society of New Zealand. This work was funded by the Marsden Fund (UOC-0705) and the Miss E.L. Hellaby Indigenous Grassland Research Trust. The funders had no role in study design, data collection and analysis, decision to publish, or preparation of the manuscript.

Competing Interests: The authors have declared that no competing interests exist.

* E-mail: scott.graham@ucf.edu

Introduction

Soils contain a pool of carbon approximately double that stored in terrestrial biomass [1]. Soil respiration (R_S), the primary pathway for return of soil carbon to the atmosphere, is increasing globally by 0.1 Pg C y^{-1} at present [2]. This increase, hypothesised to be a result of global warming, is concerning as temperatures are expected to rise by as much as 6.4°C over the next century [3]. Coupled climate models indicate a likely soil-driven positive feedback to climate change, although uncertainty remains in the magnitude of this feedback [4,5].

Numerous warming experiments have investigated the impacts of long-term climate warming on carbon cycling, suggesting that, on average, warming of 0.3–6.0°C increases soil respiration (R_S) by 20% [6]. However, several notable examples have shown the effect of warming on R_S to be only transient [7,8]. Mechanisms for this acclimation of R_S to prolonged warming include depletion of labile carbon substrates [8–10], changes to the microbial community structure [7,11], physiological acclimation of soil microbes [12], reduction in root biomass [13] and reduction in the specific root respiration rate [14]. Acclimation of soil respiration may limit potential soil carbon loss as a result of climate warming.

Global change scenarios also suggest that nitrogen cycling in terrestrial ecosystems will be altered. Nitrogen deposition due to crop fertilisation and fossil fuel combustion currently exceeds terrestrial nitrogen fixation and is expected to increase in the future [15]. As warming also increases nitrogen mineralization [6], there exists the possibility for synergistic effects of warming and anthropogenic nitrogen deposition on plant-available nitrogen.

While warming-induced increases in R_S represent a likely positive feedback to rising atmospheric CO_2 concentration, enhanced nitrogen deposition has been suggested as a possible mitigating factor due to negative impacts of nitrogen addition on R_S [16]. The findings from nitrogen addition experiments in forests suggest that reduction in R_S may represent a carbon offset equivalent to the nitrogen fertilisation effect on primary production. As well, reductions in R_S have been observed in grasslands as a result of nitrogen addition [17,18]. Due to feedbacks between the nitrogen and carbon cycles, nitrogen availability will likely influence the magnitude of the terrestrial feedback to rising atmospheric CO_2 concentration [19].

The net response of R_S to warming and nitrogen addition depends largely on the combined response of its components, autotrophic soil respiration (R_A) and heterotrophic soil respiration (R_H), which are likely to have different responses to environmental drivers. Autotrophic respiration refers to respiratory activity of roots and associated rhizosphere microbes, while R_H refers to soil organic matter decomposition by soil microbes [20]. The important distinction between R_A and R_H is that the former represents respiration of carbon recently assimilated by plants, whereas the latter releases carbon that may have residence times in the soil reaching millennia [21].

Heterotrophic respiration is widely expected to increase under warming scenarios [22,23]. Several field warming experiments demonstrated such increases [24,25]. While warming generally increases R_H, nitrogen addition can decrease microbial biomass [26], potentially reducing R_H. This reduction in R_H may explain the overall decrease in R_S observed in response to nitrogen addition [16].

In this study, we investigated the impacts of soil warming and nitrogen addition, as well as their interaction, on R_S and its components, R_A and R_H. Such multifactor experiments are important to improve the predictive ability of coupled-climate models, as single factor experiments may fail to predict interactive effects of global change drivers [27,28]. Likewise, partitioning the autotrophic and heterotrophic components of R_S can lead to greater mechanistic understanding of the response of R_S to environmental drivers [29].

Native tussock grassland was selected as a model system because grasslands are a widespread and important store of carbon in New Zealand [30], and globally [31]. Soil respiration and R_H were measured over a period of 27 months with the objective of determining the likely feedback effect that increases in R_S in grasslands will have on rising atmospheric CO_2 concentration in response to soil warming and nitrogen addition.

Methods

Study site

This study was conducted at the Cass Warming Experiment at the University of Canterbury Cass Field Station in central South Island, New Zealand (43.03° S, 171.75° E, 590 m a.s.l.). The site was constructed in January 2009 in an area of tussock grassland. Soils at the site are classified as acidic allophane brown by New Zealand Soil Classification (Typic Dystrochrept by USDA) [32,33]. Prior to this study, vegetation and the top 200 mm of topsoil were removed, twenty 12.25 m^2 plots were laid out and 90 m of resistance heating cable (Argus Heating, Ltd., Christchurch, New Zealand) were arranged in rows with 140 mm spacing between cables in each plot to achieve a heating density of 76 W m^{-2} [34]. Dummy cables were arranged similarly in unheated plots. The cables were then covered with 200 mm of topsoil and the native New Zealand tussock grasses *Festuca novae-zelandiae* (50 individuals per plot), *Poa cita* (50 per plot), *Chionochloa rigida* (22 per plot), and *Chionochloa flavescens* (12 per plot) were planted.

Five plots were assigned to each of four treatments: control, warming only, nitrogen addition only and combined warming and nitrogen addition (Appendix S1). In each of 10 plots designated for warming, three thermocouples (Type-E, Campbell Scientific, Logan, UT, USA) were buried to a depth of 100 mm in a stratified design which captured a range of horizontal distances from heating cables (directly above, one quarter of the distance between two cables and the midpoint). In each of the control plots, one thermocouple was buried to 100 mm soil depth. The heating cables were switched on and off to maintain a 3°C difference between the average of the three thermocouples in warmed plots and the nearest un-warmed plot. Warming was controlled by a datalogger (CR1000X, Campbell Scientific, Logan, UT, USA) and hourly average plot soil temperatures were recorded. An auxiliary weather station measured hourly average air temperature, soil temperature and volumetric water content at 100 mm depth.

Nitrogen addition began in February 2009. Nitrogen was applied as calcium ammonium nitrate at a rate of 10 kg N ha^{-1}

five times throughout the growing season to achieve a total amendment of 50 kg N ha^{-1} y^{-1}. For each plot, the fertiliser was dissolved in 4 L water and distributed using a watering can over both plants and soil. The continuous 3°C warming treatment was started in July 2009. Two plots, one each of the warming only and combined warming and nitrogen addition treatments, were subsequently dropped from analyses due to malfunction of the heating cables.

Respiration measurements

Measurements of soil respiration were carried out over a 27 month period beginning in August 2009 (winter) and continuing through October 2011 (spring). Six 100 mm diameter polyvinyl chloride measurement collars were installed to a soil depth of 70 mm in each plot. The rate of soil respiration in each collar was measured at 2–4 week intervals using a portable respiration system (SRC-1 and EGM-4, PP Systems, Amesbury, MA, USA). An additional two measurement collars were installed in each plot to a soil depth of 300 mm in order to exclude roots and provide an estimate of heterotrophic respiration (R_H). These deep collars extended into the clay subsoil, limiting potential root growth into the soil beneath the collar. In contrast, the shallow collars were inserted to a depth that would allow root infiltration beneath the collar while providing a seal with the soil surface. The collars remained in place for the duration of the experiment to avoid soil or root disturbance. Measurement using the deep collars began in January 2010. Simultaneous with each soil respiration measurement, soil temperature and soil water content at 50 mm depth were measured using a thermocouple (Type-E, Omega Engineering, Ltd, Stamford, CT, USA) and a soil moisture sensor (Theta Probe type ML1 and ML2, DeltaT Devices, Cambridge, UK), respectively.

Substrate addition

Availability of labile substrates in the soil is important in regulating R_S [35]. In order to assess levels of substrate limitation induced by warming and nitrogen addition treatments and the presence of roots, a substrate addition experiment was carried out in late-October 2011. In each of 16 plots representing four replicates for each treatment, two pairs of soil respiration measurement collars were selected: one pair with roots present and another pair with roots excluded. One collar from each pair was selected randomly for substrate addition. All collars were measured immediately prior to substrate addition. Subsequent to initial measurement, the two collars from each plot selected for substrate addition were amended with 20 ml of 0.2 M sucrose solution (an amount approximately equivalent to 10 days of carbon losses from R_H). In order to ensure that the sucrose solution infiltrated beyond the soil surface, 5 ml were injected with a syringe to a depth of 25 mm at four locations within each collar. The collars designated as controls were treated similarly with water. Soil respiration was then measured in each collar at 30 min, 1 h, 2.5 h, 4 h and then at 4–8 h intervals until the substrate response was no longer evident. Substrate-induced respiration (S_I) was calculated for each pair of collars as the difference between respiration rates of the control and substrate-added collars, as a proportion of the rate for the control treatment.

Soil analyses

Soils were sampled in January 2010, March 2011 and March 2012. Three 54 mm diameter soil cores were taken to a depth of 100 mm in each plot and the soil was homogenized into a single sample. Roots were removed by 8 mm sieve and dried at 60°C. As the grass roots were very fine, in 2012 a subsample was removed from the whole sample and washed over a 650 μm sieve to obtain root biomass. Microbial biomass was estimated using the fumigation-extraction technique adapted from Vance et al. [36]. Remaining soil was dried at 60°C, passed through a 2 mm sieve to remove remaining roots and ground in a ball mill. Samples were then analysed for total carbon and nitrogen concentration on an elemental analyser (CNS2000, Leco, St. Joseph, MO, USA).

Between 20 September 2011 and 22 October 2011, plant available nitrogen was estimated using ion exchange membranes (PRS probes, Western Ag Innovations, Saskatoon, Canada). The PRS probes were installed to a depth of 100 mm at three locations in each plot. Following a one-month burial period, probes were removed, rinsed with deionized water and returned to Western Ag Innovations for analysis of NH_4^+ and NO_3^-.

Statistical analyses

The effects of warming and nitrogen addition on seasonal measurements of R_S, R_H and soil water content were assessed using linear mixed-effects models conducted in the 'nlme' package [37] in R version 2.12.1 [38]. Warming, nitrogen addition and measurement date, along with their interactions were included as fixed effects, while measurement collars nested within plots were included as random effects to account for the non-independence of multiple samples through time and multiple collars per plot. Residual analyses were undertaken and log transformation was used for R_S and R_H, to correct for heteroscedasticity. The effects of warming and nitrogen on the proportion of R_S constituted by R_H (f_{RH}) were similarly assessed by treating plot averages of f_{RH} on a given date as a sample and evaluating random effects at the plot level.

Temperature responses of R_S and R_H were fitted to an Arrhenius-type curve [39], modified with a soil water content response function [40]:

$$R_S = R_{10} * \exp E_0 \left(\frac{1}{56.02} - \frac{1}{T_S - 227.13} \right)$$
$$* \exp\left(-\exp\left(a - b * \theta \right) \right) \quad (1)$$

where T_S is soil temperature (K), R_{10} is the basal respiration rate at 10°C, E_0 is the activation energy of enzymatic reactions, θ is the soil volumetric water content and a and b are parameters that determine the shape of a sigmoidal response of respiration to soil water content.

Nonlinear mixed-effects models (also conducted in the 'nlme' package for R) were used to fit Equation (1) initially to measurements of R_S, and subsequently to R_H. First, the effect of roots on the temperature response of R_S was investigated by testing how root presence (as a fixed effect) altered parameter values for R_{10} and E_0. Subsequently, temperature responses of R_S and R_H were investigated in separate models, with the latter substituting R_H in place of R_S in Equation (1). Warming and nitrogen addition, as well as their interactions, were investigated as fixed effects on R_{10} and E_0 for both the R_S and R_H models. For all the above nonlinear models, measurement collars nested within plots were evaluated as random effects. Fixed and random effects on a and b were not evaluated, as few measurement dates occurred under water-limited conditions, and a generic water content response curve which limited respiration when soil water content was less than 0.2 $m^3 m^{-3}$ was deemed appropriate based on analysis of residuals of a temperature-only model.

The final fixed effects structure was determined by first constructing a maximal model which included the presence of plant roots, warming, nitrogen and all interactions. A power variance function was fitted in order to correct for heteroscedasticity [41]. To account for autocorrelation in repeated measurements of the same collar, a first order autoregressive structure was used [42]. Fixed effects and interactions were removed iteratively based initially on their p-values and the best fit model was selected. During each step of this procedure, model comparisons were undertaken using a likelihood ratio test and selection of the best-fitting model was achieved through minimisation of Akaikie's Information Criterion (AIC). To test for potential acclimation of R_S to warming, the data were bisected such that the first full year of measurement was separated from the second. The interaction between measurement year and the warming treatment was added as a fixed effect on both R_{10} and E_0 parameters. This interaction was tested for significance to determine whether the warming effect was consistent across measurement years, with a significant negative interaction term for either parameter indicating acclimation.

Soil carbon content, soil nitrogen content, microbial biomass, plant available nitrogen and substrate induced respiration (S_I) were all assessed by multi-way ANOVA, with temperature and nitrogen treatments, as well as their interaction, as factors. For S_I, the maximum value recorded for each pair off collars over the measurement period was tested. For those variables that were measured repeatedly (soil carbon, soil nitrogen, microbial biomass carbon), a separate ANOVA was conducted for each time point.

Results

Seasonal variation of soil temperature and water content

Average soil temperature over the entire 27 month measurement period was 9.6°C. Warming increased soil temperature by an average (±SE) of 3.1±0.2°C over the course of the experiment (Fig. 1). Soil water content varied seasonally, falling below 0.2 m^3 m^{-3} periodically during summer and remaining above 0.40 m^3 m^{-3} during the winter months. Warming significantly reduced soil volumetric water content (p = 0.001, Table S1) by an average 0.01 m^3 m^{-3}. This reduction in soil water content was most evident in summer when water was limiting, with maximum reduction in soil water content of 0.04 m^3 m^{-3} observed in February 2010 and 0.07 m^3 m^{-3} in January 2011.

Seasonality of R_S

Soil respiration showed a seasonal pattern driven primarily by seasonal temperature (Fig. 1, Fig. 2A). As such, measurement date had a significant effect in the linear mixed-effects model (p<0.0001, Table S2). Heterotrophic respiration showed a similar response to seasonal temperature and, as such, measurement date was significant in predicting R_H (p<0.0001, Table S3, Fig. 2B). The average proportion of R_S constituted by R_H (i.e., f_{RH}) was 71% (Fig. 2C). No apparent seasonal pattern was observed in f_{RH}. Both R_S and R_H were sensitive to soil water content, with a reduction in respiration rate observed below 0.2 m^3 m^{-3} soil water content. This was particularly evident on 5 March 2010 and 12 December 2010, the driest measurement dates (Fig. 2).

Effects of warming and nitrogen addition on R_S

On average, R_S was increased by 41% due to warming (p<0.0001) and by 12% due to nitrogen addition (p = 0.004). The treatments combined additively, as no significant interaction was observed. Warming significantly increased R_H by 37% (p = 0.014), though nitrogen did not significantly affect R_H (p = 0.798), nor was there a significant interaction between the warming and nitrogen treatments. The proportional contribution of R_H to R_S was reduced to an average of 59% by nitrogen addition, although this

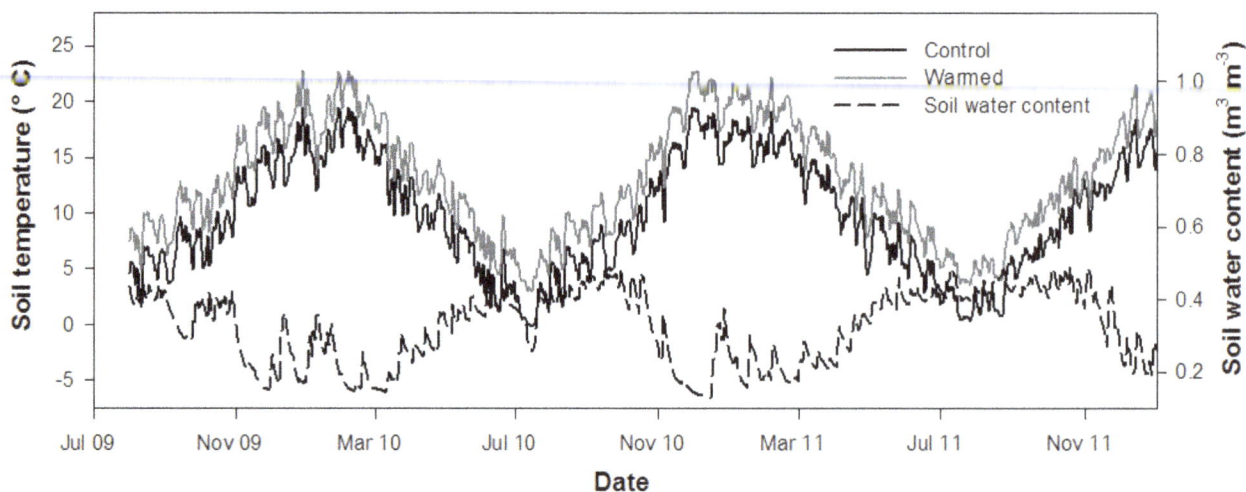

Figure 1. Seasonal variations in soil variables. Measured soil temperature, T_S (°C), and soil water content, θ, at 100 mm depth.

reduction was of marginal significance (p = 0.051, Table S4) due to high variability in f_{RH}. Warming had no significant impact on f_{RH}.

Responses of R_S and R_H to added substrate were highly variable across replicates, with values of S_I ranging from a 0.8 to 4.75 fold increase in respiration. There was no effect of the presence of roots on S_I (p = 0.487). For R_S, maximum S_I was decreased significantly by warming (p = 0.040, Table 1). In contrast, nitrogen addition did not affect S_I significantly (p = 0.146), nor was there any significant interaction with warming (p = 0.835).

Modelled temperature responses of R_S

The presence of roots increased R_{10} significantly (p<0.0001, Table S5), but had no effect on E_0 (p = 0.340). As both R_S and R_H were shown to be sensitive to soil water content, addition of the soil water content response function in Equation (1) resulted in a significant improvement in model fit over a temperature-only model (ΔAIC = 1350, p<0.0001). The best-fit model of the temperature response of R_S included warming and nitrogen addition as fixed effects on R_{10}, which was increased significantly by both warming (p<0.0001, Table S6) and nitrogen addition (p<0.0001, Table 2). E_0 was unaffected by either warming or nitrogen and there were no significant interactions between warming and nitrogen for either R_{10} or E_0, so they were removed from the model. The inclusion of a warming by measurement year interaction for R_{10} resulted in small (0.02 μmol m^{-2} s^{-1}) and marginally significant decrease in R_{10} during the second year of warming (p = 0.065, Table S7). Inclusion of the temperature by year interaction for E_0 resulted in a non-significant interaction term (p = 0.347), indicating little effect of treatment time on the temperature response of R_S. For R_H, all treatments exhibited a single temperature response curve regardless of treatment (Table 2).

Variation of soil properties and microbial biomass

The average soil carbon concentration was 43 g kg^{-1} and this was not affected significantly by either warming or nitrogen addition (Table 1). Likewise, total nitrogen concentration, which averaged 3.4 g kg^{-1}, was unaffected by any of the treatments. The average microbial biomass carbon was 646 mg kg^{-1}. This did not change significantly under the main effects of warming and nitrogen addition. However, a significant negative interaction between warming and nitrogen addition was observed on the final

soil sampling date (p = 0.023) indicating a reduced microbial biomass under combined warming and nitrogen addition. Cumulative exchange of plant available nitrogen, as estimated from the one month burial of PRS probes, was 8 mg N m^{-2} on average and was not significantly different between treatments. At the time of the final soil sample, average ($\pm$ SE) root biomass in the top 100 mm of soil was 465$\pm$26 g m^{-2}. Root biomass was highly variable and no treatment differences were detected.

Discussion

The average ($\pm$ SE) soil respiration rate measured in the control plots over the course of this study was 0.96$\pm$0.09 μmol m^{-2} s^{-1}. This value falls well within the range reported for temperate grasslands [43]. The relative contribution of R_H to R_S of 71% for control plots was very close to the average for non-forest ecosystems of 63% [20] and also agreed well with another temperate grassland warming experiment where R_H contributed 56–72% of R_S annually [25]. The 3°C warming treatment led to an increased rate of both R_S and R_H over the 27 month measurement period. The average 41% increase in R_S due to warming falls well within the reported range of a 25% reduction to a 45% increase in R_S, as a result of experimental warming [6]. Likewise, warming-induced enhancement of R_H has been observed in other grassland warming experiments [25]. Warming treatment had no significant impact on f_{RH} indicating that R_A and R_H were similarly sensitive to temperature. This is supported by the results of the temperature response curve fitting, in which similar values of E_0 were obtained for both R_S and R_H. However, the temperature response analysis did reveal a slightly higher basal respiration rate, R_{10}, for warmed soils. As this increase in R_{10} was only evident in R_S, and not R_H, we must assume that R_A is responsible for the increase in R_{10}. While an increase in the basal rate of R_A as a result of warming would appear to contradict the previous finding that f_{RH} was unaffected by warming, measurements of f_{RH} were highly variable and, as they were calculated from plot averages, subject to within plot temperature variation. Thus, our results are consistent with a slight increase in root activity in the warmed plots. As no differences in root biomass were observed as a result of warming, this increase in R_A may be due to increased specific root respiration. Further, long-term warming has been associated with increased root exudation [44], and labile carbon in these exudates may have stimulated

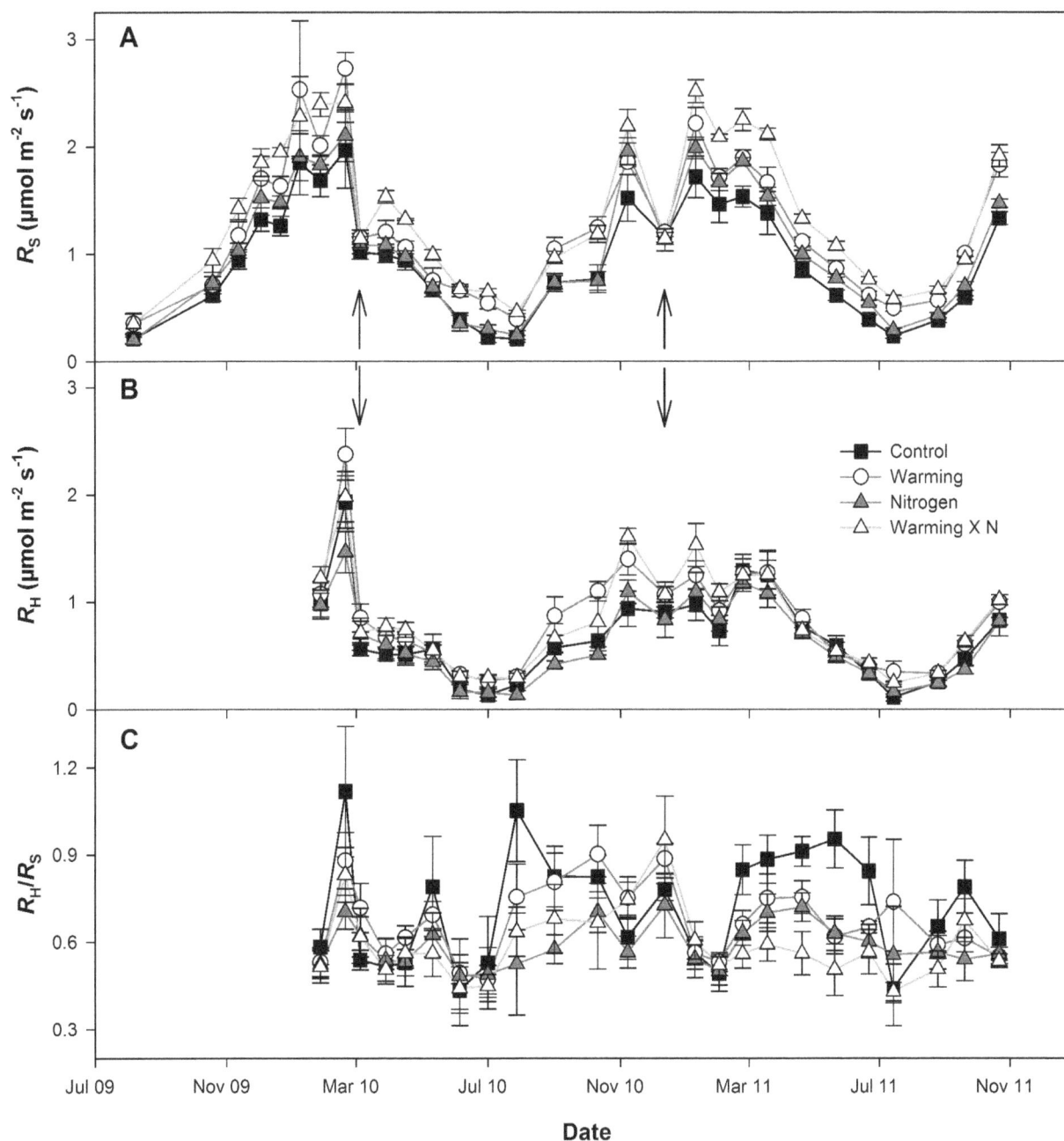

Figure 2. Seasonal rate of soil respiration. Mean $\pm$ SE soil respiration R_S (A), hetrotrophic respiration, R_H (B), and the proportion of total R_S contributed by R_H (f_{RH}) (C) by date; arrows indicate dates for which soil water content was below 0.2 m^3 m^{-3}.

rhizosphere microbial activity leading to an increase in R_A, as measured by the root exclusion approach.

The acclimation of R_S to warming frequently observed in many long-running soil warming experiments [7,8] was absent in our study. We expected that acclimation would result in a significant, negative warming by measurement year interaction (i.e., a decrease in R_{10} or E_0 relative to the control during the second year of warming). However, we observed only a small, marginally significant treatment by measurement year interaction effect for R_{10}. One possible explanation for this lack of acclimation is the relatively short duration of this experiment. Physiological acclimation of roots and soil microbes should occur rapidly compared with the duration of our experiment, though changes to biomass

and soil carbon pools may take longer. Another explanation is that acclimation has been linked to depletion of labile carbon substrates [9,10]. As the study site was recently cleared of vegetation and soil structure was disturbed, it is likely that the size of the labile carbon pool was reduced as a result of the disturbance. Our measurements would then reflect the temperature response of decomposition of more recalcitrant soil organic matter in the absence of a large labile carbon pool to which size adjustments can occur. As the system advances and labile carbon accumulates, acclimation may become evident. However, S_I was significantly higher in the control treatment, indicating that labile substrates represent a greater limitation to R_S in the control plots. This is consistent with observations of other grassland warming experiments which

Table 1. Mean ± SE of soil carbon concentration, soil nitrogen concentration, microbial biomass, plant available nitrogen and substrate induced respiration (S_I) by treatment.

Variable	Treatment			
	Control	Warming	Nitrogen	Warming ×N
Total carbon (g kg^{-1})	41.7±2.4	43.2±0.3	44.9±1.0	44.1±0.7
Total nitrogen (g kg^{-1})	3.3±0.2	3.4±0.1	3.5±0.1	3.5±0.1
Microbial biomass carbon (mg kg^{-1})	639±34	741±47	655±24	**604±25**
Plant-available nitrogen (mg m^{-2})	7.9±1.8	3.8±0.5	11.8±3.8	10.0±3.9
Root Biomass (g m^{-2})	389±51	454±27	518±83	482±30
S_I	1.91±0.38	**1.27±0.14**	1.46±0.28	0.93±0.12

Bold indicates a significance level of: $p \leq 0.05$.

showed higher labile carbon content in warmed soils due to greater belowground allocation and turnover of roots [11].

The significant reduction in soil volumetric water content as a result of soil warming has potential implications for the effects of soil warming on R_S. Both R_S and R_H were observed to be water limited below a soil volumetric water content value of 0.2 m^3 m^{-3}. Thus, warming-induced soil drying may serve to mitigate warming-enhanced carbon losses to R_S, as other warming experiments have shown [45]. However, the soil-drying effects observed here were small except when water was already limiting in all treatments. We suggest that soil drying effects of warming are contributing little to the mitigation of warming effects at this site, due to the frequency of rainfall and the relatively short duration of water limited periods.

The 12% increase in R_S with nitrogen addition is consistent with findings for temperate grasslands [46]. In young and severely nitrogen limited ecosystems, added nitrogen may increase the amount of photosynthate allocated belowground [16]. This may be a reasonable explanation for increased R_S in our experimental tussock grassland, which was planted shortly before measurements began and has very low levels of plant-available nitrogen.

The increase in R_S due to nitrogen addition can be attributed entirely to R_A, as R_H remained unaffected by nitrogen addition. The analysis of f_{RH} confirmed that autotrophic contribution to R_S increased with addition of nitrogen. Likewise, nitrogen increased R_{10} for R_S, but had no effect on R_H. Similar to warming, we found no significant increase in root biomass in the treatment with added

nitrogen, though there was a trend for higher root biomass. Specific root respiration has been shown to increase with increasing root tissue nitrogen concentration in grasslands [47]. Thus, it is likely that increased specific root respiration rate as a result of nitrogen addition contributed to this increase in R_S.

Contrary to expectations, plant-available nitrogen in the soil was not increased by warming or nitrogen addition. There may be several factors contributing to this result. First, nitrogen was applied to both the plant and soil. As a result, a portion of the nitrogen was likely intercepted by foliar uptake [48]. Additionally, the tussock grassland soils are subject to heavy leaching, likely decreasing the residence time of added nitrogen in soils. Further, the PRS probes used to estimate plant available nitrogen were inserted into soil with roots. Strong competition for nitrogen amongst roots may have contributed to the low level of plant-available nitrogen in all treatments.

No interactive effects of warming and nitrogen addition were observed for R_S, R_H or their respective temperature responses. This indicates that the effects of these separate global change drivers are additive. It has been suggested that global change drivers may interact, resulting in smaller effect sizes than those

Table 2. Mean ± SE parameter values for the temperature response of soil respiration, R_S, and heterotrophic soil respiration, R_H, generated by fitting Equation (1) to measured data using a nonlinear mixed-effects models; parameters supplied represent significant fixed effects in the final model.

	Treatment	R_{10} (μmol m^{-2} s^{-1})	E_0 (kJ mol^{-1})	a	b
R_S	Control	0.77±0.03	326±6	0.62±0.05	12.36±0.79
	Warming	0.88±0.04	-	-	-
	Nitrogen	0.93±0.04	-	-	-
	Warming ×N	1.05±0.04	-	-	-
R_H	Control	0.56±0.03	331±12	0.62±0.05†	12.36±0.79†

†fixed value, not fitted in the model.

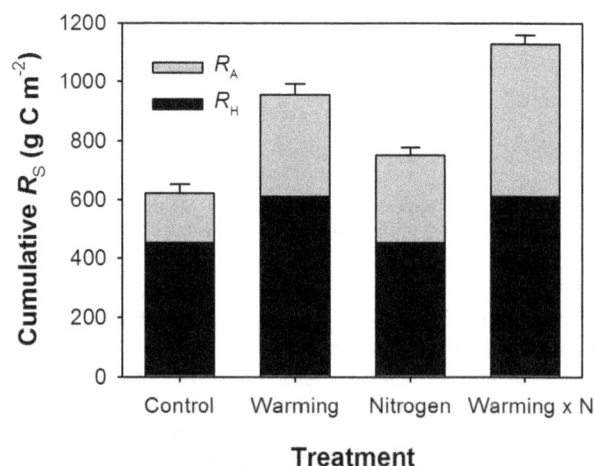

Figure 3. Cumulative soil respiration. Cumulative soil respiration, R_S, for the entire 27 month study period partitioned between autotrophic, R_A, and heterotrophic respiration, R_H. Cumulative estimates were obtained from Equation (1) and parameter values from Table 2. Error bars were estimated from the 95% confidence interval of the linear relationship between measured and modelled R_S.

reported for single drivers [28]. However, few such instances have been noted for R_S [49,50]. The only significant interaction observed in this study was the negative interaction between warming and nitrogen addition on microbial biomass. Nitrogen addition generally results in decreased microbial biomass [26]. While we found no significant decrease of microbial biomass under nitrogen addition alone, a decrease was observed under combined warming and nitrogen addition. However, this decrease did not result in reduced R_H.

Absent from this study is the inclusion of the rhizosphere priming effect in our estimate of R_H. This refers to the effect that living roots have on R_H as a result of their impact on the physical and chemical environment within the soil [51]. Priming effects can influence both the rate and temperature response of R_H [52–54]. This may represent a potential source of error in our determination of the contribution of R_H to R_S. A previous study in tussock grassland soils showed a dampening of the short-term response of R_H to temperature when plants were present [54]. However, in that study, priming effects were absent when plant and soil were held at a constant temperature of 15°C. Only when the soil temperature was perturbed from the constant incubation temperature over a period of hours were priming effects observed. As such, use of the root exclusion method may be appropriate for evaluating longer-term, seasonal temperature responses of R_H, as in the present study.

Our results highlight the potential impacts of warming and nitrogen addition on the global carbon cycle. Over the course of the 27 month experiment, simulated cumulative CO_2 emissions, based on measured temperature response curves of R_S, were 621 g C m^{-2} for the control treatment (Fig. 3). Warming increased cumulative R_S to 953 g C m^{-2}, nitrogen addition resulted in cumulative emissions of 750 g C m^{-2} and the combined effect resulted in emissions from R_S of 1127 g C m^{-2}. While these represent substantial differences in emissions, the contrasting responses of autotrophic and heterotrophic respiration to the treatments must be considered. While increases in R_A may have consequences for the carbon economy of plants, they are likely to be offset by increased primary production. However, increases in R_H due to warming present the potential for sustained loss of stored soil carbon. Extrapolation of our results to the 4.3 Mha of tussock grassland in New Zealand suggests the additional 70 g C m^{-2} y^{-1} carbon losses to R_H as a result of 3 °C warming would amount to a positive feedback to rising atmospheric CO_2 concentration equivalent to 30% of New Zealand's current annual fossil fuel emissions.

Supporting Information

Table S1 F-values for fixed effects in the best-fit linear mixed-effects model of soil volumetric water content.

Table S2 F-values for fixed effects in the best-fit linear mixed-effects model of soil respiration.

Table S3 F-values for fixed effects in the best-fit linear mixed-effects model of heterotrophic respiration.

Table S4 F-values for fixed effects in the best-fit linear mixed-effects model of the proportional contribution of heterotrophic respiration to total soil respiration.

Table S5 F-values for fixed effects in a nonlinear mixed-effects model of soil respiration including the effect of roots on R_{10} and E_0 parameters.

Table S6 F-values for fixed effects in the best-fit nonlinear mixed-effects model of soil respiration.

Table S7 F-values for fixed effects in a nonlinear mixed-effects model of soil respiration including the interaction between the warming treatment and measurement year as a fixed effect on R_{10} and E_0 parameters.

Appendix S1 Experimental layout for the Cass Warming Experiment.

Acknowledgments

We thank Jennifer Peters for help with lab and field measurements, Graeme Rogers for assistance in the field, Claudio de Sassi for design and initial setup of the warming experiment, Guy Forrester for assistance with statistical analyses and Brian Benscoter for comments on the manuscript.

Author Contributions

Conceived and designed the experiments: SLG JEH PM JMT DW. Performed the experiments: SLG TM. Analyzed the data: SLG. Contributed reagents/materials/analysis tools: SLG JMT DW. Wrote the paper: SLG.

References

1. Schlesinger WH, Andrews JA (2000) Soil respiration and the global carbon cycle. Biogeochemistry 48: 7–20.
2. Bond-Lamberty B, Thomson A (2010) Temperature-associated increases in the global soil respiration record. Nature 464: 579–582.
3. IPCC (2007) Climate Change 2007: Synthesis Report.
4. Cox PM, Betts RA, Jones CD, Spall SA, Totterdell IJ (2000) Acceleration of global warming due to carbon-cycle feedbacks in a coupled climate model. Nature 408: 184–187.
5. Sitch S, Huntingford C, Gedney N, Levy PE, Lomas M, et al. (2008) Evaluation of the terrestrial carbon cycle, future plant geography and climate-carbon cycle feedbacks using five Dynamic Global Vegetation Models (DGVMs). Global Change Biology 14: 2015–2039.
6. Rustad LE (2001) A Meta-Analysis of the response of soil respiration, net nitrogen mineralization, and aboveground plant growth to experimental ecosystem warming. Oecologia 126: 543–562.
7. Luo Y, Wan S, Hui D, Wallace LL (2001) Acclimatization of soil respiration to warming in a tall grass prairie. Nature 413: 622–625.
8. Melillo JM, Steudler PA, Aber JD, Newkirk K, Lux H, et al. (2002) Soil warming and carbon-cycle feedbacks to the climate system. Science 298: 2173–2176.
9. Kirschbaum MUF (2004) Soil respiration under prolonged soil warming: are rate reductions caused by acclimation or substrate loss? Global Change Biology 10: 1870–1877.
10. Hartley IP, Heinemeyer A, Ineson P (2007) Effects of three years of soil warming and shading on the rate of soil respiration: substrate availability and not thermal acclimation mediates observed response. Global Change Biology 13: 1761–1770.
11. Belay-Tedla A, Zhou X, Su B, Wan S, Luo Y (2009) Labile, recalcitrant, and microbial carbon and nitrogen pools of a tallgrass prairie soil in the US Great Plains subjected to experimental warming and clipping. Soil Biology and Biochemistry 41: 110–116.
12. Bradford MA, Davies CA, Frey SD, Maddox TR, Melillo JM, et al. (2008) Thermal adaptation of soil microbial respiration to elevated temperature. Ecology Letters 11: 1316–1327.
13. Zhou Y, Tang J, Melillo JM, Butler S, Mohan JE (2011) Root standing crop and chemistry after six years of soil warming in a temperate forest. Tree Physiology 31: 707–717.
14. Burton AJ, Melillo JM, Frey SD (2008) Adjustment of forest ecosystem root respiration as temperature warms. Journal of Integrative Plant Biology 50: 1467–1483.
15. Gruber N, Galloway JN (2008) An Earth-system perspective of the global nitrogen cycle. Nature 451: 293–296.

16. Janssens IA, Dieleman W, Luyssaert S, Subke JA, Reichstein M, et al. (2010) Reduction of forest soil respiration in response to nitrogen deposition. Nature Geosci 3: 315–322.

17. de Jong E, Schappert HJV, MacDonald KB (1974) Carbon dioxide evolution from virgin and clutivated soil as affected by management practices and climate. Canadian Journal of Soil Science 54: 299–307.

18. Yan L, Chen S, Huang J, Lin G (2010) Differential responses of auto- and heterotrophic soil respiration to water and nitrogen addition in a semiarid temperate steppe. Global Change Biology 16: 2345–2357.

19. Melillo JM, Butler S, Johnson J, Mohan J, Steudler P, et al. (2011) Soil warming, carbon–nitrogen interactions, and forest carbon budgets. Proceedings of the National Academy of Sciences 108: 9508–9512.

20. Hanson PJ, Edwards NT, Garten CT, Andrews JA (2000) Separating root and soil microbial contributions to soil respiration: A review of methods and observations. Biogeochemistry 48: 115–146.

21. Trumbore S (2000) Age of soil organic matter and soil respiration: radiocarbon constraints on belowground dynamics. Ecological Applications 10: 399–411.

22. Kirschbaum MUF (1995) The temperature dependence of soil organic matter decomposition, and the effect of global warming on soil organic C storage. Soil Biology and Biochemistry 27: 753–760.

23. Davidson EA, Janssens IA (2006) Temperature sensitivity of soil carbon decomposition and feedbacks to climate change. Nature 440: 165–173.

24. Schindlbacher A, Zechmeister-Boltenstern S, Jandl R (2009) Carbon losses due to soil warming: Do autotrophic and heterotrophic soil respiration respond equally? Global Change Biology 15: 901–913.

25. Zhou X, Wan S, Luo Y (2007) Source components and interannual variability of soil CO_2 efflux under experimental warming and clipping in a grassland ecosystem. Global Change Biology 13: 761–775.

26. Treseder KK (2008) Nitrogen additions and microbial biomass: a meta-analysis of ecosystem studies. Ecology Letters 11: 1111–1120.

27. Norby RJ, Luo Y (2004) Evaluating ecosystem responses to rising atmospheric CO_2 and global warming in a multi-factor world. New Phytologist 162: 281–293.

28. Leuzinger S, Luo Y, Beier C, Dieleman W, Vicca S, et al. (2011) Do global change experiments overestimate impacts on terrestrial ecosystems? Trends in Ecology & Evolution 26: 236–241.

29. Chen X, Post W, Norby R, Classen A (2011) Modeling soil respiration and variations in source components using a multi-factor global climate change experiment. Climatic Change 107: 459–480.

30. Trotter CM, Tate KR, Saggar S, Scott NA, Sutherland MA (2004) A multi-scale analysis of a national terrestrial carbon budget: uncertainty reduction and the effects of land-use change. In: Shiyomi M, Kawahata H, Tsuda A, Away Y, editors. Global Environmental Change in the Ocean and on Land. Tokyo: Terrapub. pp. 311–342.

31. Scurlock JMO, Hall DO (1998) The global carbon sink: a grassland perspective. Global Change Biology 4: 229–233.

32. Hewitt AE (2010) New Zealand Soil Classification. Lincoln, New Zealand: Manaaki Whenua Press.

33. Soil Survey Staff (2006) Keys to soil taxonomy, 10th ed.: USDA Natural Resources Conservation Service, Washington, DC.

34. Peterjohn W, Melillo J, Bowles F, Steudler P (1993) Soil warming and trace gas fluxes: experimental design and preliminary flux results. Oecologia 93: 18–24.

35. Davidson EA, Janssens IA, Luo Y (2006) On the variability of respiration in terrestrial ecosystems: moving beyond Q_{10}. Global Change Biology 12: 154–164.

36. Vance ED, Brookes PC, Jenkinson DS (1987) Microbial biomass measurements in forest soils: The use of the chloroform fumigation-incubation method in strongly acid soils. Soil Biology and Biochemistry 19: 697–702.

37. Pinheiro J, Bates D, DebRoy S, Sarkar D, R Development Core Team (2012) nlme: Linear and Nonlinear Mixed Effects Models. R package version 3.1–105.

38. R Development Core Team (2010) R: A Language and Environment for Statistical Computing. 2.12.1 ed. Vienna, Austria: R Foundation for Statistical Computing.

39. Lloyd J, Taylor JA (1994) On the temperature dependence of soil respiration. Functional Ecology 8: 315–323.

40. Bahn M, Rodeghiero M, Anderson-Dunn M, Dore S, Gimeno C, et al. (2008) Soil respiration in European grasslands in relation to climate and assimilate supply. Ecosystems 11: 1352–1367.

41. Pinheiro J, Bates D (2000) Mixed-Effects Models in S and S-PLUS; Chambers JE, W; . Hardle, W; . Sheather, S; . Tierney, L, editor. New York: Springer-Verlag.

42. Crawley MJ (2007) The R Book. Chichester, West Sussex, UK: John Wiley and Sons.

43. Raich JW, Schlesinger WH (1992) The global carbon dioxide flux in soil respiration and its relationship to vegetation and climate. Tellus B 44: 81–99.

44. Uselman S, Qualls R, Thomas R (2000) Effects of increased atmospheric CO_2, temperature, and soil N availability on root exudation of dissolved organic carbon by a N-fixing tree (*Robinia pseudoacacia L.*). Plant and Soil 222: 191–202.

45. Schindlbacher A, Wunderlich S, Borken W, Kitzler B, Zechmeister-Boltenstern S, et al. (2012) Soil respiration under climate change: prolonged summer drought offsets soil warming effects. Global Change Biology 18: 2270–2279.

46. Craine JM, Wedin DA, Reich PB (2001) The response of soil CO_2 flux to changes in atmospheric CO_2, nitrogen supply and plant diversity. Global Change Biology 7: 947–953.

47. Bahn M, Knapp M, Garajova Z, Pfahringer N, Cernusca A (2006) Root respiration in temperate mountain grasslands differing in land use. Global Change Biology 12: 995–1006.

48. Sparks J (2009) Ecological ramifications of the direct foliar uptake of nitrogen. Oecologia 159: 1–13.

49. Wan S, Norby RJ, Ledford J, Weltzin JF (2007) Responses of soil respiration to elevated CO_2, air warming, and changing soil water availability in a model old-field grassland. Global Change Biology 13: 2411–2424.

50. Contosta AR, Frey SD, Cooper AB (2011) Seasonal dynamics of soil respiration and N mineralization in chronically warmed and fertilized soils. Ecosphere 2: art36.

51. Kuzyakov Y (2002) Review: Factors affecting rhizosphere priming effects. Journal of Plant Nutrition and Soil Science 165: 382–396.

52. Zhu B, Cheng W (2011) Rhizosphere priming effect increases the temperature sensitivity of soil organic matter decomposition. Global Change Biology 17: 2172–2183.

53. Uchida Y, Hunt J, Barbour M, Clough T, Kelliher F, et al. (2010) Soil properties and presence of plants affect the temperature sensitivity of carbon dioxide production by soils. Plant and Soil 337: 375–387.

54. Graham SL, Millard P, Hunt JE, Rogers GND, Whitehead D (2012) Roots affect the response of heterotrophic soil respiration to temperature in tussock grass microcosms. Annals of Botany 110: 253–258.

5

Comparison of the Effects of Air Pollution on Outpatient and Inpatient Visits for Asthma: A Population-Based Study in Taiwan

Hui-Hsien Pan[1,2]❾, Chun-Tzu Chen[1,3]❾, Hai-Lun Sun[1,2], Min-Sho Ku[1,2], Pei-Fen Liao[1], Ko-Hsiu Lu[2], Ji-Nan Sheu[1,2], Jing-Yang Huang[4], Jar-Yuan Pai[3], Ko-Huang Lue[1,2]*

1 Department of Pediatrics, Chung Shan Medical University Hospital, Taichung City, Taiwan R.O.C, 2 School of Medicine, Chung Shan Medical University, Taichung City, Taiwan R.O.C, 3 Department of Health Policy and Management, Chung Shan Medical University, Taichung City, Taiwan R.O.C, 4 Institute of Public Health, Department of Public Health, Chung Shan Medical University, Taichung City, Taiwan R.O.C

Abstract

Background: A nationwide asthma survey on the effects of air pollution is lacking in Taiwan. The purpose of this study was to evaluate the time trend and the relationship between air pollution and health care services for asthma in Taiwan.

Methods: Health care services for asthma and ambient air pollution data were obtained from the National Health Insurance Research database and Environmental Protection Administration from 2000 through 2009, respectively. Health care services, including those related to the outpatient and inpatient visits were compared according to the concentration of air pollutants.

Results: The number of asthma-patient visits to health-care facilities continue to increase in Taiwan. Relative to the respective lowest quartile of air pollutants, the adjusted relative risks (RRs) of the outpatient visits in the highest quartile were 1.10 (P-trend = 0.013) for carbon monoxide (CO), 1.10 (P-trend = 0.015) for nitrogen dioxide (NO_2), and 1.20 (P-trend < 0.0001) for particulate matter with an aerodynamic diameter $\leq 10\mu m$ (PM_{10}) in the child group (aged 0–18). For adults aged 19–44, the RRs of outpatient visits were 1.13 (P-trend = 0.078) for CO, 1.17 (P-trend = 0.002) for NO_2, and 1.13 (P-trend < 0.0001) for PM_{10}. For adults aged 45–64, the RRs of outpatient visits were 1.15 (P-trend = 0.003) for CO, 1.19 (P-trend = 0.0002) for NO_2, and 1.10 (P-trend = 0.001) for PM_{10}. For the elderly (aged $\geq$ 65), the RRs of outpatient visits in were 1.12 (P-trend = 0.003) for NO_2 and 1.10 (P-trend = 0.006) for PM_{10}. For inpatient visits, the RRs across quartiles of CO level were 1.00, 1.70, 1.92, and 1.86 (P-trend = 0.0001) in the child group. There were no significant linear associations between inpatient visits and air pollutants in other groups.

Conclusions: There were positive associations between CO levels and childhood inpatient visits as well as NO_2, CO and PM_{10} and outpatient visits.

Editor: Stephania Ann Cormier, University of Tennessee Health Science Center, United States of America

Funding: The authors have no support or funding to report.

Competing Interests: The authors have declared that no competing interests exist.

* E-mail: cshy095@csh.org.tw

❾ These authors contributed equally to this work.

Introduction

Asthma is a common chronic inflammatory respiratory disease that affects 300 million people of all ages and all ethnic backgrounds and accounts for about 1 in every 250 deaths worldwide [1]. In Taiwan, the prevalence of asthma increased from 5.07% in 1985 to 11.9% in 2007 [2,3]. The risk factors for asthma include many external determinants such as mites, dust, mold, indoor and outdoor air pollution, and season variations [4–6]. Although air pollution has not been shown as the sole cause of respiratory illnesses, there is evidence that air pollution episodes lead to respiratory irritation, increased use of asthma medications and hospitalizations [7,8]. Traffic and industry-related pollutants, nitrogen dioxide (NO_2) and carbon monoxide (CO), were

associated with asthma hospitalizations and outpatient visits [9,10]. Elevated levels of ozone (O_3), sulfur dioxide (SO_2) and particulate matter with an aerodynamic diameter $\leq 10\mu m$ (PM_{10}) were reported to be related with increased asthma emergency room visits and admissions [11–13]. It has been reported that the rise in air pollution has increased respiratory and cardiovascular complications leading to elevated risk of death [14].

If air pollution is responsible for the observed enhanced respiratory complications and mortality, one would also expect to see an impact on clinic visits, outpatient visits, emergency department (ED) visits and hospitalization rates for asthma. However, there are no data regarding a population based survey with seasonal and air pollutants in asthma outpatient and inpatient visits in Taiwan. The objective of this study was to assess asthma-related outpatient and inpatient patterns of visits in different age groups based on the National Health Insurance Research Database (NHIRD) in Taiwan, and to compare the association with utilization of health care services and concentrations of air pollutants.

Materials and Methods

Database

The data were obtained from the NHIRD released by the National Health Research Institute (NHRI) in Taiwan. The National Health Insurance Program finances compulsory universal health care for 99% of all of residents of Taiwan [15]. The database contains demographic data, all health-care encounters, expenditure and dates of enrollment and withdrawal. To facilitate research, the NHRI randomly sampled a representative database of one million subjects enrolled in the National Health Insurance program in the year 2005 by a systematic sampling method. This one-million sample was validated to be representative of the entire insured population as reported by the NHRI. The identification numbers and personal information of all individuals in the NHRID were erased to protect the privacy of the individuals. This study was approved by the Institutional Review Board of the Chung-Shan Medical University Hospital, Taiwan.

Study Population

Cases of asthma were ascertained by the service claim for either outpatient or inpatient visit with a primary diagnosis of asthma (ICD-9-CM code 493.xx). Daily counts of clinic visits, outpatient visits, ED visits and hospital admissions for asthma were extracted from the medical insurance file for the period of 2000–2009. The outpatient visit was defined as a patient visit to a physician's office, clinic, or hospital outpatient department. The inpatient visits include ED visits and admissions. The analyses covered 54 municipalities, each with its own air quality monitoring station. We identified 306628 men and 315178 women who lived in municipalities with air quality monitoring stations during the study period. There were 33934 men and 34527 women with asthma. The patient's exact addresses were not available from the database. Therefore, we assumed that the municipality where a patient visit occurred was regarded as the same area where the patient was most likely exposed to air pollutants. Each asthma outpatient or inpatient visit was matched with the municipality's seasonal average pollutant concentrations. Each occurrence, limited to municipality with air quality monitoring stations, was counted as one visit. The outpatient and inpatient visits were further analyzed by seasons (spring as February, March and April; summer as May, June and July; autumn as August, September and October; winter as November, December and January) and four age groups (0–18, 19–44, 45–64 and ≥65 years).

Outdoor Air Pollution Monitoring

There were 76 air quality monitoring stations in Taiwan, established by the Environmental Protection Administration. Data from background and industrial air quality monitoring stations were excluded to avoid extreme levels of air pollutants. Finally, data from 54 air quality monitoring stations were enrolled (24 municipalities in northern, 9 municipalities in central, 19 municipalities in southern and 2 municipalities in eastern Taiwan). Complete monitoring data for the air pollutants included PM_{10}, SO_2, CO, O_3, and NO_2. The average seasonal concentration of each air pollutant from the Taiwan Environmental Protection Administration's air quality monitoring stations was calculated for further analysis from 2000 to 2009.

Statistical Analysis

All analyses were done by using the SAS ver. 9.3 software package (SAS Institute, Cary, NC, USA). Multivariate Poisson Regression was made in order to determine the relative risk (RR) of asthma inpatient and outpatient visits by sex, age, year and seasons. For use of health services in relation to air pollutants, the municipalities were defined as the observed units, and the level of air pollutants and counts of patient visits were collected in each municipality for each season from 2000 to 2009. As multiple visits by the same patient in a season in the municipality who lived, the count of the visits was as one visit per season. Generalized estimating equations (PROC GENMOD with repeated statement by the SAS Institute) were used to analyze of levels of pollutants by stratifying municipalities and to compare whether the inpatient and outpatient visits in those municipalities correlated with levels of pollutants. A p value <0.05 was considered statistically significant.

Results

The descriptive statistics for outpatient and inpatient visits and corresponding period and season data are shown in Table 1. Compared with women, the RRs for inpatient and outpatient visits in men were 1.31 (p<0.0001) and 1.15 (p<0.0001), respectively. Furthermore, when divided into 4 age groups, the highest RRs for inpatient and outpatient visits were among the elderly group (i.e. ≥65 years) followed by the child group (0–18), the adults aged 45–64 group and the adult aged 19–44 group. For the period effect, there were increased outpatient visits (RR: 1.29, P<0.0001) since 2005 and inpatient visits (RR: 3.30, P<0.0001) since 2006 when compared with patient visits in 2000. The peak seasons for asthma inpatient and outpatient visits in Taiwan were spring (5.1/10,000 person-season) and winter (149.6/10,000 person-season), respectively.

Table 2 shows the RRs of outpatient health service use with respect to exposure to air pollutants in the four age groups following adjustments for sex, period and quartiles of air pollutants by the generalized estimating equations model. Relative to the respective lowest quartile of air pollutants, the adjusted RRs of the outpatient visits in the highest quartile were 1.10 (95% CI: 1.04, 1.16; P-trend = 0.013) for CO, 1.10 (95% CI: 1.01, 1.18; P-trend = 0.015) for NO_2, 0.94 (95% CI: 0.88, 0.99; P-trend = 0.016) for O_3, and 1.20 (95% CI:1.13, 1.27; P-trend <0.0001) for PM_{10} in the child group. For adults aged 19–44, the adjusted RRs of the outpatient visits were 1.13 (95% CI: 1.03, 1.23; P-trend = 0.078) for CO, 1.17 (95% CI: 1.05, 1.31; P-trend = 0.002) for NO_2, 0.88 (95% CI: 0.83, 0.94; P-trend<0.0001) for O_3, and 1.13 (95% CI: 1.05, 1.21; P-trend <0.0001) for PM_{10}. The adjusted RRs of the outpatient visits in the adults aged 45–64 were 1.15 (95% CI: 1.05, 1.26; P-trend = 0.003) for CO, 1.19 (95% CI: 1.08, 1.30; P-trend = 0.0002) for NO_2, 0.93 (95% CI: 0.89, 0.98; P-trend = 0.028) for O_3, and 1.10 (95% CI: 1.02, 1.18; P-trend = 0.001) for PM_{10}. The adjusted RRs for the elderly outpatient visits across quartiles of air pollutant level were 1.00, 1.02, 1.07, and 1.12 (P-trend = 0.003) for NO_2 and 1.00, 1.04, 1.06, 1.10 (P-trend = 0.006) for PM_{10}. There was no significant association between SO_2 and outpatient visits in any group.

The use of inpatient health services in relation to levels of air pollutants by different age groups is shown in Table 3. The adjusted RRs for the child inpatient visits across quartiles of CO level were 1.00, 1.70, 1.92, and 1.86 (P-trend = 0.0001) after multivariate adjustment by sex, year and other air pollutants. There were no significant linear associations between inpatient visits and air pollutants in other age groups.

Table 1. Inpatient and Outpatient Visits Stratified by Sex, Age, Year and Season in Taiwan.

	Person-seasons	Inpatient visits[a]				Outpatient visits[b]			
		Visits per season	Visit rate per 10000 person-season	RR	P value	Visits per season	Visit rate per 10000 person-season	RR	P value
Sex									
Female	8297518	2970	3.6	-		104358	125.7	-	
Male	7904684	3711	4.7	1.31	<0.0001	113915	144.0	1.15	<0.0001
Age									
0–18	3930291	2658	6.8	-		93580	237.9	-	
19–44	7373364	1083	1.5	0.22	<0.0001	40932	55.5	0.23	<0.0001
45–64	3674627	1317	3.6	0.53	<0.0001	46489	126.5	0.53	<0.0001
≥65	1223920	1623	13.3	1.96	<0.0001	37272	304.3	1.28	<0.0001
Year									
2000	1611185	369	2.3	-		18604	115.4	-	
2001	1611426	400	2.5	1.08	0.265	19932	123.6	1.07	<0.0001
2002	1628261	393	2.4	1.05	0.469	20624	126.6	1.10	<0.0001
2003	1635688	264	1.6	0.71	<0.0001	19501	119.2	1.03	0.002
2004	1692152	323	1.9	0.83	0.017	23061	136.2	1.18	<0.0001
2005	1719118	239	1.4	0.61	<0.0001	25529	148.4	1.29	<0.0001
2006	1699621	1284	7.6	3.30	<0.0001	24301	143.0	1.24	<0.0001
2007	1685165	1380	8.2	3.58	<0.0001	25208	149.6	1.30	<0.0001
2008	1672283	1212	7.3	3.17	<0.0001	23951	143.2	1.24	<0.0001
2009	1247303	817	6.6	2.86	<0.0001	17562	140.8	1.22	<0.0001
Season									
Spring (Feb-Apr)	4159829	2127	5.1	-		60648	145.7	-	
Summer(May-Jul)	4151745	1416	3.4	0.67	<0.0001	50591	121.8	0.84	<0.0001
Autumn(Aug-Oct)	4140026	1502	3.6	0.71	<0.0001	50887	122.9	0.84	<0.0001
Winter (Nov-Jan)	3750602	1636	4.4	0.85	<0.0001	56147	149.6	1.03	<0.0001

RR, relative risk.

[a]Inpatient visits include emergency department visits and hospitalizations.

[b]Outpatient visits include physician's office, clinic, or hospital outpatient department visits.

Table 2. Outpatient Visits in Relation to Air Pollutants by Different Age Groups.

	Outpatient visits[a]							
	Age group							
	0–18		19–44		45–64		≥65	
Variables	RR	95%CI	RR	95%CI	RR	95%CI	RR	95%CI
Men/women	**1.43**	1.37–1.48	**0.84**	0.79–0.90	**0.77**	0.72–0.82	**1.20**	1.12–1.29
Year	**1.06**	1.05–1.07	**1.03**	1.02–1.04	1.00	0.99–1.02	1.01	0.99–1.02
CO (ppm)								
0.19–0.43	-		-		-		-	
0.43–0.55	**1.07**	1.03–1.11	**1.13**	1.07–1.20	**1.06**	1.01–1.12	1.02	0.97–1.06
0.55–0.69	**1.06**	1.01–1.12	**1.12**	1.03–1.21	**1.11**	1.04–1.20	1.02	0.96–1.09
0.69–1.22	**1.10**	1.04–1.16	**1.13**	1.03–1.23	**1.15**	1.05–1.26	1.04	0.95–1.13
P-Trend	**0.013**		0.078		**0.003**		0.450	
NO₂ (ppm)								
4.45–14.01	-		-		-		-	
14.01–18.64	1.02	0.97–1.08	1.03	0.95–1.11	**1.07**	1.01–1.14	1.02	0.97–1.07
18.64–24.04	1.05	0.98–1.12	1.08	0.98–1.19	**1.14**	1.06–1.23	**1.07**	1.00–1.13
24.04–47.84	**1.10**	1.01–1.18	**1.17**	1.05–1.31	**1.19**	1.08–1.3	**1.12**	1.03–1.21
P-Trend	**0.015**		**0.002**		**0.0002**		**0.003**	
O₃ (ppm)								
14.23–23.83	-		-		-		-	
23.83–27.23	0.99	0.96–1.03	**0.96**	0.93–0.99	**0.96**	0.93–0.99	0.98	0.94–1.02
27.23–30.80	**0.96**	0.92–1.00	**0.94**	0.90–0.98	0.98	0.94–1.01	0.99	0.94–1.04
30.80–50.00	**0.94**	0.88–0.99	**0.88**	0.83–0.94	**0.93**	0.89–0.98	**0.95**	0.90–0.99
P-Trend	**0.016**		**<0.0001**		**0.028**		0.103	
PM₁₀ (µg/m³)								
23.00–43.67	-		-		-		-	
43.67–54.00	**1.06**	1.02–1.10	**1.07**	1.03–1.11	1.02	0.98–1.07	**1.04**	1.01–1.08
54.00–71.00	**1.11**	1.06–1.17	**1.14**	1.09–1.19	**1.09**	1.03–1.14	**1.06**	1.01–1.11
71.00–141.67	**1.20**	1.13–1.27	**1.13**	1.05–1.21	**1.10**	1.02–1.18	**1.10**	1.03–1.17
P-Trend	**<0.0001**		**<0.0001**		**0.001**		**0.006**	
SO₂ (ppm)								
0.20–2.80	-		-		-		-	
2.80–3.77	1.03	0.97–1.08	0.99	0.93–1.05	1.00	0.94–1.05	1.04	0.99–1.09
3.77–5.00	1.05	0.98–1.11	0.96	0.90–1.03	0.96	0.90–1.03	1.05	0.99–1.10
5.00–19.17	1.04	0.97–1.13	0.96	0.89–1.04	0.97	0.90–1.05	1.01	0.95–1.08
P-Trend	0.201		0.216		0.340		0.590	

Abbreviation: CO, carbon monoxide; NO₂, nitrogen dioxide; O₃, ozone; PM₁₀, particulate matter with an aerodynamic diameter ≦10µm; RR, relative risk; SO₂, sulfur dioxide.
[a]Outpatient visits include physician's office, clinic, or hospital outpatient department visits.

Table 3. Inpatient Visits in Relation to Air Pollutants by Different Age Groups.

Variables	Inpatient visits[a] Age group 0–18 RR	95%CI	19–44 RR	95%CI	45–64 RR	95%CI	≥65 RR	95%CI
Men/women	**1.86**	1.67–2.07	0.98	0.83–1.16	0.82	0.66–1.01	1.14	0.96–1.34
Year	**1.34**	1.28–1.39	**1.22**	1.17–1.27	**1.12**	1.08–1.15	**1.15**	1.10–1.20
CO (ppm)								
0.19–0.43	-		-		-		-	
0.43–0.55	**1.70**	1.40–2.06	1.08	0.83–1.40	1.17	0.92–1.49	1.28	0.98–1.66
0.55–0.69	**1.92**	1.53–2.40	1.07	0.81–1.42	1.18	0.90–1.55	1.35	0.93–1.97
0.69–1.22	**1.86**	1.39–2.48	0.94	0.64–1.38	1.21	0.87–1.68	1.42	0.94–2.15
P-Trend	**0.0001**		0.856		0.343		0.106	
NO$_2$ (ppm)								
4.45–14.01	-		-		-		-	
14.01–18.64	0.85	0.70–1.03	0.86	0.65–1.13	0.85	0.68–1.07	1.09	0.85–1.40
18.64–24.04	0.85	0.66–1.08	1.08	0.75–1.55	1.09	0.83–1.44	1.22	0.85–1.76
24.04–47.84	0.90	0.67–1.20	1.33	0.92–1.94	1.08	0.76–1.53	1.17	0.77–1.76
P-Trend	0.615		0.068		0.447		0.434	
O$_3$ (ppm)								
14.23–23.83	-		-		-		-	
23.83–27.23	**0.86**	0.74–0.99	1.02	0.90–1.16	1.00	0.81–1.23	1.14	0.99–1.31
27.23–30.80	0.87	0.74–1.02	1.08	0.87–1.32	1.13	0.88–1.44	**1.18**	1.01–1.39
30.80–50.00	0.83	0.66–1.05	1.04	0.79–1.35	1.06	0.80–1.40	1.20	0.99–1.45
P-Trend	0.136		0.548		0.450		0.070	
PM$_{10}$ (µg/m^3)								
23.00–43.67	-		-		-		-	
43.67–54.00	**0.84**	0.73–0.97	1.04	0.83–1.30	**0.82**	0.69–0.97	0.89	0.74–1.07
54.00–71.00	0.97	0.83–1.13	1.25	0.96–1.63	1.01	0.82–1.22	1.00	0.81–1.23
71.00–141.67	0.89	0.71–1.12	1.05	0.77–1.44	1.04	0.78–1.37	0.93	0.70–1.23
P-Trend	0.757		0.336		0.393		0.900	
SO$_2$ (ppm)								
0.20–2.80	-		-		-		-	
2.80–3.77	0.95	0.77–1.18	1.13	0.83–1.52	0.90	0.69–1.17	0.95	0.72–1.26
3.77–5.00	1.03	0.79–1.33	1.26	0.96–1.65	0.91	0.70–1.19	1.00	0.76–1.31
5.00–19.17	0.82	0.63–1.07	0.96	0.71–1.29	0.81	0.61–1.08	0.78	0.57–1.05
P-Trend	0.369		0.882		0.204		0.223	

Abbreviation: CO, carbon monoxide; NO$_2$, nitrogen dioxide; O$_3$, ozone; PM$_{10}$, particulate matter with an aerodynamic diameter $\leq$10µm; RR, relative risk; SO$_2$, sulfur dioxide.
[a]Inpatient visits include emergency department visits and hospitalizations.

The use of inpatient health services increased with time as shown in Fig 1. In general, there were trends of increasing asthma outpatient visits in the children's group (Fig. 1A & B). In 2003, there was an outbreak of Severe Acute Respiratory Syndrome (SARS) in Hong Kong and nearly became a pandemic event [16]. During this period, inpatient visits dramatically decreased in all age groups. After the SARS outbreak, the rates of inpatient visits have increased with time since 2006, especially in the child and the elderly groups (Fig 1C & D).

Discussion

Among patients with asthma, air pollutant exposure causes increased asthma morbidity. Little is known about changes over time in air pollutant exposure among patients with asthma in a national sample. During the study period, the inpatient and outpatient visits by men were higher than in women. The peak seasons of asthma inpatient and outpatient visits for the total population were spring and winter, respectively. The inpatient and outpatient visits of asthmatics have not reached a plateau and have continued to increase. Inpatient visits for asthma increased with increased levels of CO in children but not for any pollutants in adults in the present study. Our study found that CO, NO_2 and PM_{10} had significant estimated associations on outpatient visits due to asthma and children are more susceptible than other age groups.

Health care visits only illustrate a small percentage and most severe inpatients of the total impacts of air pollution. Our study used the count of health care visits from NHRI databases as the measure of morbidity in the population. Clinic visits, outpatient visits, ED visits and admissions are types of health care service, but also possessed of potentially important disparities [17,18]. Because asthma is a chronic disease, patients with asthma were taught to deal with their symptoms and discomfort of an asthma attack [19,20]. When the concentrations of air pollutants rise, patients with asthma may have treated symptoms by themselves or visited neighborhood clinics and hospital outpatient departments for medical treatment. Subsequently, if patients did not receive any treatment or if the condition was deteriorated or ineffective after an outpatient visit, they would then visit hospital emergency departments for assistance. This may explain why there was no increased risk for inpatient visits for adults with increasing levels of air pollutants.

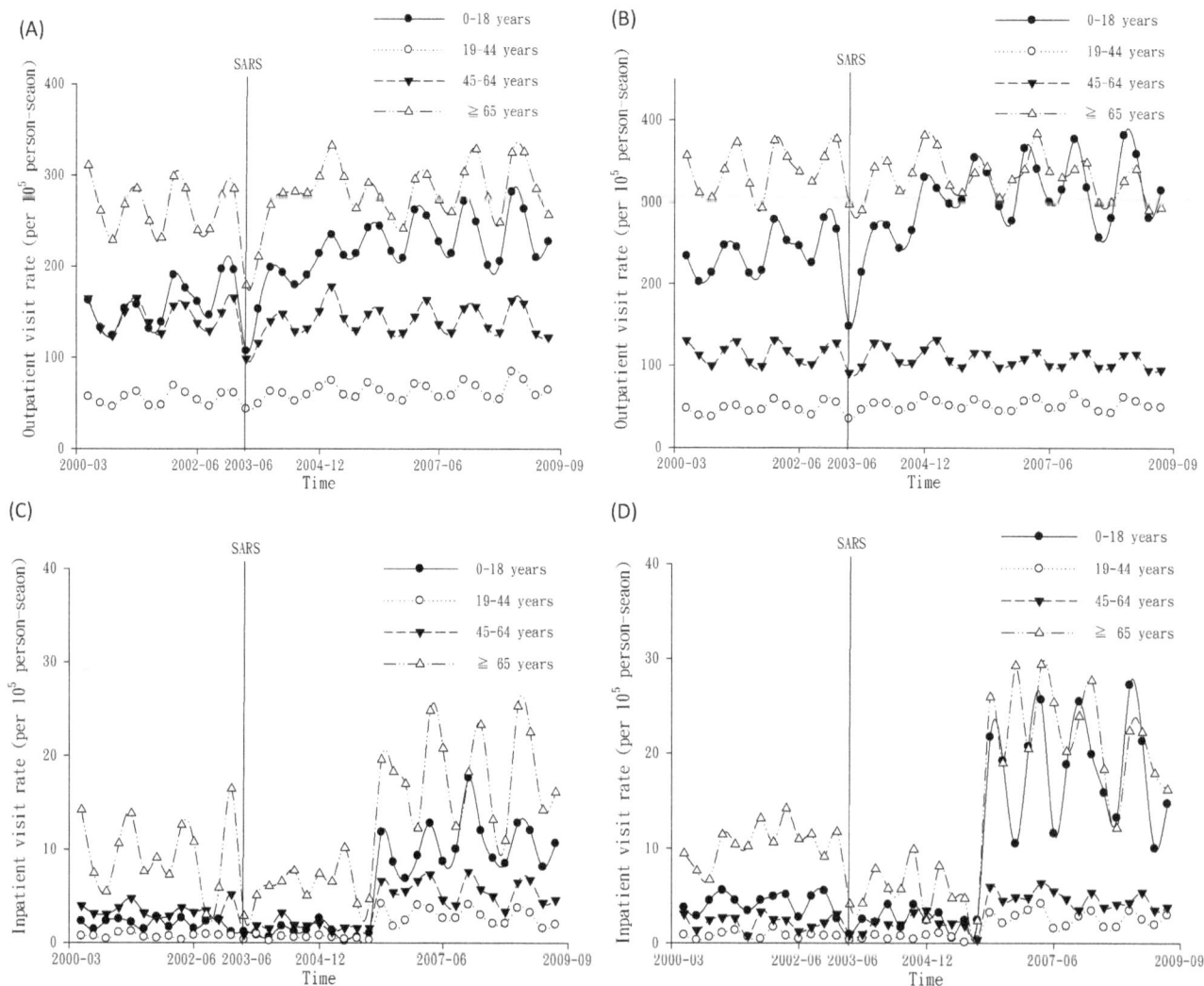

Figure 1. Outpatient visit rates for asthma in 4 age groups in women (A) and men (B), and inpatient visit rates in 4 age groups in women (C) and men (D) during 2000–2009. SARS, Severe Acute Respiratory Syndrome.

In response to the increasing prevalence, mortality rates, and medical cost of asthma, the Bureau of National Health Insurance initiated a Healthcare Quality Improvement Program for patients with asthma since November, 2001 [21]. The characteristics of the program were registry development, adherence to guidelines, patient education, and nursing care management. The Bureau offered financial incentives that motivated the nurses and physicians to change their practice patterns by following clinical guidelines; thereby, the asthma care team support from healthcare organizations was able to promote and enable patients to effectively self-manage their asthma with reduced healthcare resource utilization. During the SARS outbreak, people avoided hospital visits to prevent themselves from nasocomial infection and the inpatient visit rate for asthma decreased in all age groups [22]. However, there was an increase of inpatient and outpatient visits after the SARS outbreak. The exacerbation of air pollution probably plays a role in the rising rate of asthma visits due to growing populations, increased economic activity, rise in vehicular traffic, as well as the increasing intensity and occurrence of dust storms originating in Mongolia and China [23].

There have been associations between daily ambient air pollution levels and acute exacerbations of respiratory diseases in many time-series studies [11,12,24]. Urban air pollution constitutes a complex mixture of several compounds. The exposure of motor vehicle air pollutants, such as NO_2, CO, SO_2 and PM, increases the incidence and prevalence of asthma and bronchitic symptoms [25]. CO was reported to be associated with asthma admission and ED visits [9,26,27]. There were associations between short-term exposure to ambient CO and risk of cardiovascular disease hospitalizations, even at low ambient CO levels [28]. Sun et al conducted a one-year observation in central Taiwan that CO played a role in acute exacerbation of asthma in children and increased the number of childhood asthma ED visits, but not in adults [29]. Villeneuve et al reported that an increase in the interquartile range of the 5-day average for CO was associated with 48% increases in the risk of an asthma ED visit for children aged 2 to 4, but the associations were less pronounced in adult aged 15 to 44 [30]. In Rome, where air pollution comes mostly from combustion products of motor vehicles, CO was associated with most of the respiratory conditions in all ages, and it remained an independent predictor in multi-pollutant analysis for all respiratory admissions [31]. In London, there were significant associations between CO and daily consultations for asthma and other lower respiratory disease in children, whereas in adults the only consistent association was with PM_{10} [32]. In our study, CO was also significantly associated with asthma exacerbation and inpatient visits in children. However, a direct association between CO and asthma lacks a biologically plausible mechanism [33]. It is possible that CO might be a surrogate for other noxious incomplete combustion products [34]. Unlike with children, the other major confounders in adult asthmatic patients include occupational exposures, smoking, stress, emotional factors, and systemic diseases, which may also partly explain why outdoor air pollution was not associated with inpatient visits in the adults in our study.

Coal- and oil-fired power plants and diesel- and gasoline-powered motor vehicle engines are the main sources of ambient NOx emissions [35]. In a meta-analysis study, inhalation of NO_2 in the air significantly increased the development of childhood asthma and symptoms of wheezing [36]. A previous study in Taipei showed that most robust associations were found for NO_2 elevation and asthma admission rates [26]. NO_2 levels were associated with childhood asthma exacerbations and ED visits in Santa Clara, California [37]. In a spatiotemporal analysis of air pollution and asthma patient visits in Taipei, elevated levels of NO_2 had a positive association on outpatient visits [38]. In summary, there was a linear response in outpatient visits on days with elevated NO_2.

Previous studies have not yielded consistent results concerning associations between O_3 and SO_2 and asthma admissions. SO_2 was least frequently mentioned in the correlation with asthma hospitalization rate. Most previous studies have not shown a significant effect of SO_2 on asthma hospitalization rates supporting our findings [29,39]. O_3 is a highly reactive gas and induces bronchial inflammation, constriction of the airways and decreased lung function [40]. Long-term exposure to outdoor O_3 increases the prevalence of bronchitic symptoms among children [41]. O_3 levels have been previously reported to be associated with asthma admission and ED visits [11,24,38]. In contrast, asthma exacerbations were not associated with O_3 levels in North America [42]. Daily general-practice consultations for respiratory conditions were unrelated to O_3 in London [32] and Taiwan [43].

PM_{10} is a heterogeneous mixture of small solid or liquid particles with varying compositions in the atmosphere. There were no consistent results concerning associations between PM_{10} and asthma admissions. Some studies reported that PM_{10} was significantly associated with asthma admissions [8,39], other studies reported a lack of association between PM_{10} and asthma admission [27]. In our study, levels of PM_{10} were associated with outpatient visits for asthma, but not associated with admissions. Hwang and Chan used data obtained from clinic records and environmental monitoring stations in Taiwan and reported that PM_{10} had an impact on outpatient visits [43]. This was consistent with our findings.

Most of the previous studies are cross-sectional and have focused air pollutants on short-term, regional area and respiratory system diseases [8,29,38]. We conducted a nationwide asthma survey on the effects of air pollution in Taiwan and evaluated the association between different air pollutants and outpatient and inpatient visits. One of the strengths of the present study is the use of a computerized database, which is population-based and is highly representative. Because we enrolled all patients diagnosed with asthma from 2000 to 2009, we can rule out the possibility of selection bias. Since the data were obtained from a historical database, which collects all information, recall bias was avoided. Besides, we analyzed levels of pollutants at different municipalities and compared whether the inpatient and outpatient visits in those municipalities correlated with levels of pollutants. Thus this study directly associated the patient visits with the levels of pollutants in those municipalities. As multiple visits by the same patient would lead to misinterpretation of the data, we used the statistic method to reduce bias.

There were still several limitations of the present study. First, although we adjusted for several potential confounders in the statistical analysis, a number of possible confounding variables, including family history of atopy, dietary habits, physical activity, occupational exposures, smoking habits, stress and emotional factors, which are associated with asthma were not included in our database. Second, we were unable to ask patients for severity of asthma because of de-identifcation. Third, self-treatment with over-the-counter medications or alternative health services was not included in the database. These data also do not include the number of asthmatic subjects who had respiratory problems but did not search for health service. Therefore, the extent of the issue may have been considerably underestimated. Fourth, potentially inaccurate data in the records could lead to possible mis-classification.

In conclusion, the present study provides evidence that exposure to the outdoor air pollutant, CO, exerted adverse effects on health and increases in the child admission. Our study also showed a linear association between NO_2, CO, and PM_{10} and outpatient visits. It is an important public health policy to monitor air quality and warn the public about atmospheric factors that could be associated with increased risks of asthma.

Acknowledgments

This study is based on data from the National Health Insurance Research Database provided by the Bureau of National Health Insurance,

Department of Health and managed by National Health Research Institutes. The descriptions or conclusions herein do not represent the viewpoints of the Bureau of National Health Insurance, Department of Health or National Health Research Institutes.

Author Contributions

Conceived and designed the experiments: K.H. Lue HHP CTC. Performed the experiments: JYH JYP. Analyzed the data: K.H. Lue HLS JNS. Wrote the paper: HHP CTC. Reviewed the manuscript and gave input to the final version: HLS MSK PFL K.H. Lu JNS.

References

1. Masoli M, Fabian D, Holt S, Beasley R (2004) The global burden of asthma: executive summary of the GINA Dissemination Committee report. Allergy 59: 469–478.
2. Hsieh KH, Shen JJ (1988) Prevalence of childhood asthma in Taipei, Taiwan, and other Asian Pacific countries. J Asthma 25: 73–82.
3. Hwang CY, Chen YJ, Lin MW, Chen TJ, Chu SY, et al. (2010) Prevalence of atopic dermatitis, allergic rhinitis and asthma in Taiwan: a national study 2000 to 2007. Acta Derm Venereol 90: 589–594.
4. Han YY, Lee YL, Guo YL (2009) Indoor environmental risk factors and seasonal variation of childhood asthma. Pediatr Allergy Immunol 20: 748–756.
5. Chiang CH, Wu KM, Wu CP, Yan HC, Perng WC (2005) Evaluation of risk factors for asthma in Taipei City. J Chin Med Assoc 68: 204–209.
6. Guo Y, Jiang F, Peng L, Zhang J, Geng F, et al. (2012) The association between cold spells and pediatric outpatient visits for asthma in Shanghai, China. PLoS One 7: e42232.
7. Yeh KW, Chang CJ, Huang JL (2011) The association of seasonal variations of asthma hospitalization with air pollution among children in Taiwan. Asian Pac J Allergy Immunol 29: 34–41.
8. Kuo HW, Lai JS, Lee MC, Tai RC, Lee MC (2002) Respiratory effects of air pollutants among asthmatics in central Taiwan. Arch Environ Health 57: 194–200.
9. Delamater PL, Finley AO, Banerjee S (2012) An analysis of asthma hospitalizations, air pollution, and weather conditions in Los Angeles County, California. Sci Total Environ 425: 110–118.
10. Wang KY, Chau TT (2013) An association between air pollution and daily outpatient visits for respiratory disease in a heavy industry area. PLoS One 8: e75220.
11. Wilson AM, Wake CP, Kelly T, Salloway JC (2005) Air pollution, weather, and respiratory emergency room visits in two northern New England cities: an ecological time-series study. Environ Res 97: 312–321.
12. Qiu H, Yu IT, Tian L, Wang X, Tse LA, et al. (2012) Effects of coarse particulate matter on emergency hospital admissions for respiratory diseases: a time-series analysis in Hong Kong. Environ Health Perspect 120: 572–576.
13. Cadelis G, Tourres R, Molinie J (2014) Short-term effects of the particulate pollutants contained in Saharan dust on the visits of children to the emergency department due to asthmatic conditions in Guadeloupe (French Archipelago of the Caribbean). PLoS One 9: e91136.
14. Wong TW, Tam WS, Yu TS, Wong AH (2002) Associations between daily mortalities from respiratory and cardiovascular diseases and air pollution in Hong Kong, China. Occup Environ Med 59: 30–35.
15. Lu JF, Hsiao WC (2003) Does universal health insurance make health care unaffordable? Lessons from Taiwan. Health Aff (Millwood) 22: 77–88.
16. Tsang KW, Ho PL, Ooi GC, Yee WK, Wang T, et al. (2003) A cluster of cases of severe acute respiratory syndrome in Hong Kong. N Engl J Med 348: 1977–1985.
17. Gold LS, Thompson P, Salvi S, Faruqi RA, Sullivan SD (2014) Level of asthma control and health care utilization in Asia-Pacific countries. Respir Med 108: 271–277.
18. Sun HL, Kao YH, Lu TH, Chou MC, Lue KH (2007) Health-care utilization and costs in Taiwanese pediatric patients with asthma. Pediatr Int 49: 48–52.
19. Centers for Disease Control and Prevention (2011) Vital signs: asthma prevalence, disease characteristics, and self-management education: United States, 2001—2009. MMWR Morb Mortal Wkly Rep 60: 547–552.
20. Weng HC (2005) Impacts of a government-sponsored outpatient-based disease management program for patients with asthma: a preliminary analysis of national data from Taiwan. Dis Manag 8: 48–58.
21. Bureau of National Health Insurance (2001) A national comprehensive disease management program. Taipei: Bureau of National Health Insurance.
22. Huang YT, Lee YC, Hsiao CJ (2009) Hospitalization for ambulatory-care-sensitive conditions in Taiwan following the SARS outbreak: a population-based interrupted time series study. J Formos Med Assoc 108: 386–394.
23. Bell ML, Levy JK, Lin Z (2008) The effect of sandstorms and air pollution on cause-specific hospital admissions in Taipei, Taiwan. Occup Environ Med 65: 104–111.
24. Winquist A, Klein M, Tolbert P, Flanders WD, Hess J, et al. (2012) Comparison of emergency department and hospital admissions data for air pollution time-series studies. Environ Health 11: 70.
25. Gasana J, Dillikar D, Mendy A, Forno E, Ramos Vieira E (2012) Motor vehicle air pollution and asthma in children: a meta-analysis. Environ Res 117: 36–45.
26. Yang CY, Chen CC, Chen CY, Kuo HW (2007) Air pollution and hospital admissions for asthma in a subtropical city: Taipei, Taiwan. J Toxicol Environ Health A 70: 111–117.
27. Slaughter JC, Kim E, Sheppard L, Sullivan JH, Larson TV, et al. (2005) Association between particulate matter and emergency room visits, hospital admissions, and mortality in Spokane, Washington. J Expo Anal Environ Epidemiol 15: 153–159.
28. Bell ML, Peng RD, Dominici F, Samet JM (2009) Emergency hospital admissions for cardiovascular diseases and ambient levels of carbon monoxide: results for 126 United States urban counties, 1999–2005. Circulation 120: 949–955.
29. Sun HL, Chou MC, Lue KH (2006) The relationship of air pollution to ED visits for asthma differ between children and adults. Am J Emerg Med 24: 709–713.
30. Villeneuve PJ, Chen L, Rowe BH, Coates F (2007) Outdoor air pollution and emergency department visits for asthma among children and adults: a case-crossover study in northern Alberta, Canada. Environ Health 6: 40.
31. Fusco D, Forastiere F, Michelozzi P, Spadea T, Ostro B, et al. (2001) Air pollution and hospital admissions for respiratory conditions in Rome, Italy. Eur Respir J 17: 1143–1150.
32. Hajat S, Haines A, Goubet SA, Atkinson RW, Anderson HR (1999) Association of air pollution with daily GP consultations for asthma and other lower respiratory conditions in London. Thorax 54: 597–605.
33. American Thoracic Society (1996) Health effects of outdoor air pollution. Committee of the Environmental and Occupational Health Assembly of the American Thoracic Society. Am J Respir Crit Care Med 153: 3–50.
34. Norris G, YoungPong SN, Koenig JQ, Larson TV, Sheppard L, et al. (1999) An association between fine particles and asthma emergency department visits for children in Seattle. Environ Health Perspect 107: 489–493.
35. Trasande L, Thurston GD (2005) The role of air pollution in asthma and other pediatric morbidities. J Allergy Clin Immunol 115: 689–699.
36. Takenoue Y, Kaneko T, Miyamae T, Mori M, Yokota S (2012) Influence of outdoor NO2 exposure on asthma in childhood: meta-analysis. Pediatr Int 54: 762–769.
37. Lipsett M, Hurley S, Ostro B (1997) Air pollution and emergency room visits for asthma in Santa Clara County, California. Environ Health Perspect 105: 216–222.
38. Chan TC, Chen ML, Lin IF, Lee CH, Chiang PH, et al. (2009) Spatiotemporal analysis of air pollution and asthma patient visits in Taipei, Taiwan. Int J Health Geogr 8: 26.
39. Lin M, Stieb DM, Chen Y (2005) Coarse particulate matter and hospitalization for respiratory infections in children younger than 15 years in Toronto: a case-crossover analysis. Pediatrics 116: e235–240.
40. Khatri SB, Holguin FC, Ryan PB, Mannino D, Erzurum SC, et al. (2009) Association of ambient ozone exposure with airway inflammation and allergy in adults with asthma. J Asthma 46: 777–785.
41. Hwang BF, Lee YL (2010) Air pollution and prevalence of bronchitic symptoms among children in Taiwan. Chest 138: 956–964.
42. Schildcrout JS, Sheppard L, Lumley T, Slaughter JC, Koenig JQ, et al. (2006) Ambient air pollution and asthma exacerbations in children: an eight-city analysis. Am J Epidemiol 164: 505–517.
43. Hwang JS, Chan CC (2002) Effects of air pollution on daily clinic visits for lower respiratory tract illness. Am J Epidemiol 155: 1–10.

6

Foliar Symptoms Triggered by Ozone Stress in Irrigated Holm Oaks from the City of Madrid, Spain

Carlos Calderón Guerrero[1,2¤], **Madeleine S. Günthardt-Goerg**[1]*, **Pierre Vollenweider**[1]

1 Forest Dynamics. Swiss Federal Research Institute WSL, Birmensdorf, Switzerland, **2** Department of Silvopasture, Faculty of Forest Engineering (EUIT Forestal), Universidad Politécnica de Madrid, Madrid, Spain

Abstract

Background: Despite abatement programs of precursors implemented in many industrialized countries, ozone remains the principal air pollutant throughout the northern hemisphere with background concentrations increasing as a consequence of economic development in former or still emerging countries and present climate change. Some of the highest ozone concentrations are measured in regions with a Mediterranean climate but the effect on the natural vegetation is alleviated by low stomatal uptake and frequent leaf xeromorphy in response to summer drought episodes characteristic of this climate. However, there is a lack of understanding of the respective role of the foliage physiology and leaf xeromorphy on the mechanistic effects of ozone in Mediterranean species. Particularly, evidence about morphological and structural changes in evergreens in response to ozone stress is missing.

Results: Our study was started after observing ozone -like injury in foliage of holm oak during the assessment of air pollution mitigation by urban trees throughout the Madrid conurbation. Our objectives were to confirm the diagnosis, investigate the extent of symptoms and analyze the ecological factors contributing to ozone injury, particularly, the site water supply. Symptoms consisted of adaxial and intercostal stippling increasing with leaf age. Underlying stippling, cells in the upper mesophyll showed HR-like reactions typical of ozone stress. The surrounding cells showed further oxidative stress markers. These morphological and micromorphological markers of ozone stress were similar to those recorded in deciduous broadleaved species. However, stippling became obvious already at an AOT40 of 21 ppm·h and was primarily found at irrigated sites. Subsequent analyses showed that irrigated trees had their stomatal conductance increased and leaf life -span reduced whereas the leaf xeromorphy remained unchanged. These findings suggest a central role of water availability *versus* leaf xeromorphy for ozone symptom expression by cell injury in holm oak.

Editor: Daniel Ballhorn, Portland State University, United States of America

Funding: The work was supported by the Social Council of the Universidad Politécnica de Madrid. http://www.upm.es/institucional/UPM/ConsejoSocial. The funders had no role in study design, data collection and analysis, decision to publish, or preparation of the manuscript.

Competing Interests: The authors have declared that no competing interests exist.

* E-mail: madeleine.goerg@wsl.ch

¤ Current address: Department of Projects and land use planning, Faculty of Forest Engineering (ETSI Montes), Universidad Politécnica de Madrid, Madrid, Spain

Introduction

Southern Europe is affected by high tropospheric ozone (O3) concentrations [1]. With 6.4 million inhabitants and 4.4 million motor vehicles, the Madrid conurbation acts as a large source of O3 precursors leading to substantial O3 pollution - especially in the Madrid outskirts [2], [3], [4], [5], [6], [7]. During the summer months on the central plateaus, the polluted air masses are re-circulated inside convective cells remaining stable for many days or even months [8], [9] making Madrid one of the regions with the highest O3 pollution in the Iberian Peninsula [3], [4], [5], [6], [7] during 2003–2008.

For several decades, visible foliar injury caused by O3 stress has been investigated in more than 75 European and 66 North American plant species and partly validated by controlled exposure experiments and microscopic analysis [6]. [11], [12], [13]. Despite a high variability, macro and micro-morphological markers of O3 stress share common structural and distribution features which can be used for identifying an O3 stress signature [14], [15], [16], [17]. These features are indicative of outbalances within the antioxidant detoxification system as a consequence of reactive oxygen species (ROS) produced in cascade after O3 uptake and synergies between O3 and photooxidative stress [18], [19], [20]. The elicited plant response and its associated structural changes in foliage can vary according to the O3 dose and levels of photooxidative stress thus leading to more than one pattern of O3 symptom expression within the same species [21]. However, O3 symptoms in broadleaved Mediterranean evergreen trees have so far seldom been documented and, to our knowledge, only one study has shown evidence of microscopic injury [22]. Holm oak (*Quercus ilex* L.) is the main tree species in many Mediterranean sclerophyll evergreen forests. Its deep rooting system, xeromorphic leaf structure and efficient stomatal control ensure tolerance to yearly summer droughts [23], [24], [25]. Compared to other sclerophylls however, it prefers rather mesic and slightly moist sites [26], [27]. In the Madrid region, holm oak is a dominant climacic species in the forest belt surrounding the city [28] and is valued as an ornamental tree in Madrid parks and streets.

The O3 sensitivity of holm oak is still controversial. In a general way and similar to other sclerophylls, this species appears to be rather O3-tolerant [1], [29], partly as a consequence of the xeromorphic traits to be found in the foliage and which are regarded as being an efficient morphological protection against O3 stress [30]. However, some of the most extreme stress reactions to O3 exposure among all sclerophyll evergreen trees so far tested were found in experiments with this species [31], [32]. Depending on the peak O3 concentration, daily irrigated holm oak seedlings thus showed photosynthesis, biomass or chlorophyll content reduction and an increase in some detoxifying enzyme activity in response to O3 exposures as low as 3.6 and 11.7 ppm•h [30], [32]. Visible leaf injury in the form of "slight stippling" [28] or "dark pigmented stipples" [31] has been observed in response to O3 exposure (AOT 40) of 59.27 ppm•h in 6 months and 79.8 ppm•h in 11 months respectively.

The present study is part of an investigation about air pollution mitigation by urban trees. During a bioindication survey, abiotic O3-like injury was identified in foliage of the holm oaks growing on an irrigated lawn strip in the center of Madrid. Given the little structural evidence available for O3 symptoms in broadleaved evergreen species, a study was undertaken in 2007 with the following objectives 1) confirm the diagnosis, 2) investigate the extent of symptoms in holm oaks growing in Madrid and 3) analyze the environmental factors contributing to O3 injury. Therefore, macro- and micromorphological markers of O3 stress were analyzed, using the aforementioned lawn strip as an intensive study site, (objective 1), 65 other urban sites with holm oaks were surveyed for similar type of leaf injury (objective 2) and data on the possible abiotic contributors, i.e. the Madrid climate, lawn strip irrigation and air pollution, were collected and analyzed (objective 3). Given the generally higher O3 sensitivity of trees growing at moist sites [33], [34] and the relative insensitivity of sclerophylls [1], [29], higher rates of stomatal conductance (first hypothesis) and reduced xeromorphic traits (second hypothesis) were hypothesized to be the principal factors determining the development of O3 injury observed at irrigated sites. Their contribution was verified by measuring gas exchanges and assessing the leaf biomass during a subsequent vegetation season (2011) at the irrigated *versus* another comparable but non-irrigated urban intensive site nearby.

Materials and Methods

All necessary permits were obtained for the described field studies at these sites and for the symptoms survey (paragraph 2.4.) by the Madrid park service (Dirección General de Patrimonio Verde del Ayuntamiento de Madrid, signed by Mr. Santigo Soria Carrera, vice-director of green spaces and urban trees) and by the municipal authority (Departamento de Calidad del Aire del Ayuntamiento de Madrid, provided by Mr. Francisco Moya, head of the air quality department).

Intensive Study Sites

The irrigated site was situated in the center of Madrid near the train station of Atocha (Fig. 1; Fig. 2B). It consisted of a green lawn strip irrigated by tap water sprayers and planted with *Quercus ilex* ssp. Ilex (a holm oak sub-species showing minor differences with the *Quercus ilex* spp. ballota native to the Madrid area) and with wide spaces between the trees. Escalonilla, the non-irrigated site, was situated 4 km west of Atocha along a paved street lined with similarly and regularly spaced trees planted on a 0.8 m2 grate and surrounded by a concrete pavement (Fig. 1; Fig. 2A). The holm oak subspecies at Escalonilla was the same as in Atocha.

Climate and Air Pollution in the Madrid Conurbation

Climate conditions and air pollution of Madrid were characterized on the basis of the 1971–2007 daily records of temperature and precipitation and the 2003–2007 hourly O3 and other air pollutant concentrations from four air quality monitoring stations (Fig. 1). The closest air monitoring station was 900 m away from Atocha at a similar elevation (650 m a.s.l.). AOT40 exposure index, expressed on a daily or yearly basis, was calculated as a cumulative dose of O3 concentrations over a threshold of 40 ppb, using the April to September hourly average data measured during daylight hours and for solar radiation above 500 W/m2 [35].

SO$_2$ concentrations were low over the reported period (yearly mean = 11 µg/m^3). NO$_2$ concentrations (yearly mean = 60 µg/m^3) do not induce visible leaf injury [13,36]. Climate, air pollution and site irrigation data was provided by the national meteorological agency (Agencia Estatal de Meteorología - AEMET), the municipal authority (Departamento de Calidad del Aire del Ayuntamiento de Madrid) and the Madrid park service (Dirección General de Patrimonio Verde del Ayuntamiento de Madrid), respectively. Irrigation data was converted to mm of precipitation per month and added to the natural precipitation to calculate the total water supply.

Macro- and Micromorphological Observations

At Atocha, three 8 m high trees with a diameter at breast height (1.3 m, dbh) of 20±2.5 cm and with up to four leaf generations (current: C+0, 1-year: C+1, 2-year: C+2, 3-year: C+3 formed in 2007, 2006, 2005 and 2004, respectively), were selected. In June 2007, four sun-exposed branches per tree were sampled at the mid crown position, assessed for abiotic visible injury using a hand lens and dried in a herbarium after excision of leaf material for microscopy (see below). In the laboratory and to determine the percentage of stippling per leaf area, individual leaves were photographed using a macro-objective, natural light and a dark background. Digital images were analyzed by means of an image analysis system (Scion Image, Scion Corporation, Frederick, Maryland, USA) [37].

For microscopic analysis, the aforementioned sample collection harvested in June was completed in October of the same year using the same trees and mid crown branches at Atocha. Disks, 1 cm in diameter, were excised from asymptomatic and symptomatic C+0, C+1 and C+2 leaves. The leaf disks were immediately fixed either in methanol or in 2.5% glutaraldehyde buffered at pH 7.0 with 0.067 M Sorensen's phosphate buffer. They were entirely infiltrated with the solution by evacuation before storage at 4°C until further processing. Histological, cytological and histochemical observations were performed using 2 µm semi-thin or 50 µm hand-microtomed cuttings. Semi-thin sections were obtained after dehydrating the fixed material with 2-methoxyethanol (three changes), ethanol, n-propanol, n- butanol [38], embedding in Technovit 7100 (Kulzer HistoTechnik) and cutting using a Supercut Reichert 2050 microtome. Sections were stained with different methods including toluidine blue O, p-phenylenediamine and acid-vanillin and subsequently mounted in inclusion medium [17], [21]. All sections were observed using a Leica microscope Leitz DM/RB, 5× to 100× objectives and diascopic light illumination. Micrographs were taken using the digital Leica DC 500 camera interfaced by the Leica DC500 TWAIN software under control of the Image Access Enterprise 5 (Imagic, Glattbrugg, Switzerland) image management system.

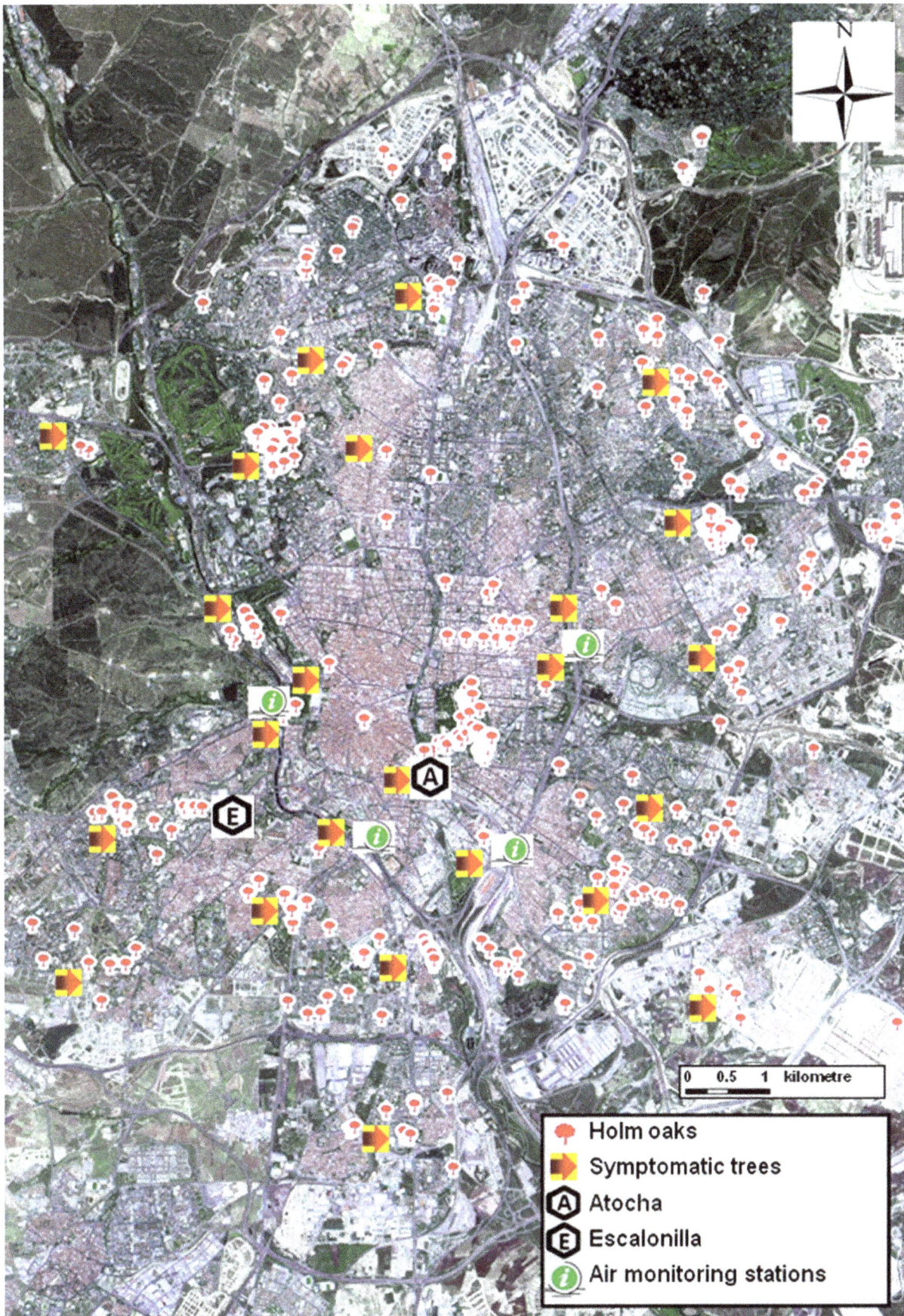

Figure 1. Localization of holm oak sites and air monitoring stations in Madrid. The Atocha (A) and Escalonilla (E) intensive study sites were located in the city centre. Sites with at least one symptomatic tree are indicated by red arrows.

Symptomatic Tree Survey

In October 2007, once all current foliage had completed its development, the extent of visible O_3 injury in Madrid's holm oaks was investigated by surveying 65 public park and street sites inventoried during a preceding tree survey [39]. 257 out of 2314 holm oaks with growth and age similar to trees at Atocha were selected on the basis of their dbh. Examining all leaf generations, the presence/absence of visible O_3 injury in foliage accessible from the ground was assessed and the proportion of trees showing symptoms per site calculated. Already in the field, it quickly became apparent that symptomatic sites had been irrigated and that the mode of water supply should be recorded.

Gas Exchange and Biomass Measurements

From February to October 2011, gas exchange and biomass measurements were carried out at the irrigated (Atocha) and non-irrigated (Escalonilla) intensive study sites using the same trees as in 2007 in Atocha and selecting trees of comparable age (40–50 years), height (8–10 m), and dbh (17.5–22.5 cm) at Escalonilla. The Atocha trees showed visible O3 injury similar to findings in 2007 with regard to the injury distribution and intensity whereas the Escalonilla holm oaks were asymptomatic. New foliage sprouted twice a year at Atocha (end of April and occasionally September, average air temperature reaching 22 and 25°C, respectively) versus only once at Escalonilla (end of May, average air temperature reaching 22.5°C); the leaves being completely developed by the onset of the summer drought (June, Fig. 3). Once a month, 5 leaves per leaf generation in 3 randomly selected branches (10–15 leaves per branch) from the mid part of the sun crown of each tree were measured (10 repetitions per leaf). Stomatal conductance (gs) photosynthetic active radiation (PAR) at leaf surface and leaf temperature (Tleaf) were measured in situ under ambient conditions between 09:00 and 15:00 (CET) using a portable infrared gas analyzer (IRGA model ADC-LCA4) equipped with a 6.25 cm2 chamber for broadleaf plants (PLC4, ADC Inc., Hoddesdon, Hertfordshire, UK). During measurements, the leaf and air temperature remained within a ±2°C range and leaf natural orientation was maintained. Daily course of gs, PAR and Tleaf was measured from dawn to dusk during two subsequent and clear days with similar weather conditions (Fig. 4), using C+1 leaves (formed in 2010).

Following measurements, the selected branches were harvested with a view to leaf area and biomass determination. Individual leaf area was ascertained using an Epson GT5000 scanner and images were analyzed using the aforementioned image analysis system. Leaves were then dried (85°C until constant weight), weighed and the leaf mass per area (LMA) determined.

Statistical Analysis

In the case of the amount of stippling assessed at the irrigated intensive study site in 2007, an estimate for a given leaf generation was calculated by averaging measurements from three leaves per branch and four branches per tree the statistical unit being the branch (n = 4). Hence, the experiment was a split-plot design with the whole plot factor *tree* and the split-plot factor *leaf generation*. Effects of these factors on the stippling intensity were tested by means of ANOVA (with post hoc pairwise Tukey's studentized range (HSD) test) using the SAS software package (SAS Institute, Inc, Cary NC). Given the incomplete randomization of whole plot

factors in a split-plot design, the factor *tree* was tested against its interaction with the *leaf generation* factor and the *leaf generation* against the residual error.

In 2011, gs and LMA estimates per leaf generation at the irrigated *versus* non-irrigated intensive study site were calculated by averaging five leaves per branch and three branches per tree but the statistical unit in this case was the tree (n = 3). The experiment was also a split-plot design with the whole plot factor *irrigation* and the split-plot factor *leaf generation* and *month*. Effects of these factors on gs and LMA were also tested by means of ANOVA followed by post hoc tests with the factor *irrigation* and *leaf generation* tested against their interaction and the *month* against the residual error.

Results

Site Conditions

The climate of Madrid (Fig. 3) is Mediterranean and continental with hot summers (on average 23.2°C), cold winters (on average 8.1°C) and little annual precipitation (436 mm), especially during the summer (precipitation minimum in August). Therefore to alleviate the summer drought many places throughout the city of Madrid are irrigated either manually or automatically, as can be seen at Atocha. At this site, artificial irrigation is supplied by sprinklers at varying levels between March and November, peaking in June, July and August and reaching overall 1027 mm per year (Fig. 3).

Ozone Pollution

The concentration and yearly course of O3 recorded at the Madrid air monitoring stations between April and September (2003–2007) was typical for an urban site. On average, and as a consequence of precursor accumulation and O3 production by road traffic and solar radiation, O3 concentration increased during the day and reached 48 ppb at 17:00 CET. In the evening and during the night, O3 concentration dropped to a minimum of 14 ppb at 09:00 CET (Fig. 5). The daily average, calculated on an hourly basis, reached 31 ppb with values ranging from 0 to 99 ppb. The exceedance of O3 threshold values (one-hour O3 concentration >180 μg/m3/92 ppb; [36]) amounted to 10 hours over 7 days in 2007, 13 hours over 8 days in 2006, 113 hours over 29 days in 2005, 74 hours over 22 days in 2004, and 165 hours over 36 days in 2003.

Regarding O3 exposure, yearly AOT40 (April to September) amounted to 15/11/9/13/8 ppm•h in 2003/2004/2005/2006/2007 and an average of 11 ppm•h for the whole period (Fig. 6). The highest daily AOT40 were recorded in 2003 and 2006 whereas rainy and cloudy weather, especially in 2007, reduced O3 exposure sizably. The cumulated O3 dose experienced by C+3 foliage, formed in 2004, amounted to 41 ppm•h.

Visible Injury

Visible O3-like injury in foliage of holm oak appeared as depressed, tiny, necrotic and intercostal stipples amid still green leaf tissue (Fig. 2I, Fig. 2J, Fig. 2K). Their small size and high frequency let the leaf appear homogeneously discoloured unless the stipples were resolved using a hand lens (Fig. 2D, Fig. 2E *versus* Fig. 2I, Fig. 2K). Stipples developed on the upper leaf side of non-shaded foliage exposed to full sun light. Leaf parts shaded by other leaves or twigs showed reduced stippling (Fig. 2F, Fig. 2G). The recently flushed foliage (C+0; Fig. 2D, Fig. 2H) was predominantly

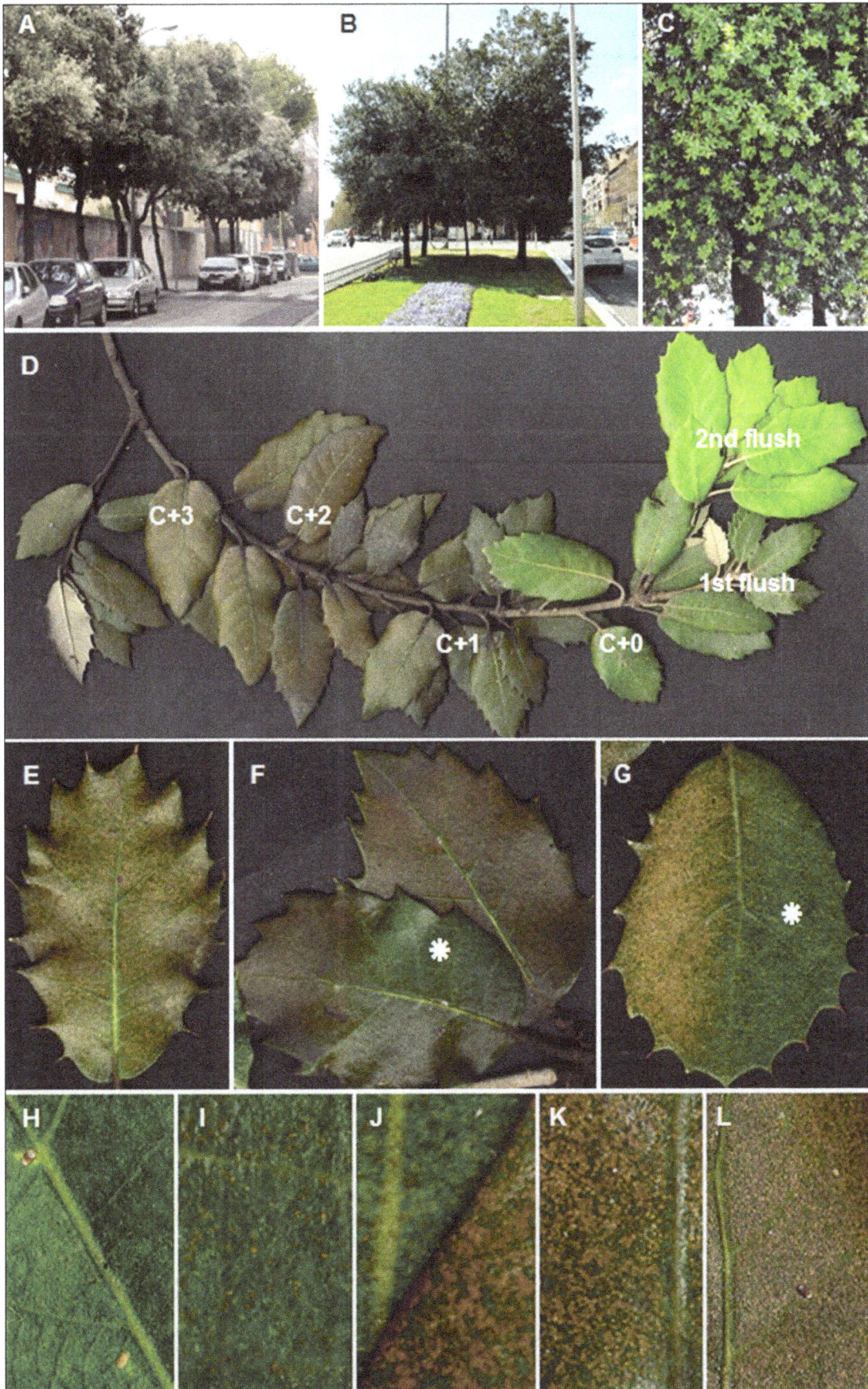

Figure 2. Visible injury caused by ozone stress in urban holm oaks from Madrid. A the non-irrigated intensive study site at Escalonilla. Trees were asymptomatic. **B, C** the irrigated intensive study site at Atocha. At tree level, the older and symptomatic foliage showed dark brownish tones whilst the newly flushed leaves were green (**C**). **D–L** visible injury in holm oak at Atocha in 2007. **D** at branch level, the symptomatic foliage showed a bronze discoloration that increased with leaf age. **E–L** at leaf level, symptoms were characterized by, tiny, slightly depressed, intercostal and necrotic adaxial stippling surrounded by still green leaf parts. The high stippling frequency gave an overall bronze appearance to the injured leaf (**E, L**). Shaded leaf parts (*) showed less injury (**F–G**). The stippling frequency increased with leaf age (asymptomatic: **H**: C+0; symptomatic: **I**: C+0, **J**: C+1, **K**: C+2, **L**: C+3; leaf formation: C+0:2007, C+1:2006, C+2:2005, C+3:2004).

asymptomatic whereas stipples generally developed in C+1 and C+2 leaves (Fig. 2J, Fig. 2K). Stippling intensity increased with leaf age (Fig. 2D) to such an extent as to give the older foliage an overall bronzed appearance (Fig. 2B, Fig. 2C, Fig. 2E, Fig. 2L). The stippling rates varied significantly between trees and increased with leaf age (P<0.02, Fig. 7). Holm oak subspecies showed similar type of visible injury (Fig. not shown).

Other visible symptoms occasionally observed and unrelated to the aforementioned stippling include 1) aphid exuviae and honeydew traces on C+0 leaves, 2) accumulation of soot and dust particles primarily trapped by hairs on the lower leaf side and nesting epiphytic communities in older foliage and 3) discretely distributed fungal infections (Fig. not shown).

Microscopic Symptoms

The leaf blade structure of the investigated holm oak leaves showed xeromorphic traits typical of a Mediterranean evergreen tree and which include a thick leaf lamina, thick-walled and lignified epidermis, thick cuticle and lower leaf side stomata protected by a thick and dense layer of hair (Fig. 8). In leaf parts with stipples, discretely distributed groups of necrotic cells were observed in the mesophyll (Fig. 8C, D versus 8A, B). Necrosis developed in the upper palisade cells and often extended into the lower assimilative layers. Stipples showed characteristic hypersensitive response-like (HR-like, [41], [42]) traits including 1) distribution of dead cells in discrete intercostal groups 2) cell collapse 3) cell content disruption and 4) cell remnant condensation (Fig. 8D versus Fig. 8B). Similar to stipples in fumigated foliage of Fraxinus ornus [21], folds and cracks in cell walls together with cell fragments leaking into the intercellular space were observed. Stipples were surrounded by degenerating cells as shown by cell wall thickening, chloroplast condensation and vacuolar accumulation of phenolics (Fig. 8C). Interestingly, the latter two markers

were also observed within dead cells belonging to stipples (Fig. 8D). In contrast to the tissue level, cell-level gradients of injury caused by varying light exposure were missing. Droplets of cell wall material protruding into the inter-cellular space were often observed in the lower leaf blade - mostly within the spongy parenchyma layers (Fig. 8F versus 8E). Elevated levels of oxidative stress were indicated by the accumulation of oligo-proanthocyanidins (OPC) inside and surrounding recently formed stipples (Fig. 8H versus Fig. 8G). In C+1 versus C+0 leaves, an increase in oxidation of the cellular material was shown by the lower OPC signal and brownish unspecific staining of stipples (Fig. 8I versus Fig. 8H).

Structural changes by fungi and bacteria or insects were detected but they were spatially distinct and causally unrelated to stipples. They included 1) cell wall thickening in cells of hairs covering the lower leaf side and trapping dust particles or 2) cell collapse, cell wall thickening and cell content degeneration in vein phloem and nearby lower leaf blade tissues as a consequence of aphid feeding (Fig. not shown).

O3 Symptoms Survey

Out of the 65 sites surveyed in the Madrid conurbation in 2007, 24 (37%) including Atocha, were symptomatic with foliage of holm oaks showing varying levels of O3 injury (Fig. 1). With the exception of one site next to a stream, all symptomatic sites had supplementary water supplied by an automated irrigation system. On average, the proportion of symptomatic trees per site amounted to 25% ±5.3 (SE; range: 11–100%). Out of the 41 asymptomatic sites, only 12 (30%) were irrigated with water supplied by drip (19%) or manual (11%) once a month, presumably with lower amounts than at symptomatic sites. Variation in the plot size, water supply, site conditions or holm oak sub-species prevented further quantification of the stippling frequency.

Stomatal Conductance

On average, during a typical summer day in 2011, and from dawn to dusk, the C+1 holm oak foliage showed higher gs at Atocha than Escalonilla (P<0.001) with values 55.5% larger at the irrigated versus non-irrigated intensive study site (Fig. 4). At both sites, gs was highest in the morning topping at 9:00/7:30 CET in Atocha/Escalonilla. Gs experienced a slight midday depression increasing moderately again from 16:00 to17:00 CET, prior to a further drop in the evening. Hence, at Atocha, besides increasing gs, irrigation delayed the midday depression by a few hours, as shown by significant differences (P = 0.05) between the sites from 9:00 to 12:00 am (Fig. 4).

The stomatal conductance varied as a function of the irrigation, leaf age and month (Table 1; Fig. 9). Particularly current year leaves C+0 followed the irrigation curve (Fig. 9 compared to Fig. 3, Pearson correlation coefficient for C+0 and irrigation = 0.88), whereas C+1 was correlated to C+2 (0.86). The site irrigation caused a significant increase in gs from May to October. During the peak irrigation period (May to September), gs values in C+0/C+1 leaves were 54%/31% higher on average at Atocha versus

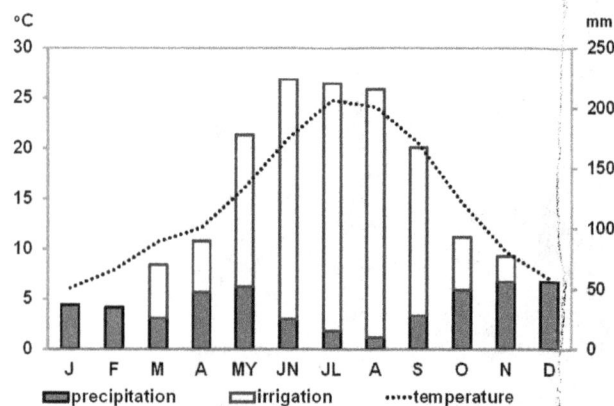

Figure 3. Climate diagram. Climatic conditions in Madrid and monthly irrigation totals at the Atocha intensive study site. Reference period for the climatic data: 1971–2007, average summer/winter temperature: 23.2°C/8.1°C, annual rainfall: 436 mm, annual irrigation: 1'027 mm.

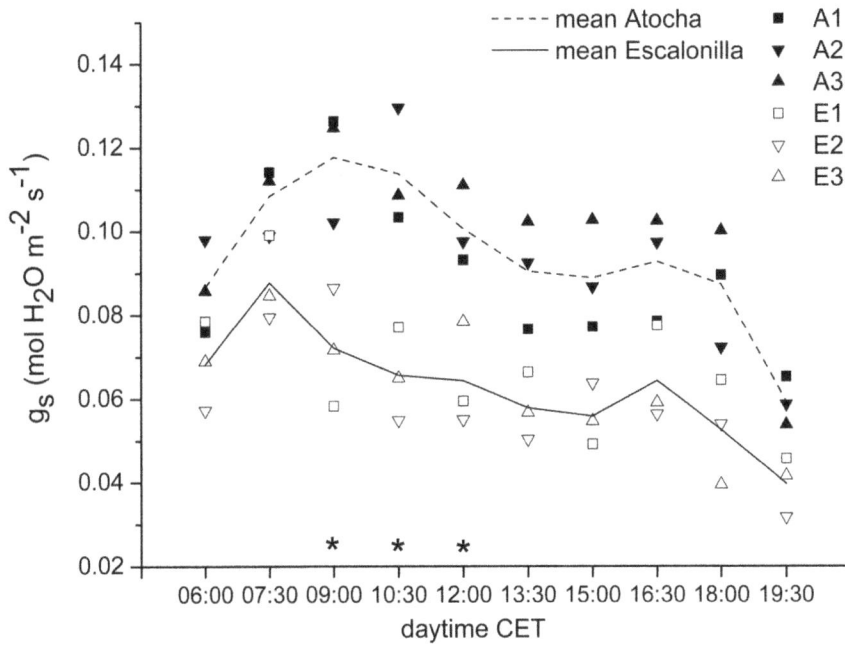

Figure 4. Daily time-course of stomatal conductance (gs). C+1 leaves (leaf formation: 2010) during a typical early summer day at the irrigated (A) Atocha (10th of June 2011; $T_{min} = 16.1°C$, $T_{med} = 20.1°C$, $T_{max} = 25.5°C$) and non-irrigated (E) Escalonilla (11th of June 2011; $T_{min} = 16.4°C$, $T_{med} = 20.5°C$, $T_{max} = 26°C$) intensive study site (means ± SE, n = 3 trees). The factors site (p>0.0001) and daytime (P<0.003) were significant. Stars indicate a significant difference between the site means (p<0.05) from 9:00 to 12:00 am.

Escalonilla. Leaf age caused a significant decrease in stomatal conductance and younger leaves were more responsive to higher water availability (significant irrigation*leaf age interaction from May to October, Table 1). The C+1 and C+2 leaves showed a decrease in gs after the new C+0 foliage had sprouted. During the vegetation season (February to October), stomatal conductance varied between months, especially regarding younger leaves (significant leaf age*month interaction from May to October).

Leaf Biomass Partition and LMA

At both intensive study sites, the youngest leaf generation formed the highest biomass fraction in the analyzed branches

(Fig. 10). However, there were differences between sites and, prior to and after the development of new C+0 leaves, older foliage made up a larger proportion of the total foliage biomass at Escalonilla than at Atocha. Hence from June to October 2011, during the highest irrigation and most O3-polluted period, the C+0 and older foliage biomass fraction amounted to 56% and 44% at Escalonilla *versus* 80% and 20% at Atocha. Finally at each site in October, the C+0 foliage's contribution to the total foliage biomass amounted to 61% and 86%, respectively.

As indicated by LMA and irrespective of the leaf generation, holm oak foliage showed a similar leaf xeromorphy at both sites (Table 2). During the 2011 vegetation season, monthly LMA

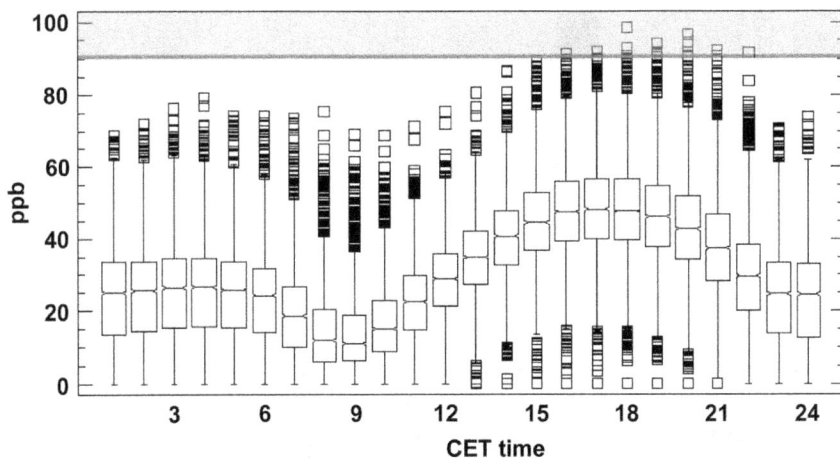

Figure 5. Boxplot of the average hourly (CET) O3 concentrations during the vegetation season. Data from April to September in Atocha for the years 2003–2007. The grey zone outlines the range of values exceeding the population warning threshold (box: interquartile range; whiskers: lower and upper quartiles; median horizontal line of boxes: median; white squares: maxima and minima).

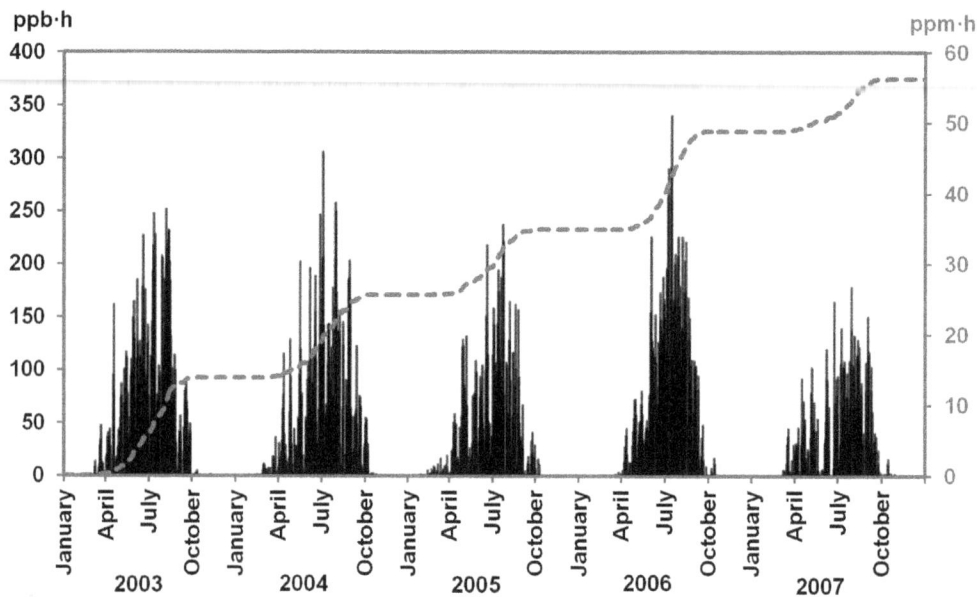

Figure 6. Daily (ppb·h, black spikes) and cumulated (ppm·h, grey line) AOT40 in Atocha from 2003 to 2007. Yearly AOT40 (April to September) in 2003/2004/2005/2006/2007 amounted to 14.55/11.26/9.0/13.45/7.55 ppm·h.

estimates for C+1 and C+2 leaves at Atocha *versus* Escalonilla were not significantly different except during new foliage development (April and May). Regarding the C+0 leaves, it took three months at Atocha *versus* four at Escalonilla (Fig. 10) until adult and comparable leaf LMA values could be achieved (75 ± 10.5 to 162 ± 5.3 mg/cm2 from April to June *versus* 74 ± 1.3 to 164 ± 3.6 mg/cm2 from May to July).

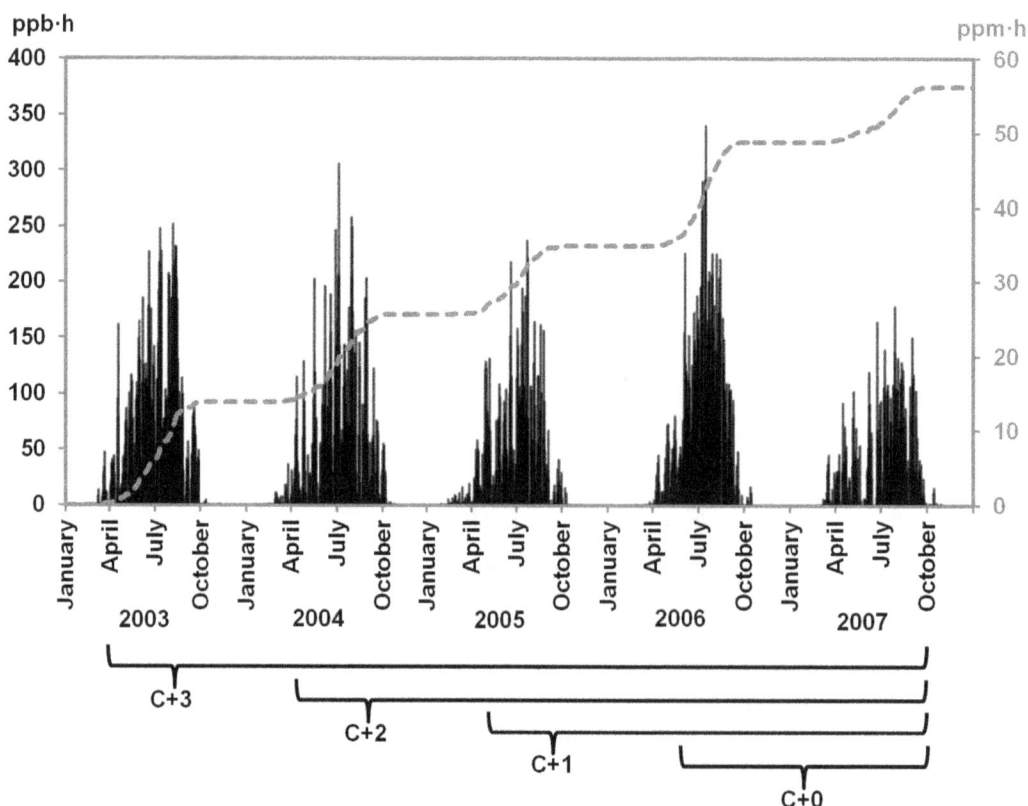

Figure 7. Mean percentage ± SE of leaf area showing adaxial stippling in holm oaks. Samples from Atocha in June 2007 (n = 4 branches per tree each with leaf age C+0, C+1, C+2, C+3). Different letters indicate significantly different percentages of symptomatic leaf area ($p \leq 0.05$).

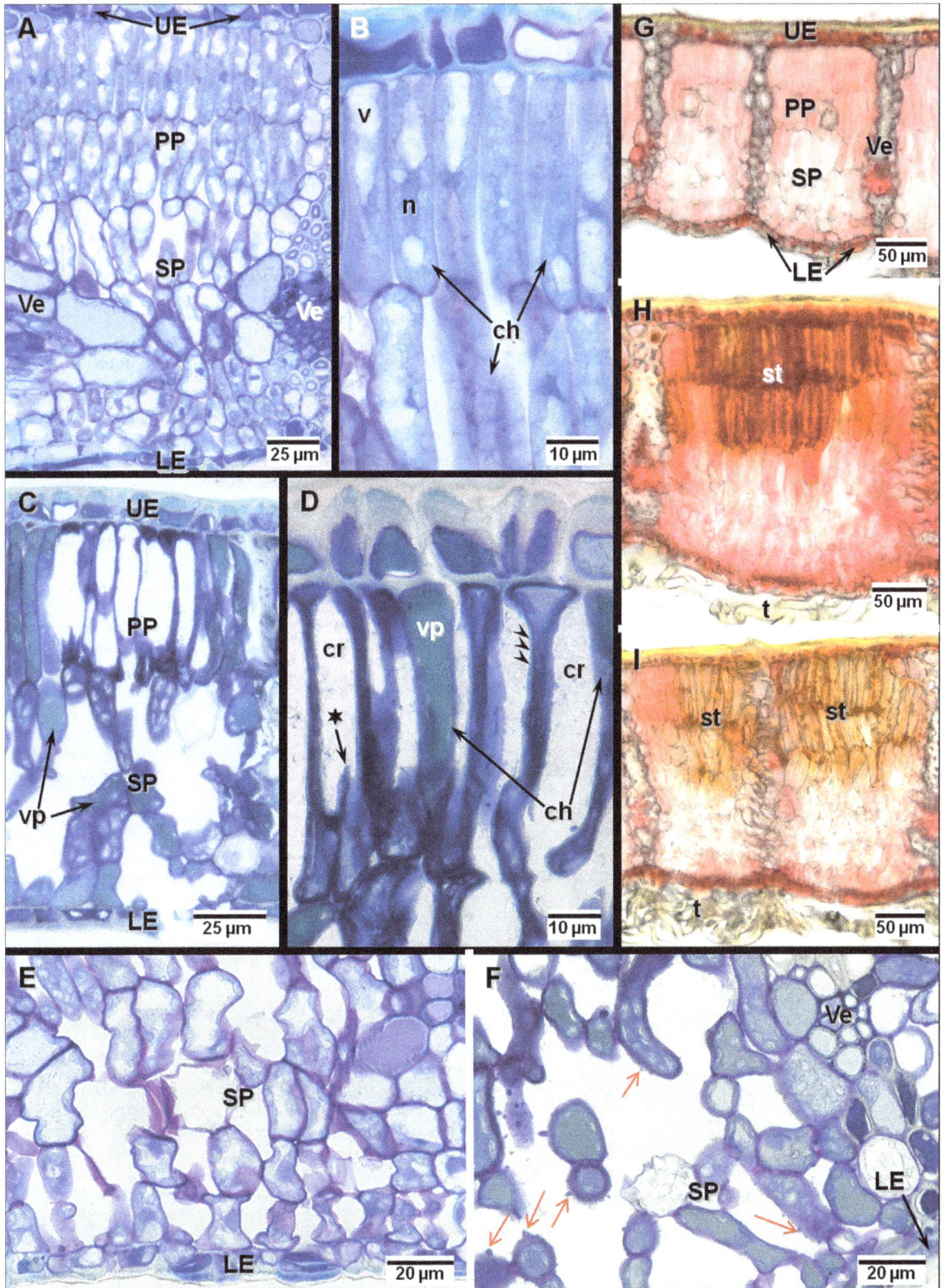

Figure 8. Structural and histochemical changes in the leaf blade. Leaf age/formation C+0/2007, (**G, H**) and C+1/2006 (**A–F, I**). Symptomatic (**C, D, F, H, I**) *versus* asymptomatic (**A, B, E, G**) foliar samples. Leaf parts with stipples in symptomatic *versus* asymptomatic (**C *versus* A**) material showed discrete groups of necrotic and collapsed palisade parenchyma (PP) cells surrounded by degenerating mesophyll tissue. At cell level (**D *versus* B**), necrotic cells showed cell wall thickening (arrowheads), cracking (*) and folding and a disrupted cell content. The intercellular space contained cellular remains (cr). Degenerating cells showed thickened cell walls, enlarged vacuoles (v) filled with phenolics (vp) and smaller and condensed chloroplasts (ch). Within the spongy parenchyma, cell wall protrusions (red arrows), the frequency of which increased in symptomatic versus asymptomatic material (**F *versus* E**), were indicative of oxidative stress in the apoplast. **G–I** Photo-oxidative stress in stipples (st) of symptomatic (**H, I**) *versus* asymptomatic (**G**) samples was shown by gradients of condensed tannin reacting with acid-vanillin (red staining) between the upper (stronger staining) and lower (weaker staining) mesophyll cell layers. In older samples (C+1, **I**) and in contrast to younger symptomatic samples (C+0, **H**), stronger oxidation of proanthocyanidins in stipples was shown by the weak reaction of condensed tannins to acid-vanillin. UE, LE upper and lower epidermis; Ve: veins; n: nucleus; t: trichomes.

Discussion

Stipples as Structural Injury Due to Ozone Stress

Within the analyzed holm oak leaves, the stipple morphology and the changes observed at cell level were typical of those associated with acute O3 stress as described for deciduous broadleaved species [13], [43]. Apparently, the thick leaf blade, a xeromorphic trait in *Q. ilex* [44], [45], did not modify the development of stipples although the distance between the O3-absorbing stomata and necrotic upper palisade parenchyma was larger than in deciduous foliage. In comparison to *Pistacia lentiscus* [22], not only degenerative changes but also necrotic stipples indicative of HR-like and resulting from defensive programmed cell death (PCD, [13], [46]) were found. Cracks in cell walls and cell content leakage similar to HR-like injuries reported for fumigated Manna ash seedlings [21] indicated a large production of and severe injury by ROS [47]. Differing from the latter species, necrotic cells in the present study showed phenolic accumulation and cell content disruption suggesting that PCD was preceded by a degenerative phase lasting more than a year according to morphological observations about stippling emergence. Indeed, PCD is ROS- concentration dependent [48] and an oxidative stress threshold thus needs to be exceeded prior to activating a PCD-program. Other oxidative and O3 stress markers in the studied holm oaks included 1) the wart-like droplets in lower mesophyll [49], [14], [16], 2) the positive reaction with acid-vanillin in and around young stipples [50] and 3) the impediment of the acid-vanillin reaction within older leaf material due to the oxidation of necrotic cell remnants [17]. The interaction between O3 and photo-oxidative stress [21] was indicated by the gradient of injury and OPC between lower and upper mesophyll in leaf parts with stipples.

Table 1. Significance (P-values) of two-way analysis of variance.

Factor	d.f.	g$_s$ February to April	g$_s$ May to October
Irrigation	1	ns	<0.001
Leaf age	3	0.001	<0.001
Month	2 or 5	0.011	ns
Irrigation * leaf age	2	ns	0.003
Irrigation * month	2	ns	ns
Leaf age * month	4 or 10	ns	0.006

Effects of the factors: irrigation (Atocha *versus* Escalonilla), leaf age/formation (C+0/2011, C+1/2010, C+2/2009, C+3/2008) and month on stomatal conductance (gs) and their interactions during spring with little irrigation (February to April) and summer with irrigation (May to October), ns not significant p≥0.05.

The visible stippling morphology and distribution, together with the observed shading effects, were typical for O3 stress [10]. The homogeneous and intercostal distribution of stipples in foliage of the sun-exposed crown, their frequency increasing with leaf age and their occurrence within large tree crown portions of several trees per site at the many sites further confirmed the diagnosis [51]. Regarding the role of other stress factors, as potential causes for the observed leaf injury, besides ozone, a contribution can be excluded based on: 1) other phytotoxic components of photo-chemical smog, such as peroxyacetyl nitrate (PAN), do not cause HR-like reaction leading to stippling symptoms [13], [52], 2) the concentration of other gaseous air pollutants, such as the aforementioned SO$_2$ and NO$_2$, were too low or not phytotoxic, 3) the detected biotic injury was spatially and causally not related to the analyzed stippling, 4) eventual nutrient deficiencies or imbalances cause specific patterns of visible injury different from those caused by ozone stress [43] and 5) eventual soil contamination with metals lead to microscopic changes primarily along the water pathway through the leaf and these microscopic symptoms are clearly different from those induced by ozone stress [13], [51].

Ozone-triggered stippling has already been observed in fumigated holm oak seedlings [31], [32] and similar visible leaf injury has been documented in other deciduous and partly evergreen Spanish oak species (*Q. faginea, Q. pyrenaica*) exposed to O3 under controlled conditions [53]. To our knowledge however, the findings presented here are the first to show the structural changes associated with O3-triggered stippling in leaves of holm oak.

Ozone Stress in the Holm Oaks of Madrid

In the center of Madrid, the 2003–2007 AOT40 average (11 ppm•h) was above both the former (10 ppm•h) and present (5 ppm•h) concentration-based critical level for European forest trees [54], [55]. Compared to other South-Western European sites, it was slightly inferior to the 2000–2002 AOT40 range (13–19 ppm•h), whereas the warmer 2003 and 2006 years fitted into the lower part of the range [56]. Gradients of O3 concentration, increasing between the Madrid center and suburbs and varying according to micro-climatic conditions [2], may relate to the higher frequency of symptomatic sites at the city's periphery. Between 2003 and 2007, Madrid experienced exceedances over the 180 µg/m3 alert threshold more often than on average in the Iberian Peninsula but less frequently than in South-Eastern France and Italy (for example [3], [4], [5], [6], [7], 2003–2008). The 21 ppm•h O3 exposure required for the appearance of visible injury in holm oak at Atocha (sum of 2007 and 2006 AOT40) was lower than that indicated for the fumigated holm oak seedlings mentioned previously [31], [32]. However, it was still largely higher than that which causes symptoms in most young deciduous broadleaved trees so far tested [57], [58], [59] or promoting functional alterations in holm oak foliage fumigated experimentally (3 ppm•h, [32]). The O3 exposure needed for the develop-

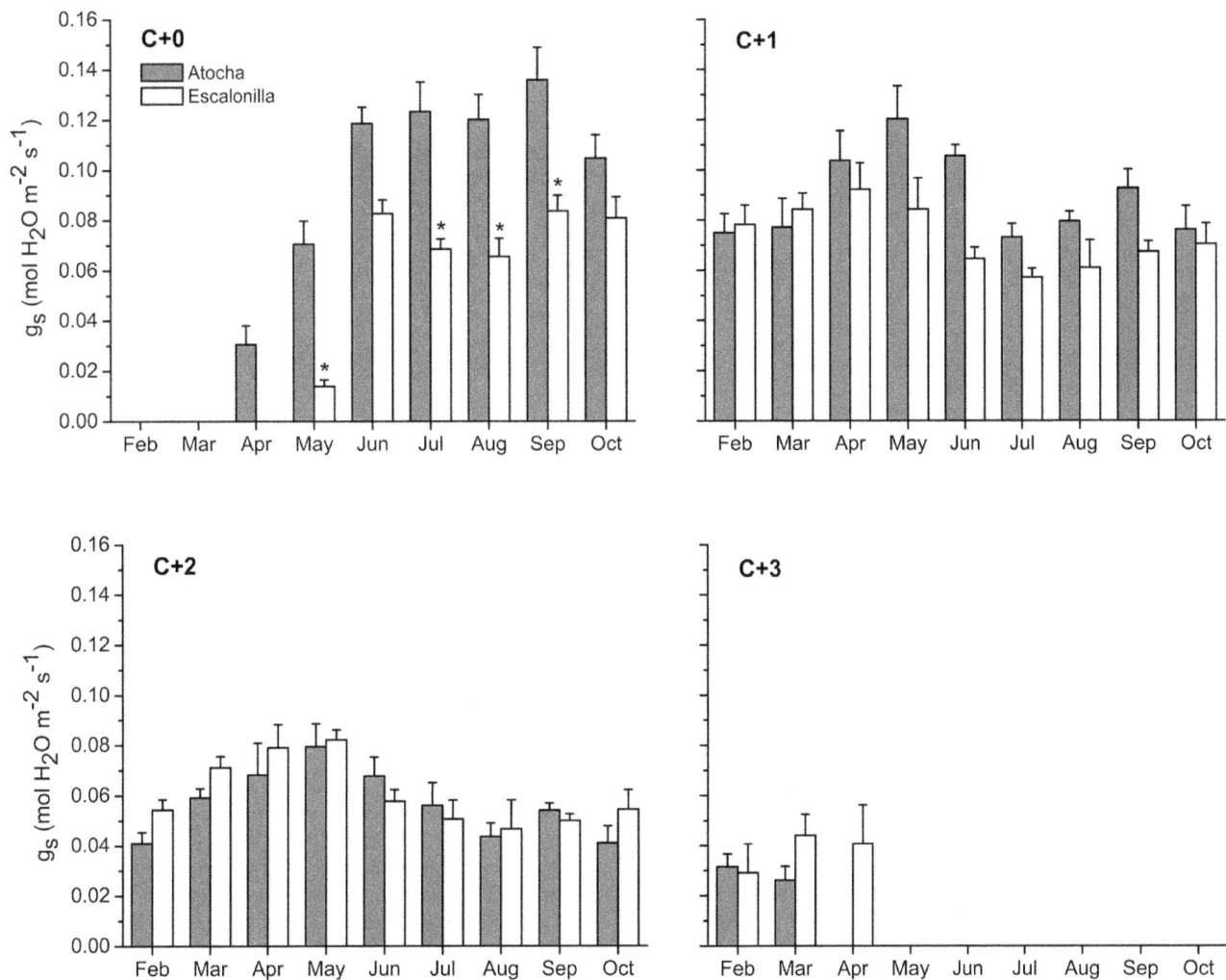

Figure 9. Seasonal variation of stomatal conductance (gs). C+0, C+1, C+2 and C+3 foliage (leaf formation: 2011, 2010, 2009, 2008, respectively) at the irrigated (Atocha, grey bars) *versus* non- irrigated (Escalonilla, white bars) intensive study site in 2011 (means ± SE, n = 3 trees). The monthly irrigation supply at Atocha is shown in Figure 2, the significance of influencing factors in Table 2.

ment of leaf injury may also change according to the years and experimental settings as shown for *Pistacia lentiscus*, another evergreen xerophyte, with injury threshold varying from 9 to 74 ppm•h [22]. Hence, the O3 exposure recorded at Atocha was considerable with respect to the Iberian average but remained, with regard to structural injury in holm oak, within the lower range of values expected to cause symptoms in a sclerophyll evergreen tree relatively insensitive to O3 stress [1], [60].

Table 2. Mean leaf mass per area.

Leaf age/ formation	C+0/2011	C+1/2010	C+2/2009	C+3/2008
Atocha	16.5±0.1	16.9±0.3	17.3±0.1	17.2±0.3
Escalonilla	16.5±0.1	16.7±0.3	17.1±0.2	17.1±0.3

LMA (mg/cm2) ± SE per leaf generation at the irrigated (Atocha) and non-irrigated (Escalonilla) site in October (C+0, C+1, C+2) and March (C+3) 2011. Differences between sites and leaf generations were not significant (p>0.05; N = 3 trees).

Increased Ozone Uptake in Holm Oak Foliage as a Trade-off for Site Irrigation

By raising gs in Atocha *versus* Escalonilla, irrigation was confirmed to increase O3 uptake during the whole day and alleviate the midday gas exchange reduction during peak O3 hours. Other studies have also documented the responsiveness of holm oak to higher soil moisture availability [61], [62] with enhanced O3 stomatal uptake at O3 polluted sites as a trade-off for an elevated water supply [63], similar to findings on other species [33]. In Madrid, findings from the leaf injury survey suggest that O3 symptoms were even conditioned to site irrigation and higher water availability. At Atocha, the highest levels of O3 exposure, site irrigation and leaf gs were recorded during the summer and these factors could synergistically contribute to an increased O3 uptake in irrigated *versus* non-irrigated holm oak. Interestingly, only the younger C+0 and C+1 leaf generations were responsive to elevated water availability. Besides shading by new foliage and lower gs with increasing leaf age [64], leaf injury might further reduce gs in older holm oak foliage as suggested by the concomitant development of stippling and reduction of gs in C+1 leaves recorded during the summer. Overall, the gs values

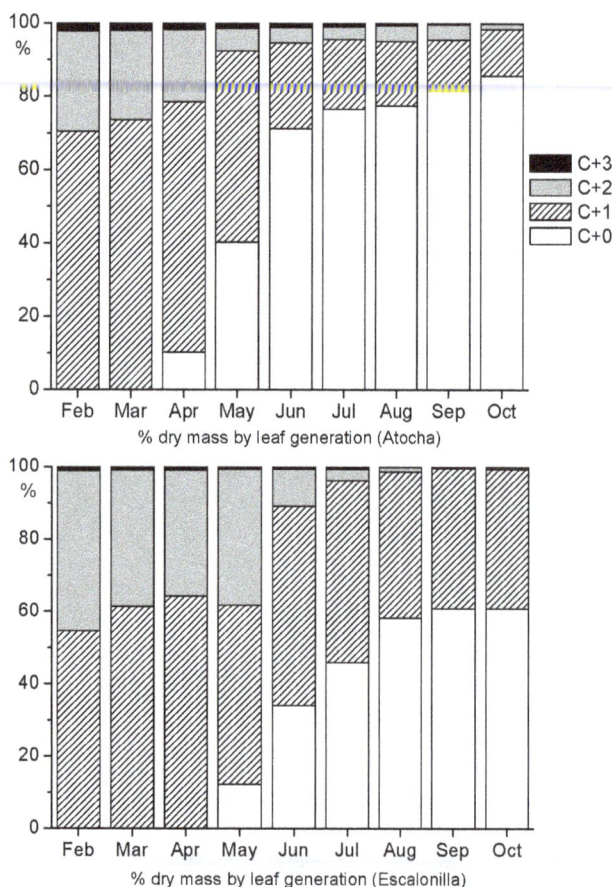

Figure 10. Monthly changes in the foliage biomass fraction (expressed in %) of each leaf generation. Leaf age/formation C+0/2011, C+1/2010, C+2/2009, C+3/2008 within holm oaks from the irrigated (Atocha) and non-irrigated (Escalonilla) intensive study site in 2011 (mean values of 3 trees).

by the concomitant decrease of gs and development of stippling. Hence and synergistically with other causes, O3 might contribute to reduced leaf life span in the irrigated and symptomatic holm oaks.

Leaf Xeromorphy and Irrigation

Whatever the leaf generation, the leaf xeromorphy was not affected by irrigation, as indicated by similar LMA at both study sites. With reference to [72], these findings were unexpected. The values found in Madrid were similar to those indicated for holm oak in other urban conditions [73] on rather mesic Italian sites [74] or under similar precipitation and temperature regimes in Catalonia, Spain [75]. As found by [45], the LMA of mature leaves did not change significantly through time. Consequently, the development of O3 injury proceeded independent of the leaf xeromorphy and primarily related to enhanced gs and higher O3 uptake.

Conclusions

In synthesis, the initial O3 symptom diagnosis was confirmed on the basis of the macro- and micro-morphological changes found in irrigated holm oak foliage (objective 1). Ozone injury similar to that detected at the intensive study site of Atocha was found throughout Madrid but predominantly at sites with automated irrigation (objective 2). O3 exposure up to a harmful level for the natural vegetation was recorded in air monitoring stations close to our intensive study sites but at levels apparently too low to cause visible injury in an evergreen tree rather insensitive to O3 stress (objective 3). On the basis of subsequent gas exchange and biomass/LMA measurements, higher rates of stomatal O3 uptake in irrigated and symptomatic trees were corroborated (first hypothesis) whereas no difference in the leaf xeromorphy between the irrigated and non- irrigated site was found and therefore the second hypothesis was rejected. Given the concomitant maximum irrigation and peak O3 pollution, the O3 tolerance of irrigated holm oaks appeared to be lowered to levels similar to those recorded for other broadleaved trees. This particular case of leaf injury by O3 stress because of the irrigation gives insight into mechanisms driving O3 symptom expression in sclerophyll evergreen trees. In holm oak at least, they suggest that stomatal closure, particularly during peak O3 pollution, can be more effective than leaf xeromorphy to reduce stomatal O3 uptake and outlines the driving contribution of soil moisture availability for O3 symptom expression in dry climates.

Acknowledgments

The authors wish to thank the Madrid park service (Dirección General de Patrimonio Verde del Ayuntamiento de Madrid), municipal authority (Departamento de Calidad del Aire del Ayuntamiento de Madrid) and national meteorological agency (Agencia Estatal de Meteorología - AEMET) for graciously providing irrigation, air pollution and climate data. They are also grateful to Dr. Rosa Inclán from CIEMAT, Spain, for sharing unpublished evidence about visible O3 symptoms in holm oak, to Terry Menard for technical assistance and scientific English correction.

Author Contributions

Conceived and designed the experiments: CCG PV MGG. Performed the experiments: CCG. Analyzed the data: CCG PV MGG. Contributed reagents/materials/analysis tools: CCG PV. Wrote the paper: CCG PV MGG.

measured in Madrid were in the range of those published for Mediterranean forests subjected to summer drought [65], [66], the highest gs rates for C+0 leaves were in line with findings by [62]. The C+0 and C+1 leaf generations with the highest gs were also those least symptomatic. This paradox probably related to the aforementioned exceedance of an oxidative stress threshold needed for triggering a PCD-program and causing visible stippling, as a consequence.

Leaf Life Span of Irrigated Foliage

Compared to Escalonilla, the leaf turn-over in Atocha's holm oak foliage was accelerated. Generally, foliage showing higher stomatal conductance, as in Atocha, is also shed earlier [67]. Furthermore, water-restricted versus water unrestricted evergreen trees tend to keep their older foliage a longer time and use it more intensively [68]. Competition between older and younger leaves might also contribute to leaf turn-over, as suggested by leaf-drop primarily after new leaf flushing instead of throughout the spring and after the summer drought period, as usual. Besides leaf physiology and competition factors, O3 accelerates leaf senescence [69], [70] which can lead to a reduction in the amount of leaf generations in evergreen trees [71]. Here, this effect is suggested

References

1. Paoletti E (2006) Impact of ozone on Mediterranean forests: A review. Environ Pollut 144: 463–474.
2. Sanz MJ, Sanz F, Sanchez-Peña G (2001) Spatial and annual temporal distribution of ozone concentrations in the Madrid basin using passive samplers. Scientific World J 1: 785.
3. EEA (2003) Air pollution by ozone in Europe in summer 2003. Overview of exceedances of EC ozone threshold values during the summer season April-August 2003 and comparisons with previous years. Copenhagen: European Environment Agency Topic report No 3/2003.
4. EEA (2005) Air pollution by ozone in Europe in summer 2004. Overview of exceedances of EC ozone threshold values for April–September 2004. Copenhagen: European Environment Agency Technical report No 3/2005.
5. EEA (2006) Air pollution by ozone in Europe in summer 2005. Overview of exceedances of EC ozone threshold values for April–September 2005. Copenhagen: European Environment Agency Technical report No 3/2006.
6. EEA (2007) Air pollution by ozone in Europe in summer 2006. Overview of exceedances of EC ozone threshold values for April–September 2006. Copenhagen: European Environment Agency Technical report No 5/2007.
7. EEA (2008) Air pollution by ozone across Europe during summer 2007. Overview of exceedances of EC ozone threshold values for April–September 2007. Copenhagen: European Environment Agency Technical report No 5/2008.
8. Millán MM, Salvador R, Mantilla E, Kallos G (1997) Photooxidant dynamics in the Mediterranean basin in summer: results from European research projects. J Geophys Res 102: 8811–8823.
9. Sanz MJ, Millán M (1998) The dynamics of aged air masses and ozone in the Western Mediterranean: relevance to forest ecosystems. Chemosphere 36: 1089–1094.
10. Innes JL, Skelly JM, Schaub M (2001) Ozone and broadleaved species. A guide to the identification of ozone-induced foliar injury. Bern: Haupt. 136 p.
11. Orendovici T, Skelly JM, Ferdinand JA, Savage JE, Sanz MJ, et al. (2003) Response of native plants of northeastern United States and southern Spain to ozone exposures; determining exposure/response relationships. Environ Pollut 125: 31–40.
12. Porter E (2003) Ozone sensitive plant species on national park service and U.S. fish and wildlife service lands: results of a June 24–25, 2003 workshop. US Department of the Interior. NPS D1522. Natural resource report NPS/ NRARD/NRR_2003/01, Available: http://www.nature.nps.gov/air/pubs/ ozone.cfm?CFID = 11417561&CFTOKEN = 98472238.
13. Günthardt-Goerg MS, Vollenweider P (2007) Linking stress with macroscopic and microscopic leaf response in trees: New diagnostic perspectives. Environ Pollut 147: 467–488.
14. Gravano EF, Bussotti F, Strasser RJ, Schaub M, Novak K, et al. (2004) Ozone symptoms in leaves of woody plants in open-top chambers: ultrastructural and physiological characteristics. Physiol Plant 121: 620–633.
15. Kivimäenpää M, Sutinen S, Karlsson PE, Sellden G (2003). Cell structural changes in the needles of Norway spruce exposed to long-term ozone and drought. Ann Bot 92: 779–93.
16. Oksanen E, Haikio E, Sober J, Karnosky DF (2003) Ozone-induced H2O2 accumulation in field-grown aspen and birch is linked to foliar ultrastructure and peroxisomal activity. New Phytol 161: 791–99.
17. Vollenweider P, Ottiger M, Günthardt-Goerg MS (2003) Validation of leaf ozone symptoms in natural vegetation using microscopical methods. Environ Pollut 124: 101–118.
18. Foyer CH, Lelandais M, Kunert KJ (1994) Photooxidative stress in plants. Physiol Plant 92: 696–717.
19. Elstner EF (1996) Die Sauerstoffaktivierung als Basis pflanzlicher Stressreaktionen. In: Brunold Ch, Rüegsegger A, Brändle R, editors. Stress bei Pflanzen. Ökologie, Physiologie, Biochemie, Molekularbiologie. Bern: Paul Haupt, 347–362.
20. Yamasaki H, Sakihama Y, Ikehara N (1997) Flavonoid-peroxidase reaction as a detoxification mechanism of plant cells against H2O2. Plant Physiol 115: 1405–1412.
21. Paoletti E, Contran N, Bernasconi P, Günthardt-Goerg MS, Vollenweider P (2009) Structural and physiological responses to ozone in Manna ash (*Fraxinus ornus* L.) leaves of seedlings and mature trees under controlled and ambient conditions. Sci Total Environ 407: 1631–43.
22. Reig-Armiñana J, Calatayud V, Cervero J, Garcia-Breijo FJ, Ibars A, et al. (2004) Effects of ozone on the foliar histology of the mastic plant (*Pistacia lentiscus* L.). Environ Pollut 132: 321–31.
23. Bombelli A, Gratani L (2003) Interspecific differences of leaf gas exchange and water relations of three evergreen Mediterranean shrub species. Photosynthetica 41: 619–625.
24. Romane F, Terradas J, editors (1992) *Quercus ilex* L. Ecosystems: Function, Dynamics and Management. Vegetatio 99/100. 376 p.
25. Turner IM (1994) Sclerophylly: primarily protective? Funct Ecol 8: 669–675.
26. Ogaya R, Peñuelas J, Martines-Vilalta J, Mangirón M (2003) Effect of drought on diameter increment of *Quercus ilex, Phillyrea latifolia* and *Arbutus unedo* in a holm oak forest of NE Spain. For Ecol Manage 180: 175–184.

27. Tetriach M (1993) Photosynthesis and transpiration of evergreen Mediterranean and deciduous trees in an ecotone during a growing season. Acta Oecol 341–360.
28. Zazo J, Calderón Guerrero C, Cornejo L (2000) Caracteres culturales y otras características de interés de algunas frondosas forestales españolas: Part I: *Quercus ilex, Quercus suber, Castanea sativa, Fagus sylvatica, Quercus robur, Quercus petraea* y *Quercus pyrenaica*. Madrid, Spain: Servicio de Publicaciones de la EUIT Forestal (UPM). 220 p.
29. Calatayud V, Marco F, Cervero J, Sanchez-Peña G, Sanz MJ (2010) Contrasting ozone sensitivity in related evergreen and deciduous shrubs. Environ Pollut 158: 3580–3587.
30. Manes F, Vitale M, Donato E, Paoletti E (1998) O3 and O3+CO2 effects on a Mediterranean evergreen broadleaf tree, holm oak (*Quercus ilex* L.). Chemosphere 36: 801–806.
31. Inclán R, Ribas A, Peñuelas J, Gimeno B (1999) The relative sensitivity of different Mediterranean plant species to ozone exposure. Water Air Soil Pollut 116: 273–277.
32. Ribas A, Peñuelas J, Elvira S Gimeno BS (2005) Contrasting effects of ozone under different water supplies in two Mediterranean tree species. Atmos Environ 39: 685–93.
33. Schaub M, Skelly JM, Steiner KC, Davis DD, Pennypacker SP, et al. (2003) Physiological and foliar injury responses of *Prunus serotina, Fraxinus americana* and *Acer rubrum* seedlings to varying soil moisture and ozone. Environ Pollut 124: 307–320.
34. Vollenweider P, Kelty MJ, Hofer RM, Woodcock H (2003) Reduction of stem growth and site dependency of leaf injury in Massachusetts black cherries exhibiting ozone symptoms. Environ Pollut 125: 467–480.
35. Fuhrer J, Skärby L, Ashmore MR (1997) Critical levels for ozone effects on vegetation in Europe. Environ Pollut 97: 91–106.
36. Günthardt-Goerg MS, Schmutz P, Matyssek R, Bucher JB (1996) Leaf and stem structure of poplar (*Populus x euramericana*) as influenced by O3, NO2, their combination and different soil N supplies. Can J for Res 26: 649–657.
37. Murakami PF, Turner MR, Van den Berg AK, Schaberg PG (2005) An instructional guide for leaf color analysis using digital imaging software. United States Department of Agriculture Publication. Tech Rep NE-327.
38. Feder N, O'Brien TP (1968) Plant microtechnique: some principles and new methods. Am J Bot 55: 123–42.
39. Calderón Guerrero C, Saiz de Omenaca González JA, Günthardt-Goerg MS (2009) Contribución del arbolado urbano y periurbano del municipio de Madrid en la mejora de la calidad del aire y sumidero de contaminantes atmosféricos como beneficio para la sociedad. In: SECF, editor. Actas del 5 Congreso Forestal Español. 1–16.
40. EU Directive 2008/50/EC. Directive on Ambient Air Quality and Cleaner Air for Europe. Available: http://eur- lex.europa.eu/LexUriServ/LexUriServ.- do?uri = CELEX:32008L0050:EN:NOT.
41. Schraudner M, Langebartels C, Sandermann HJr (1996) Plant defense systems and ozone. Biochem Soc Transact 24: 456–461.
42. Sandermann HJ, Ernst D, Heller W, Langebartels C (1998) Ozone: an abiotic elicitor of plant defence reactions. Trends Plant Sci - Reviews 3: 47–50.
43. Fink S (1999). Pathological and regenerative plant anatomy. Encyclopedia of plant anatomy Vol. XIV/6. Berlin: Gebrüder Bornträger. 1095 p.
44. Gratani L (1995). Structural and ecophysiological plasticity of some evergreen species of the Mediterranean maquis in response to climate. Photosynthetica 31: 335–343.
45. Cunningham SA, Summerhayes B, Westoby M (1999). Evolutionary divergences in leaf structure and chemistry, comparing rainfall and soil nutrient gradients. Ecology 69: 569–588.
46. Sandermann H (2004) Molecular ecotoxicology of plants. Trends Plant Sci 9: 406–413.
47. Foyer CH, Noctor G (2005). Oxidant and antioxidant signalling in plants: a re-evaluation of the concept of oxidative stress in a physiological context. Plant Cell Environ 28: 1056–1071.
48. Rao MV Davis KR (2001). The physiology of ozone induced cell death. Planta 213: 682–690.
49. Günthardt-Goerg MS, McQuattie C, Scheidegger C, Rhiner C, Matyssek R (1997) Ozone-induced cytochemical and ultrastructural changes in leaf mesophyll cell walls. Can J For Res 27: 453–463.
50. Bussotti F, Agati G, Desotgiu R, Matteini P, Tani C (2005). Ozone foliar symptoms in woody plant species assessed with ultrastructural and fluorescence analysis. New Phytol 166: 941–955.
51. Vollenweider P, Günthardt-Goerg MS (2006). Diagnosis of abiotic and biotic stress factors using the visible symptoms in foliage. Environ Pollut 140: 562–571.
52. Davis DD (1977) Responses of ponderosa pine primary needles to separate and simultaneous ozone and PAN exposures. Plant Dis Rep 61: 640–644.
53. Sanz MJ, Calatayud V (2012). Ozone injury in European Forest Species. Available: http://www.ozoneinjury.org.
54. Kärenlampi L, Skärby L (1996) Critical Levels for Ozone in Europe: Testing and Finalizing the Concepts. UN-ECE Workshop Report. University of Kuopio, Dept. of Ecology and Environmental Science. 363 p.

55. ICP (2004). Modelling and Mapping. Manual on methodologies and criteria for modeling and mapping critical loads and levels and air pollution effects, risks and trends. UBA-Texte 52/04.

56. Ferretti M, Bussotti F, Rocchini D (2007) Estimates of ozone AOT40 from passive sampling in forest sites in South-Western Europe. Environ Pollut 145: 629–35.

57. Günthardt-Goerg MS, McQuattie C, Maurer S, Frey B (2000) Visible and microscopic injury in leaves of five deciduous tree species related to current critical ozone levels. Environ Pollut 109: 489–500.

58. VanderHeyden D, Skelly J, Innes J, Hug C, Zhang J, et al. (2001) Ozone exposure thresholds and foliar injury on forest plants in Switzerland. Environ Pollut 111: 321–331.

59. Novak K, Skelly JM, Schaub M, Kraeuchi N, Hug C, et al. (2003) Ozone air pollution and foliar injury development on native plants of Switzerland. Environ Pollut 125: 41–52.

60. Calatayud V, Cervero J, Calvo E, García-Breijo FJ, Reig-Armiñana J, et al. (2011) Responses of evergreen and deciduous Quercus species to enhanced ozone levels. Environ Pollut 159: 55–63.

61. Galle A, Florez-Sarasa I, Aououad H E, Flexas J (2011) The Mediterranean evergreen *Quercus ilex* and the semi-deciduous *Cistus albidus* differ in their leaf gas exchange regulation and acclimation to repeated drought and re-watering cycles. J Exp Bot 62: 5207–5216.

62. Pardos M, Royo A, Pardos J A (2005) Growth, nutrient, water relations and gas exchange in a holm oak plantation in response to irrigation and fertilization. New Forests 30: 75–94.

63. Gerosa G, Vitale M, Finco A, Manes F, Denti AB, et al. (2005) Ozone uptake by an evergreen Mediterranean Forest (*Quercus ilex*) in Italy. Part I: Micrometeo-rological flux measurements and flux partitioning. Atmos Environ 39: 3255–3266.

64. Niinemets Ü, Cescatti A, Rodeghiero M, Tosens T (2005). Leaf internal diffusion conductance limits photosynthesis more strongly in older leaves of Mediterranean evergreen broad-leaved species. Plant Cell Environ 28: 1552–1566.

65. Asensio D, Peñuelas J, Ogaya R, Llusià J (2007) Seasonal soil and leaf CO2 exchange rates in a Mediterranean holm oak forest and their responses to drought conditions. Atmos Environ 41: 2447–2455.

66. Gulías J, Cifre J, Jonasson S, Medrano H, Flexas J (2009). Seasonal and inter-annual variations of gas exchange in thirteen woody species along a climatic gradient in the Mediterranean island of Mallorca. Flora - Morphology, Distribution, Functional Ecology of Plants 204: 169–181.

67. Reich PB, Ellsworth DS, Walters MB, Vose JM, Gresham C, et al. (1999) Generality of leaf trait relationships: A test across six biomes. Ecology 80: 1955–1969.

68. Niinemets Ü, Lukjanova A (2003) Total foliar area and average leaf age may be more strongly associated with branching frequency than with leaf longevity in temperate conifers. New Phytol 158: 75–89.

69. Bortier K, Ceulemans R, de Temmerman L (2000) Effects of tropospheric ozone on woody plants. In: Agrawal SB, Agrawal M, editors. Environmental pollution and plant responses. Boca Raton: Lewis Publishers. 153–173.

70. Karnosky DF, Pregitzer KS, Zak DR, Kubiske ME, Hendrey GR, et al. (2005) Scaling ozone responses of forest trees to the ecosystem level in a changing climate. Plant Cell Environ 28: 965–981.

71. Miller PR, Arbaugh MJ, Temple PJ (1997) Ozone and its known and potential effects on forests in Western United States. In: Sandermann JrH, Wellburn AR, Heath RL, editors. Forest decline and ozone. Ecological Studies 127. Berlin: Springer. 39–67.

72. Gratani L, Varone L (2006) Long-time variations in leaf mass and area of Mediterranean evergreen broad-leaf and narrow-leaf maquis species. Photo-synthetica 44: 161–168.

73. Gratani L, Pesoli P, Crescente MF (1998) Relationship between photosynthetic activity and chlorophyll content in an Isolated *Quercus Ilex* L. tree during the year. Photosynthetica 35: 445–451.

74. Bussotti F, Bettini D, Grossoni P, Mansuino S, Nibbi R, et al. (2002) Structural and functional traits of *Quercus ilex* in response to water availability. Environ Exp Bot 47: 11–23.

75. Ogaya R, Peñuelas J (2007) Leaf mass per area ratio in *Quercus ilex* leaves under a wide range of climatic conditions. The importance of low temperatures. Acta Oecol 31: 168–17.

7

Sensitivity of Global and Regional Terrestrial Carbon Storage to the Direct CO_2 Effect and Climate Change Based on the CMIP5 Model Intercomparison

Jing Peng[1], Li Dan[1]*, Mei Huang[2]

1 START Temperate East Asia Regional Center and Key Laboratory of Regional Climate-Environment for Temperate East Asia, Institute of Atmospheric Physics, Chinese Academy of Sciences, Beijing, China, **2** Key Laboratory of Ecosystem Network Observation and Modeling, Institute of Geographical Sciences and Natural Resources Research, Chinese Academy of Sciences, Beijing, China

Abstract

Global and regional land carbon storage has been significantly affected by increasing atmospheric CO_2 concentration and climate change. Based on fully coupled climate-carbon-cycle simulations from the Coupled Model Intercomparison Project Phase 5 (CMIP5), we investigate sensitivities of land carbon storage to rising atmospheric CO_2 concentration and climate change over the world and 21 regions during the 130 years. Overall, the simulations suggest that consistently spatial positive effects of the increasing CO_2 concentrations on land carbon storage are expressed with a multi-model averaged value of 1.04PgC per ppm. The stronger positive values are mainly located in the broad areas of temperate and tropical forest, especially in Amazon basin and western Africa. However, large heterogeneity distributed for sensitivities of land carbon storage to climate change. Climate change causes decrease in land carbon storage in most tropics and the Southern Hemisphere. In these regions, decrease in soil moisture (MRSO) and enhanced drought somewhat contribute to such a decrease accompanied with rising temperature. Conversely, an increase in land carbon storage has been observed in high latitude and altitude regions (e.g., northern Asia and Tibet). The model simulations also suggest that global negative impacts of climate change on land carbon storage are predominantly attributed to decrease in land carbon storage in tropics. Although current warming can lead to an increase in land storage of high latitudes of Northern Hemisphere due to elevated vegetation growth, a risk of exacerbated future climate change may be induced due to release of carbon from tropics.

Editor: Vanesa Magar, Centro de Investigacion Cientifica y Educacion Superior de Ensenada, Mexico

Funding: This study was supported by the project of National Natural Science Foundation of China (grant no. 41275082 and 41305070), the CAS Strategic Priority Research Program (grant no. XDA05110103), the Knowledge Innovation Program of the Chinese Academy of Sciences (KZCX2-EW-QN208 and 7-122158), the National Basic Research Program of China (grant no. 2010CB950500). The funders had no role in study design, data collection and analysis, decision to publish, or preparation of the manuscript.

Competing Interests: The authors have declared that no competing interests exist.

* E-mail: danli@tea.ac.cn

Introduction

Variations in the whole terrestrial carbon cycle are accompanied with increasing atmospheric CO_2 concentration and climate change. The physical climate can influence the terrestrial carbon storage and the carbon exchange between atmosphere and land [1–5], and the sequent changes in atmospheric concentration of CO_2 simultaneously affect the physical climate system [1,6]. By the end of the twenty-first century, there is an additional CO_2 change between 20 and 200 ppm considering warming alone, while the higher CO_2 values can lead to an additional climate warming ranging between 0.1° and 1.5°C [1]. Zeng et al. (2004) [7] also suggested a positive feedback to global warming from the interactive carbon cycle introduces an additional increase of 90 ppm in the atmospheric CO_2 and then 0.6 degree additional warming during the period of 1860–2100 with the prescribed IPCC-SRES-A1B emission scenario. Regionally, using the Coupled Climate Carbon Cycle Model Intercomparison Project (C^4MIP), Cox et al. (2013) [8] estimates that warming alone will release 53 ± 17 gigatonnes K^{-1} of carbon to the atmosphere over

tropical land from 30° north to 30° south. On the other side, increasing CO_2 concentrations would influence land carbon balance, such as the stimulation of carbon storage, increases in gross primary production, net primary productivity and heterotrophic respiration [6,8–12]. With increasing CO_2 concentrations alone there is a widely distributed terrestrial carbon sink of 1.4–3.8 $PgCyr^{-1}$ during the 1990s, rising to 3.7–8.6 $PgCyr^{-1}$ a century later [13]. Such an enhancement of land carbon storage was accessed due to CO_2 fertilization effects with a rate of $0.07PgCyr^{-1}$ using simulation from Organizing Carbon and Hydrology considering fire disturbance [12].

As sensitivities of land carbon storage are largely disturbed and altered by both increasing CO_2 concentrations and climate change [12,14–18], previous studies have focused on it [1,14]. Generally, when climate change is only accounted for, a reduction is suggested for the efficiency of the terrestrial ecosystems to absorb anthropogenic carbon emissions [2,19,20] and general negative sensitivities of land carbon storage have been revealed from the earth system models (ESMs)[14] and Dynamic Global Vegetation Models (DGVMs) [19]. Arora et al. (2013) [14] estimated earth system models (ESMs) simulations from CMIP5, which represent the interactions between the carbon cycle and physical climate system. Negative impacts of continuous warming on terrestrial carbon storage have been suggested. Sitch et al. (2008) [19] used the five Dynamic Global Vegetation Models (DGVMs) and IPSL GCM to evaluate the terrestrial carbon cycle and climate-carbon cycle feedbacks, and found the similar sign of sensitivities of terrestrial land carbon to the changes in atmospheric CO_2 and climate. Generally, terrestrial carbon balance and storage is discovered to be sensitive to rising atmospheric CO_2 concentration and climate change [17,21,22]. Land carbon sensitivities to rising CO_2 are positive implying increases under enhanced atmospheric CO_2 concentration about 1.19~1.32 PgC ppm^{-1}, while its sensitivities to temperature are negative ranging between $-137 \sim -85$ PgCK^{-1}[23].

Although both the common results suggest the positive response of terrestrial carbon storage to increasing CO_2 concentration and its negative response to climate change [1,8,23], many potentially uncertainties are remained [14,19], such as the response magnitudes and regional pattern, often because they are poorly understood [24], or scantily quantified at scales relevant for models [1]. Several previous studies have been carried in order to evaluate and predict land carbon sensitivities to CO_2 and temperature [1,14,23]. But most of them have not distinguished the regional variability due to impacts of rising atmospheric CO_2 concentration or climate change on carbon storage in order to identify the sensitive regions, respectively. On the basis of this point, we focus on the sensitivities of land carbon storage to the rising atmospheric CO_2 concentrations and climate change at global and regional scales by multi-models, and compare the simulated variations between models that represent a potential source of uncertainty. Among regions which are more fragile to climate change, we quantify sensitivities of land carbon storage at regional scales. It is important for mitigation and adaptation of terrestrial ecosystems to climate change. In this study, we use simulations from earth system models offered by the CMIP5 to quantify and compare the carbon sensitivities to changes in CO_2 and climate for the period from 1860 to 1989. The disturbances of none-CO_2 greenhouse gases, land use and fire are totally ignored.

Experiments and Methods

The fifth phase of Coupled Model Intercomparison Project (CMIP5) can provide a common framework [25] to compare and assess land carbon storage responses in the context of climate simulations [14]. There are two experiments used in this analysis, which have been downloaded from http://cmip100pcmdi.llnl.gov/cmip5/forcing.html. These chosen experiments from CMIP5 have been described by Peng et al.(2013; 2014) [26,27]. They are detailed as follows: (1) in the experiment of only considering the single effect of atmospheric increasing CO_2 concentrations, biogeochemistry only responds to the increasing CO_2 concentrations in land models while the radiative forcing is fixed at the pre-industrial values in the atmospheric modules; (2) in the experiment of only thinking of the single effect of climate change, the radiative forcing responds to the increasing atmospheric CO_2 concentration (prescribed by 1% increase of CO_2 concentration per year in atmosphere from the pre-industrial values to the quadruple) while the biogeochemistry remains at the pre-industrial values in the biogeochemistry modules. In addition, in order to estimate and compare the sensitivities of land carbon storage to increasing CO_2 concentration and climate change, we use the simulations from six of the fully carbon-climate coupled ESMs, which participate in the CMIP5 intercomparison project. These models include Had-

GEM2-ES [28], IPSL-CM5A-LR [29], CESM1-BGC [14], MPI-ESM-LR [30,31], CanESM2 [32] and BCC-CSM1-1 [33].

The ESMs, which are used to analyze the sensitivities, except for HadGEM2-ES and MPI-ESM-LR, do not include dynamical vegetation cover. Their spatial distribution is controlled by the competition between different PFTs (Plant functional types) [14]. A patch-based representation of vegetation structure competition has been made [34]. HadGEM2-ES can simulate the transient shifts and geographic patterns of vegetation. Its atmosphere/land component resolution is 1.6° by 1.6°. A mixture of 12 PFTs is shown in HadGEM2-ES [28]. Over each patch a transient dynamics of vegetation also has be presented by MPI-ESM-LR. It represents the resolution of atmospheric component of 1.9° by 1.9° [31].

The IPSL-CM5A-LR [29] is the new generation Earth System Model developed by the Institute Pierre Simon Laplace. A resolution of 3.6° by 1.8° at latitudes and longitudes is used in the land and atmospheric components. There are 39 vertical levels of atmosphere. With a daily time step land component ORCHIDEE [35] simulates processes of photosynthesis, carbon allocation, litter decomposition, soil carbon dynamics, maintenance and growth respiration and phenology for 13 different plant functional types [34].

Version 1 of the Community Earth System Model (CESM1) is the successor to version 4 of the Community Climate System Model (CCSM4). In this model, global climate model consisting of land, atmosphere, ocean, and sea-ice components has been fully coupled [36].The terrestrial carbon cycle is represented by the CLM4 land surface scheme. And it contains 16 different PFTs. Coupled carbon-nitrogen dynamics are included and expressed as CLM4CN [37,38].

CanESM2 is the new generation Earth System Model developed from the first generation Canadian earth system model (CanESM1) [21,39]. It is produced and developed by the Canadian Centre for Climate Modelling and Analysis (CCCma) and described by Arora et al. (2013) [14]. The atmospheric component has horizontal resolution of about 2.8° in a grid cell. The Canadian Terrestrial Ecosystem Model (CTEM) [40] is used to model terrestrial ecosystem processes. It simulates terrestrial carbon supported by three live vegetation pools (leaves, stem, and root) and two dead pools (litter and soil organic carbon) for PFTs (e.g. needle leaf evergreen and deciduous trees, broadleaf evergreen and cold and dry deciduous trees, and C3 and C4 crops and grasses) [14].

BCC-CSM1.1 is a fully coupled global climate–carbon model including interactive vegetation and global carbon cycle [33]. The atmospheric component BCC-AGCM2.1 is a global spectral model with a horizontal resolution of 2.81 degrees, with vertical 26 levels of atmosphere. For the land surface processes, the land surface component Atmosphere-Vegetation Interaction Model (AVIM) [41,42] is incorporated into the biogeophysical frame of NCAR/CLM3. With 15 PFTs including natural vegetation and crop, each a grid cell can contains up to four PFTs. Terrestrial carbon cycle components simulates biochemical and physiological processes. For example, photosynthesis and respiration of vegetation, allocation of carbohydrate to leaves, stem, and root tissues, carbon loss due to turnover and mortality of vegetation can be modeled as described by previous work [43].

In order to analyze the outputs simulated by the global coupled models, a criteria of regions has been employed by Peng et al. (2013; 2014)[26,27] in the regional scale performance (Table 1 and Figure 1). It has been described by Giorgi et al (2000) [44] as follows: (1) the size of the regions vary in the range of a few thousand to several thousand km in each direction, so that each

Table 1. List of 21 regions.

ID	Region	Abbreviation
1	Australia	AUS
2	Amazon basin	AMZ
3	southern South America	SSA
4	Central America	CAM
5	western North America	WNA
6	central North America	CNA
7	eastern North America	ENA
8	Alaska	ALA
9	Greenland	GRL
10	Mediterranean basin	MED
11	northern Europe	NEU
12	western Africa	WAF
13	eastern Africa	EAF
14	southern Africa	SAF
15	Sahara	SAH
16	southeastern Asia	SEA
17	eastern Asia	EAS
18	southern Asia	SAS
19	central Asia	CAS
20	Tibet	TIB
21	northern Asia	NAS

region includes at least a few Earth system models (ESM) grid points and thus contains the smallest wavelength of ESM solutions; (2) all land areas in the World with a number of regional management simple shapes have been approximately covered; (3) selection of specific regions is in order to represent different climatic regimes and terrain settings. On the basis of such criteria of regional selection, we calculate the sensitivities of global and regional land carbon storage to climate change and increasing atmospheric CO_2 concentrations. These sensitivities to direct CO_2 effects and climate change are described as β_{Li} (Pg C per ppm) and γ_{Li} (Pg C K^{-1}) as the following equations, respectively. The calculation method can be referred to a previous study reported by Cox et al. (2013) [8].

$$\beta_{Li} = \frac{\Delta C_{Li}^{co2} - \gamma_{Li}\Delta T_{Li}^{co2}}{\Delta C_a^{co2}} \tag{1}$$

$$\gamma_{Li} = \frac{\Delta C_{Li}^{C\lim ate}}{\Delta T_{Li}^{C\lim ate}} \tag{2}$$

Where $\Delta C_{Li}^{CO_2}$ is the change in global or regional land carbon storage (in PgC), $\Delta T_{Li}^{CO_2}$ is the change in mean global and regional atmospheric temperature (in K) and $\Delta C_a^{CO_2}$ is the change in atmospheric CO_2 concentration (in ppm), in response to increasing atmospheric CO_2 concentrations. $\Delta T_{Li}^{C\lim ate}$ is the change in mean global and regional atmospheric temperature (in K) and $\Delta C_{Li}^{C\lim ate}$ is the change in global and regional land carbon storage (in PgC), in response to climate change. These changes in temperature, land carbon storage and atmospheric CO_2 concentration averaged over the period of the last 30 years of 130-years relative to the period of the first 30 years in all cases of calculation for the sensitivities responding the increasing CO_2 concentration (γ_{Li}) and climate change (β_{Li}).

In this analysis, R is a correlation coefficient between the annual variable (e.g., temperature, precipitation) and the natural sequence 1, 2, 3,..., n, at a given year [45]. R is calculated by the following formula:

$$R_{xt} = \frac{\sum_{k=1}^{n} (x_k - \bar{x})(k - \bar{t})}{\sqrt{\sum_{k=1}^{n} (x_k - \bar{x})^2 \sum_{k=1}^{n} (k - \bar{t})^2}} \tag{3}$$

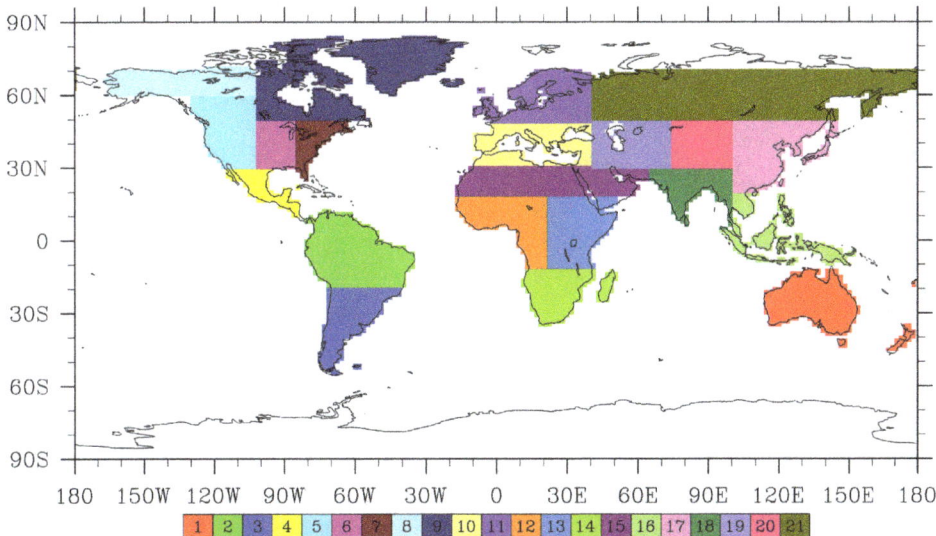

Figure 1. Map of 21 regions.

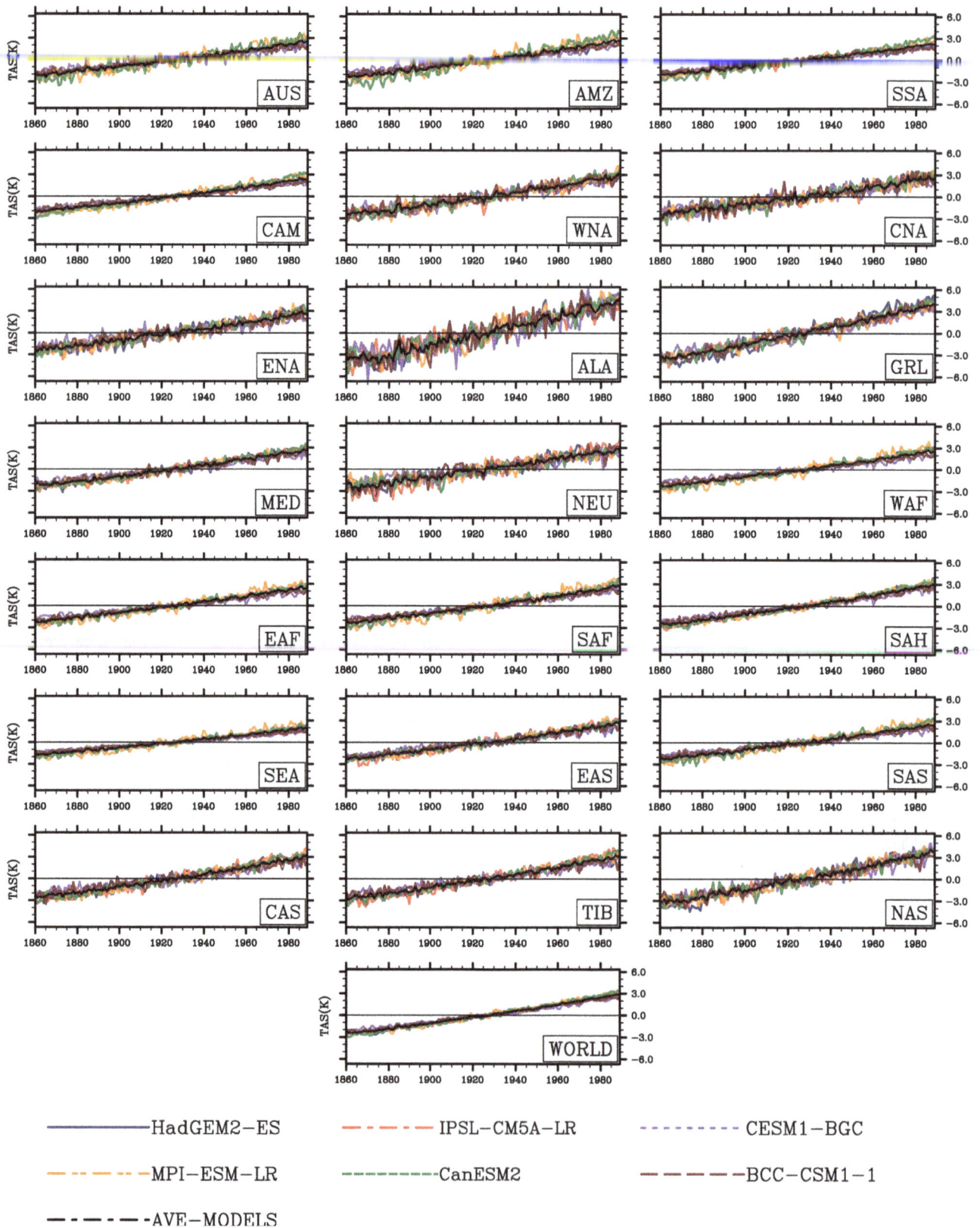

Figure 2. Variability in anomalies of annual mean surface air temperature (K) at regional and global scales considering climate change alone.

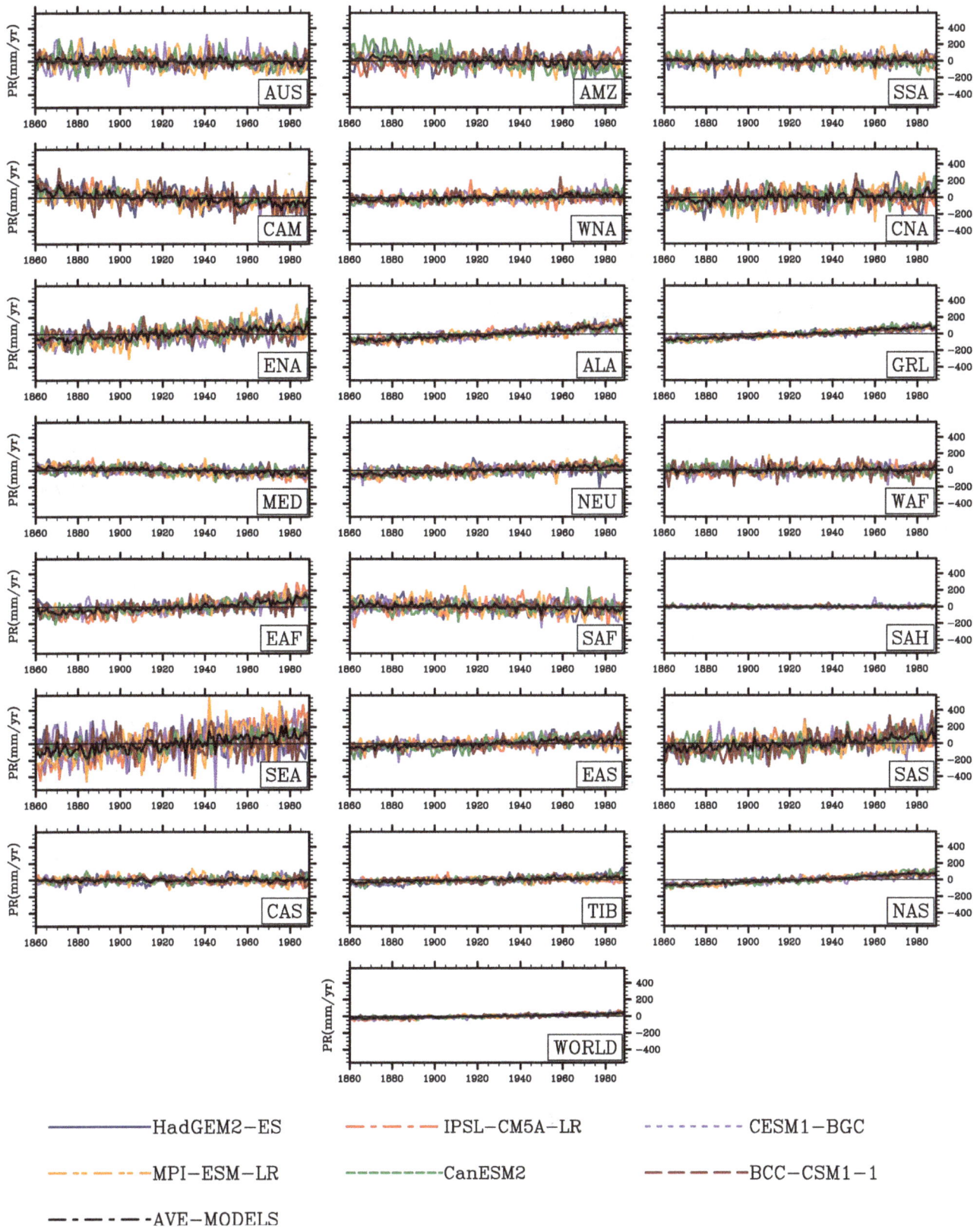

Figure 3. Same as Figure 2, but for annual precipitation (mm yr^{-1}).

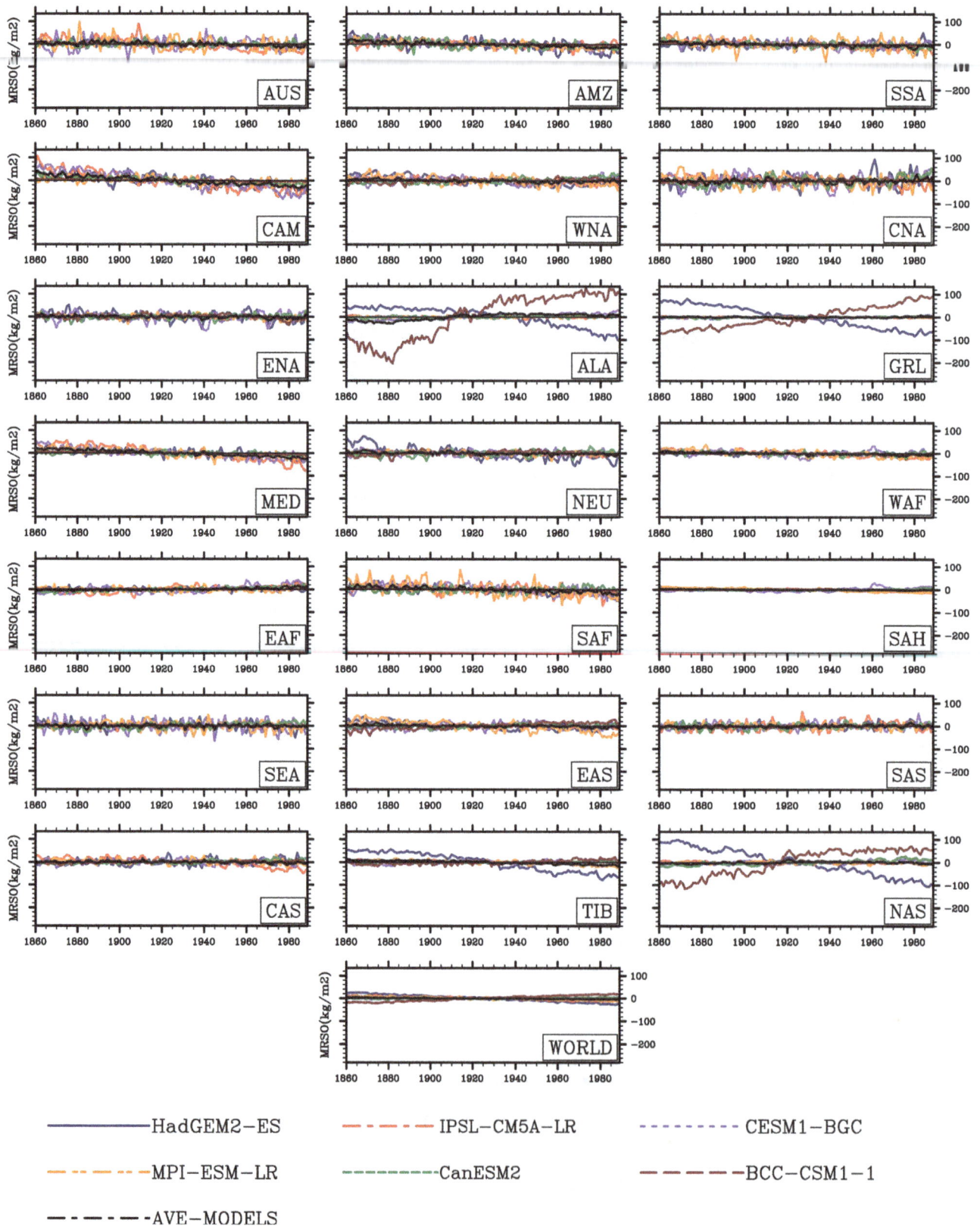

Figure 4. Same as Figure 2, but for soil moisture (kg m^{-2}).

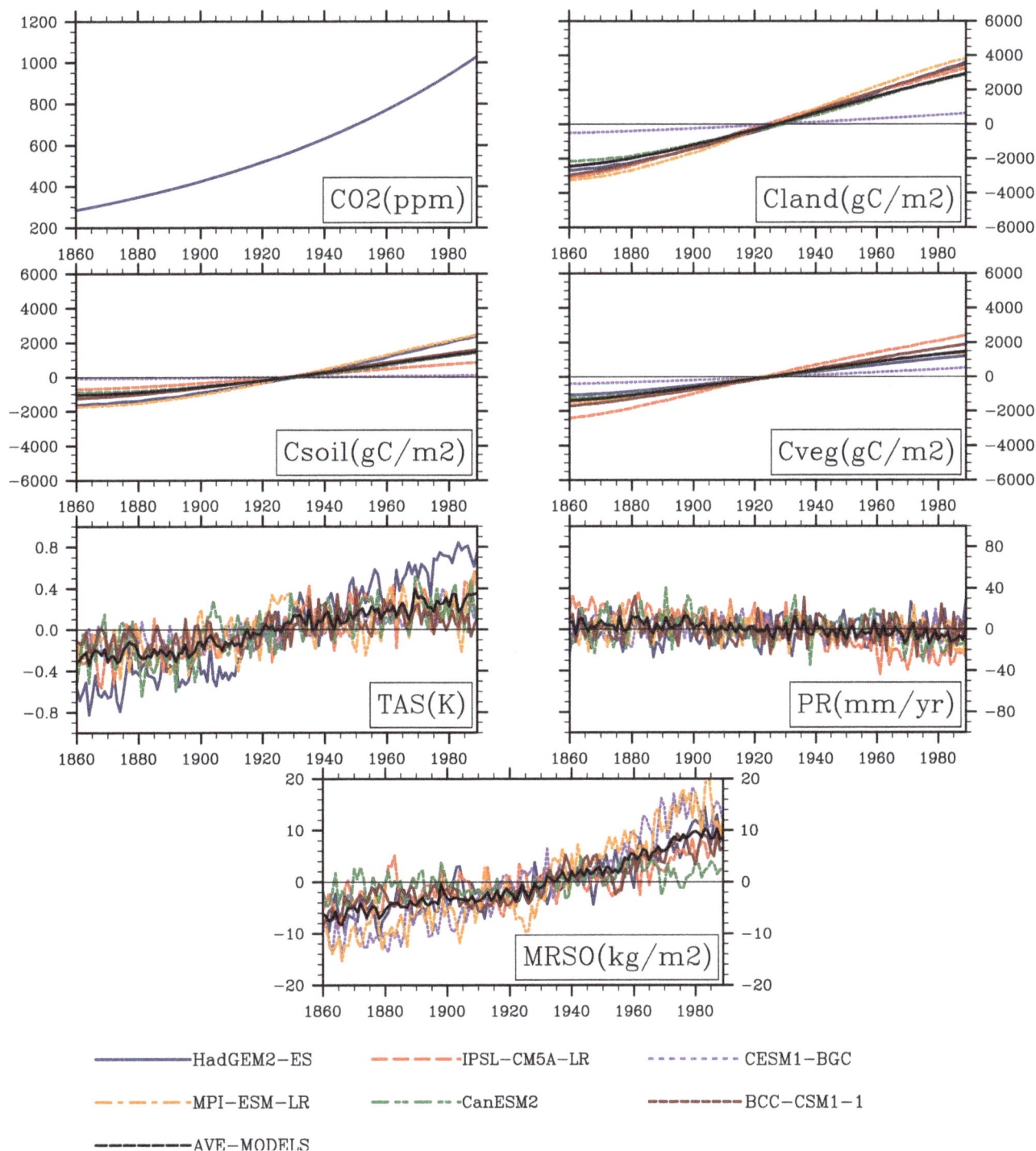

Figure 5. Variability in atmospheric CO$_2$ concentration (ppm) at rate of 1% per year from pre-industrial values until concentration quadruples for a 130-year long simulation. variability in anomalies of global land carbon storage (Cland, gC m^{-2}), soil carbon storage (Csoil, gC m^{-2}), vegetation carbon storage (Cveg, gC m^{-2}), annual mean temperature (TAS, K), annual precipitation (PR, mm yr^{-1}) and soil moisture (MRSO, kg m^{-2}) only allowing for the direct CO$_2$ effect.

Where x_k is the annual variable in the time k, n is the sequential year, and $\bar{x}$ is the multi-year mean for this variable; $\bar{t}$ is equal to the mean of 1 and n. The linear trend of the variable is presented within the period of 1860–1989. When a "significant" linear trend for a variable is shown, R must pass the significance level (e.g., $P < 0.05$) using the student t-test.

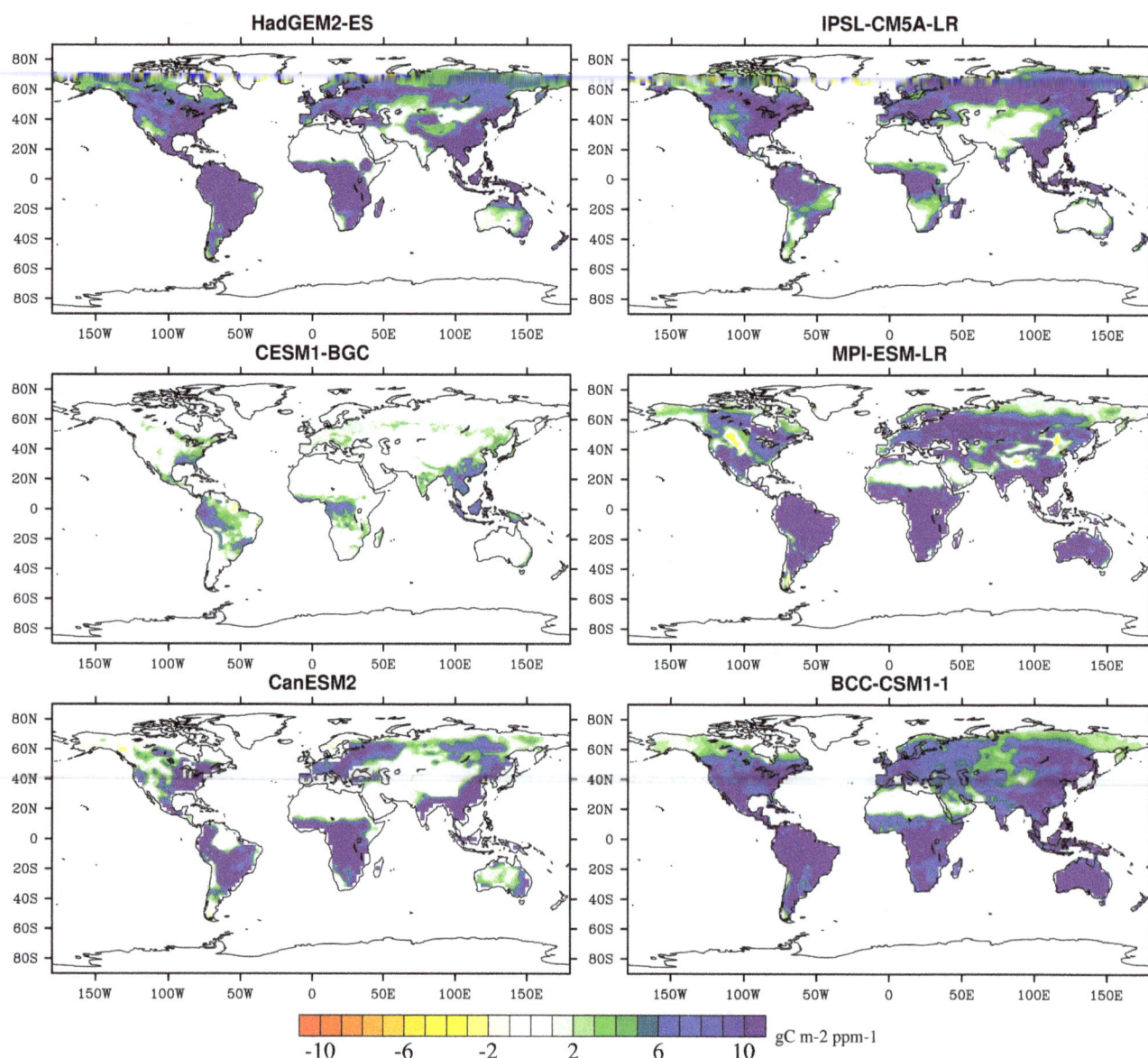

Figure 6. Spatial distribution in sensitivity of terrestrial carbon storage to upward atmospheric CO$_2$ concentration simulated from six the fifth Coupled Model Intercomparison Project (CMIP5) models. units: gC m^{-2} ppm^{-1}.

Results

The global and regional climate consistently show increasing trends in temperature (0.05 °C yr^{-1}, R = 0.95, P<0.001), but no consensus trends in precipitation (PR) and soil moisture are presented at regional scales (Figures 2–4). As shown in Figure 3, increases in precipitation in high-latitude regions are shown (e.g., in northern Asia with 1.17mm yr^{-2}, R = 0.97, P<0.01), while decreases in precipitation are located in part of the low latitudes such as Amazon basin with −0.50mm yr^{-2} (R = 0.50, P<0.01) and Central America with −1.28 mmyr^{-2} (R = 0.79, P<0.01). These spatial divergences on changes in soil moisture are also shown in Figure 4. Only considering climate change, the increases in soil moisture is mainly distributed in high latitudes, while the area principally in Amazon basin exhibits a considerable decrease of soil moisture with −0.23 kg yr^{-1} (R = 0.86, P<0.001).

Figure 5 shows the interannual variability in modeled global annual land carbon storage anomaly, soil carbon storage anomaly, vegetation carbon storage anomaly, mean temperature anomaly, annual precipitation anomaly and soil moisture anomaly only responding to the CO$_2$ increases. As a result, the increase in global annual land carbon storage has been presented across the whole globe. Such an increase in turn affects the soil carbon storage with increasing rate of 20.7 gC yr^{-1} (R = 0.989, P<0.001).

The geographical distribution of in the direct CO$_2$ impacts on the terrestrial and regional land carbon storage is shown in Figures 6 and Table 2. Sensitivities of land carbon storage, considering the CO$_2$ fertilization effect alone, are fairly positive in the most terrestrial ecosystem, which means the terrestrial biosphere acting as enhanced carbon storage by 1.0 Pg C per ppm. The most dramatic increases caused by the increased CO$_2$ concentrations are mostly located in the regions such as Amazon

Table 2. The impact of increasing atmospheric CO_2 concentration on global and regional land carbon storage (units: PgC ppm^{-1}).

ID	Acronym	HadGEM2-ES	IPSL-CM5A-LR	CESM1-BGC	MPI-ESM-LR	CanESM2	BCC-CSM1-1
1	AUS	0.04	0.02	0.01	0.1	0.04	0.1
2	AMZ	0.22	0.19	0.04	0.24	0.12	0.21
3	SSA	0.08	0.03	0.01	0.06	0.06	0.06
4	CAM	0.04	0.02	0.01	0.03	0.02	0.03
5	WNA	0.04	0.04	0	0.03	0.02	0.05
6	CNA	0.04	0.04	0.01	0.02	0.03	0.03
7	ENA	0.03	0.03	0.01	0.02	0.05	0.03
8	ALA	0.02	0.01	0	0.01	0	0.01
9	GRL	0.02	0.04	0	0.02	0.01	0.01
10	MED	0.03	0.03	0	0.05	0.03	0.03
11	NEU	0.04	0.05	0	0.04	0.03	0.03
12	WAF	0.09	0.08	0.02	0.12	0.12	0.08
13	EAF	0.09	0.06	0.01	0.14	0.11	0.1
14	SAF	0.09	0.03	0.01	0.12	0.08	0.06
15	SAH	0	0	0	0.01	0	0.01
16	SEA	0.07	0.13	0.03	0.09	0.13	0.08
17	EAS	0.09	0.08	0.03	0.06	0.1	0.09
18	SAS	0.04	0.04	0.02	0.07	0.07	0.05
19	CAS	0.02	0.01	0	0.06	0.01	0.04
20	TIB	0.03	0.01	0	0.03	0	0.03
21	NAS	0.1	0.15	0.01	0.1	0.06	0.09
22	WLD	1.21	1.08	0.21	1.42	1.09	1.23

basin, northern Asia, western Africa and eastern Africa, which have larger vegetation biomass than other regions. However, the warming greatly reduces the land carbon storage and the carbon sequestration. Excluding Tibet, northern Asia, Alaska and Greenland, the impacts of the warming on land carbon storage are negative in the terrestrial ecosystems with about -41.6 PgCK^{-1}. Additionally, it should be noted that these simulations forced by both increasing CO_2 concentration and climate change are properly different among six ESMs and regions. The multi-model values across the globe range from 0.2 PgC per ppm in CESM1-BGC to 1.4 PgC per ppm in MPI-ESM-LR responding the direct CO_2 effects and from -17.5 PgC K^{-1} in CESM1-BGC to -58.6 PgC K^{-1} in CanESM2 considering climate change alone across the whole terrestrial ecosystem (Tables 2–3 and Figures 6–7).

Figure 7 exhibits the spatial pattern of changes in land carbon storage only responding to warming. In response to such a warming, the largest loss in land carbon storage is mainly distributed in central Amazon basin, southeastern Asia, Central America and eastern Africa (Table 3 and Figure 7). Generally, negative sensitivities of land carbon storage mostly located in the tropics and Southern Hemisphere. Inversely, areas in northern high latitudes primarily exhibit a considerable increase of land carbon storage. For example, a significantly enhanced carbon storage appears in Tibet with 3.26 gC m^{-2} yr^{-1} (R = 0.98, P<0.001). Increases in land carbon storage in northern Asia and Greenland Alaska are also detected (Figure 8). The simulation shows a consistent pattern (a negative sensitivity for the land carbon storage to temperature across the major land) modeled by the major models in terms of the sign of sensitivities of land carbon

storage to temperature. Spatial discrepancies among regions are evidently presented as mentioned above: increases in land carbon storage in the mainly temperature-limited regions, but decrease in higher temperature regions. On the other side, in comparison with different regions in the Figures 7, it's worth noting that the multi-model simulations provide fairly consistent results in most tropical areas and Southern Hemisphere (e.g. Australia, Eastern Africa, Southern Africa, Western Africa, Southeast Asia, Amazon Basin), but inconsistent results in a part of northern latitudes (e.g. some show positive values; others show negative values). The inconsistency is particularly evident in high latitudes of Northern Hemisphere, such as Alaska and part of northern Europe.

Spatially, analysis of responses of land carbon storage to climate change also reveals major pattern: negative impacts of rising temperature at the low and middle latitudes of the World are simulated by the major models in terms of the sign of response of land carbon storage; it's positive impacts at the high latitudes is presented. The changes are defined as averaged values of the period 1960–1989 relative to 1860–1889 in order to calculate these sensitivities as mentioned above. In Figure 9, in response to rising atmospheric temperature, average zonal-annual (sensitivities averaged over the regions and year) patterns and the agreement between simulations are deduced: ≥80% of areas across the whole terrestrial ecosystems shows the same sign of land carbon storage sensitivities modeled by a majority of simulations (≥4/6 simulations). A majority of the simulations agrees on the negative impact of temperature in tropics and positive impact in temperature-limited regions in response to climate change. For example, 99.6% of Amazon basin and 99.2% of western Africa show the negative impacts of increasing temperature on the land carbon storage. In

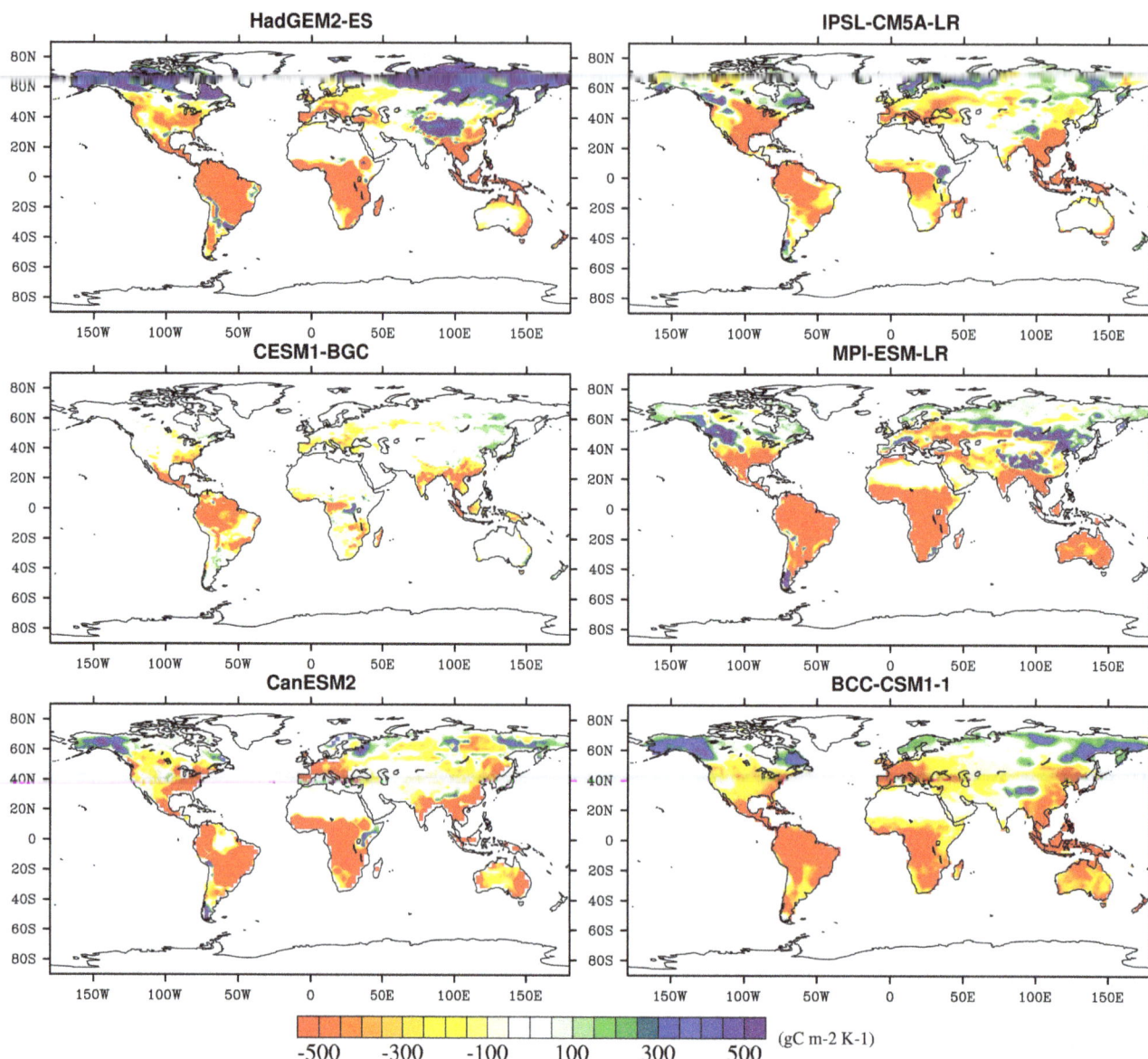

Figure 7. Same as Figure 9, but to the warming (gC m^{-2} K^{-1}).

contrast, the reverse sensitivities to warming are shown both quantitatively and qualitatively in high latitudes of Northern Hemisphere. In the most areas, there is little agreement among simulations for the sensitivities to increasing temperature around 60°N and Greenland.

Discussion

In terms of regions (Figures 2–4), the spatial patterns of temperature, precipitation and soil moisture are so different. In high latitudes and altitudes, temperature and radiation impose a complex and varying limitations on vegetation activity [60]. For example, in the Western Europe, solar radiation is an evident limited factor to the vegetation growth. In the eastern of Tibetan plateau temperature change is the main cause to affect the vegetation growth and NPP [47]. Accompanied with such warming, increases in soil moisture have been reported by Peng et al. (2013) [26] in these regions. However, the tropical change estimated from the simulations is currently different from that of temperature-limited regions. Indeed water limitation is enhanced (e.g. increased droughts) in some of these tropical regions. Regionally, the CO_2-induced climatic change enhances a decrease in soil moisture most in Amazon basin and part of arid and semi-arid regions (e.g. Mediterranean basin, Sahara and central Asia). In some cases, only considering the radiative CO_2 forcing, the increasing atmospheric CO_2 can influence the climate factors (e.g., rising temperature). Although the increasing atmospheric CO_2 only influences the climate but not the biogeochemistry, the climatic change in turn affects the biogeochemistry. It can be to say biogeochemistry can respond to rising temperature and changes in other climatic variables. For example, associated warming, the increased evapotranspiration produced by increasing temperature can also exacerbate drought in Amazon basin [48].

Table 3. The impact of rising temperature on global and regional land carbon storage (units: $PgCK^{-1}$).

ID	Acronym	HadGEM2-ES	IPSL-CM5A-LR	CESM1-BGC	MPI-ESM-LR	CanESM2	BCC-CSM1-1
1	AUS	−1.93	−0.99	−0.19	−6.25	−3.42	−3.15
2	AMZ	−13.88	−9.99	-7.45	−12.85	−10.58	−8.49
3	SSA	−2.78	−1.25	−0.84	−2.61	−3.85	−2.48
4	CAM	−2.04	−1.53	−1.49	−1.9	−2.01	−1.52
5	WNA	−1.03	−0.6	−0.1	−0.27	−1.13	−0.54
6	CNA	−0.86	−2.64	−0.33	−1.53	−1.15	−0.74
7	ENA	−0.57	−0.79	−0.29	−0.49	−1.62	−0.83
8	ALA	1.51	−0.35	0.01	0.19	0.55	0.61
9	GRL	1.44	0.39	−0.02	0.39	0.11	0.33
10	MED	−1.37	−1.62	−0.56	−2.01	−2.14	−1.51
11	NEU	−0.59	−0.35	−0.19	−0.76	−1	−0.71
12	WAF	−4.38	−3.14	−0.97	−6.17	−7.61	−3.82
13	EAF	−4.71	−0.97	−0.22	−6.22	−4.27	−3.18
14	SAF	−4.67	−1.61	−0.88	−6.21	−4.94	−2.72
15	SAH	−0.03	0	−0.01	−0.37	−0.07	−0.24
16	SEA	−5.26	−5.13	−1.28	−3.9	−6.54	−3.45
17	EAS	−1.29	−3.05	−0.85	0.45	−3.74	−2.93
18	SAS	−0.86	−1.85	−1.44	−3.19	−3.53	−0.75
19	CAS	−0.4	−0.49	−0.1	−1.9	−0.29	−0.84
20	TIB	1.65	0.07	−0.01	0.67	−0.3	0.21
21	NAS	4.94	0.09	−0.05	0.97	−0.66	1.24
22	WLD	−36.52	−35.92	−17.46	−53.78	−58.59	−35.49

As a consequence, enhanced drought can cause reduction in captured carbon in living biomass.

There is current lack of information on the accurate magnitude of the response of terrestrial carbon storage and the affecting causes, and thus large differences exist among different ESMs. Such differences among models are determined by the used approaches (e.g., the models using a biogeochemical approach to calculate the terrestrial photosynthesis) [14], which provides an indication of the key potential processes controlling the CO_2-induced and climate-induced carbon uptake/storage [4,22]. For example, CESM-BGC simulates the smallest sensitivities of land carbon storage to direct CO_2 effects. As this model includes the CO_2 fertilization effect constrained by nitrogen limitation compared with other models. Other factors potentially influencing land carbon storage include warming resulting in the changes in land carbon storage, especially at the high latitudes. In the high latitudes and altitudes of Northern Hemisphere, the positive impact of the increasing temperature on land carbon storage has been shown in Figure 7. This is partly due to, in these regions, the enhanced growth caused by the elevated temperatures [49]. Although the intense warming there increases soil organic matter decomposition, soil organic carbon storage has been observed to continue to increase [2]. This change is partly caused by the increase in vegetation productivity, leading to more turnover (litterfall) into the soil [2]. Nevertheless, a qualitative modeled intercomparison of changes in total land carbon storage to increasing temperature, both increasing vegetation storages and soil carbon storage provide useful insights in these temperature-limited regions. Reversely, multi-modeled simulations conformably agree on the negative response of terrestrial ecosystems

mainly coming from the reduction in carbon storage in tropics and most Southern Hemisphere.

The temperature in this study accounts for a highly important role in the land carbon storage for high latitudes of Northern Hemisphere. Associated with global warming, changes in carbon storage fairly depend on the balance between the input of carbon as the net primary production (NPP) and the loss of carbon as heterotropic respiration (RH) [46]. Over the past two decades, an increase in terrestrial photosynthetic activity has been documented across high latitudes of Northern Hemisphere [39,50,51]. NPP has an increase with a rate of about $18.4 TgC \ yr^{-2}$ (Peng et al unpublished data) in association with the warming in the regions including Alaska, Greenland, northern Europe, Tibet and northern Asia. Thus, accompanied with warming global vegetation growth is significantly elevated in these high latitudes of Northern Hemisphere. Simultaneously, in these regions covered by boreal forest, the lengthened plant growing period has been observed (Linderholm, 2006; Zhu et al., 2012). The longer growing season somewhat contributes such change, based on both the earlier onset of spring [52] and the later ending of autumn [50]. Overall, both experimental and modelling studies suggest greening [53] and increased NPP [2,50], which infer a upward trend of land carbon storage [2], responding to global warming. This is consistent with our results in high latitudes of Northern Hemisphere. In contrast, in the hot regions such as tropics, the sensitivities of land carbon storage would be generally negative responding to the warming. Such a negative response of carbon storage [54] and reduced carbon sequestration [55] have been also reported in the previous studies. Indeed, unlike in temperature-limited regions, our results suggest that total carbon storage can

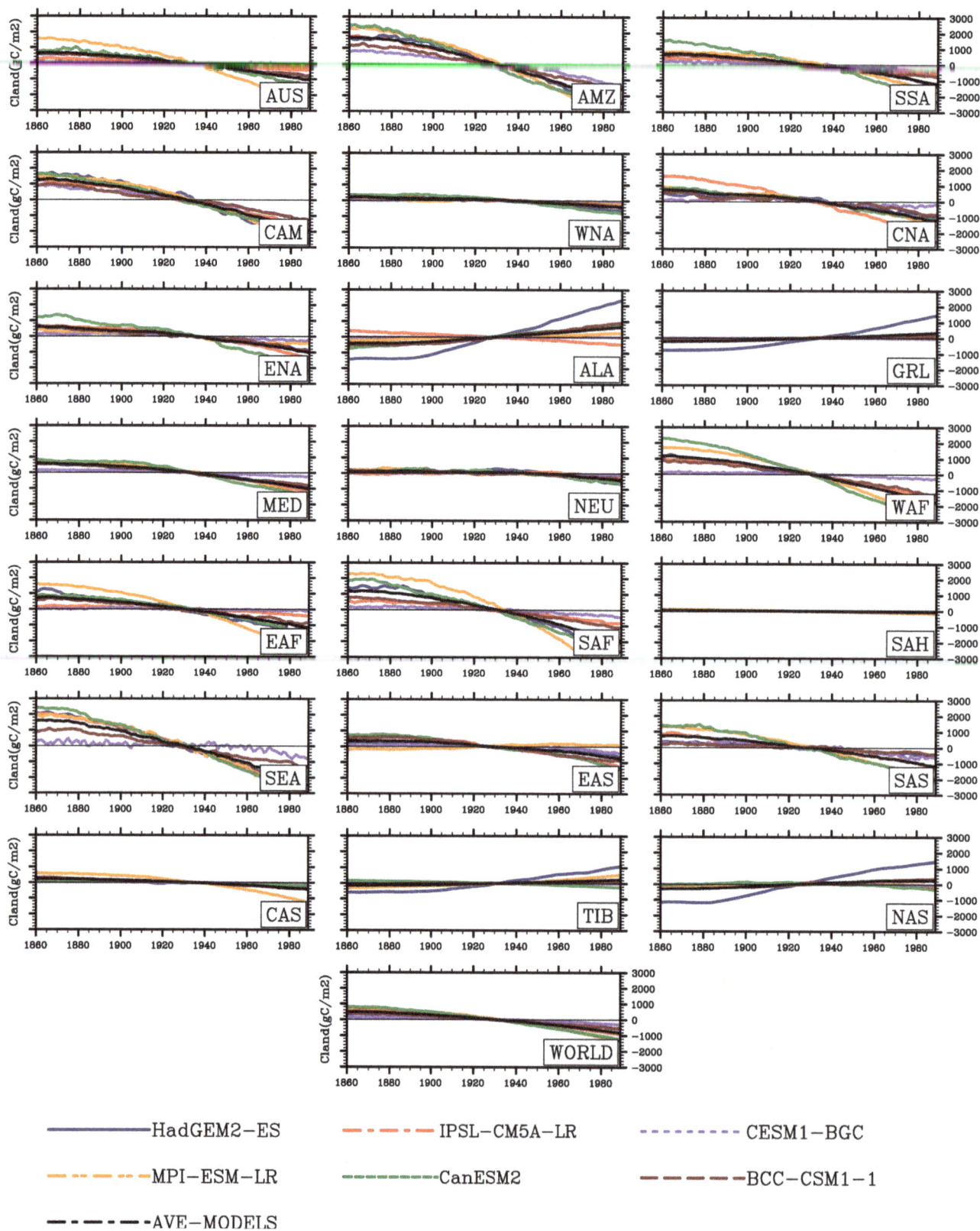

Figure 8. Same as Figure 2, but for land carbon storage (Cland, gC m⁻²).

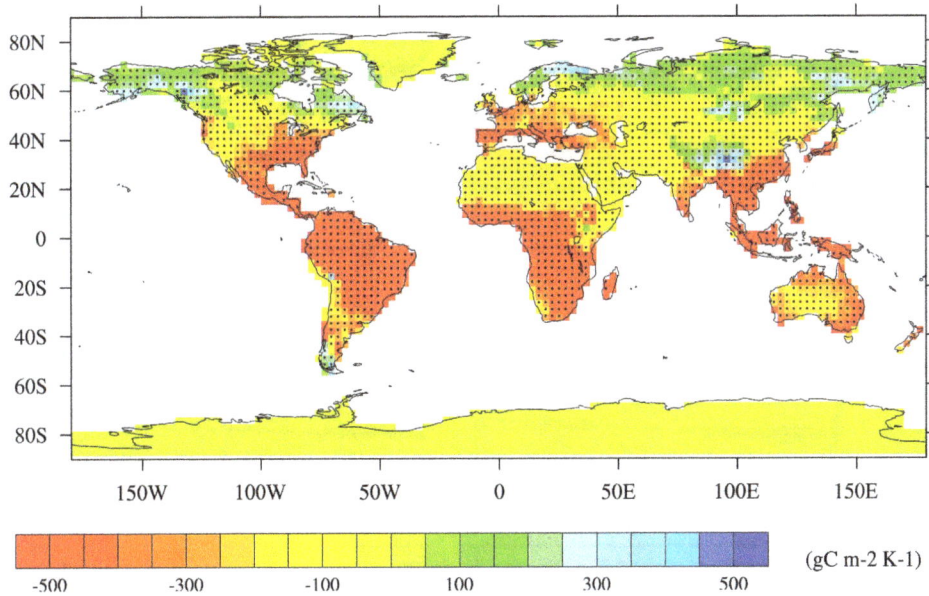

Figure 9. Six model average of the sensitivities of land carbon storage to climate change and agreement between simulations. Hatched areas means 4 or more of models agree on the same sign in the change in sensitivities to the rising temperature.

not be enhanced by the just rising temperature in current high temperature regions (e.g., Amazon basin) (Figures 10–11). This may lead to a net carbon source of the terrestrial ecosystems tend into under the global warming environment. Such a change may be attributed to changes in climatic factors. For example, the tropics exhibit higher increase in temperature and decreases in precipitation and soil moisture as mentioned above. Consequently, reduction in precipitation and soil moisture can result in the diminution in carbon storage and thus such negative response is presented. This is further evident from the 2005 and 2010 drought in Amazon basin [56,57] leading to reduction in land carbon storage through decreased vegetation productivity and/or increased respiration [16]. In addition, heat and drought can introduce increase in forest dieback [58] and thus a consequence of decreased vegetation production and carbon storage [59]. Compared with change in the vegetation storage produced by this change in vegetation production, a reduction in soil carbon storage can be found due to the decreased litter supply from vegetation (leaves, stems and roots) to soil [60]. On the other hand, the decreased soil storage is commonly generated by the greatly accelerated microbial decomposition with the sequentially rising temperature. Particularly, the larger differences in the responses of total carbon storage to climate change are presented in these regions among the ESMs (e.g., from -13.9 PgCK^{-1} in Had-GEM2-ES to -7.5 PgCK^{-1} in CESM1-BGC in Amazon basin), which suggests that the estimation of changes in temperature and precipitation is especially important in evaluating response of carbon cycle of the biosphere. In the same region, the study of Cramer et al. (2001) also suggested the importance of climatic variables for the captured carbon in biomass [13]. Climatic system itself is liable to reduce total carbon storage of terrestrial ecosystems. Hence, improvement in knowledge about it is very important to reduce and even remove uncertainties in sensitivities of land carbon storage at global and regional scales, especially in the areas of around 60°N covered by boreal forest.

Moreover, we compared our results with previous works to clarify the magnitudes and spatial pattern of sensitivities of land carbon storage to rising atmospheric CO$_2$ concentration and climate change at both regional and global scales. The used simulations are forced by increasing 1% yr^{-1} until quadruple CO$_2$ concentration with/without the carbon-cycle feedback or radiative CO$_2$ forcing. Ranges of multi-model sensitivities commonly fall into C^{4}MIP model range (sensitivities to increasing CO$_2$ between $0.2\sim2.8$ PgCppm^{-1} and to rising temperature between $-177\sim-20$ PgCK^{-1}) [19]. The multi-model positive response to increasing CO$_2$ concentrations suggests that the carbon sequestration is improved owning to CO$_2$ fertilization effect, especially in broad areas of forests. Such an increase can be explained by the vegetation increased photosynthesis accompanied with the rising atmospheric CO$_2$ concentration. For example, the increase of 23.6 gC yr^{-1} (R^2=0.99, P<0.001) in vegetation carbon storage has been shown at the global scale (Figure 5), in association with the increased GPP with a rate of about 5.0 g C m^{-2} yr^{-2}(R^2=0.998, P<0.01) (Peng et al unpublished data). For the high latitudes of Northern Hemisphere such as northern Europe and at tropical latitudes such as western Africa, which contain broad areas of forest, great accumulations of carbon into plant biomass appear and the consistent result has be found by Cramer et al.(2001), Sitch et al. (2008) and Ito (2005) [13,19,60]. In addition, simulated sensitivities of land carbon storage responding to climate change have large uncertainties. For example, the magnitudes of these sensitivities to temperature in CanESM2 and MPI-ESM-LR are significantly larger than that in CESM-BGC. Compared with the estimates by Arora et al. (2013) [14], the range among multi-model sensitivities to rising CO$_2$ is larger, while the range between sensitivities to temperature is smaller. Also such an extent to temperature is smaller than the range assessed by Friedlingstein et al. (2006) [1] and Zickfeld et al. (2011) [23]. Such differences straightforwardly lie in the discrepancy forcing scenario, the used approach and different but plausible representations of the underlying physical and chemical processes [1,13,23]. In our study, in all cases sensitivities are calculated based on changes from the last 30-yr results relative to the first 30-yr results of 130-yr simulations. Agreed on simulations from both the multi-model differences in dynamic global vegetation models (DGVMs)[13,19] and Earth system models

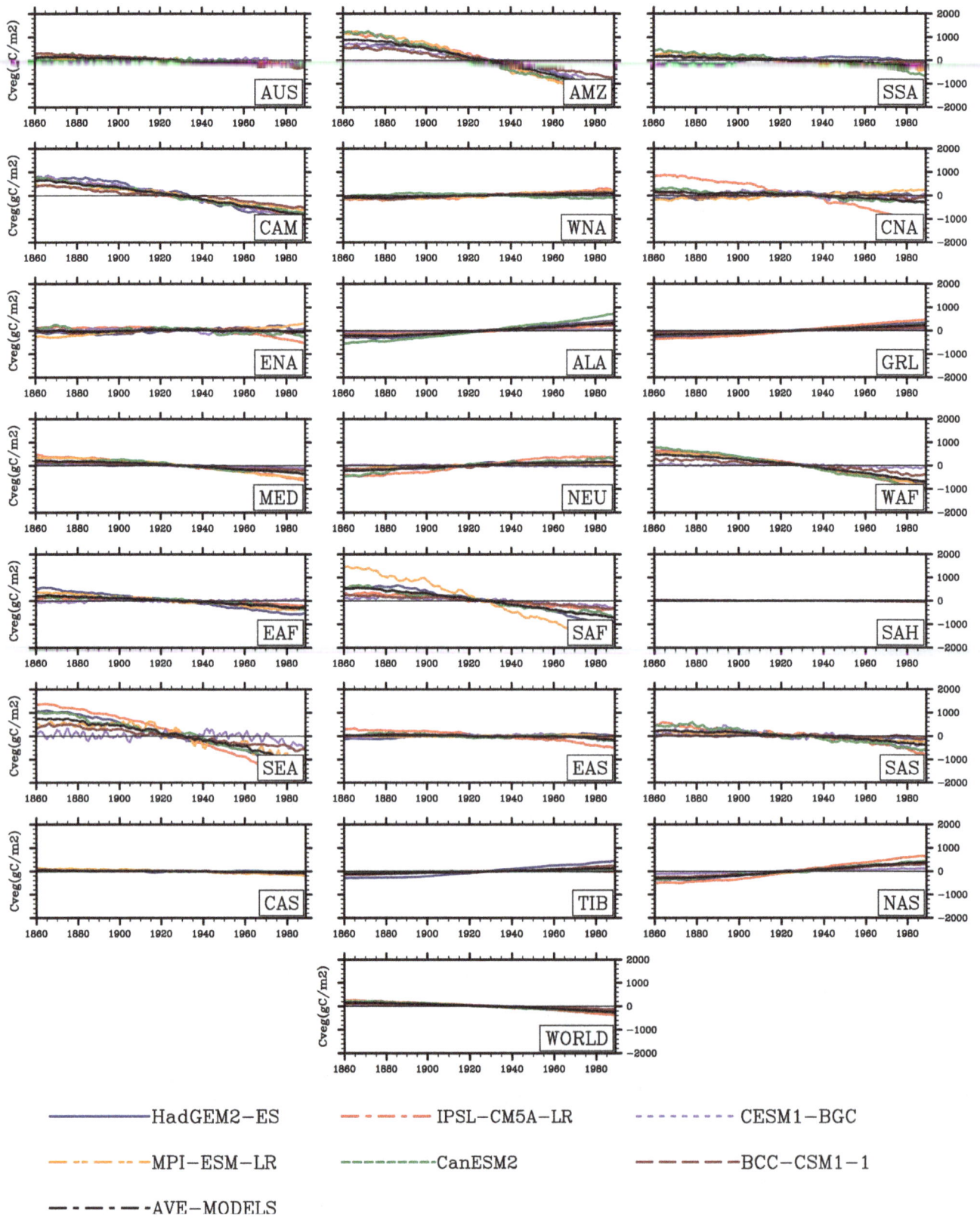

Figure 10. Same as Figure 2, but for vegetation carbon storage (Cveg, gC m^{-2}).

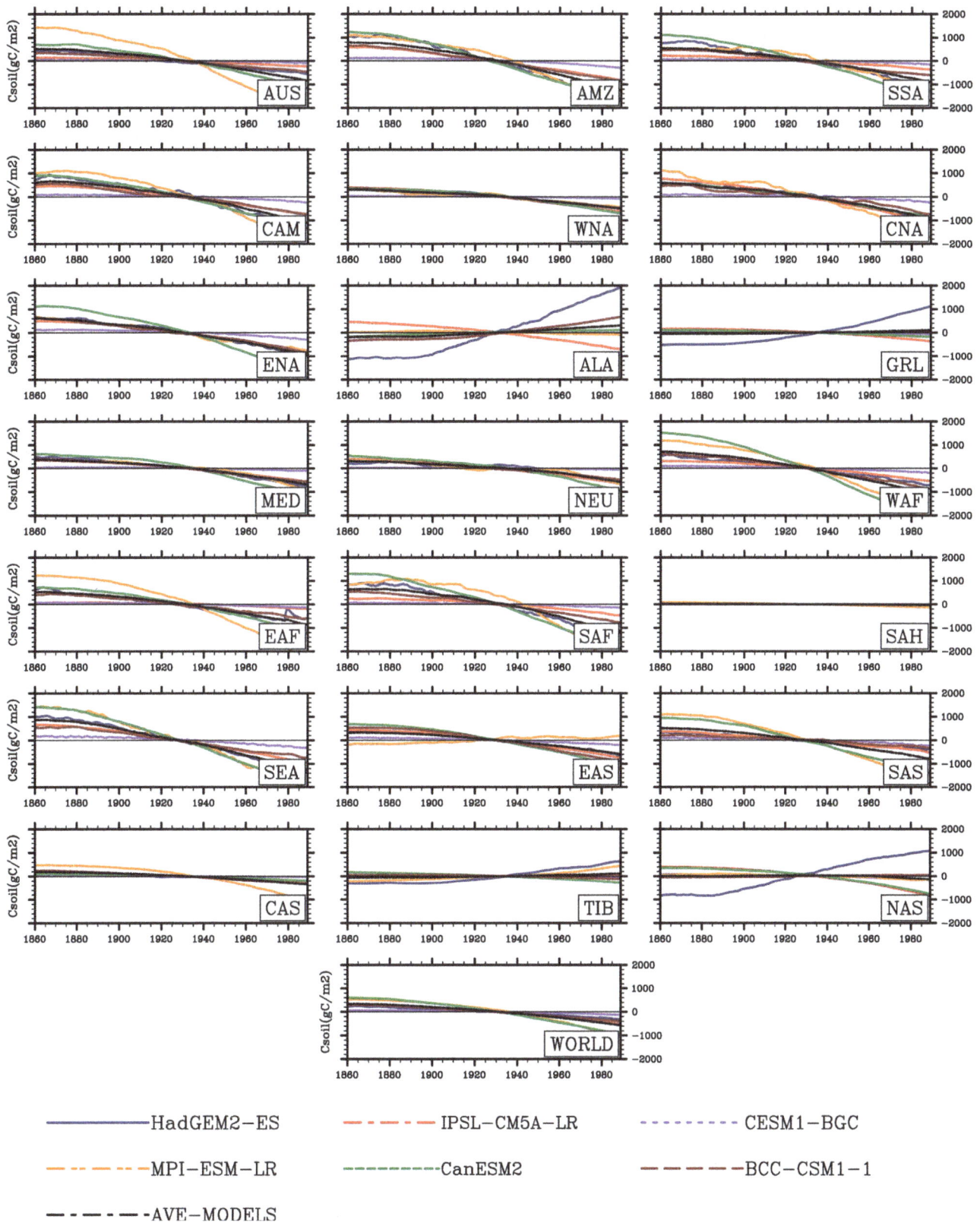

Figure 11. Same as Figure 2, but for soil carbon storage (Csoil, gC m^{-2}).

(ESMs) [1,14], the changes of land carbon storage across the terrestrial ecosystem caused by CO_2 increases of about 563.4 ppm account for about 247% of the changes induced by warming of about 4.3 K. Generally, the magnitude of responses of NPP versus RH to climate change [19,61] is still debated and strength of dynamical responses of vegetation growth to increasing CO_2 concentrations is qualitatively different among different models. Such magnitude of NPP has been documented to depend on the local water availability [62]. These changes in water availability depend critically upon uncertain regional aspects of climate change projections and are therefore likely to be another dominant source of uncertainty [14,19]. Hence the simulated divergences in precipitation and soil moisture can partly contribute to the variances in land carbon storage from the modeled simulations. Overall, these uncertainties can be concluded in the processes: temperature dependent RH [63] and strength of NPP affected by drought and CO_2 fertilization [62,64,65], and captured carbon in the forest aboveground biomass [66], especially in the Amazon forest [67,68]. Hence, it is urgently needed to pay more attention on key processes (e.g. quantifying the strength of CO_2 fertilization effect and removing or lessening the uncertainty in climate change) and critical regions. For example, for tropical regions, which play extremely important role in the terrestrial carbon cycle, great uncertainties are even maintained in the competition between the direct CO_2 effects and climate change due to lacking fully understanding driven mechanism in dynamic biosphere component.

Conclusion

The results of this study show sensitivities of global and regional land carbon storage to rising atmospheric CO_2 concentration and climate change, which are directly based on simulations from the CMIP5. Positive impacts of increasing CO_2 concentration on land carbon storage have been exhibited over the majority of whole terrestrial ecosystems, which are attributed to CO_2 fertilization effect. At regional scale, the strongest positive impacts mainly occur in broad areas covered by tropical and temperate forests (e.g., Amazon basin, western Africa, southern Asia and southeastern Asia). Great spatial divergence of responses of land carbon storage to warming has been suggested among multi-model simulations. In high latitudes and altitudes, positive effects of increasing temperature are introduced in association with an extended growing season length and enhanced photosynthesis. Current global warming has already accelerated carbon storage loss in most tropics and Southern Hemisphere. This change is partly attributed to local reduction in soil moisture and decline in precipitation. Decreases in land carbon storage of areas including Amazon basin, southern South America, western Africa, eastern Africa, southern Africa and southeastern Asia account for 61.4% simulated by IPSL-CM5A-LR to 97.7% simulated by HadGEM2-ES of the decreases in global land carbon storage responding to the rising temperature. Further, majority of the simulations ($\geq 4/6$) agree on the sign of its negative effects of climate change on land carbon storage across low latitudes and Southern Hemisphere. Conversely, across the areas of around $60°N$, there is less agreement on effects of climate change among models.

Author Contributions

Conceived and designed the experiments: LD JP MH. Performed the experiments: JP LD. Analyzed the data: LD JP. Contributed reagents/materials/analysis tools: JP. Wrote the paper: JP LD.

References

1. Friedlingstein P, Cox P, Betts R, Bopp L, Von Bloh W, et al. (2006) Climate-carbon cycle feedback analysis: Results from the C4MIP model intercomparison. Journal of Climate 19: 3337–3353.
2. Qian H, Joseph R, Zeng N (2010) Enhanced terrestrial carbon uptake in the northern high latitudes in the 21st century from the coupled carbon cycle climate model intercomparison project model projections. Global Change Biology 16: 641–656.
3. Randerson JT, Hoffman FM, Thornton PE, Mahowald NM, Lindsay K, et al. (2009) Systematic assessment of terrestrial biogeochemistry in coupled climate–carbon models. Global Change Biology 15: 2462–2484.
4. Schuur EAG, Vogel JG, Crummer KG, Lee H, Sickman JO, et al. (2009) The effect of permafrost thaw on old carbon release and net carbon exchange from tundra. Nature 459: 556–559.
5. Yi C, Ricciuto D, Li R, Wolbeck J, Xu X, et al. (2010) Climate control of terrestrial carbon exchange across biomes and continents. Environmental Research Letters 5: 034007.
6. Koven CD, Ringeval B, Friedlingstein P, Ciais P, Cadule P, et al. (2011) Permafrost carbon-climate feedbacks accelerate global warming. Proceedings of the National Academy of Sciences 108: 14769–14774.
7. Zeng N, Qian H, Munoz E, Iacono R (2004) How strong is carbon cycle-climate feedback under global warming? Geophysical Research Letters 31: L20203.
8. Cox PM, Pearson D, Booth BB, Friedlingstein P, Huntingford C, et al. (2013) Sensitivity of tropical carbon to climate change constrained by carbon dioxide variability. Nature, doi:10.1038/nature11882.
9. McGuire AD, Sitch S, Clein JS, Dargaville R, Esser G, et al. (2001) Carbon balance of the terrestrial biosphere in the Twentieth Century: Analyses of CO2, climate and land use effects with four process-based ecosystem models. Global Biogeochemical Cycles 15: 183–206.
10. Zak DR, Pregitzer KS, Kubiske ME, Burton AJ (2011) Forest productivity under elevated CO2 and O3: positive feedbacks to soil N cycling sustain decade-long net primary productivity enhancement by CO2. Ecology letters 14: 1220–1226.
11. Peñuelas J, Canadell JG, Ogaya R (2011) Increased water-use efficiency during the 20th century did not translate into enhanced tree growth. Global Ecology and Biogeography 20: 597–608.
12. Piao S, Ciais P, Friedlingstein P, de Noblet-Ducoudré N, Cadule P, et al. (2009) Spatiotemporal patterns of terrestrial carbon cycle during the 20th century. Global Biogeochemical Cycles, doi: 10.1029/2008GB003339.
13. Cramer W, Bondeau A, Woodward FI, Prentice IC, Betts RA, et al. (2001) Global response of terrestrial ecosystem structure and function to CO2 and climate change: results from six dynamic global vegetation models. Global change biology 7: 357–373.
14. Arora VK, Boer GJ, Friedlingstein P, Eby M, Jones CD, et al. (2013) Carbon-concentration and carbon-climate feedbacks in CMIP5 Earth system models. Journal of Climate 26, 5289–5314.
15. Hemming D, Betts R, Collins M (2011) Sensitivity and uncertainty of modelled terrestrial net primary productivity to doubled CO2 and associated climate change for a relatively large perturbed physics ensemble. Agricultural and Forest Meteorology 170 (15): 79–88.
16. Heimann M, Reichstein M (2008) Terrestrial ecosystem carbon dynamics and climate feedbacks. Nature 451: 289–292.
17. Boer G, Arora V (2013) Feedbacks in Emission-Driven and Concentration-Driven Global Carbon Budgets. Journal of Climate 26: 3326–3341.
18. Schimel D, Melillo J, Tian H, McGuire AD, Kicklighter D, et al. (2000) Contribution of increasing CO2 and climate to carbon storage by ecosystems in the United States. Science 287: 2004–2006.
19. Sitch S, Huntingford C, Gedney N, Levy P, Lomas M, et al. (2008) Evaluation of the terrestrial carbon cycle, future plant geography and climate-carbon cycle feedbacks using five Dynamic Global Vegetation Models (DGVMs). Global Change Biology 14: 2015–2039.
20. Roeckner E, Giorgetta M, Crueger T, Esch M, Pongratz J (2011) Historical and future anthropogenic emission pathways derived from coupled climate–carbon cycle simulations. Climatic Change 105: 91–108.
21. Arora V, Boer G, Christian J, Curry C, Denman K, et al. (2009) The effect of terrestrial photosynthesis down regulation on the twentieth-century carbon budget simulated with the CCCma earth system model. Journal of Climate 22: 6066–6088.
22. Boer G, Arora V (2010) Geographic aspects of temperature and concentration feedbacks in the carbon budget. Journal of Climate 23: 775–784.
23. Zickfeld K, Eby M, Matthews HD, Schmittner A, Weaver AJ (2011) Nonlinearity of carbon cycle feedbacks. Journal of Climate 24: 4255–4275.
24. Arneth A, Harrison SP, Zaehle S, Tsigaridis K, Menon S, et al. (2010) Terrestrial biogeochemical feedbacks in the climate system. Nature Geoscience 3: 525–532.
25. Taylor KE, Stouffer RJ, Meehl GA (2012) An overview of CMIP5 and the experiment design. Bulletin of the American Meteorological Society 93: 485–498.

26. Peng J, Dong W, Yuan W, Chou J, Zhang Y, et al. (2013) Effects of increased CO2 on land water balance from 1850 to 1989. Theoretical and Applied Climatology 111: 483–495.

27. Peng J, Dan L, Dong W (2014) Are there interactive effects of physiological and radiative forcing produced by increased CO_2 concentration on changes of land hydrological cycle? Global and Planetary Change 112: 64–78.

28. Collins W, Bellouin N, Doutriaux-Boucher M, Gedney N, Halloran P, et al. (2011) Development and evaluation of an Earth-system model–HadGEM2. Geoscientific Model Development Discussions 4: 997–1062.

29. Dufresne J-L, Foujols M-A, Denvil S, Caubel A, Marti O, et al. (2013) Climate change projections using the IPSL-CM5 Earth System Model: from CMIP3 to CMIP5. Climate Dynamics: 1–43.

30. Girardin MP, Bernier PY, Raulier F, Tardif JC, Conciatori F, et al. (2011) Testing for a CO2 fertilization effect on growth of Canadian boreal forests. Journal of Geophysical Research doi: 10.1029/2010JG001287.

31. Raddatz T, Reick C, Knorr W, Kattge J, Roeckner E, et al. (2007) Will the tropical land biosphere dominate the climate–carbon cycle feedback during the twenty-first century? Climate Dynamics 29: 565–574.

32. Todd-Brown K, Randerson J, Post W, Hoffman F, Tarnocai C, et al. (2013) Causes of variation in soil carbon simulations from CMIP5 Earth system models and comparison with observations. Biogeosciences 10: 1717–1736.

33. Wu T, Li W, Ji J, Xin X, Li L, et al. (2013) Global carbon budgets simulated by the Beijing Climate Center Climate System Model for the last century. Journal of Geophysical Research doi: 10.1002/jgrd.5032.

34. Brovkin V, Boysen L, Arora V, Boisier J, Cadule P, et al. (2013) Effect of anthropogenic land-use and land cover changes on climate and land carbon storage in CMIP5 projections for the 21st century. Journal of Climate 26 (18): 6859–6881.

35. Krinner G, Viovy N, de Noblet-Ducoudré N, Ogée J, Polcher J, et al. (2005) A dynamic global vegetation model for studies of the coupled atmosphere-biosphere system. Global Biogeochemical Cycles doi: 10.1029/2003GB002199.

36. Gent PR, Danabasoglu G, Donner LJ, Holland MM, Hunke EC, et al. (2011) The community climate system model version 4. Journal of Climate 24: 4973–4991.

37. Thornton PE, Doney SC, Lindsay K, Moore JK, Mahowald N, et al. (2009) Carbon-nitrogen interactions regulate climate-carbon cycle feedbacks: results from an atmosphere-ocean general circulation model. Biogeosciences 6: 2099–2120.

38. Thornton PE, Lamarque JF, Rosenbloom NA, Mahowald NM (2007) Influence of carbon-nitrogen cycle coupling on land model response to CO_2 fertilization and climate variability. Global Biogeochemical Cycles 21 doi: 10.1029/2006GB002868.

39. Christian J, Arora V, Boer G, Curry C, Zahariev K, et al. (2010) The global carbon cycle in the Canadian Earth system model (CanESM1): Preindustrial control simulation. Journal of Geophysical Research 115: G03014.

40. Arora VK, Boer GJ (2005) A parameterization of leaf phenology for the terrestrial ecosystem component of climate models. Global Change Biology 11: 39–59.

41. Ji J (1995) A climate-vegetation interaction model: Simulating physical and biological processes at the surface. Journal of Biogeography: 445–451.

42. Dan L, Ji J, Li Y (2005) Climatic and biological simulations in a two-way coupled atmosphere–biosphere model (CABM). Global and Planetary Change 47: 153–169.

43. Dan L, Ji J (2007) The surface energy, water, carbon flux and their intercorrelated seasonality in a global climate-vegetation coupled model. Tellus B 59: 425–438.

44. Giorgi F, Francisco R (2000) Uncertainties in regional climate change prediction: a regional analysis of ensemble simulations with the HADCM2 coupled AOGCM. Climate Dynamics 16: 169–182.

45. Gao Q, Li Y, Wan Y, Qin X, Jiangcun W, et al. (2009) Dynamics of alpine grassland NPP and its response to climate change in Northern Tibet. Climatic change 97: 515–528.

46. Post WM, King AW, Wullschleger SD (1997) Historical variations in terrestrial biospheric carbon storage. Global Biogeochemical Cycles 11: 99–109.

47. Piao S, Cui M, Chen A, Wang X, Ciais P, et al. (2011) Altitude and temperature dependence of change in the spring vegetation green-up date from 1982 to 2006 in the Qinghai-Xizang Plateau. Agricultural and Forest Meteorology 151: 1599–1608.

48. Nepstad D, Lefebvre P, Lopes da Silva U, Tomasella J, Schlesinger P, et al. (2004) Amazon drought and its implications for forest flammability and tree growth: A basin-wide analysis. Global Change Biology 10: 704–717.

49. Way DA, Oren R (2010) Differential responses to changes in growth temperature between trees from different functional groups and biomes: a review and synthesis of data. Tree Physiology 30: 669–688.

50. Zhu W, Tian H, Xu X, Pan Y, Chen G, et al. (2012) Extension of the growing season due to delayed autumn over mid and high latitudes in North America during 1982–2006. Global Ecology and Biogeography 21: 260–271.

51. Myneni RB, Keeling C, Tucker C, Asrar G, Nemani R (1997) Increased plant growth in the northern high latitudes from 1981 to 1991. Nature 386: 698–702.

52. Christidis N, Stott PA, Brown S, Karoly DJ, Caesar J (2007) Human contribution to the lengthening of the growing season during 1950-99. Journal of Climate 20: 5441–5454.

53. Alcaraz-Segura D, Chuvieco E, Epstein HE, Kasischke ES, Trishchenko A (2010) Debating the greening vs. browning of the North American boreal forest: differences between satellite datasets. Global Change Biology 16: 760–770.

54. Fisher JB, Sikka M, Sitch S, Ciais P, Poulter B, et al. (2013) African tropical rainforest net carbon dioxide fluxes in the twentieth century. Philosophical Transactions of the Royal Society B: Biological Sciences 368: 20120376.

55. Boyero L, Pearson RG, Gessner MO, Barmuta LA, Ferreira V, et al. (2011) A global experiment suggests climate warming will not accelerate litter decomposition in streams but might reduce carbon sequestration. Ecology Letters 14: 289–294.

56. Phillips OL, Aragão LE, Lewis SL, Fisher JB, Lloyd J, et al. (2009) Drought sensitivity of the Amazon rainforest. Science 323: 1344–1347.

57. Lewis SL, Brando PM, Phillips OL, van der Heijden GM, Nepstad D (2011) The 2010 amazon drought. Science 331: 554–554.

58. Bonan GB (2008) Forests and Climate Change: Forcings, Feedbacks, and the Climate Benefits of Forests. Science 320: 1444–1449.

59. Cox PM, Betts RA, Collins M, Harris PP, Huntingford C, et al. (2004) Amazonian forest dieback under climate-carbon cycle projections for the 21st century. Theoretical and Applied Climatology 78: 137–156.

60. Ito A (2005) Climate-related uncertainties in projections of the twenty-first century terrestrial carbon budget: off-line model experiments using IPCC greenhouse-gas scenarios and AOGCM climate projections. Climate Dynamics 24: 435–448.

61. Nemani RR, Keeling CD, Hashimoto H, Jolly WM, Piper SC, et al. (2003) Climate-driven increases in global terrestrial net primary production from 1982 to 1999. Science 300: 1560–1563.

62. Zhao M, Running SW (2010) Drought-induced reduction in global terrestrial net primary production from 2000 through 2009. Science 329: 940–943.

63. Tuomi M, Vanhala P, Karhu K, Fritze H, Liski J (2008) Heterotrophic soil respiration—comparison of different models describing its temperature dependence. Ecological Modelling 211: 182–190.

64. Berthelot M, Friedlingstein P, Ciais P, Dufresne JL, Monfray P (2005) How uncertainties in future climate change predictions translate into future terrestrial carbon fluxes. Global Change Biology 11: 959–970.

65. Hickler T, Smith B, Prentice IC, Mjöfors K, Miller P, et al. (2008) CO2 fertilization in temperate FACE experiments not representative of boreal and tropical forests. Global Change Biology 14: 1531–1542.

66. Houghton R (2005) Aboveground forest biomass and the global carbon balance. Global Change Biology 11: 945–958.

67. Saatchi S, Houghton R, Dos Santos Alvala R, Soares J, Yu Y (2007) Distribution of aboveground live biomass in the Amazon basin. Global Change Biology 13: 816–837.

68. Martin AR, Thomas SC (2011) A reassessment of carbon content in tropical trees. PLoS One 6: e23533.

8

Spatial Prediction of N$_2$O Emissions in Pasture: A Bayesian Model Averaging Analysis

Xiaodong Huang[1,2]*, Peter Grace[2], Wenbiao Hu[3], David Rowlings[2], Kerrie Mengersen[1]

1 Mathematical Sciences, Queensland University of Technology, Brisbane, Australia, **2** Institute of Sustainable Resources, Queensland University of Technology, Brisbane, Australia, **3** School of Population Health, The University of Queensland, Brisbane, Australia

Abstract

Nitrous oxide (N$_2$O) is one of the greenhouse gases that can contribute to global warming. Spatial variability of N$_2$O can lead to large uncertainties in prediction. However, previous studies have often ignored the spatial dependency to quantify the N$_2$O – environmental factors relationships. Few researches have examined the impacts of various spatial correlation structures (e.g. independence, distance-based and neighbourhood based) on spatial prediction of N$_2$O emissions. This study aimed to assess the impact of three spatial correlation structures on spatial predictions and calibrate the spatial prediction using Bayesian model averaging (BMA) based on replicated, irregular point-referenced data. The data were measured in 17 chambers randomly placed across a 271 m^2 field between October 2007 and September 2008 in the southeast of Australia. We used a Bayesian geostatistical model and a Bayesian spatial conditional autoregressive (CAR) model to investigate and accommodate spatial dependency, and to estimate the effects of environmental variables on N$_2$O emissions across the study site. We compared these with a Bayesian regression model with independent errors. The three approaches resulted in different derived maps of spatial prediction of N$_2$O emissions. We found that incorporating spatial dependency in the model not only substantially improved predictions of N$_2$O emission from soil, but also better quantified uncertainties of soil parameters in the study. The hybrid model structure obtained by BMA improved the accuracy of spatial prediction of N$_2$O emissions across this study region.

Editor: Matteo Convertino, University of Florida, United States of America

Funding: The authors thank the Institute of Sustainable Resources (ISR), Queensland University of Technology (QUT), and the Australian Research Council (ARC) for funding this study. The funders had no role in study design, data collection and analysis, decision to publish, or preparation of the manuscript.

Competing Interests: The authors have declared that no competing interests exist.

* E-mail: xiaodong.huang@student.qut.edu.au

Introduction

Soils have been considered as an important source for nitrous oxide (N$_2$O), a well-known greenhouse gas [1]. N$_2$O fluxes often exhibit spatial autocorrelation at multiple scales due to the distribution of soil properties and topography. It is difficult to precisely estimate annual N$_2$O emissions at a field scale level because of high spatial variability within the field [2]. In light of these large uncertainties in prediction, spatial variation should be an explicit consideration in any analysis of N$_2$O emissions [3–5].

To date, the relationship between N$_2$O emissions and environmental covariates has largely been quantified by aggregating over all sites and assuming independent observations in multiple linear regression models. However, the presence of spatial correlation can render these models invalid since they can lead to biased estimates and incorrect inferences [6,7].

In the past decade, a variety of models that take into account the spatial nature of data have been developed [8,9] and are widely applied in ecology, epidemiology, economics and so on. These models can help to better identify and explore influential factors and guide more informed inferences, as well as improve further experimental design in order to obtain more precise estimates [10].

Bayesian spatial conditional autoregressive (CAR) models are appropriate for all locations that have a similar size and are regularly arranged [11], whereas geostatistical models are more suitable for spatial data with unidentified neighbourhoods [7]. Most published research on the comparison of spatial models has been based on areal data with identified neighbours or point data with a regular sampling pattern [7–9,12,13]. However, differences between the CAR model and geostatistical model with respect to parameter estimation and predicted spatial distribution based on point-referenced data with an irregular sampling interval and undetermined boundaries are not well understood.

One concern with spatial models is that different representations of the spatial correlation based on the same dataset might give different estimated effect sizes, inferences about significant parameters or estimated error structures [7,9]. Many candidate spatial correlation structures are available in spatial analysis. It is often difficult to determine the best spatial correlation structure based on standard information-based criteria. However, Bayesian model averaging (BMA) can take account of such model uncertainty and provide better average predictive performance [14,15]. For example, Boone and Bullock [16] used BMA to pool information from four spatial candidate structures in the analysis of a loblolly pine dataset.

In this study, we consider three spatial correlation structures (independence, distance-based and neighbourhood-based) in spatial analyses of N$_2$O emission for point data obtained from irregular sampling intervals in pasture. All models are developed under a hierarchical Bayesian inferential framework. Key attributes of Bayesian approaches are the use of probability for quantifying uncertainty in inferences, formal accommodation of parameter uncertainty [17], and flexibility of model description [8,18]. The deviance information criterion (DIC) is used to compare the various models [19] and provide weights for BMA [20]. The aims of this study are to assess the effects of various spatial dependencies on spatial prediction, to calibrate spatial predictions of N_2O by BMA across the study region based on the environment-N_2O relationships obtained from the three models.

Materials and Methods

Study Site and Data Collection

The study site is located at Mooloolah ($26°38'40''$ S., $152°56'23''$E.) on the Sunshine Coast in Australia. N_2O (ug N_2O-N m^{-2} hr^{-1}) emissions and 7 potential independent variables (gravimetric soil moisture (%), soil temperature (°C), soil NO_3^- concentration (Kg N ha^{-1}), soil pH; soil sand, silt and clay content (%)) were measured at 17 chambers randomly placed across a 271 m^2 subtropical pasture at monthly intervals between October 2007 and September 2008.

The pasture was a mixture of the tropical grass *Setaria sphacelata* and the legumes Silverleaf Desmodium (*Desmodium uncinatum*) and White Clover (*Trifolium repens*). No nitrogenous fertilizer had been applied to the pasture site for over 20 years. The soil was classified according to the Australian Soil Classification as a Haplic, Eutrophic, Black Dermosol [21] and had a bulk density of 1.0 g cm^{-3} (0–10 cm) and an organic carbon content of 2.8%. Average soil texture across the site was classified as a loam.

The closed static chamber technique was used for measurements of N_2O emissions. Chambers were 200 mm high (diameter 200 mm) inserted 100 mm into the soil, allowing a headspace of 80–100 mm. Chambers remained in situ throughout the length of the experiment. Chambers were closed for one h and sampled using 12 ml evacuated glass vials (Exetainer; Labco, High Wycombe, Buckinghamshire, UK) at zero (0) min and 60 min. Full details of chamber method and site climate are described in Rowlings et al. [22].

Statistical Analysis

The observed data can be defined as point-referenced data [23]. In order to assess the effects of different covariance structures on estimates of spatial variation in N_2O fluxes and compare the estimated parameters among three Bayesian spatial models, we used a Thiessen-polygon approach to convert point-referenced data to areal data. This method creates a polygon enclosing each original point, such that each point has its own polygon. The defined boundaries of the Thiessen polygons can be used to establish a neighbourhood weight matrix for each data point [24]. In this study, we focused on a linear regression model with three different correlation structures: 1) independent model (no spatial correlation structure), 2) geostatistical (EXP) model (spatial correlation described as the exponential decay function of the distance between pairs of points), and 3) conditional autoregressive model (spatial correlation described as first-order neighbourhood). Although other spatial correlation models, for example, simultaneous autoregressive model (SAR) and geostatistical models with other Matérn correlation functions are available, CAR and EXP models are most commonly used in practice [23].

In all of the following models, let y_{ir} be the observed N_2O fluxes at location i for replicate r, ($i = 1,\ldots,Q$, $r = 1,\ldots,M$; $Q = 17$, $M = 13$). The vector $Y_i = [y_{i1},y_{i2},y_{i3},\ldots,y_{iM}]$ represents N_2O fluxes at the ith location. Let X_{ir} be a vector of length $K = 7$, representing the covariates comprising soil moisture, soil temperature, NO_3^-, soil

texture (including sand, silt and clay) and soil pH at the ith site for replicate r. Measurements of N_2O fluxes exhibited skewness, so were log-transformed to better approximate a normal distribution.

Bayesian Linear Regression Model (Independent Structure)

In this first model, we assumed that locations were independent and that N_2O emissions were affected by the nominated covariates independently at each location so that:

$$y_{ir} = \beta_0 + \sum_{k=1}^{7} \beta_k x_{irk} + \varepsilon_{ir} \qquad (1)$$

where β_k are the regression coefficients and $\varepsilon_{ir} \sim N(0,\sigma^2)$ is the residual under the independence and normality assumptions [25]. In a Bayesian framework, the posterior distribution for the parameters of interest is thus given by:

$$p(\beta,\sigma^2|Y) \propto p(Y|\beta,\sigma^2)P(\beta)P(\sigma^2)$$

where $p(Y|\beta,\sigma^2) = N(Y|\beta,\sigma^2)$. Diffuse priors were imposed for the regression parameters, so that $\beta \sim \mathcal{N}(0.0, 1.0E6)$ and $\sigma \sim U(0,5)$.

Bayesian Geostatistical Model (EXP Model)

The second model considered is an extension of the normal linear regression described above, with an additional term to account for spatial correlation between the experimental sites. The additional term is modelled as a random effect with the variance reflecting the spatial correlation. Letting $s = (s_i; i = 1,\ldots,17)$ be the vector of site-specific spatial Gaussian random effects, equation (1) is extended as follows:

$$y_{ir} = \beta_0 + \sum_{k=1}^{7} \beta_k x_{irk} + s_i + \varepsilon'_{ir} \qquad (2)$$

$$Y|(\beta,\sigma^2,S) \sim N(X\beta + S, \sigma^2 I)$$

$$S|(\sigma_S^2,\theta) \sim N(0,\sigma_S^2 \Phi(\theta))$$

$$\Phi_{ij} = f(d_{ij},\theta,\delta), i,j = 1,\ldots,17$$

$$f(d_{ij},\theta,\delta) = \exp[-(\theta d_{ij})^\delta], 0 < \delta < 2; \theta > 0$$

Here, $\varepsilon'_{ir} \sim N(0,\sigma^2)$ is a spatially uncorrelated error term; I is a $r \times r$ identity matrix and $S \sim N(0,\sigma_S^2 \Phi(\theta))$ is assumed to be a stationary, isotropic Gaussian process with mean zero and correlation matrix Φ with elements $\Phi_{ij} = f(d_{ij},\theta,\delta)$ between s_i and s_j [23]. The pairwise correlations Φ_{ij} are usually described as a parametric function of the distance d_{ij} between each pair of sites i and j. The exponential decay function $f(d_{ij},\theta,\delta) = exp[-(\theta d_{ij})^\delta]$ [17] is the most popular. Here θ is the rate of decrease in spatial correlation per unit of distance, with a large value of θ indicating that the spatial correlation decreases rapidly [6]. The prior distribution for θ was specified as Uniform with lower and upper bounds corresponding to a correlation of 0.05 the maximum distance (25.35 m) and minimum distance (0.75 m), respectively, between any pair of locations across the study site [26]. Covariate coefficients were modelled with diffuse normal prior distributions $\beta \sim \mathcal{N}(0.0, 1.0E6)$. The parameter δ controls the amount of spatial smoothing. Thomas et al. [27] advise a value of $\delta = 1$. The standard deviation σ was described by a uniform prior $\sigma \sim U(0,5)$.

Bayesian Spatial Intrinsic Conditional Autoregressive Model (CAR)

The third model considered employed a different representation of the spatial nature of the data. Here equation (1) is extended as follows:

$$y_{ir} = \beta_0 + \sum_{k=1}^{7} \beta_k x_{irk} + u_i + \varepsilon''_{ir} \qquad (3)$$

$$Y|(\beta,\sigma^2,U) \sim N(X\beta+U,\sigma^2 I)$$

$$u_i|u_{-i} \sim N(\sum \frac{w_{ij}}{w_{i+}}u_{-i}, \frac{\sigma_u^2}{w_{i+}}) \qquad (4)$$

Here $\varepsilon''_{ir} \sim N(0,\sigma^2)$ is the within-site residual variation. A conditional autoregressive (CAR) model was used to describe the spatial component. This is represented by the term U, with elements u_i denoting the local dependence at site i as a function of the site's neighbours u_{-i}, where $u_{-i} = [u_1, u_2, ..., u_{i-1}, u_{i+1},...,u_Q]$ [28]. The local neighbourhood relationship is represented as a symmetric $n \times n$ matrix W of spatial weights with elements w_{ij}, and $w_{i+} = \sum_{j=1}^{Q} w_{ij}$. This representation allows a great deal of flexibility in describing the spatial correlation. For example the spatial neighbourhood may be specified only as first-order neighbourhood for each site, in which case $w_{ij} = 1$ if sites i and j share a boundary, and zero otherwise. As before, all covariate coefficients had diffuse normal priors, given by $\beta \sim \mathcal{N}(0.0, 1.0E6)$, and σ_u and σ had uniform priors, $\sigma_u \sim U(0,10)$ and $\sigma \sim U(0,5)$.

Bayesian Model Averaging (BMA)

Bayesian model averaging can account for model uncertainty by taking a weighted average of models over a given model space [14]. Let M be the model space, comprising $L \geq 1$ model structures M_l with parameter set π_l based on data (D). Let Δ be the quantity of interest; this could represent, for example, the posterior predictive distribution of y. Hence the posterior distribution of Δ given data D is [14]:

$$p(\Delta|D) = \sum_{l=1}^{L} p(\Delta|M_l,D)p(M_l|D)$$

The posterior probability for M_l is given by:

$$p(M_l|D) = \frac{p(D|M_l)p(M_l)}{\sum_{q=1}^{L} p(D|M_q)p(M_q)}$$

where $p(D|M_l) = \int p(D|\pi_l, M_l)p(\pi_l|M_l)d\pi_l$.

Here, $p(D|M_l)$ is the marginal likelihood of the data D given model M_l and $p(\pi_l \mid M_l)$ is the prior density of π_l given model M_l. $p(M_l)$ is the prior probability for model M_l when M_l is regarded as the true model [14]. A Laplace approximation, typically the Bayesian information criterion (BIC) [29] can be used to approximate $p(D|M_l)$ [14,30,31]:

$$\log\{(p(D|M_l)\} \approx \log\{p(D|\widehat{\pi}_l, M_l)\} - d_l \log(n)$$

$$BIC = -2 \log\{p(D|\widehat{\pi}_l, M_l)\} + d_l \log(n)$$

Here $\log\{p(D|\widehat{\pi}_l, M_l)\}$ is the maximized log-likelihood of model l, which estimates goodness of fit; d_l is the number of parameters in model l, and n is the sample size. In the absence of other information, it is common to assume equal prior model probabilities $p(M_l)$ for the candidate models [16,31]. Hence the BMA weights are approximately

$$w_l = \exp(-0.5 * BIC)$$

The posterior probability for M_l is calculated as

$$p(M_l|D) = \frac{w_l}{\sum_{q=1}^{L} w_q}$$

Other information criterion can be used instead of the BIC. For example Akaike's information criterion $(AIC = -2 \log\{p(D|\widehat{\pi}_l, M_l)\} + 2d_l)$ [32] was suggested by Jackson et al. [31]. In the present study, candidate models were compared and combined using the deviance information criterion (DIC) [19]. The DIC is based on the posterior expectation of the deviance $\bar{D}$ and the effective number of parameters p_D in the model, and is expressed as:

$$DIC = P_D + \bar{D}$$

Deviance is defined as $-2 \log\{p(D|\pi_l, M_l)\}$ where p_D is the difference between the expected deviance and the deviance value for the posterior expectation. The DIC is easily computed from the samples generated through MCMC [23]. A smaller DIC value indicates a better model fit, accounting for model parsimony. In the BMA analysis, we let $p(M_l)$ be 1/3, indicating no prior preference for any of the three correlation structures considered in this study.

Bayesian Analysis and Spatial Interpolation

Markov chain Monte Carlo (MCMC) was used to obtain distributions and corresponding posterior structures of means, standard deviations and quantiles for parameters of interest. Convergence was assessed by checking the trace and the autocorrelation plots for the sample of each chain [33]. For each model we ran a single MCMC chain for 150,000 iterations, discarding the first 50,000 iterations as burn-in. The MCMC analysis was undertaken using WinBUGS software version 1.4 [34].

The posterior predictions of N_2O obtained from the three models and hybrid model developed by BMA were mapped across the study site using GS+ software [35]. If input values are available across the study site, the model can be used to provide predictions between the experimental locations. In our case, these values were not available, so kriging was used for spatial interpolation of the predicted N_2O values.

Results

Summary statistics for observed N_2O and covariates were provided in Table 1. The overall means were 27.4 ug N_2O-N m^{-2} hr^{-1}, 19.0 Kg N ha^{-1}, 35.6%, 22.2°C and 5.5 for N_2O, NO_3^{-}, soil moisture, soil temperature and soil pH under 17 sampling chambers, respectively. N_2O tended to be more variable. For soil texture across 17 chambers, the overall means were 18.4%, 44.3% and 37.3%, with range 9.7 to 23.3, 34.4 to 60.9 and 22.8 to 50.9% for soil clay, soil silt and soil sand, respectively. Soil clay had lower percentages in soil texture.

The DIC values, measuring goodness of fit of each model, are shown in Table 2. The DIC was obviously smaller for the CAR and EXP models compared with the independent model, indicating the value of including spatial dependency in describing the N_2O emissions in this dataset. The DIC values were similar for the CAR model and EXP model, indicating little difference in overall goodness of fit between the two representations of spatial variation. The results also showed that there were 2.7%, 1.8% and 1.4% of observed values that did not fall within the 95% posterior predictive intervals for the linear regression model, CAR and EXP model, respectively. The sum of the squared residuals from the geostatistical model was 317.12, while the sum of the squared residuals from the CAR model was 319.5.

Table 2 shows the posterior means and 95% credible intervals (CI) of parameters for the three models. Soil moisture and soil temperature had a substantive positive relationship with N_2O emissions in all three models. Only the two spatial models showed a negative relationship between N_2O emissions and NO_3^{-} in the presence of the other variables in the model. Soil pH and soil texture, such as clay, silt and sand, were not substantial influential factors for N_2O emissions in the three models in this study.

The spatial patterns of predicted N_2O using the CAR and EXP models were similar to the observed spatial pattern, particularly the CAR model (Figure 1). However, there were slight errors for

classifications of areas into different emission level groups for the two spatial models, particularly the EXP model. The results of the independent model could not match the observed spatial distribution of N_2O emission.

Figure 2 shows the distributions of the posterior means of the spatial variation in N_2O emissions which were obtained using the two spatial models and BMA model. The three maps of posterior spatial variation show similar patterns. However, the CAR model displayed slightly larger areas for high or low spatial variation in N_2O than those from the EXP model (Figure 2).

The CAR model had the highest posterior probability of 5.434E-1, whereas the independent model had a negligible probability of 0.000E-1. The EXP model had a posterior probability of 4.566E-1. The map of the averaged posterior predicted N_2O emissions across the three structures was much more similar to the map of observed N_2O emissions comparison to the maps of the CAR, EXP and independent models (Figure 1). The map of averaged spatial predictions of N_2O displayed better performance in high or low emission areas than that obtained from the EXP model and also it improved the accuracy of the spatial prediction of N_2O on left side of the map compared with the CAR model. There were also slight changes in the map of the distribution of averaged posterior spatial variation (Figure 2).

Discussion

The CAR and EXP models are popular approaches for describing spatially correlated data and are widely used in many areas of scientific research. In this case study, we applied these two models and a baseline model that ignored the spatial correlation altogether. In order to gain some insight into the effects of different assumed spatial correlation structures on parameter estimation and spatial prediction of N_2O emission for the same point data on an irregular grid, and to account for the uncertainty in evaluating spatial variability of N_2O emissions using Bayesian model averaging.

All three models identified soil temperature and soil moisture as potentially important influential factors positively associated with N_2O emissions. This is supported by a large body of previous research [5,36–42]. In this study, the average of soil moisture was around 36% in the pasture. Nitrification occurs when soil water-filled pore space is <60% [43]. Our result supported that increasing soil moisture and soil temperature increased N_2O emissions via the nitrification pathway [36,39,44]. Only the CAR and EXP models yielded a significant coefficient for NO_3^{-}. The inverse relationships between N_2O and NO_3^{-} from nitrification have been found in grass-clover pasture and laboratory study [36,45]. The results showed that allowing for spatial dependence in the model affected not only the scale of posterior mean but also affected the determination of significant factors in the model for the pasture data. Moreover, the different spatial correlation structures in the models resulted in differences in the magnitude of the corresponding coefficients. Hence, the selection of an appropriate model structure is a critical step [9].

The sum of the squared residuals showed that the geostatistical model was slightly better than the CAR model. However, the plots of spatial interpolation of the predicted N_2O values by kriging showed that the geostatistical model tended to oversmooth high N_2O emission areas in comparison to the results of CAR model. We found that the predicted N_2O values of the locations which were close to the highest emission site were underestimated by the geostatistical model in comparison to the CAR model and the observed data. The tendency of the EXP model to oversmooth is supported by Best [8]. On the other hand, mapping the spatial prediction of interest is often an important aim of developing a spatial model. Spatial interpolation is a straightforward approach

Table 1. Summary statistics of observed variables for the 17 chambers over the sampling period from a subtropical pasture at Mooloolah, Queensland.

Variables	Mean	SD	Minimum	Maximum
N_2O (µg N_2O-N m^{-2} hr^{-1})	27.2	39.4	0.0	280.4
NO_3^{-} (kg N ha^{-1})	18.98	14.1	0.0	90.34
Gravimetric soil moisture (%)	35.57	9.27	12.37	70
Soil temperature (°C)	22.16	3.07	14.8	27.3
pH	5.47	0.29	5.2	6.4
Sand (%)	37.25	7.74	22.75	50.89
Silt (%)	44.34	7.2	34.44	60.94
Clay (%)	18.4	3.07	9.65	23.34

Table 2. Posterior means and 95% credible intervals of parameters for three models for pasture.

Parameter	CAR model	EXP model	Independent model
	Mean	Mean	Mean
β_0	−49.2 (−100–6.89)	162.6 (−1728–2049)	−5.84 (−40.62–30.2)
$\beta_{soil\ moisture}$	0.055 (0.034–0.075)	0.054 (0.034–0.075)	0.039 (0.02–0.06)
$\beta_{soil\ temperature}$	0.16 (0.1–0.22)	0.16 (0.1–0.22)	0.15 (0.08–0.21)
$\beta_{NO_3^-}$	−0.018 (−0.031– −0.004)	−0.017 (−0.031– −0.003)	−0.006 (−0.02–0.008)
β_{Ph}	0.4 (−0.73–1.55)	0.32 (−1.19–1.84)	0.21 (−0.42–0.91)
β_{sand}	0.45 (−0.11–1.0)	−1.67 (−20.54–17.22)	0.023 (−0.35–0.37)
β_{silt}	0.46 (−0.096–0.978)	−1.66 (−20.52–17.24)	0.033 (−0.34–0.38)
β_{clay}	0.4 (−0.16–0.94)	−1.7 (−20.57–17.21)	0.004 (−0.37–0.36)
σ^2	1.59 (1.31–1.92)	1.59 (1.30–1.93)	2.0 (1.66–2.43)
σ_u^2, σ_s^2	1.78 (0.46–4.81)	0.76 (0.25–1.84)	
DIC	745.75	746.1	786.69

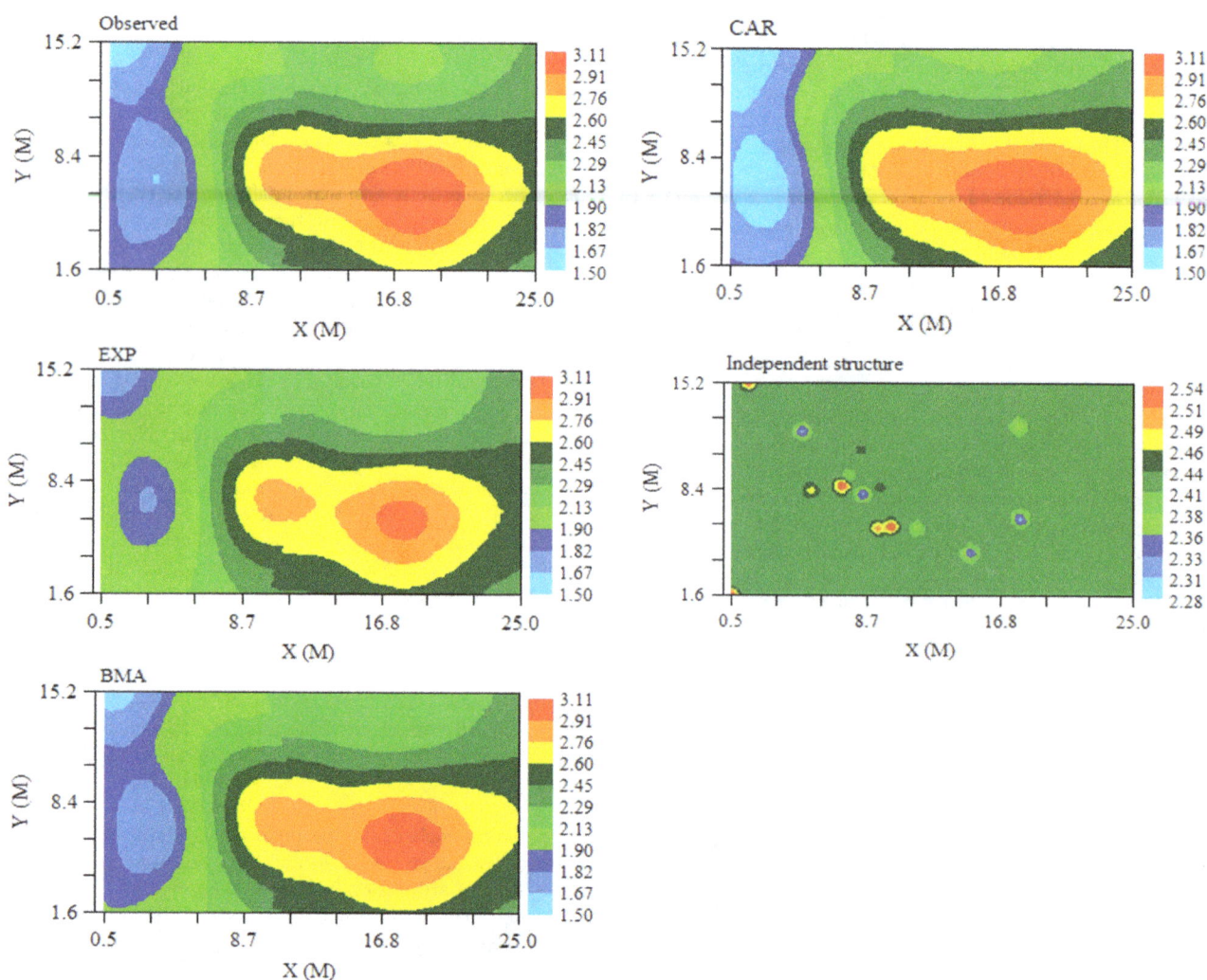

Figure 1. Maps of observed and posterior mean Ln(N$_2$O) (ug N$_2$O-N m^{-2} hr^{-1}) from the CAR, EXP, BMA and linear regression models across the study site in pasture.

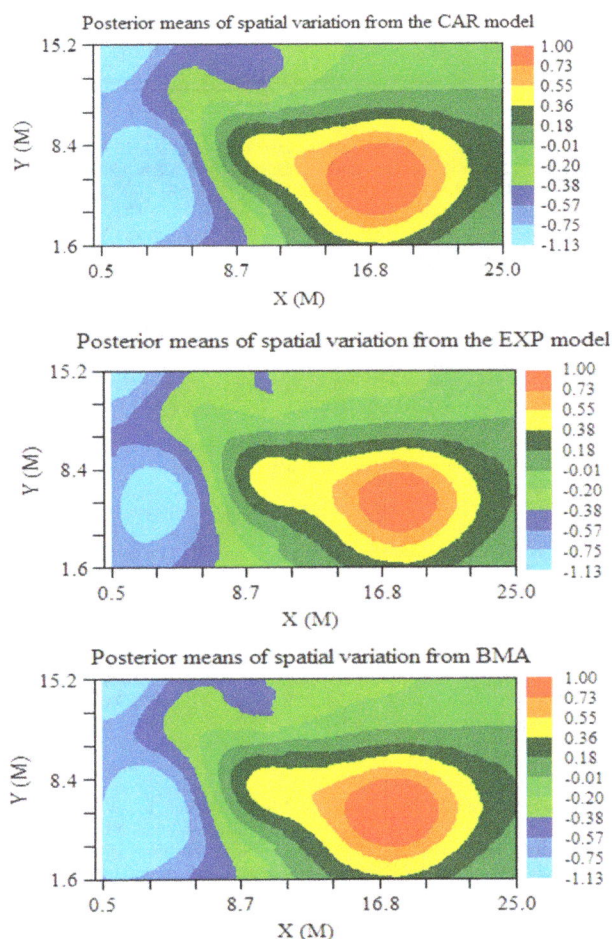

Figure 2. Maps of the posterior means of spatial variation in Ln(N$_2$O) (ug N$_2$O-N m^{-2} hr^{-1}) emission using two spatial models and Bayesian model averaging in pasture.

for spatial prediction. The map (Figure 1) based on the CAR model was visually quite similar to the map based on observed data, in that the patch of predicted high emissions under the CAR model matched the observed high emission locations well. The map for the linear regression model indicated that it oversmoothed the study region and poorly predicted the spatial distribution, in that it did not capture some of the regions with low or high emissions. We suggest that the CAR model is better at capturing the distribution of high N$_2$O emissions areas in this study.

The posterior spatial variation in N$_2$O emissions using the CAR model based on a first-order neighbourhood function tended to be slightly greater than these of the EXP model based on an exponential distance decay function across the study site (Figure 2). This indicated that the CAR model gave more weight to the random effect than did the EXP model. However, the two spatial correlation structures largely revealed similar natural phenomena associated with geographic variation in this study. Finally, the N$_2$O distribution could be predicted well across the survey region based on the environmental covariates-N$_2$O relationship only, after adjusting for spatial autocorrelation in the models in this study.

Both the CAR model and the EXP model yielded similar parameter estimates for N$_2$O emissions underlying point-referenced data with irregular sampling intervals. Our results support previous assertions that the CAR model is comparatively more

flexible, and hence potentially more accurate and precise, for such data structures in that it can better represent geographic phenomena and accommodate more complex spatial structures [9]. Finally, the computing time of the CAR model was much faster than that of geostatistical model due to the different representations of the weight matrices [6].

The best spatial correlation structure was unclear based on the DIC values associated with the CAR and EXP models in this study. The two spatial models showed differences in spatial prediction of N$_2$O distribution. This justified model averaging across the three structures via BMA. In this article, the posterior model probability was approximated by the DIC, which can be considered as a Bayesian analogue of the AIC suitable for hierarchical models with random effects [19]. Jackson et al. [31] suggested that it was worth investigating the use of the DIC as a basis for model averaging, given the increasing popularity of Bayesian hierarchical models. Our results clearly indicated that the spatial prediction of N$_2$O from the hybrid structure could better capture the observed N$_2$O distribution across the study region than any of the individual component models. We therefore concluded that Bayesian model averaging was a potentially useful method to take account of uncertainty of different spatial correlation structures and could improve the accuracy of spatial prediction of N$_2$O emissions.

In this research, it is acknowledged that spatial analysis not only improves prediction but also highlights clustering and probabilistic uncertainties. The maps of the spatial distribution and the spatial variation in N$_2$O emissions may help to guide cultivation practices and determine emission reduction strategies. More attention should still be paid to how to select appropriate spatial correlation structure to improve the accuracy and precision of spatial prediction of N$_2$O in study regions such as the one described here.

Conclusions

Our study showed that soil temperature, soil moisture and NO$_3^-$ were important influential factors in N$_2$O emissions in pasture across this study region. High emission areas were accompanied by high uncertainties after taking soil moisture, soil temperature, NO$_3^-$, soil pH and soil texture into account. It was important to incorporate spatial dependency in the model when quantifying the relationship between N$_2$O emissions and environmental factors for this pasture. Allowing for spatial dependency in the model could yield accurate spatial prediction of N$_2$O. The CAR and EXP models yielded similar parameter estimates based on point-referenced data with irregular sampling intervals. The CAR model was better at capturing high N$_2$O emissions areas. The hybrid model structure obtained by BMA could improve the accuracy of mapping the spatial prediction of N$_2$O emissions in pasture across this study region. The maps of spatial distribution and spatial variation in N$_2$O can improve subsequent experiment design and investigate unknown influential environmental covariates in a study site.

Acknowledgments

We thank the Institute of Sustainable Resources (ISR), Queensland University of Technology (QUT) for assistance on data collection.

Author Contributions

Conceived and designed the experiments: XH KM PG. Performed the experiments: XH KM PG DR. Analyzed the data: XH KM WH. Contributed reagents/materials/analysis tools: XH KM PG WH DR. Wrote the paper: XH KM PG DR WH.

References

1. Bouwman AF (1998) Nitrogen oxides and tropical agriculture. Nature 392: 866–867.

2. Saggar S, Andrew RM, Tate KR, Hedley CB, Rodda NJ, et al. (2004) Modelling nitrous oxide emissions from dairy-grazed pasture. Nutr Cycl Agroecosys 68: 243–255.

3. Bouwman AF, Vanderhoek KW, Olivier JGJ (1995) Uncertainties in the global source distribution of nitrous-oxide. J Geophys Res-Atmos 100: 2785–2800.

4. Bouwman AF (1996) Direct emission of nitrous oxide from agriculture soils. Nutr Cycl Agroecosys 46: 53–70.

5. Dalal R, Wang W, Robertson G, Parton W (2003) Nitrous oxide emission from Australian agricultural lands and mitigation options: a review. Soil Res 41: 65–195.

6. Cressie N (1993) Statistics for Spatial Data. 2nd ed. New York: Wiley.

7. Keitt TH, Bjørnstad OT, Dixon PH, Citron-Pousty S (2002) Accounting for spatial pattern when modelling organism-environment interactions. Ecography 25: 616–625.

8. Best N, Richardson S (2005) A comparison of Bayesian spatial models for disease mapping. Stat methods med res 14: 35–59.

9. Dormann CF, McPherson JM, Araújo MB, Bivand R, Bolliger J, et al. (2007) Methods to account for spatial autocorrelation in the analysis of species distributional data: a review. Ecography 30: 609–628.

10. Huang XD, Grace P, Mengersen K, Weier K (2011) Spatio-temporal variation in soil derived nitrous oxide emissions under sugarcane. Sci Total Environ 409: 4572–4578.

11. Kelsall J, Wakefield J (1999) Discussion of "Bayesian models for spatially correlated disease and exposure data", by Best et al.. In: Bernardo J, Berger J, Dawid A, Smith A, editors. Bayesian Statistics 6. Oxford: Oxford University Press.

12. Wall MM (2004) A close look at the spatial structure implied by the CAR and SAR models. J Stat Plan Infer 121: 311–324.

13. Goovaerts P, Gebreab S (2008) How does Poisson Kriging compare to the popular BYM model for mapping disease risks? International journal of health geographics 7:6 doi:10.1186/1476–072X-7-6.

14. Hoeting JA, Madigan D, Raftery AE, Volinsky CT (1999) Bayesian Model Averaging: A Tutorial. Stat Sci 14: 382–417.

15. Draper D (1995) Assessment and propagation of model uncertainty. J. Roy. Statist. Soc. Ser. B 57: 45–97.

16. Boone EL, Bullock BP (2008) Spatial correlation matrix selection using Bayesian model averaging to characterize inter-tree competition in loblolly pine trees. J Appl Stat 35: 967–977.

17. Diggle PJ, Tawn JA, Moyeed RA (1998) Model-based geostatistics. ApplStatist 47: 299–350.

18. Gelman A, Carlin JB, Stern HS, Rubin DB (2004) Bayesian data anlysis. New York: Chapman&Hall/CRC.

19. Spiegelhalter D, Best NG, Carlin BP, van der Linde A (2002) Bayesian Measures of Model Complexity and Fit (with discussion). J R Stat Soc B Sta 64: 583–639.

20. Borgonovo E (2006) Measuring uncertainty importance: investigation and comparison of alternative approaches. Risk Analysis 26: 1349–1361.

21. Isbell RF (2002) The Australian soil classification. Collingwood: CSIRO Press.

22. Rowlings DW, Grace PR, Kiese R, Weier KL (2012) Environmental factors controlling temporal and spatial variability in the soil-atmosphere exchange of CO_2, CH_4 and N_2O from an Australian subtropical rainforest. Global Change Biology 18: 726–738.

23. Banerjee S, Carlin BP, Gelfand AE (2004) Hierarchical Modeling and Analysis for Spatial Data. USA: CHAPMAN&HALL/CRC.

24. Anselin L (2003) GeoDa 0.9 User's Guide. Spatial Analysis Laboratory. Available: www.unc.edu/~emch/gisph/geoda099.pdf. Accessed 2011 Apr 19.

25. Wackerly D, Mendenhall III W, Scheaffer R (2002) Mathematical Statistics with Applications, Sixth Edition. USA: Duxbury.

26. Thomas A, Best N, Arnold R, Spiegelhalter D (2002) GeoBUGS User Manual Version 1.1. Available: wwwmrc- bsucamacuk/bugs/winbugs/manualgeopdf. Accessed 2011 May 2.

27. Thomas A, Best N, Lunn D, Arnold R, DS (2004) GeoBUGS User Manual Version 1.2. Available: wwwmrc- sucamacuk/bugs/winbugs/geobugs12manualpdf. Accessed 2011 Apr 26.

28. Besag J, York J, Mollie A (1991) Bayesian image restoration, with two applications in spatial statistics. Ann I Stat Math 43: 1–59.

29. Schwarz G (1978) Estimating the dimension of a model. Ann Stat 6: 461–464.

30. Clyde M (2000) Model uncertainty and health effect studies for particulate matter. Environmetrics 11: 745–763.

31. Jackson CH, Thompson SG, Sharples LD (2009) Accounting for uncertainty in health economic decision models by using model averaging. J R Stat Soc A Sta 172: 383–404.

32. Akaike H (1973) Information theory and an extension of the maximum likelihood. 2nd International Symposium on Information Theory. Budapest: Akademia Kaido. 267–281.

33. Gelman A, Rubin DB (1992) Inference from iterative simulation using multiple sequences. Stat Sci 7: 457–472.

34. Spiegelhalter D, Thomas A, Best N, Lunn D (2003) WinBUGS User Manual,Version 1.4. Available: wwwmrc-bsucamacuk/bugs/winbugs/manual14pdf. Accessed 2010 Jun 12.

35. Robertson GP (2008) GS+: Geostatistics for the Environmental Sciences. Michigan USA: Gamma Design Software.

36. Goodroad LL, Keeney DR (1984) Nitrous oxide production in aerobic soils under varying pH, temperature and water content. Soil Biol Biochem 16: 39–43.

37. Williams E, Hutchinson G, Fehsenfeld F (1992) NOx and N_2O emissions from soil. Global Biogeochem Cy 6: 351–388.

38. Kamp T, Steindl H, Hantschel R, Beese F, Munch J (1998) Nitrous oxide emissions from a fallow and wheat field as affected by increased soil temperatures. Biol Fer Soils 27: 307–314.

39. Maag M, Vinther FP (1996) Nitrous oxide emission by nitrification and denitrification in different soil types and at different soil moisture contents and temperatures. Appl Soil Ecol 4: 5–14.

40. Yan Y, Sha L, Cao M, Zheng Z, Tang J, et al. (2008) Fluxes of CH_4 and N_2O from soil under a tropical seasonal rain forest in Xishuangbanna, Southwest China. J Environ Sci 20: 207–215.

41. Machefert S, Dise N, Goulding K, Whitehead P (2004) Nitrous oxide emissions from two riparian ecosystems: key controlling variables. Water, Air, Soil Poll 4: 427–436.

42. Huang X, Grace P, Rowlings D, Mengersen K (2013) A flexible Bayesian model for describing temporal variability of N_2O emissions from an Australian pasture. Sci Total Environ 454: 206–210.

43. Davidson E (1991) Fluxes of nitrous oxide and nitric oxide from terrestrial ecosystems. In: Rogers J, Whitman W, editors. Microbial Production and Consumption of Greenhouse Gases: Methane, Nitrogen Oxides, and Halomethanes. Washington, DC: Am Soc Microbiol.

44. Huang X, Grace P, Weier K, Mengersen K (2012) Nitrous oxide emissions from subtropical horticultural soils: a time series analysis. Soil Research 50: 596–606.

45. Ambus P (2005) Relationship between gross nitrogen cycling and nitrous oxide emission in grass-clover pasture. Nutr Cycl Agroecosys 72: 189–199.

9

Projected Carbon Dioxide to Increase Grass Pollen and Allergen Exposure Despite Higher Ozone Levels

Jennifer M. Albertine[1]*, William J. Manning[2], Michelle DaCosta[2], Kristina A. Stinson[3], Michael L. Muilenberg[4], Christine A. Rogers[4]*

1 Harvard Forest, Harvard University, Petersham, MA 01366, United States of America, 2 Stockbridge School of Agriculture, University of Massachusetts, Amherst, MA 01003, United States of America, 3 Department of Environmental Conservation, University of Massachusetts, Amherst, MA 01003, United States of America, 4 Environmental Health Sciences, School of Public Health and Health Sciences, University of Massachusetts, Amherst, MA 01003, United States of America

Abstract

One expected effect of climate change on human health is increasing allergic and asthmatic symptoms through changes in pollen biology. Allergic diseases have a large impact on human health globally, with 10–30% of the population affected by allergic rhinitis and more than 300 million affected by asthma. Pollen from grass species, which are highly allergenic and occur worldwide, elicits allergic responses in 20% of the general population and 40% of atopic individuals. Here we examine the effects of elevated levels of two greenhouse gases, carbon dioxide (CO_2), a growth and reproductive stimulator of plants, and ozone (O_3), a repressor, on pollen and allergen production in Timothy grass (*Phleum pratense* L.). We conducted a fully factorial experiment in which plants were grown at ambient and/or elevated levels of O_3 and CO_2, to simulate present and projected levels of both gases and their potential interactive effects. We captured and counted pollen from flowers in each treatment and assayed for concentrations of the allergen protein, Phl p 5. We found that elevated levels of CO_2 increased the amount of grass pollen produced by ~50% per flower, regardless of O_3 levels. Elevated O_3 significantly reduced the Phl p 5 content of the pollen but the net effect of rising pollen numbers with elevated CO_2 indicate increased allergen exposure under elevated levels of both greenhouse gases. Using quantitative estimates of increased pollen production and number of flowering plants per treatment, we estimated that airborne grass pollen concentrations will increase in the future up to ~200%. Due to the widespread existence of grasses and the particular importance of *P. pratense* in eliciting allergic responses, our findings provide evidence for significant impacts on human health worldwide as a result of future climate change.

Editor: Tai Wang, Institute of Botany, Chinese Academy of Sciences, China

Funding: The authors have no support or funding to report.

Competing Interests: The authors have declared that no competing interests exist.

* Email: jalbertine@fas.harvard.edu (JMA); car@ehs.umass.edu (CAR)

Introduction

One expected effect of climate change on human health is increasing allergic and asthmatic symptoms through changes in pollen biology [1–5]. Allergic diseases have a large impact on human health globally, with 10–30% of the population affected by allergic rhinitis and more than 300 million affected by asthma [6]. Pollen from grass species, which are highly allergenic and occur worldwide, elicits allergic responses in 20% of the general population and 40% of atopic individuals [7].

Climate change is noticeably affecting plant, animal and human systems and is anticipated to have large impacts on human health [4], [5], [8], [9]. A major concern is that wind-borne pollen, a primary cause of allergic rhinitis, may change in timing, amount, and allergenicity with future climate change, and may increase both the symptom severity and number of people affected [1], [3], [5], [9], [10], [11]. This concern is particularly relevant for grasses, which are widely distributed over the globe and affect a large number of sensitized individuals [1], [4], [12]. In fact, peaks in atmospheric grass pollen have been directly correlated to ambulance calls by patients under respiratory stress [13] and ER visits for asthma and wheeze [14].

Here we examine the effects of elevated levels of two greenhouse gases, carbon dioxide (CO_2), a growth and reproductive stimulator of plants, and ozone (O_3), a repressor, on pollen and allergen production in Timothy grass (*Phleum pratense* L.).

Atmospheric CO_2 levels resulting from anthropogenic sources, a major driver of climate change, are expected to rise from around 400 ppm currently to 730–1020 ppm by the year 2100 [15]. Likewise, tropospheric O_3 has been predicted to rise from current background levels of about 30–40 ppb to 42–84 ppb by 2100 [15], [16]. These greenhouse gases have been shown to affect plant growth in contrasting ways. Carbon dioxide has been well characterized as stimulating plant growth through increased photosynthetic carbon assimilation [17]. Ozone has been shown to decrease growth due to oxidative damage of photosynthetic components [18]. Together these two gases act antagonistically on plants, the degree to which depends highly on plant species and other environmental factors [18].

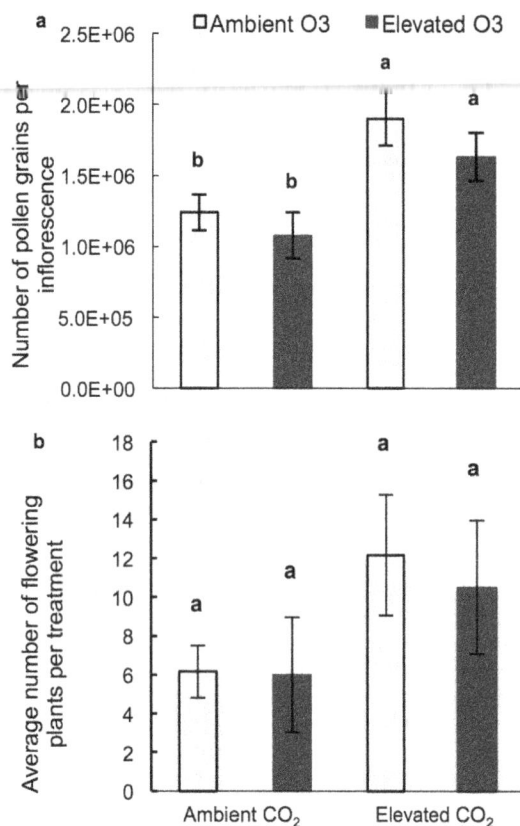

Figure 1. Effects on flowering and pollen production. a) Average pollen number per inflorescence. **b)** Average number of flowering plants per treatment. Ambient $O_3 = 30$ ppb ozone; Elevated $O_3 = 80$ ppb ozone; Ambient $CO_2 = 400$ ppm carbon dioxide; Elevated $CO_2 = 800$ ppm carbon dioxide. Significant differences ($p<0.05$) denoted by letters above bars determined by the Tukey-Kramer test.

Figure 2. Effect on inflorescence size. a) inflorescence length (cm) **b)** Inflorescence weight (g) Ambient $O_3 = 30$ ppb ozone; Elevated $O_3 = 80$ ppb ozone; Ambient $CO_2 = 400$ ppm carbon dioxide; Elevated $CO_2 = 800$ ppm carbon dioxide. Significant differences ($p<0.05$) denoted by letters above bars were determined using the Tukey-Kramer test.

The effects of elevated CO_2 on the production and allergen content of grass pollen have not been examined to date [4], although increases in pollen production 55–90% have been reported in other allergenic plants, such as ragweed (*Ambrosia artemisiifolia* L.), at elevated CO_2 [9], [19], [20]. Elevated CO_2 has also been shown to increase the reproductive output of ragweed populations through increasing the reproduction of subordinate plants in a competitive stand [21]. In a meta-analysis, C_3 and C_4 grass species demonstrated 44% and 33% growth stimulation, respectively, in response to elevated CO_2 [22]. It is therefore likely that elevated CO_2 could also increase pollen output in grasses. *In vivo* exposure to O_3 has been shown to reduce amount of viable pollen by preventing pollen maturation through reduced anther starch content in perennial rye grass (*Lolium perenne* L.) and increase the group 5 allergen content of the pollen [23], [24]. Ozone has also been shown to decrease pollen viability of *in vitro* exposed Timothy grass pollen through disruption of the cell membrane; however, this disruption could increase allergen exposure by releasing cytoplasmic allergen-containing granules from within pollen [25], [26]. Finally, allergenicity has been found to decrease in Timothy grass pollen from reduced IgE recognition due to mechanical damage and post translational modifications when pollen was exposed *in vitro* to the mixture of air pollutants: ozone, sulfur dioxide, and nitrogen dioxide [26]. While these studies suggest that there could be a reduction in pollen viability and change in allergen quality at

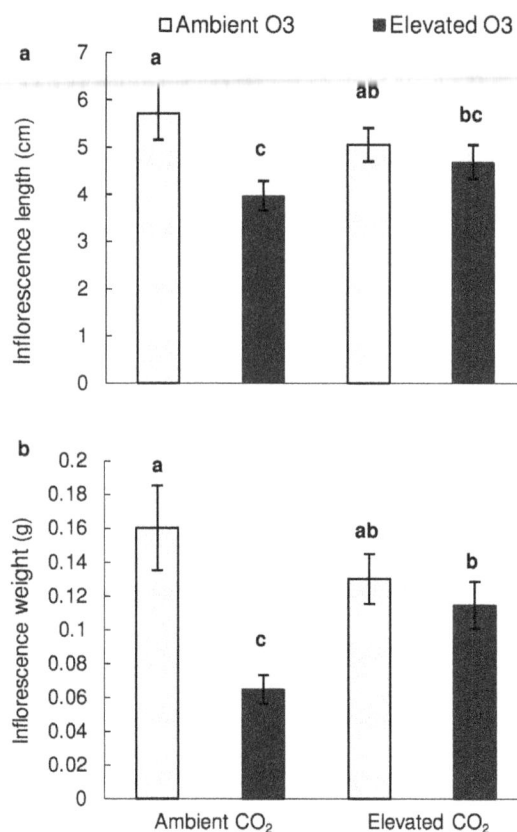

projected O_3 levels, it is not clear how these effects would interact with expected stimulatory effects of CO_2 on growth and reproduction in grasses.

Timothy grass (*Phleum pratense* L.) is a widespread perennial C_3 grass species used in agriculture and found growing naturally throughout temperate zones of North America and Europe. It produces a single inflorescence per plant with abundant amounts of easily aerosolized pollen, and is a major cause of early summer allergies. The allergens in Timothy grass pollen are similar to, and cross-reactive with, allergens from many other grass taxa; hence Timothy grass pollen extracts are often used in skin testing for broadly diagnosing grass allergy [7], [12]. The processes that affect these proteins in Timothy grass may also be representative of responses in other grass taxa.

In this study we determined the interactive effects of CO_2 and O_3 at current levels (400 ppm CO_2, 30 ppb O_3) and at projected elevated levels (800 ppm CO_2, 80 ppb O_3) in a full factorial design, on both the amount of pollen produced and the concentration of Phl p 5 protein, the major allergen in Timothy grass pollen. Plant exposure to gas treatments was entirely *in vivo* in order to assess whole plant responses. On each plant we measured: number of pollen grains per inflorescence, concentration of Phl p 5 per inflorescence and per pollen grain, inflorescence weight, inflorescence length, and number of inflorescences produced. Using generalized linear model approaches, we examined the effects of the factorial treatments on each parameter, and on the relationships between parameters.

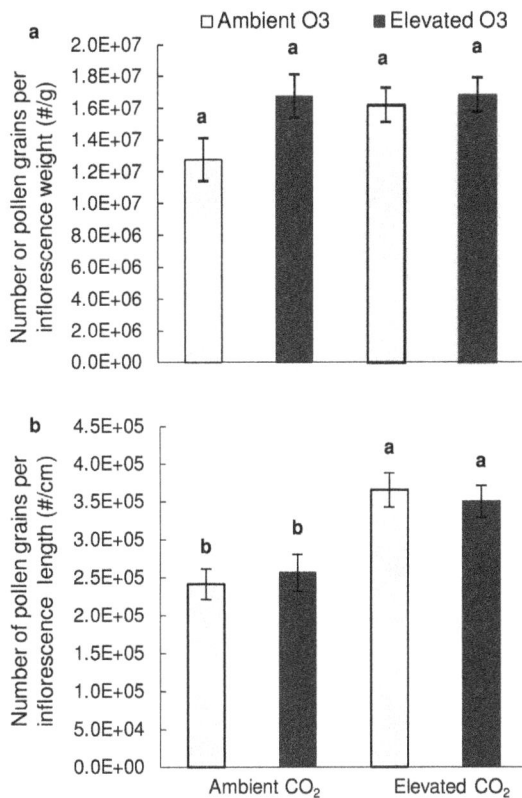

Figure 3. Relationship between pollen amount and inflorescence size. a) Pollen Number per length of inflorescence **b)** Pollen number per gram of inflorescence. Ambient $O_3 = 30$ ppb ozone; Elevated $O_3 = 80$ ppb ozone; Ambient $CO_2 = 400$ ppm carbon dioxide; Elevated $CO_2 = 800$ ppm carbon dioxide. Significant differences (p< 0.05) denoted by letters above bars were determined using the Tukey-Kramer test.

Figure 4. Phl p 5 allergen content. a) Average Phl p 5 concentration per inflorescence **b)** Average Phl p 5 concentration per pollen Ambient $O_3 = 30$ ppb ozone; Elevated $O_3 = 80$ ppb ozone; Ambient $CO_2 = 400$ ppm carbon dioxide; Elevated $CO_2 = 800$ ppm carbon dioxide. Significant differences (p<0.05) denoted by letters above bars determined by the Tukey-Kramer test.

Methods

Seeds of *Phleum pratense* L. var. CLIMAX were sown in 10.5 cm×10.5 cm×35 cm pots (Treepots, Hummert International, Missouri, USA) at the rate of 20 seeds per pot using MM 200 growing medium (SunGro, Washington, USA). At seeding, equal numbers of pots were placed per treatment in continuously-stirred tank reactor (CSTRs) chambers [27] and plants were allowed to germinate and grow through maturity. Grass was fertilized weekly with a dilute concentration of Peat Light Special (1/3 concentration; 15-16-17; Peter's Professional; Scotts, Ohio USA) and watered as needed. Plants were exposed to natural light since chambers were located in a single greenhouse, however light was also supplemented with 400W metal Halide lights (Metal Arc, Sylvania, Massachusetts, USA) which were on 12- hours/day for the first five weeks, then on for 16-hours/day for the remainder of the experiment. Greenhouse temperatures ranged from 15.5°C–26°C. Chambers were monitored for temperature and relative humidity levels using Hoboware data loggers (Onset Computer Corp; Massachusetts, USA).

Eight CSTRs [27] chambers were assigned to factorial treatments with either 30 ppb or 80 ppb ozone and either 400 ppm or 800 ppm carbon dioxide, to simulate present and future projected levels of both gases, respectively. Carbon dioxide treatments were administered continuously while ozone treatments were applied 9:00–16:00 daily to better simulate natural exposure

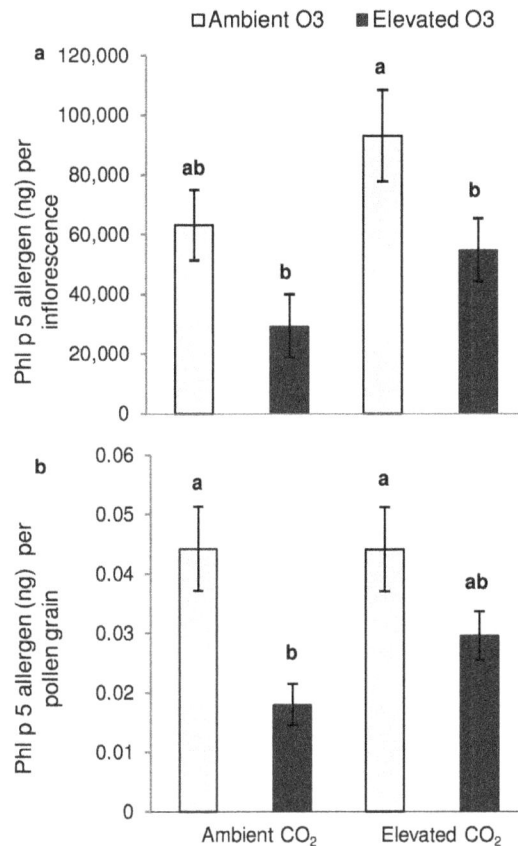

cycles. Experiments were repeated three times for a total of six replications (two replications per experiment); each replication was randomized in chambers and blocked by experiment (two block per experiment) in analysis to account for changes in environmental variables (light, temperature) across repeated experiments.

To capture pollen, polyethylene bags were placed over flowers upon emergence and held open. Bag and flower spike were removed following complete dehiscence of pollen and stored at −20°C until analysis.

Bagged flowers were washed three times with PBS-T (Phosphate Buffered Saline with 0.05% Tween 20 pH 7.4) to remove pollen; the amount of PBS-T was determined based on flower length and was measured precisely using a micro-pipette. Protein was extracted from the pollen by incubating at room temperature in the PBS-T solution for 2 hours. Pollen was separated by centrifugation at RCF 2,200×g. The extracted protein supernatant was stored at −20°C until allergen analysis. The pelleted pollen was suspended in a precisely measured volume of PBS-T solution and counted using a hemocytometer (3 times and averaged).

The Phl p 5 allergen content was determined using a monoclonal sandwich enzyme-linked immunosorbent assay (ELISA) (Indoor Biotechnologies, Inc., Charlottesville, VA, USA) using a mouse monoclonal IgG1 primary antibody to Phl p 5a&b, a 1% bovine serum albumin in PBS-T blocking agent, and a biotinolated mouse IgG1 secondary detection antibody to Phl p

5a&b. Concentrations were determined calorimetrically using streptavidin-peroxidase and substrate 1 mM ABTS (2,2'-azino-di-(3 ethylbenzthiazoline sulfonic acid)) in 70 mM citrate- phosphate buffer (pH 4.2). Serial dilutions of protein extracts were compared to a standard curve to quantify Phl p 5 allergen in each sample.

Analysis of variance of experimental data was used to account for treatment differences and was performed using a generalized linear model in SAS 9.3 (SAS Institute Inc, North Carolina, USA). Differences between treatment means were determined using Tukey- Kramer post-hoc test to account for uneven replications in response to treatments. Linear regressions were performed using analysis of covariance (ANCOVA) to evaluate relationship between co-factors.

Results

Elevated CO_2 (800 ppm) increased the amount of pollen produced by individual inflorescences by ~53% whereas elevated O_3 had no effect on the amount of pollen produced (Figure 1a; File S1). Since each plant only produces one inflorescence, the pollen output per plant was increased. The number of plants flowering per chamber was increased by elevated CO_2, with and without elevated O_3, during the experiment but was not statistically significant (Figure 1b).

Elevated CO_2 did not stimulate an increase in inflorescence length or weight as has been found in ragweed (Figure 2a) [19]. However, our findings suggest that stimulation by CO_2 offsets ozone damage and partially ameliorates O_3–induced reductions in inflorescence size (Figure 2a, 2b). Furthermore, plants grown at elevated CO_2 and O_3 did not differ in the amount of pollen per weight (gram) of inflorescence than the control (Figure 3a). However, inflorescences of plants grown at elevated CO_2 regardless of O_3 level produced significantly more pollen per length of inflorescence (Figure 3b).

Concentrations of Phl p 5 allergen per inflorescence and per pollen grain were not affected by CO_2 levels but were reduced by elevated O_3. (Figure 4a, 4b). Again, elevated CO_2 partially ameliorated this reduction by elevated O_3 (Figure 4b).

Finally, to determine the overall changes in pollen and allergen production between current and projected CO_2 and O_3, we multiplied the average number of inflorescences produced per treatment by the number of pollen grains produced per inflorescence in that treatment (Figure 5a). Linear regression analysis determined a significant increase in pollen production of approximately 200% of current levels of CO_2 to future predicted levels for both O_3 treatments. The increase was only slightly affected by elevated O_3 (202% at ambient O_3, 165% at elevated O_3), indicating that we can expect a large increase in pollen production due to increased CO_2 regardless of future ozone levels.

Likewise, for Phl p 5 allergen levels, we multiplied the concentration of Phl p 5 per inflorescence by the average number of inflorescences per treatment. Linear regression analysis indicated a significant allergen increase at high CO_2 concentrations with a larger 190% increase in Phl p 5 allergen production at ambient O_3 than the 48% increase at elevated O_3 (Figure 5b). Although the rate of increase of Phl p 5 concentration in response to elevated CO_2 was slowed by elevated O_3, the strong CO_2-stimuation of pollen production suggests increased exposure to Timothy grass allergen overall.

Discussion

We provide the first evidence that pollen production is significantly stimulated by elevated carbon dioxide in grasses, a finding that expands earlier work on the annual forb, common ragweed [9], [19] to another important allergenic plant taxonomic

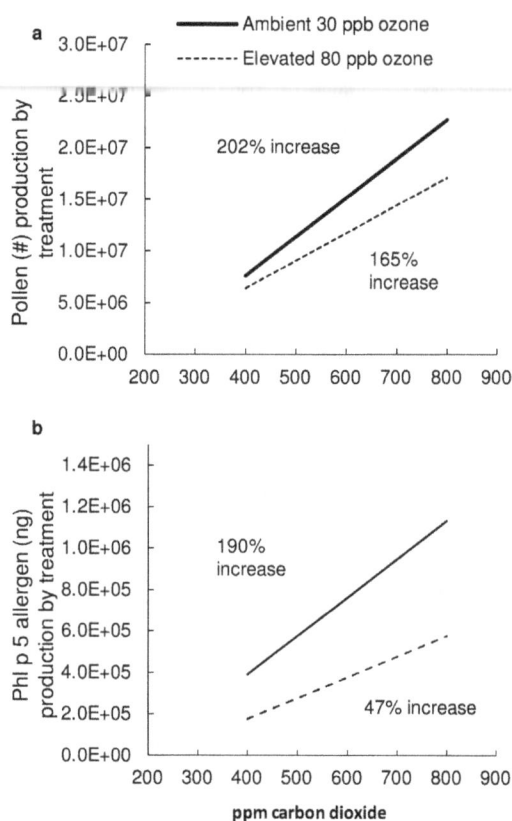

Figure 5. Predicted increases in pollen and allergen. a) Predicted increase in pollen production calculated by multiplying the average number of inflorescences in each treatment by the number of pollen produced per inflorescence. CO_2 by O_3 interaction is not significant indicating equal rates of increase with increasing CO_2 **b)** Predicted increase in Phl p 5 allergen production calculated by multiplying the average number of inflorescences per treatment by the concentration of Phl p 5 per Inflorescence. The CO_2 by O_3 interaction was significant (ANCOVA p<0.05) indicating that elevated O_3 reduced the rate of increase of Phl p 5 in response to elevated CO_2. Solid line (−) determines trend at current ambient level of ozone (30 ppb). Dashed line (- - -) is future elevated level of ozone (80 ppb). Percentage number on graph indicates the percent increase in production with increasing CO_2 levels.

group. The inflorescences produced more pollen without increasing inflorescence size, indicating a greater number of male flowers per inflorescence and/or that male flowers in the inflorescence produced more pollen per anther.

The number of plants flowering per chamber was increased by elevated O_3 or elevated CO_2 during the experiment but was not statistically significant (Figure 1b). However, others have found an increased number of plants flowering in Timothy grass in response to elevated CO_2, while elevated O_3 had no impact on flowering [28]. In our study, the trend toward more plants flowering at elevated CO_2 in Figure 1b indicates, however, that stimulation of flowering in Timothy grass warrants further research. Moreover, rising temperatures related to elevated CO_2 may lengthen the flowering season thereby further increasing atmospheric pollen loads [19], [29], [30].

Biomass allocation in response to elevated carbon dioxide may be an important factor in pollen production [9], [20], [21]. In a concurrent study, biomass allocation was analyzed at different growth stages over the Timothy grass life cycle [31]. It was found that the highest level of biomass stimulation, particularly in the photosynthetic leaf tissue, occurred in the tillering stages leading up to the flowering stage, which could have contributed to increased energy available for pollen production during flowering.

In a previous study, ragweed increased Amb a 1 production, the major allergen in the pollen coat, in response to elevated levels of CO_2 [10]. Ragweed differs in physiology from grasses; for instance, it is an annual plant, a dicot, and can be self-compatible, while grasses are perennial, monocots, and self-incompatible. The physiological importance of these two allergens may differ across these species [32]. Another explanation for lack of stimulation on Phl p 5 production could be that our work was conducted using moderate levels of nitrogen ($\sim$5% N applied in solution once a week) and Phl p 5 protein production was potentially limited by nitrogen availability [17], [33].

Our experiment differed from other work on ozone in that we exposed our plants from seeding through dehiscence as would occur naturally with background O_3 levels, while other experiments used more artificial exposures and durations [23–26]. In our study, the number of pollen grains produced per inflorescence was unaffected by O_3; however, we did not test the viability of the pollen nor quantitate the maturity of the pollen, so we do not know the impacts on its function. There is a chance that our study underestimates the treatment effects on allergen concentration due to a change in recognition of the allergen by the monoclonal antibodies in our ELISA assay. Since ozone is a strong oxidant known to quickly degrade fatty acids and double bonds found in proteins [34], oxidative damage could have altered the Phl p 5 protein resulting in reduced recognition by the mouse IgG antibodies used in ELISA. Rogerieux et al. [26] came to similar conclusions, using IgE antibodies, when looking at multiple individual allergens found in Timothy grass. This degradation of allergen by ozone would also happen in the natural environment impacting recognition of allergens by human immune systems and hence is not exclusively an artifact of the experiment.

The implications of increasing CO_2 for human health are clear. Stimulation of pollen production will increase airborne concentrations and increase exposure and suffering in pollen allergic individuals. The additional effects of increasing ozone are more complex. Ozone alone impacts human health negatively and has been shown to exacerbate the allergic airway response through irritation of the mucus membrane in respiratory pathways [35].

The projected O_3 levels are nearing the current US National Ambient Air Quality standard (80 ppb pre-2008 and 75 ppb post-2008) used to protect human health [36]. Hence elevated levels would likely elicit negative respiratory health effects independent of any additional indirect health effects as a result of elevated CO_2. The IPCC AR5 predicts lower O_3 levels in some areas in the future due to air pollution mitigation strategies to reduce precursors to O_3 and also climate feedbacks that affect O_3 formation [37]. While increasing levels would increase airway irritation by ozone, lower levels of O_3 would mean sustained levels of Phl p 5 allergen with higher Timothy grass pollen production resulting in increased allergen exposure and the potential for further worsening of allergic symptoms.

Conclusion

Taken together, our results strongly suggest significant increases in grass pollen production and allergen exposure under future predicted levels of CO_2 and O_3 of 165–202%. Due to the widespread existence of grasses and their importance in eliciting allergic responses, these results indicate there will be a significant impact on human health worldwide as a result of future climate change.

Acknowledgments

A special thanks to Gaurav Dhawan, MPH and undergraduates Matt Seeley and Erin Stockman who helped with the tedious process of collecting, preparing, and counting millions of pollen grains.

Author Contributions

Conceived and designed the experiments: JMA WJM MD MLM CAR. Performed the experiments: JMA. Analyzed the data: JMA KAS CAR. Contributed reagents/materials/analysis tools: CAR WJM KAS. Contributed to the writing of the manuscript: JMA CAR KAS MLM MD WJM.

References

1. Beggs PJ (2004) Impacts of climate change on aeroallergens: past and future. Clin Exp Allergy 34: 1507–1513.
2. Beggs JP, Bambrick HJ (2005) Is the global rise of asthma an early impact of anthropogenic climate change? Environ Health Perspect 113: 915–919.
3. Cecchi L, D'Amato G, Ayres JG, Galan C, Forastiere F, et al. (2010) Projections on the effects of climate change on allergenic asthma: the contributions of aerobiology. Allergy 651: 1073–1081.
4. Gamble JL, Reid CE, Post E, Sacks J (2008) A Review of the Impacts of Climate Variability and Change on Aeroallergens and their Associated Effects- Final Report of the Global Change Research Program: US EPA.
5. Ziska LH, Epstein PR, Rogers CA (2008) Climate change, aerobiology, and public health in the Northeast United States. Mitig Adapt Strat Glob Change 13: 607–613.
6. Pawankar R, Canonica GW, Holgate ST, Lockey RF (2011) World Allergy Organization White Book on Allergy 2011-2012: Executive Summary. World Allergy Organization.
7. Andersson K, Lidholm J (2003) Characteristics and Immunobiology of Grass Pollen Allergens. Int Arch Allergy Immunol 130: 87–107.
8. Shea KM, Truckner RT, Weber RW, Peden DB (2008) Climate Change and allergic disease. Clin Rev Allergy Immunol 122: 443–451.
9. Ziska LH, Caulfield FA (2000) Rising CO_2 and pollen production of common ragweed (Ambrosia artemisiifolia L.), a known allergy- inducing species; implications for public health. Aust J Plant Physiol 27: 893–898.
10. Singer BD, Ziska LH, Frenz DA, Gebhard DE (2005) Increasing Amb a 1 content in common ragweed (Ambrosia artemisiifolia L.) as a function of rising atmospheric CO_2. Funct Plant Biol 32: 667–670.
11. Ziello C, Sparks TH, Estrella N, Belmonte J, Bergmann KC, et al. (2012). Changes to airborne pollen counts across Europe. PLoS ONE 7(4): e34076.
12. White JF, Bernstein DI (2003) Key pollen allergens in North America. Ann Allergy Asthma Immunol 91: 425–435.
13. Héguy L, Garneau M, Goldberg MS, Raphoz M, Guay F, et al. (2008) Associations between grass and weed pollen and emergency department visits for asthma among children in Montreal. Environmental Research 106: 203–211.
14. Darrow LA, Hess J, Rogers CA, Tolbert PE, Klein M, et al. (2012) Ambient pollen concentrations and emergency department visits for asthma and wheeze. J Allergy Clin Immunol 130: 630–638.
15. Meehl GA, Stocker TF, Collins WD, Friedlingstein P, Gaye AT, et al. 2007. Global climate projections. In: Solomon S, Qin C, Manning M, Chen Z, Marquis M, Averyt KB, et al., Editors. Climate Change 2007: The Physical Science Basis. Contribution of Working Group I to the Fourth Assessment Report of the Intergovernmental Panel on Climate Change. New York: Cambridge University Press, pp. 47–847.
16. Vingarzan R (2004) A review of surface ozone background levels and trends. Atmos Environ 38: 3431–3442.
17. Leakey AD, Ainsworth EA, Bernacchi CJ, Rogers A, Long SP, et al. (2009) Elevated CO_2 effects on plant carbon, nitrogen and water relations: six important lessons form FACE. J Exp Bot 60: 2859–2876.

18. Ainsworth EA, Yendrek CR, Sitch S, Collins WJ, Emberson LD (2012) The effects of tropospheric ozone on net primary productivity and implications for climate change. Annu Review of Plant Biol 63: 637–661.

19. Rogers CA, Wayne PM, Macklin EA, Muilenberg ML, Wagner CJ, et al. (2006) Interaction of the onset of spring and elevated atmospheric CO₂ on ragweed (*Ambrosia artemisiifolia* L.) pollen production. Environ Health Perspect 114: 865–869.

20. Wayne P, Foster S, Connolly J, Bazzaz F, Epstein P (2002) Production of allergenic pollen by ragweed (*Ambrosia artemisiifolia* L.) in increased in CO₂-enriched atmospheres. Ann Allergy Asthma Immunol 88: 279–280.

21. Stinson KA, Bazzaz FA (2006) CO₂ enrichment reduces reproductive dominance in compering stands of *Ambrosia artemisiifolia* L. (common ragweed). Oecologia 147: 155–163.

22. Wand SJE, Midgley GF, Jones MH, Curtis PS (1999) Responses of wild C4 and C3 grass (Poaceae) species to elevated atmospheric CO₂ concentration: a meta-analytic test of current theories and perceptions. Glob Change Biol 5: 723–741.

23. Masuch G, Franz J-Th, Shcoene K, Musken H, Bergmann K-Ch (1997) Ozone increases group 5 allergen content of *Lolium perenne*. Allergy 52: 874–875.

24. Schoene K, Franz J-Th, Masuch G (2004) The effect of ozone on pollen development in *Lolium perenne* L. Environ Pollut 131: 347–354.

25. Motta AC, Marliere M, Peltre G, Sterenberg PA, Lacroix G (2006) Traffic-related air pollutants induce the release of allergen- containing cytoplasmic granules from grass pollen. Int Arch Allergy Immunol 139: 294–298.

26. Rogerieux F, Godfrin D, Senechal H, Motta AC, Marliere M, et al. (2007) Modifications of *Phleum pratense* grass pollen allergens following artificial exposure to gaseous air pollutants (O₃, NO₂, SO₂). Int Arch Allergy Immunol 143: 127–134.

27. Manning WJ, Krupa SV (1992) Experimental methodology for studying the effects of ambient ozone on crops and trees. In: Lefohn AS, editor. Surface Level Ozone Exposures and their Effects on Vegetation. Michigan: Lewis Publishing, pp. 111–119.

28. Johnson BG, Hale BA, Ormrod DP (1996) Carbon dioxide and ozone effects on growth in a legume- grass mixture. J Environ Qual 25: 908–916.

29. Frei T (1998) The effects of climate change in Switzerland 1969–1996 on airborne pollen quantities from hazel, birch, and grass. Grana 37: 172–179.

30. Ziska LH, Knowlton K, Rogers CA, Dalan D, Tierney N, et al. (2011) Recent warming by latitude associated with increase of ragweed pollen in central North America. Proceedings of the National Academy of Sciences 108: 4248–4251.

31. Albertine JM (2013) Understanding the links between Human Health and Climate Change: Agricultural productivity and Allergenic Pollen Production of Timothy Grass (Phleum pratense L.) under future predicted levels of Carbon Dioxide and Ozone. PhD Dissertation, The University of Massachusetts-Amherst. Available: http://scholarworks.umass.edu/cgi/viewcontent.cgi?article=1782&context=open_access_dissertations. Accessed 15 September 2014.

32. Knox RB, Suphioglu C (1996) Pollen allergens: development and function. Sex Plant Reprod 9: 318–323.

33. Townsend AR, Howarth RW, Bazzaz FA, Booth MS, Cleveland CC et al. (2003) Human health effects of a changing global nitrogen cycle. Front Ecol Environ 1: 240–246.

34. Roshchina VV, Roshchina VD (2003) Ozone and Plant Cell. Boston: Kluwer Academic Press, pp. 55–102.

35. Peden DB, Woodrow Setzer R, Devlin RB (1995) Ozone exposure has both a priming effect on allergen-induced responses and an intrinsic inflammatory action in the nasal airways on perennially allergic asthmatics. Am J Respir Crit Care Med 151: 1336–1345.

36. EPA NAAQS (Environmental Protection Agency National Ambient Air Quality Standards). 2014. NAAQS homepage. Available: http://www.epa.gov/air/criteria.html. Accessed 1 May 2014.

37. Kirtman B, Power SB, Adedoyin JA, Boer GJ, Bojariu R, et al. (2013) Near-term Climate Change: Projections and Predictability. In: Stocker TF, Qin D, Plattner G-K, Tignor M, Allen SK, Boschung J, et al., editors. Climate Change 2013: The Physical Science Basis. Contribution of Working Group I to the Fifth Assessment Report of the Intergovernmental Panel on Climate Change. New York: Cambridge University Press, pp. 953–1028.

10

Recent Widespread Tree Growth Decline Despite Increasing Atmospheric CO_2

Lucas C. R. Silva, Madhur Anand*, Mark D. Leithead

Global Ecological Change Laboratory, School of Environmental Sciences, University of Guelph, Guelph, Ontario, Canada

Abstract

Background: The synergetic effects of recent rising atmospheric CO_2 and temperature are expected to favor tree growth in boreal and temperate forests. However, recent dendrochronological studies have shown site-specific unprecedented growth enhancements or declines. The question of whether either of these trends is caused by changes in the atmosphere remains unanswered because dendrochronology alone has not been able to clarify the physiological basis of such trends.

Methodology/Principal Findings: Here we combined standard dendrochronological methods with carbon isotopic analysis to investigate whether atmospheric changes enhanced water use efficiency (WUE) and growth of two deciduous and two coniferous tree species along a 9° latitudinal gradient across temperate and boreal forests in Ontario, Canada. Our results show that although trees have had around 53% increases in WUE over the past century, growth decline (measured as a decrease in basal area increment – BAI) has been the prevalent response in recent decades irrespective of species identity and latitude. Since the 1950s, tree BAI was predominantly negatively correlated with warmer climates and/or positively correlated with precipitation, suggesting warming induced water stress. However, where growth declines were not explained by climate, WUE and BAI were linearly and positively correlated, showing that declines are not always attributable to warming induced stress and additional stressors may exist.

Conclusions: Our results show an unexpected widespread tree growth decline in temperate and boreal forests due to warming induced stress but are also suggestive of additional stressors. Rising atmospheric CO_2 levels during the past century resulted in consistent increases in water use efficiency, but this did not prevent growth decline. These findings challenge current predictions of increasing terrestrial carbon stocks under climate change scenarios.

Editor: Tamara Natasha Romanuk, Dalhousie University, Canada

Funding: This work was funded by Inter-American Institute for Global Change Research, the Natural Sciences and Engineering Council of Canada, the Canadian Foundation for Innovation, and the University of Guelph grants to M.A. who holds the Canada Research Chair in Global Ecological Change. The funders had no role in study design, data collection and analysis, decision to publish, or preparation of the manuscript.

Competing Interests: The authors have declared that no competing interests exist.

* E-mail: manand@uoguelph.ca

Introduction

According to the principles of plant physiology, higher CO_2 concentrations generally increase the ratio between carboxilation and water transpired - water use efficiency (WUE) - enhancing productivity [1,2]. In temperate and boreal forests the synergetic effects of recent changes in climate and rising atmospheric CO_2 are expected to further stimulate primary production [3–6]. Recent dendrochronological studies have identified unprecedented tree growth [7]; however, it is not clear whether this can be attributed to CO_2 fertilization or recent changes in climate. Results from ecosystem level experiments have generally supported the CO_2 fertilization hypothesis [8,9], but they are limited in their ability to assess long-term mechanistic relationships between plants and the atmosphere. This, along with the fact that warming induced stress can cause tree growth decline [10,11], leave the question of how vegetation will respond to current atmospheric changes open for debate.

Combined with isotopic analyses, dendrochronology can be used to address the limitations of the above-mentioned approaches. For example, ratios of carbon isotopes in tree rings ($\delta^{13}C$) allow us to determine atmospheric effects on stomatal conductance and plant gas exchange through time, while tree growth, measured by tree-ring width converted into basal area increment (BAI), is a reliable proxy for total carbon uptake [1,12]. Together these two lines of evidence offer a physiologically based tool useful to decipher the past and to predict future trends of vegetation/atmosphere interactions.

Increases in WUE may be caused by either greater photosynthetic rates (carbon uptake) or by reductions in stomatal conductance and, consequently, lower transpiration [1]. In either case, changes in WUE yield changes in $\delta^{13}C$ of the bulk biomass [12]. Therefore, if there is an ongoing CO_2 fertilization effect, we should be able to detect continuous increments in WUE through changes in $\delta^{13}C$ across tree-ring chronologies. Moreover, along with such increases in WUE an increasingly greater BAI should also be observed. Conversely, if BAI declines while WUE increases, it indicates that any photosynthetic advantage conferred by higher CO_2 concentrations has not been enough to overcome warming-induced stress [13,14]. In these cases, changes in climate should explain growth decline.

Here we determine whether systematic changes in tree growth and WUE have been occurring in temperate and boreal forests in

Ontario, Canada. We sampled mature individuals of two deciduous (red oak - *Quercus rubra* L. and red maple - *Acer rubrum* L.) and two coniferous (black spruce - *Picea mariana* Mill. B. S. P. and red pine - *Pinus resinosa* Ait) species. We determined ring width and annual BAI through standard dendrochronological methods. In mature trees, ring width declines with age; thus, declining growth may be impossible to detect based on changes in ring width alone. The conversion of ring width to BAI overcomes this problem. Unlike width, age-related trends in unstandardized BAI are generally positive and this can be maintained for many decades after trees reach maturity [15]. We estimated changes in WUE by measuring carbon isotope abundances in tree rings [12]. We used multiple regression models to identify significant correlations between BAI, WUE, and climatic variables over the past century.

Results

Trees typically showed a period of early growth suppression before a release phase, which was followed by growth decline (Fig. 1b). Overall, growth patterns for the relatively young trees resembled those of the oldest trees (Fig. 2) and cannot be solely attributed to aging. Exceptions were young deciduous trees in the lowest studied latitudes, where growth decline was not evident. We found significant correlations between BAI and recent changes in climate (temperature and/or precipitation) in most cases (Fig. 2).

We observed that growth of black spruce in the northernmost site and red oak and red maple in the southernmost site was positively correlated with temperature, but we found negative or non-significant growth responses to recent warmer conditions in all other cases (Fig. 2). We only observed continuously positive trends in growth for black spruce in the northernmost studied site, with declines occurring progressively sooner toward lower latitudes (Fig. 1b). We did not observe consistent latitudinal patterns for the other species. When significant correlations were found with precipitation, BAI was generally greater during wetter periods. Red pine and red oak in the northernmost sites were exceptions with BAI negatively correlated with precipitation. The combined effect of temperature and precipitation explained growth patterns only for younger trees growing at the southern end of the latitudinal gradient.

Carbon isotope abundances in the tree rings confirmed that in the vast majority of cases, changes in BAI were significantly correlated with changes in WUE (Fig. 3). We found two types of relationships between WUE and BAI: linear positive and negative polynomial. When BAI responses to temperature were positive or not significant, we found linear correlations between BAI and WUE (Figs. 2 and 3). Conversely, where relationships with temperature were negative and/or precipitation had a positive influence over tree growth, WUE correlations with BAI were non-linear. Due to these different relationships, increasing atmospheric CO_2 levels during the past century did not produce a consistent

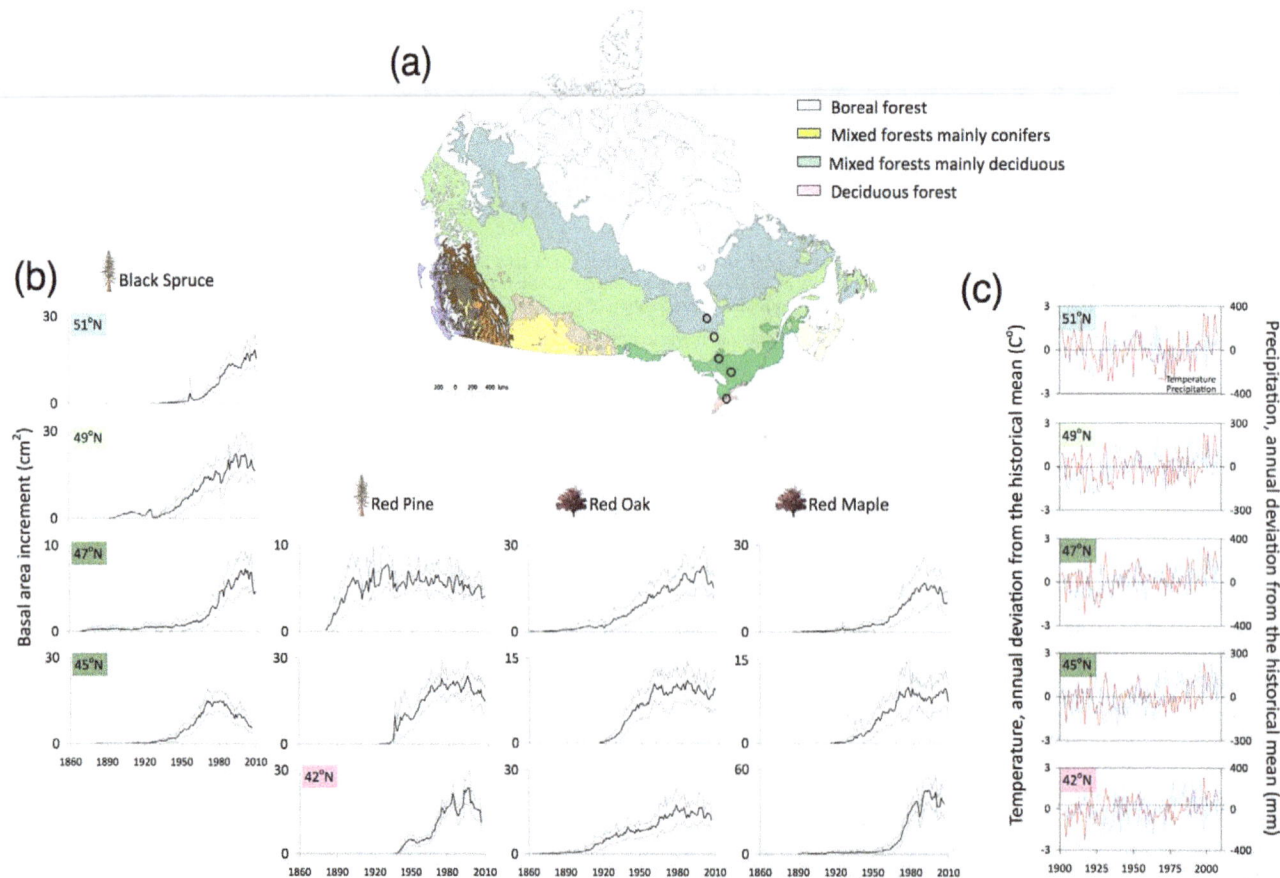

Figure 1. (**a**) Vegetation map of Canada and sampling sites (black circles); (**b**) average annual basal area increment (BAI) for black spruce - *Picea mariana* (Mill.) B. S. P.; red pine - *Pinus resinosa* Ait.; red oak - *Quercus rubra* L. and red maple - *Acer rubrum* L.; (**c**) past century annual temperature and precipitation deviations from the historical average mean (since 1900) for each sampling site. Species were sampled according to their range. Gray lines represent one standard deviation.

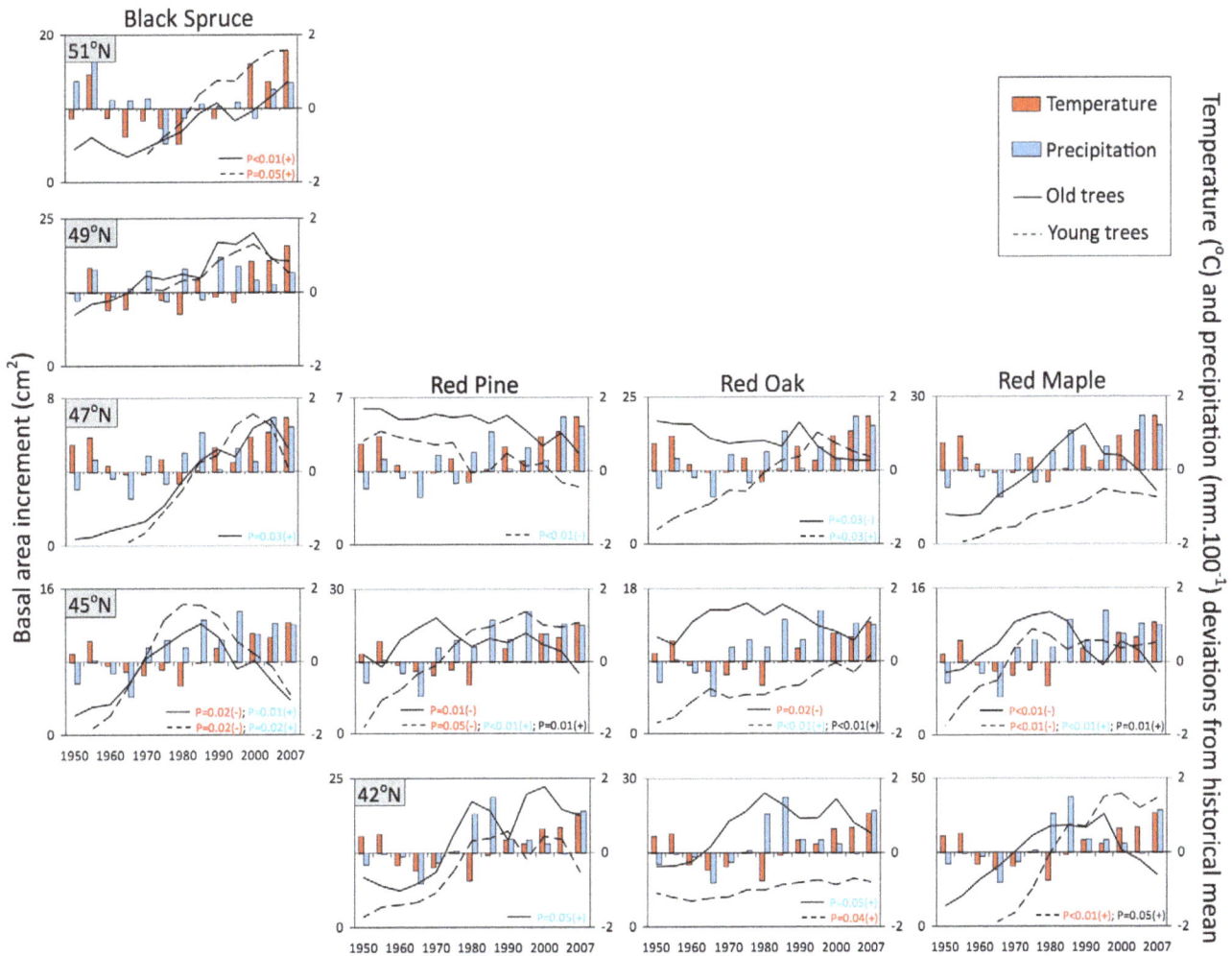

Figure 2. Five years averaged annual mean precipitation and temperature since 1950, deviations from the historical mean (1900–2007) and basal area increment (BAI) for the five youngest (dashed lines) and five oldest (solid lines) trees for black spruce - *Picea mariana* **(Mill.) B. S. P.; red pine -** *Pinus resinosa* **Ait.; red oak -** *Quercus rubra* **L. and red maple -** *Acer rubrum* **L., along a latitudinal gradient in Ontario, Canada.** P values represent significant correlations between BAI and temperature (red), precipitation (blue) or their interaction (black).

increase in BAI despite its positive influence on WUE (Fig. 4). Correlations between WUE and BAI tended to be similar among young and old trees at each studied site confirming that our findings cannot be solely attributed to aging (Fig. 3). Since WUE represents the ratio between carbon assimilated and water transpired, higher WUE could be due to either greater carbon assimilation in response to increased atmospheric CO_2 or lower transpiration in response to water stress. The former would lead to an increase in BAI but the latter would lead to a decrease in BAI. When tree growth was negatively correlated with temperature and/or positively correlated with precipitation, BAI declined in spite of long-term increases in WUE (Fig. 3), suggesting water stress [13,14].

Discussion

We found recent declines in BAI for all study species and latitudes in spite of consistent long-term increases in WUE. Initial tree growth is strongly affected by stand level variations in light, nutrients, water availability and disturbance regimes [15,16]. Accordingly, because site histories are not identical, BAI

developed differently in different sites until the trees reached maturity. The growth release phase observed prior to BAI decline may have been accelerated by past century changes in the atmosphere. However, stability in BAI after growth release is the expected trend for mature trees, which should not show a decrease in BAI until they begin to senesce [15]. Since BAI and WUE trends for the relatively young trees follow the same pattern as that of the oldest trees, the recent growth decline observed here cannot be attributed to aging.

Changes in isotope ratios with ontogeny linked to developmental and microclimatic effects, also known as "age related effects", have received attention in recent studies [12]. Such effects however are usually related to initial tree growth and cannot explain BAI decline in mature trees. Age-related effects might represent part of the WUE increases that we observed here, but ontogenic processes are known to be minor when compared with long-term effects of changes in atmospheric CO_2 and climate [12–14]. Co-occurring ontogenic related influences include: assimilation of $\delta^{13}C$-depleted air near the forest floor [17], changes in irradiance and photosynthetic capacity [1], and changes in the vapor pressure deficit with height in the canopy [18], none of

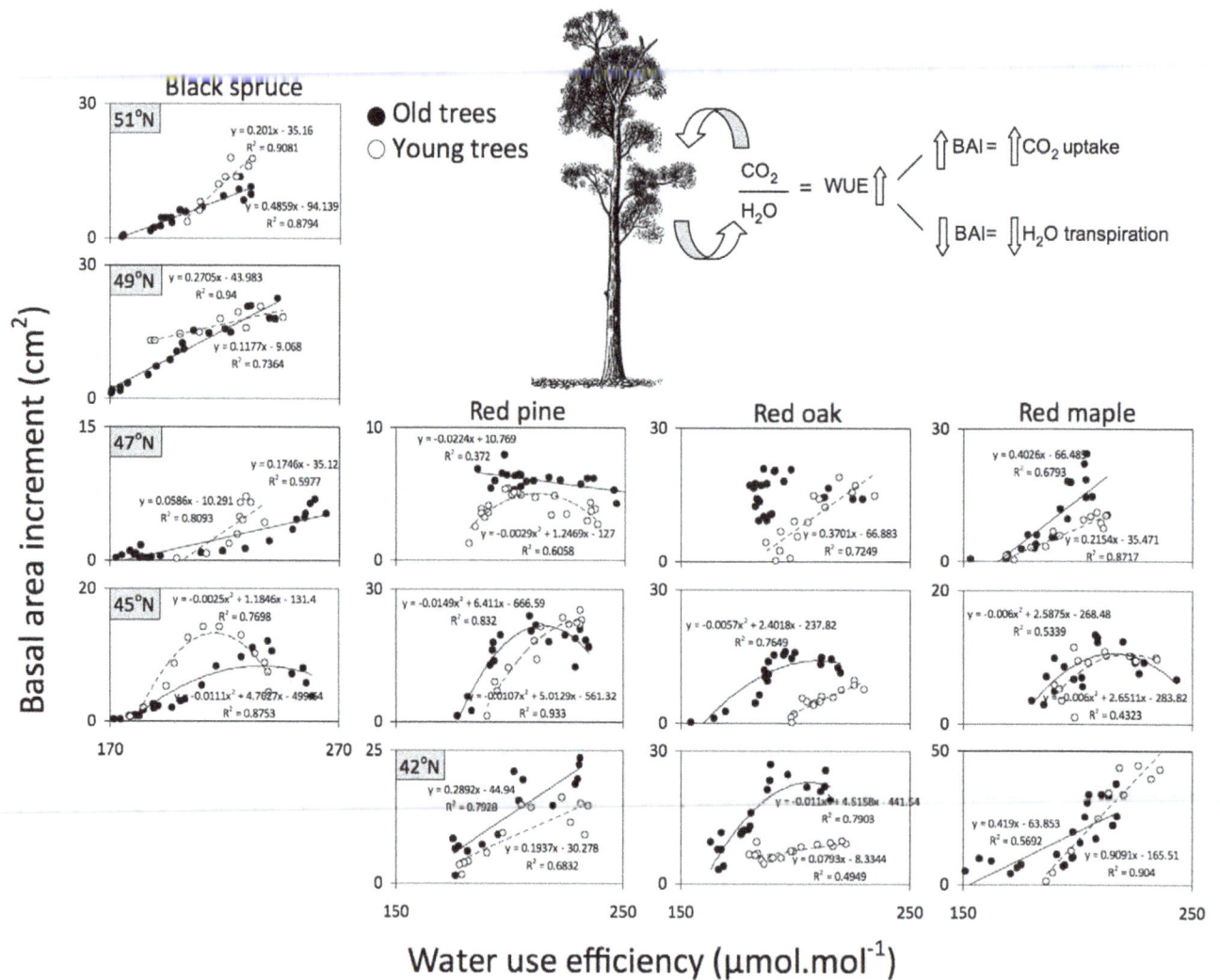

Figure 3. Correlations between water use efficiency (WUE) and basal area increment (BAI) for the five youngest (white circles) and five oldest (black circles) trees during the past century. WUE values were obtained from 5-year pooled rings and BAI values represent 5-year increment averages. Solid lines show significant correlation for old trees and dashed lines significant correlations for young trees (P<0.05). Positive linear relationships correspond to greater carbon uptake, while polynomial functions correspond to water stress.

which can explain a ~53% WUE change observed during the past century. Similar increases in WUE were recently reported for species of tropical, temperate and boreal trees, consistent with theorized CO_2 fertilization effects [13,14,19,20,21]; however, our observations show a recent decrease in BAI in spite of long-term increases in WUE.

It is possible that changes in BAI are not related to CO_2 fertilization effects, but driven by warmer temperatures. Strong increases in tree growth in colder conditions above (but not below) the tree line and growth declines associated with warmer climates at lower altitudes have been recently reported [7,13]. In our study we observed continuously positive trends in growth only for black spruce in the northernmost (coldest) studied site, with the onset of declines occurring progressively sooner (1970 at 43°N) towards lower latitudes. Deciduous species did not show the same latitudinal pattern and tended to have lower decline rates than conifers. Growth of black spruce in the northernmost site and red oak and red maple in the southernmost sites represented the only cases where we did observe significant positive growth in response to recent higher temperature (Fig. 2). Combined effects of

temperature and precipitation were significantly related to growth patterns, but only for young trees growing at low latitudes (Fig 2). BAI declines were not evident for young deciduous trees at the lower end of the latitudinal gradient (Fig 2) and accordingly they represent the outliers observed in Fig 4. If this trend persists, a reconfiguration of species ranges, as previously theorized through northward migration of the temperate/boreal ecotone [9,22,23] may hold true.

In some sites WUE and BAI were linearly and positively related, suggesting that growth decline is not always related to water stress. These were the very same sites where growth trends were not negatively related with warmer conditions and/or positively related with precipitation. In these cases BAI decline would be necessarily linked with reductions in WUE, meaning loss of sensitivity to CO_2. Secondary effects of changes in climate, such as changes in seasonality, snowmelt time and differential growth/climate relationships could explain this loss of sensitivity [11]. In addition, warmer climates may also cause a deceleration in tree growth by increasing rates of respiration. However, since growth is not limited by carbon and acclimation of respiration is likely

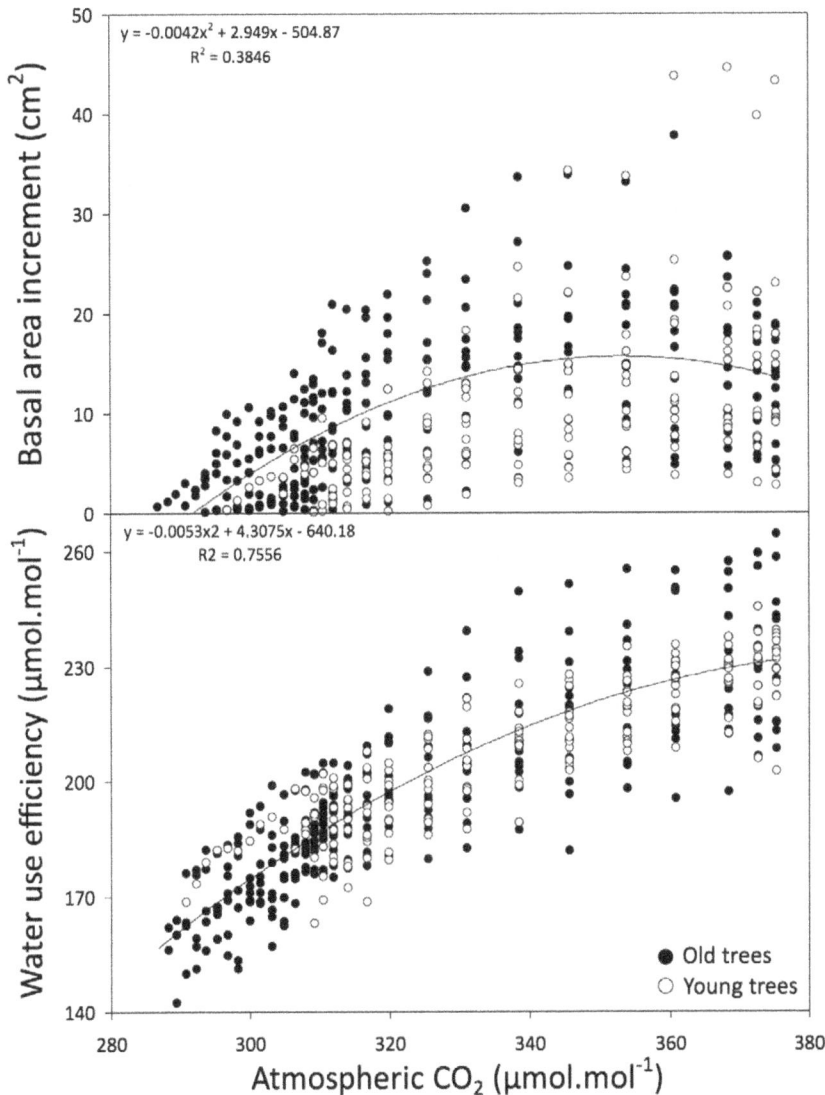

Figure 4. Correlations between atmospheric CO$_2$ and average basal area increment (BAI) and water use efficiency (WUE) for the five youngest (white circles) and five oldest (black circles) trees from all species and sites. Atmospheric concentrations of CO$_2$ summarize actual measurements and estimated values from ice cores in Antarctica from 1850 to the present [12].

occurring, respiration cannot explain growth declines [24]. It is also possible that warming-induced canopy dieback, recently observed in western US temperate forests [25], could produce such effects. However, our sampling criteria focused on healthy trees without signs of canopy damage.

There are several non-climatic explanations for tree decline, some of which we can rule out for our study. Autogenic succession and endogenous increasing light competition are unlikely explanations because we sampled canopy trees and the species studied represent diverse successional groups. Rising tropospheric ozone concentrations have been shown to damage the photosynthetic apparatus, reducing plant sensitivity to CO$_2$ and promoting growth decline, but this would require extremely high concentrations of O$_3$ near the canopy. Additionally, deleterious effects of O$_3$ on carboxilation rates seem to be restricted to young seedlings [26] and O$_3$-induced effects at the leaf level do not necessarily correspond to reduced tree growth or WUE [27].

A possible explanation for growth decline in the cases where we did not find evidence of water stress is progressive nutrient limitation. Recent ecosystem level experiments have demonstrated this process in US temperate forests [8,28]. Photosynthetic capacity is mostly limited by nutrient limitations, which may reduce WUE and BAI [1,24]. Although tree growth is generally consistent with species niche differentiation, soil-plant feedbacks can have a greater influence on growth than species-specific differences at the ecosystem level [29,30] and yield convergent signals such as the ones observed here. Nitrogen has been considered to have a preeminent role in regulating tree growth and other aspects of ecosystems beyond the species level [8,28]. The relative importance however of nitrogen compared to other nutrients may have decreased because of atmospheric deposition and other nutrients (e.g., calcium and phosphorus) appear now to be of equal importance as nitrogen in determining limits to ecosystem productivity [31] and tree growth rates [16,32].

Most of our study sites are located within the Canadian Shield that is characterized by very thin organic soil lying on granite bedrock with many bare rock outcrops formed from glacial retreat. Trees growing on these soils should have higher sensitivity to drought stress and nutrient limitation than on deeper, well-structured soils found elsewhere. For instance, the common subsurface water runoff immediately after rainfall events within the Shield can lead to water stress. Additionally, despite high levels of inorganic nutrients in throughfall, subsurface runoff losses are nearly inexistent, showing that biological demand for these nutrients is greater than that available in soils [33]. Thus, favorable water balance and a rapid nutrient turnover from biomass to labile forms are essential to maintain steady productivity in these environments.

In conclusion, we suggest that either drought stress or nutrient limitation could explain the recent growth decline reported here. Demand for other resources may have a greater effect on tree growth than a positive influence of new atmospheric conditions such as elevated CO_2. Long-term experiments to test soil-plant feedbacks and the effects of atmospheric changes on tree physiology should provide a more in-depth understanding of why expected increases in tree growth are not found. This, in combination with new forest models that take into account multiple stressors will improve our knowledge of forest dynamics as well as our estimates of terrestrial carbon stocks under climate change scenarios.

Materials and Methods

Sampling design

We chose mature stands in preserved areas with no recent history of anthropogenic disturbance. We sampled between twenty and thirty-five living trees of black spruce (*Picea mariana* Mill. B. S. P.), red pine (*Pinus resinosa* Ait.), red oak (*Quercus rubra* L.) and red maple (*Acer rubrum* L.) at five different latitudes according to their range (Fig. 1a). This latitudinal gradient represents the transition between the southern deciduous (Carolinian) forest to the south and the boreal forest to the north in Ontario, Canada. We sampled trees at: Long Point Waterfowl & Wetland Research Station (42°42′N; 80°21′W); Algonquin Provincial Park (45°35′N; 78°21′W); Wolf Lake Forest Reserve and Fairbanks Provincial park (46°47′/51′N; 80°/81°37′/43′W); Esker Lakes Research Station (49°10′N; 80°5 9′W) and a black spruce dominated forest a few kilometers west of Moosonee municipality (51°17′N; 80°38′W).

Analysis of wood cores

We obtained two to four wood cores for each tree with an increment borer (5.1 mm). We mounted the cores on supports where they were air-dried and polished with sandpaper from 60 to 600 grains. We counted and measured the growth rings and dated the cores using the software Windendro. We assigned the calendar age of the growth rings according to Schulman's criteria [34]. In order to check dating accuracy, we compared ring-width time series within trees at each site. This gives a similar pattern of growth in a given population, allowing us to identify annual rings for each tree and to correct primary errors caused by partial or false rings. Dendrochronological measurements were performed at the Climate and Ecosystem Dynamics Research (CEDaR) Laboratory of the University of Guelph.

In mature trees, ring width declines with age and therefore cannot be considered an accurate measure of tree growth [12]. For instance, if a declining growth trend is suspected, it may be impossible to detect it on the basis of changes in width alone. The conversion of ring width into BAI overcomes this problem. Unlike width, age-related trends in unstandardized BAI are generally positive. Therefore mature trees should yield linearly positive trends in BAI over the past few decades [15]. Here, the conversion

of ring width into BAI is shown annually (1 year = 1 full ring, late and early wood), assuming that increment was uniform along each ring:

$$BAI = \pi(R_n^2 - R_{n-1}^2) \qquad (1)$$

where R is the tree radius and n is the year of tree ring formation.

We further investigated all individual trees for changes in WUE as shown by changes in wood carbon isotope composition through time. For isotopic analysis we used the five youngest and five oldest trees of each species at each site, resulting in a total of 130 trees. For each of these trees we pooled rings (5-year blocks) using a microscope and separated them using a thin sharp blade. Combining rings for isotopic analysis is a common practice that allows adequate sample homogenization, yields enough material for the analysis and reduces processing time. We coarsely ground wood samples in a microgrinder until the particles achieved a thickness of 1.5 mm and transferred them to Eppendorf tubes. We took an aliquot from each of the Eppendorf tubes and weighed each on a microbalance (weight range 2–3 mg). We transferred each aliquot to a tin stain capsule (3 mm diameter and 8 mm height, Elemental Micro-analysis, Milan, Italy), sealed by compressing the cup and then stored in a plate ready for analysis. We prepared one standardized aliquot at every 10th aliquot and a "blank" (empty tin cup) every 40th aliquot. We prepared standards in the same way as the wood samples by weighing between 1 and 2 mg of semolina. We placed all samples in an automated elemental analyzer (Euro-EA-Elemental Analyzer, Eurovector, Milan, Italy) connected to a continuous flow isotope ratio mass spectrometer (Isoprime, GV, Manchester, England) generating results with 0.1‰ of accuracy.

To investigate WUE using this type of analysis we rely on the natural existence of both carbon 12 (C^{12}) and carbon 13 (C^{13}) stable isotopes in the atmosphere. The CO_2 molecules contain these isotopes in the proportion of 98.89% C^{12} and 1.11% C^{13}; however in all plant tissues carbon isotope ratios are variable and the C^{13} abundance relative to C^{12} is usually expressed as $\delta^{13}C$:

$$\delta^{13}C(‰) = (R_{sample}/RPDB) - 1) \times 1000 \qquad (2)$$

in which R_{sample} and $RPDB$ represent the $^{13}C/^{12}C$ ratios of the sample and PeeDee international standard respectively [1].

We compiled historic atmospheric $\delta^{13}C$ levels, and estimated values for the recent atmospheric concentration of CO_2, from McCarroll & Loader [12] who summarized high precision records of the long-term atmospheric $\delta^{13}C$ measured from an ice core in Antarctica. We determined the C_i (intercellular CO_2 concentration), eliminating changes in atmospheric isotopic composition effects on plant carbon, using Francey & Farquhar's [1] equation:

$$C_i = Ca((\delta^{13}C_{plant} - \delta^{13}C_{atmosphere} + a/(a-b)) \qquad (3)$$

where $\delta^{13}C_{plant}$ and $\delta^{13}C_{atmosphere}$ are the plant and atmospheric isotopic carbon ratios, respectively, a is the diffusion fractionation across the boundary layer and the stomata (≈4.4‰), b is the RuBisCo enzymatic biologic fractionation (≈27.0‰), C_i and C_a are the internal and external partial pressure of CO_2, respectively.

Carboxilation rates (total carbon uptake) and plant water loss (transpiration) are, like carbon isotope discrimination, controlled by photosynthesis and stomatal conductance (see Fick's first law for further explanations). Since the ratio of carbon fixed to water loss is the definition of water use efficiency, we expressed the carbon ratio values in terms of changes in water use efficiency [12] by the following equation:

$$WUE = A/g + C_a(1 - (C_i/C_a)0.625) \qquad (4)$$

where the intrinsic water use efficiency (WUE) or the rate of CO_2 assimilation by the plant (A) to its stomatal conductance (g) is a function of C_i and C_a, internal and external partial pressure of CO_2, considered that g for CO_2 molecules is 0.625 times g for leaf conductance to water vapor. All the isotopic analyses were performed at the Laboratory of Stable Isotope Ecology (LSIETE) of the University of Miami.

Data treatment and statistical analysis

We calculated the average annual BAI and standard deviation values for each species at each site. Growth patterns were consistent regardless of changes in tree age and/or tree size. The five youngest and five oldest trees of each species at each site showed similar radial growth (BAI) patterns through time. We used multiple regression models and analyses of variance (ANOVAs) to identify significant correlations between BAI (dependent variable) and temperature, precipitation and their interaction (independent variables) since 1950, when most trees had already reached maturity. As typically used to interpret climate data, changes in temperature and precipitation are presented as the deviation from the historical mean (1901–2007) (Figs. 1c, 2). Historical data on precipitation and temperature were

provided by the Ministry of Natural Resources of Canada and can be accessed through the website: http://cfs.nrcan.gc.ca/subsite/ glfc-climate/meansurfaces.

We observed that changes in BAI were linked with changes in WUE and tested the significance ($P<0.05$) of these relationships for the youngest and oldest trees of each species at each site during the past century using least-squares and non-linear regressions. Using non-linear regressions we determined BAI and WUE responses to increasing atmospheric CO_2 for young and old trees for all species across sites. We performed all regressions using 5-year BAI average values to match the WUE values calculated form pooled rings (5-years blocks) as previously described. We performed all statistical analysis using JMP software for Macintosh, version 8.0.

Acknowledgments

We thank Ze'ev Gedalof (director of the Climate and Ecosystem Dynamic Research Lab, University of Guelph), Daniel McKenney and Wayne Bell (Ministry of Natural Resources, Canada) and Andrew Gordon and Paul Sibley (School of Environmental Sciences, University of Guelph) for logistic support and/or valuable comments.

Author Contributions

Conceived and designed the experiments: LS MA. Performed the experiments: LS ML. Analyzed the data: LS ML. Contributed reagents/materials/analysis tools: MA. Wrote the paper: LS MA ML.

References

1. Francey R, Farquhar G (1982) An explanation of 13C/12C variations in tree rings. Nature 297: 28–31.
2. Lamarche VC, Graybill DA, Fritts HC, Rose MR (1984) Increasing atmospheric carbon-dioxide - tree-ring evidence for growth enhancement in natural vegetation. Science 225: 1019–1021.
3. Bonan GB (2008) Forests and climate change: Forcings, feedbacks, and the climate benefits of forests. Science 320: 1444–1449.
4. Huang JG, Bergeron Y, Denneler B, Berninger F, Tardif J (2007) Response of forest trees to increased atmospheric CO2. Critical Reviews in Plant Sciences 26: 265–283.
5. Luo Y, Hui D, Zhang D (2006) Elevated CO2 stimulates net accumulations of carbon and nitrogen in land ecosystems: a meta-analysis. Ecology 87: 53–63.
6. Guiot J, Corona C ESCARSEL members (2010) Growing season temperatures in Europe and climate forcings over the past 1400 years. PLoS ONE 5.
7. Salzer M, Hughes M, Bunn A, Kipfmueller K (2009) Recent unprecedented tree-ring growth in bristlecone pine at the highest elevations and possible causes. Proceedings of the National Academy of Sciences 106: 20348.
8. Finzi A, Moore D, DeLucia E, Lichter J, Hofmockel K, et al. (2006) Progressive nitrogen limitation of ecosystem processes under elevated CO2 in a warm-temperate forest. Ecology 87: 15–25.
9. Soja A, Tchebakova N, French N, Flannigan M, Shugart H, et al. (2007) Climate-induced boreal forest change: Predictions versus current observations. Global and Planetary Change 56: 274–296.
10. Loarie S, Carter B, Hayhoe K, McMahon S, Moe R, et al. (2008) Climate change and the future of California's endemic flora. PLoS ONE 3.
11. D'arrigo R, Wilson R, Liepert B, Cherubini P (2008) On the 'Divergence Problem' in Northern Forests: A review of the tree-ring evidence and possible causes. Global and Planetary Change 60: 289–305.
12. McCarroll D, Loader NJ (2004) Stable isotopes in tree rings. Quaternary Science Reviews 23: 771–801.
13. Peñuelas J, Hunt J, Ogaya R, Jump A (2008) Twentieth century changes of tree-ring 13C at the southern range-edge of Fagus sylvatica: increasing water-use efficiency does not avoid the growth decline induced by warming at low altitudes. Global Change Biology 14: 1076–1088.
14. Silva LCR, Anand M, Oliveira JM, Pillar VD (2009) Past century changes in Araucaria angustifolia (Bertol.) Kuntze water use efficiency and growth in forest and grassland ecosystems of southern Brazil: implications for forest expansion. Global Change Biology 15: 2387–2396.
15. Weiner J, Thomas SC (2001) The nature of tree growth and the "age-related decline in forest productivity". Oikos 94: 374–376.
16. Bigelow S, Canham C (2007) Nutrient limitation of juvenile trees in a northern hardwood forest: calcium and nitrate are preeminent. Forest Ecology and Management 243: 310–319.
17. Schleser G, Jayasekera R (1985) 13 C-variations of leaves in forests as an indication of reassimilated CO2 from the soil. Oecologia 65: 536–542.
18. Sternberg L, Mulkey S, Joseph Wright S (1989) Oxygen isotope ratio stratification in a tropical moist forest. Oecologia 81: 51–56.
19. Hietz P, Wanek W, Dunisch O (2005) Long-term trends in cellulose delta 13 C and water-use efficiency of tropical Cedrela and Swietenia from Brazil. Tree Physiology 25: 745–752.
20. Linares J, Delgado-Huertas A, Julio Camarero J, Merino J, Carreira J (2009) Competition and drought limit the response of water-use efficiency to rising atmospheric carbon dioxide in the Mediterranean fir Abies pinsapo. Oecologia 161: 611–624.
21. Saurer M, Siegwolf R, Schweingruber F (2004) Carbon isotope discrimination indicates improving water-use efficiency of trees in northern Eurasia over the last 100 years. Global Change Biology 10: 2109–2120.
22. McLachlan J, Clark J, Manos P (2005) Molecular indicators of tree migration capacity under rapid climate change. Ecology 86: 2088–2098.
23. Caplat P, Anand M, Bauch C (2008) Interactions between climate change, competition, dispersal and disturbances in a tree migration model. Theoretical Ecology 1: 209–220.
24. Lloyd J, Farquhar G (2008) Effects of rising temperatures and CO2 on the physiology of tropical forest trees. Philosophical Transactions B 363: 1811–1817.
25. van Mantgem PJ, Stephenson NL, Byrne JC, Daniels LD, Franklin JF, et al. (2009) Widespread increase of tree mortality rates in the western United States. Science 323: 521–524.
26. Manning W (2005) Establishing a cause and effect relationship for ambient ozone exposure and tree growth in the forest: progress and an experimental approach. Environmental Pollution 137: 443–454.
27. Novak K, Cherubini P, Saurer M, Fuhrer J, Skelly J, et al. (2007) Ozone air pollution effects on tree-ring growth, delta 13C, visible foliar injury and leaf gas exchange in three ozone-sensitive woody plant species. Tree physiology 27: 941–949.
28. Reich P, Hobbie S, Lee T, Ellsworth D, West J, et al. (2006) Nitrogen limitation constrains sustainability of ecosystem response to CO2. Nature 440: 922–925.
29. DeLuca T, Zackrisson O, Gundale M, Nilsson M (2008) Ecosystem feedbacks and nitrogen fixation in boreal forests. Science 320: 1181.
30. Léotard G, Debout G, Dalecky A, Guillot S, Gaume L, et al. (2009) Range expansion drives dispersal evolution in an equatorial three-species symbiosis. PLoS ONE 4.
31. Jeziorski A, Yan N, Paterson A, DeSellas A, Turner M, et al. (2008) The widespread threat of calcium decline in fresh waters. Science 322: 1374–1377.
32. Gradowski T, Thomas S (2006) Phosphorus limitation of sugar maple growth in central Ontario. Forest Ecology and Management 226: 104–109.
33. Hill A, Kemp W, Buttle J, Goodyear D (1999) Nitrogen chemistry of subsurface storm runoff on forested Canadian Shield hillslopes. Water Resources Research 35: 811–821.
34. Schulman E (1956) Dendroclimatic changes in semiarid America: Tucson, Arizona (Estados Unidos).

11

Effect of Dielectric and Liquid on Plasma Sterilization Using Dielectric Barrier Discharge Plasma

Navya Mastanaiah[1], Judith A. Johnson[1,2], Subrata Roy[1]*

1 Applied Physics Research Group (APRG), Department of Mechanical and Aerospace Engineering, University of Florida, Gainesville, Florida, United States of America,
2 Department of Pathology, Immunology and Laboratory Medicine, College of Medicine and Emerging Pathogens Institute, University of Florida, Gainesville, Florida, United States of America

Abstract

Plasma sterilization offers a faster, less toxic and versatile alternative to conventional sterilization methods. Using a relatively small, low temperature, atmospheric, dielectric barrier discharge surface plasma generator, we achieved ≥ 6 log reduction in concentration of vegetative bacterial and yeast cells within 4 minutes and ≥ 6 log reduction of *Geobacillus stearothermophilus spores* within 20 minutes. Plasma sterilization is influenced by a wide variety of factors. Two factors studied in this particular paper are the effect of using different dielectric substrates and the significance of the amount of liquid on the dielectric surface. Of the two dielectric substrates tested (FR4 and semi-ceramic (SC)), it is noted that the FR4 is more efficient in terms of time taken for complete inactivation. FR4 is more efficient at generating plasma as shown by the intensity of spectral peaks, amount of ozone generated, the power used and the speed of killing vegetative cells. The surface temperature during plasma generation is also higher in the case of FR4. An inoculated FR4 or SC device produces less ozone than the respective clean devices. Temperature studies show that the surface temperatures reached during plasma generation are in the range of 30°C–66°C (for FR4) and 20°C–49°C (for SC). Surface temperatures during plasma generation of inoculated devices are lower than the corresponding temperatures of clean devices. pH studies indicate a slight reduction in pH value due to plasma generation, which implies that while temperature and acidification may play a minor role in DBD plasma sterilization, the presence of the liquid on the dielectric surface hampers sterilization and as the liquid evaporates, sterilization improves.

Editor: Mohammed Yousfi, University Paul Sabatier, France

Funding: The authors are highly grateful to Sestar Medical for their generous financial support in funding this research. Sestar Medical had no role in study design, data collection and analysis, decision to publish or preparation of the manuscript.

Competing Interests: The authors are highly grateful to Sestar medical for their generous financial support in funding this research.

* E-mail: roy@ufl.edu

Introduction

Plasma makes up the majority of the universe. Natural and fabricated plasmas occur over a wide range of pressures, temperatures and electron number densities. Fabricated plasmas are ionized gases, made up of ions, electrons and neutrals. These are commonly categorized based on either temperature or electron number density. In this paper, we are working with low temperature plasmas generated from room air.

Depending on the applied voltage and discharge current, different types of plasma discharges can be obtained [1]. A corona discharge occurs in regions of high electric field near sharp points in gases prior to electrical breakdown. The transition from the Townsend and corona discharge regime to the sub-normal and normal glow discharge regime is accompanied by a decrease in voltage and increase in current. The 'glow discharge' owes its name to the luminous glow seen during plasma generation as seen in Figure 1 below. For the purpose of this paper, this device is denoted as the *'sawtooth electrode'*. The dielectric barrier discharge (DBD) surface plasma, which is the type of plasma discussed in this paper, occurs in the transition between corona and normal glow discharge.

DBD plasmas are a special class of plasmas that operate at pressures of 0.1–10 atm. They are effective ozonizers and are used in a number of additional applications such as surface modification, plasma chemical vapor deposition and most popularly, in large plasma display panels used in television [2]. In its simplest configuration, DBD is the gas-discharge between two electrodes, separated by one or more dielectric layers. Gap between electrodes is typically of the order of millimeters. A broad range of voltages (1–100 kV) and frequencies (50 Hz- 1 MHz) are required to sustain such a discharge. The presence of the dielectric barrier inhibits the transition from glow to arc, thus ensuring stable, non-thermal plasma. DBD plasmas can be classified into two configurations, as shown below in Figure 2.

The volume plasma configuration, shown in Figure 2 (A), consists of two electrodes separated in between by one or more dielectric layers and a discharge gap. Plasma is generated within this discharge gap. Much of the work with plasma sterilization has focused on this type of setup that sterilizes items within a chamber and there has been difficulty in attaining sterilization with air plasmas using this configuration [3]. The surface plasma configuration, shown in Figure 2 (B), differs from the volume plasma configuration, in that there is no discharge gap. Plasma is generated on the surface of the device. This paper focuses on

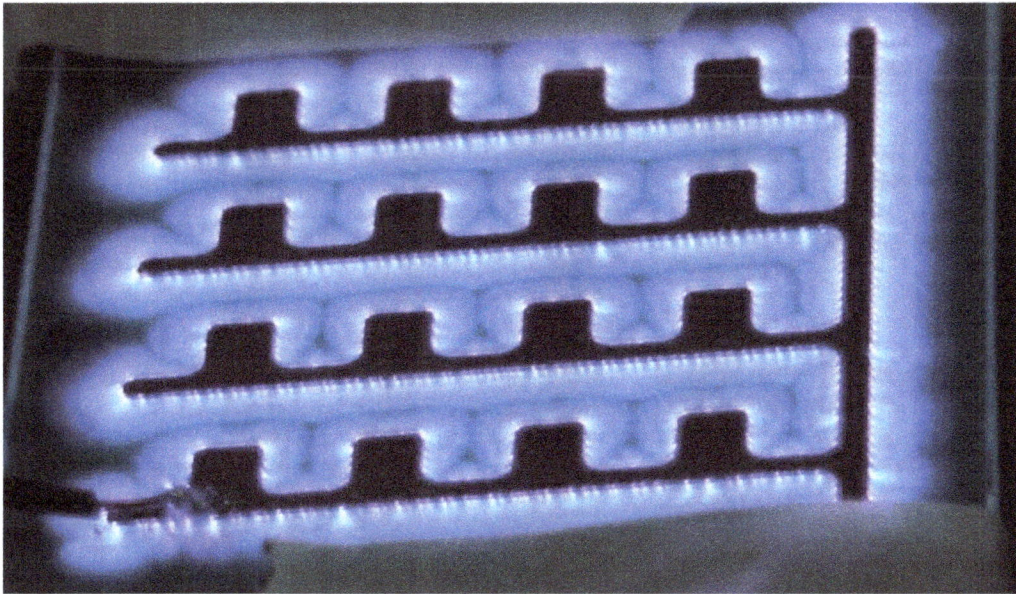

Figure 1. Plasma device (sawtooth electrode) used for earlier experiments in this paper.

surface plasma sterilization for which the discharge configuration consists of a dielectric layer, whose either side is embedded with electrodes such that plasma is generated atop a surface of the dielectric layer. The plasma is concentrated in a thin layer near the surface that sterilizes the surface containing the electrode as well as nearby surfaces. If the electrode is shaped to allow close proximity, this concentration may aide in sterilization.

Sterilization refers to any process that results in the complete elimination or destruction of all living microorganisms. Conventional methods of sterilization such as autoclaving, dry heat, ethylene oxide (EtO) fumigation and γ-irradiation, while established as effective methods, do have their disadvantages, especially damage to heat sensitive polymers, long processing times and/or the need for expensive and potentially dangerous equipment. The ideal sterilant as defined by Moisan et.al. [3] should provide (a)

short sterilization (b) low processing temperatures (c) versatility of operation and (d) harmless operation for patients, operators and materials. Plasma sterilization provides advantages in all these criteria. Very short times have been reported by literature [4–5]. DBD plasma operates almost at room temperature. Plasma generated from ambient air produces a variety of reactive species such as oxygen and nitrogen ions as well as UV photons. Since most of these chemical species disappear milliseconds after the discharge is switched off, they do not leave any toxic residue. However, DBD devices are also known ozone generators, which contribute to killing, and any sterilization setup using these devices must be equipped with measures to control the excessive amount of ozone produced.

The origins of plasma sterilization can be traced back to a patent filed by Menashi [6] in 1968. Research in plasma

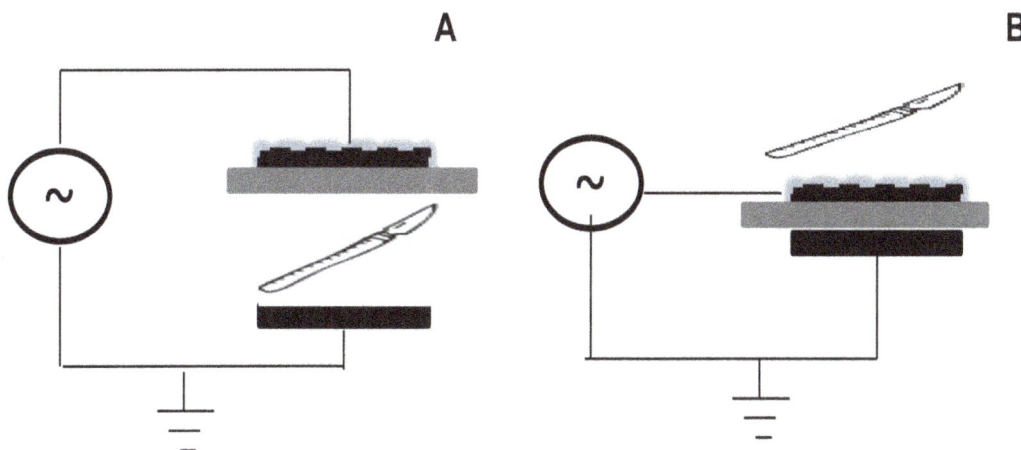

Figure 2. Schematic showing the difference between (A) Volume Plasma Configuration (B) Surface Plasma Configuration. In both (A) and (B), the powered electrode is the dashed, black surface on top. The grounded electrode is the solid, black surface on the bottom. The grey surface in between is the dielectric barrier. In (A), a small scalpel that needs to be sterilized would be placed in the discharge gap between the dielectric barrier and grounded electrode since this is where plasma is generated. In (B), the same small scalpel would be placed on top of the dashed black surface, since this is where the plasma would be generated.

sterilization can be traced along two parallel paths: 1) Determination of the optimum set of parameters for fast, safe plasma sterilization 2) Understanding the underlying mechanism of plasma sterilization. Early research in plasma sterilization was aimed at the former. Hury et.al. [7] conducted a parametric study wherein they tested the destruction efficiency of a 2.45 GHz plasma, generated in a cylindrical reactor using oxygen as the discharge gas, on *Bacillus subtilis* spores. Their studies confirmed previous assertions that oxygen plasmas achieved more killing than argon plasmas, with H_2O_2 and CO_2 plasmas achieving high destruction efficiencies. Lerouge et.al. [8] conducted studies using a large volume microwave plasma reactor, wherein different gas compositions were compared in terms of spore destruction efficiencies. They found that O_2/CF_4 plasma achieved most effective sterilization, due to the combined etching action of both oxygen and fluoride atoms. Moreau et.al. [9] used the flowing afterglow of a 2.45 GHz microwave plasma to inactivate *B. subtilis* spores within 40 minutes.

Most of the papers cited above fall in the low pressure regime. A lot of recent work has been published in the atmospheric pressure regime. This recent body of research ranges in scope from the exploration of new plasma sources for sterilization to usage of diagnostic and microbiological tools to investigate the mechanism of plasma sterilization. Ying et.al. [10] compared yeast inactivation in helium(He), Air and nitrogen (N_2) DBD (volume) plasma at atmospheric pressure. Working at a frequency of 0–20 kHz and an input voltage of 40 kV p-p for a treatment time of 5 minutes, they reported a 5-log reduction with N_2, 6-log reduction with air and 7 log reduction with He. Sladek et.al. [11] reported atmospheric plasma interaction with *S. mutans* biofilms and concluded that a single plasma treatment for 1 minute on biofilms cultured without sucrose caused no re-growth within the observation period. Kalghatgi et.al. [12] took a more fundamental route in assessing the damge due to plasma exposure. They concluded that the effect of plasma ranges from increasing cell proliferation to inducing apoptosis (programmed cell death). Joshi et.al. [13] used anti-oxidants (compounds that protect bacteria from oxidative stress) to prove that when these agents were used to scavenge the reactive oxygen species produced during plasma generation, membrane lipid peroxidation and oxidative DNA damage was significantly inhibited, proving that the ROS causing oxidative DNA damage is a major mechanism involved in DBD plasma sterilization.

In spite of many years of study, plasma sterilization has still not become widely used. Optimization of plasma killing has been difficult due to the complexity of plasma and limited understanding of how it interacts with microbes. Moisan et.al. [3] wrote about the uncertainty concerning the role of UV in the process of plasma sterilization. While earlier experiments [14], mostly conducted in the low-pressure regime, believed that UV (especially in the VUV range (<200 nm)) was a primary factor in sterilization, later experiments conducted at higher pressure suggested that UV radiation was of less importance. Laroussi et.al. [15] used a DBD setup in the volume plasma configuration to record the UV spectrum of air plasma in which they noted there was no significant UV emission below wavelengths of 285 nm. Similarly, Dobrynin et.al. [16] reported experiments wherein they used a quartz filter (transparent to UV photons of >200 nm) during plasma treatment of bacteria They noted from these experiments that there was no visible effect on bacteria by UV/VUV radiation. However, they end with the note that the role of VUV should not be discounted completely.

Dobrynin et.al. also explored the plasma dosage required for bacterial inactivation in cases with and without water. Their results showed that the plasma dosage required for complete bacterial inactivation in cases with water was lower than that required for cases without water. They also concluded from other experiments in the same paper that the presence of water and direct plasma treatment were both required to achieve fast inactivation and this inactivation was highly dependent on the amount of water. Other approaches in understanding plasma sterilization have also included using various protein-detection assays to detect the leakage of a particular protein that might indicate the rupture of the cell wall [17]. Numerical models for plasma sterilization have also been proposed taking into account sterilization times and reaction constants from existing empirical data [5], [18].

This paper describes the implementation of a high-frequency DBD plasma source (operating at 14 kHz and low input power) to sterilize vegetative microbes and spores on a surface with applied electrodes. We compare two different dielectric substrates: FR4 (which is commonly used in manufacturing printed circuit boards and has a mean dielectric constant (ε) of 4.15) and a semi-ceramic laminate (RO3003®, which has a dielectric constant of 3.00 ± 0.04). Plasma is characterized by the spectral signature and ozone levels produced. Further, we evaluate the effect of the liquid portion of the test culture on efficacy of sterilization. These experiments begin the process of optimizing plasma sterilization.

Materials and Methods

Experimental Setup

Figure 3 above shows the schematic of the experimental setup used in plasma generation and testing. A function generator (Agilent ® 33120A) is used to generate a sinusoidal RF signal of frequency 14 kHz. The power of this signal is then amplified using an amplifier (model Crown CDi4000). This amplified signal is then passed through a step-up transformer. The input power from the transformer is fed to the powered electrode (shown in red) of the device via a metal connector. This electrode configuration can also be flipped without affecting sterilization effectiveness, i.e. the red electrode can be grounded and the blue one powered. This allows one to design devices with a reduced risk of electrical shock due to touching of the powered electrode. The powered and grounded electrodes are separated by a sheet of dielectric material, about 1.6 mm thick. In this paper, two types of dielectric material are considered and compared: FR4 (Flame Retardant 4) and Rogers®3003 semi-ceramic (SC) dielectric with $\varepsilon = 3.00 \pm 0.04$. FR4, which is commonly used for making printed circuit boards, has a dielectric constant (ε) of 3.8–4.5 (mean 4.15). Both dielectric sheets are overlaid with a copper layer and are etched out, according to the requisite electrode pattern.

To compare characteristics of the plasma generated using these two dielectric materials, input voltage and current are measured using an Agilent ® DSO1004 Oscilloscope and a current probe. The final input signal into the plasma device has a power $\sim$7–10 W and an input voltage of 12 kV peak-to-peak (p-p). The other electrode (shown in blue) of the device is connected to an electrically grounded bench, atop which the device sits. Before any experiment, the experimental bench was swabbed with 70% proof ethyl or isopropyl alcohol and allowed to completely dry so that a clean testing environment was maintained. Following sterilization trials, plasma devices were removed from the bench and placed in sample bags for microbiological testing.

The plasma device consists of a dielectric square that measures 3.5×3.5 cm^2. This dielectric is embedded with the bottom and top electrodes (on either side). The bottom (grounded) electrode is a square sheet of metal, measuring 2.4×2.4 cm^2. The powered electrode has a *comb-like* design that has the same surface area as

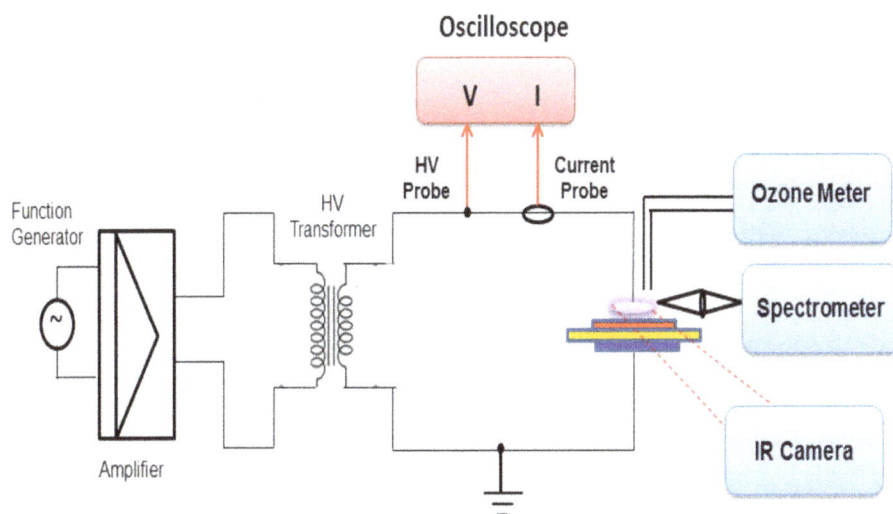

Figure 3. Schematic of the Experimental Setup used.

the grounded electrode. Earlier configurations of these plasma devices also incorporated a *sawtooth* design for the powered electrode.

Two electrode configurations are used in this paper. Figure 4(A) above shows the type of device that has been used for most of the experiments in this paper. For the purpose of this paper, this device is denoted as the *'comb electrode'*. Figure 4(B) shows the same device; powered and producing glow discharge. Note that the plasma seen in this figure covers the entire electrode surface area, contrary to the *'sawtooth electrode'* show in Figure 1. In the *sawtooth electrode*, parts of the electrode surface area are enveloped by the plasma glow and parts of it (the electrodes themselves) are not. The significance of the plasma glow enveloping the entire electrode surface area will be described later while discussing the sterilization curves.

The spectroscopic signature of the generated DBD plasma is determined using the Ocean Optics® USB 2000+ spectrometer. This spectrometer has a detector range of 200–1100 nm, an optical resolution of ~0.3–10 nm (FWHM), a dynamic range of 1300:1 for a single scan and is fitted with a custom-made grating designed to be sensitive to wavelengths between 200–650 nm. An uncoated UV Fused Silica Plano-Convex Lens ($\Phi 2''$, f = 75 mm) is used to collect and focus the incident plasma glow from the plasma device, which is then detected by the spectrometer via a fiber-optic probe. Baseline data for each device was collected with the device powered for 2 minutes while its spectral signature was recorded every 10s. Readings were also taken during sterilization experiments.

A 2B Tech® Ozone meter is used to measure the ozone levels at fixed time intervals within a closed chamber. This ozone meter operates on the principle that the maximum absorption of ozone

Figure 4. The plasma device used for sterilization experiments in this paper (A) un-powered (B) powered (generating plasma).

takes place at 254 nm. Air is sampled every 10s and the sampled ozone levels are saved to a computer via a LabView ® Interface. Surface temperature of the dielectric substrate during plasma generation is measured using an infrared camera (FLIR A320®). The A320 operates at a spectral range of 7.5–13 µm and has a pixel resolution of 320×240 pixels. The distance between the plasma device and infrared camera, ambient temperature and humidity and the emissivity of the FR4, SC dielectric are measured to be 0.2667±0.0127 m, 24.4±2.3°C, 59±3% RH and 0.9097±0.03, 0.929±0.03 respectively.

Microbiological Testing

Cultures were maintained frozen at −80°C in broth with 25% glycerol and inoculated onto fresh plates weekly. *Saccharomyces cerevisiae* (Fleischmann's Baker's Yeast) was grown on Sabouroud's (SAB) agar at 30°C overnight. A single colony was inoculated into SAB broth and incubated at 30°C with shaking. *Escherichia coli C600* was grown on Luria-Bertani (LB) agar or broth at 37°C. Purified *Geobacillus stearothermophilus* spores (SCM Biotech, Bozeman, MT) at $3.1×10^7$ spores/ml, in alcohol, were stored at −20°C and grown on trypticase soy (T-soy) agar or broth at 50°C. Before each experiment, the optical density (OD) of the microbial sample was measured using an Ultrospec 10 cell density meter (GE Healthcare Bio-Sciences Corp., Piscataway, NJ) to estimate the density of the culture. An OD of 1 corresponds to approximately $5×10^8$ colony forming units (CFU) for *E. coli* and $1×10^8$ CFU for *S. cerevisiae*. Cultures were diluted if needed to ensure that approximately 10^6 CFU were inoculated on the device.

For each experiment, the plasma devices were inoculated with 20 µl of the bacterial sample (unless otherwise noted) spread uniformly over the entire surface area of the top electrode, using a sterile inoculating loop. A separate plasma device was used for each time-point. After each experiment, plasma devices were either autoclaved (for spore experiments) or disinfected with 70% ethyl alcohol and sealed in sterile bags. Once the experiment was completed, each tested device was deposited in a sterile bag with 5 ml of appropriate culture broth. The bag was sealed and agitated thoroughly using a Fisher Scientific ® Mini Vortexer Lab Mixer to detach any microorganisms clinging to the device. Serial dilutions were spread on appropriate plates that were then incubated at the required temperature for 24–48 hours and counted. Plate counts were also performed on the inoculum to determine the exact concentration of organisms and an inoculated device not exposed to plasma was processed to control for loss of viable counts due to drying or adherence to the device. Experiments were performed in triplicate unless otherwise noted.

Results

Sterilization Curves Using Several Vegetative Microbes and Spores as Test Pathogens

In testing out our in-house DBD plasma sterilization setup, baker's yeast (*S. cerevisiae*) was used for preliminary tests of killing of vegetative cells by plasma. The sterilization plots from these trials are shown below in Figure 5(A). These tests were conducted using an input frequency of 14 kHz and an input voltage of 12 kVp-p for plasma generation. The DBD devices used had the *sawtooth electrode* configuration, shown in Figure 1.

While these tests showed a 5 log reduction in 75s, subsequent tests started yielding inconsistent data. It was soon realized that the electrode configuration was at fault. As is seen in Figure 1, when the device is powered, the electrode itself is not covered by plasma (enveloped by the bluish glow) unlike the rest of the dielectric surface. This led to contaminated areas atop the electrode surface

that did not seem to be completely sterilized by the generated plasma as was confirmed by placing the electrode, inoculated surface face down on a SAB agar plate following plasma exposure. Colonies only grew on the area covered by the electrode (data not shown). Hence, it was realized that thinner electrodes placed with an optimum gap in between led to uniform plasma coverage over a surface, thus helping uniform sterilization. This led to the second electrode configuration (*comb-like*) shown in Figure 4, where the generated plasma is seen enveloping the entire electrode and dielectric surface. Using the new electrode configuration, consistent sterilization times of 60s–90s were observed in the case of yeast (as shown above in Figure 5(B)).

To test for killing of Gram-negative vegetative bacterial cells, *E. coli C600* was plasma treated for different time intervals and the sterilization plot is shown in Figure 6(A) below. 10^7 cfu were killed within 90s using the FR4 dielectric plasma devices. The same sterilization trials using the semi-ceramic (SC) dielectric plasma devices resulted in 10^7 cfu being killed within 120s. This result is also shown below in Figure 6(B). Both sterilization plots show a phase with little or no loss of viability followed by a rapid killing of the test sample.

Spores are tough, dormant, non-reproductive organisms produced by some bacteria as a survival mechanism when threatened by harsh conditions. Sterilization, by definition, requires the ability to kill bacterial spores. Purified *G. stearothermophilus* spores were used as the spore challenge. In this case, the inoculation volume used was 40 µl. As shown in Figure 7 below, the sterilization curve for spores shows a triphasic pattern with a 2-log reduction in the first 5 minutes, a slow killing period, and complete inactivation (6-log reduction) within 20 minutes.

Spectroscopic Studies

Spectral signatures of the devices were recorded before and during sterilization experiments. For the spectral signatures obtained in Figure 8 below, both clean and inoculated FR4 plasma devices were powered for a total of 2 minutes.

Intensity peaks observed at particular wavelengths can be compared to existing literature [19] in order to identify the molecular species responsible for the respective intensity peaks. Two dominant intensity peaks are observed at wavelengths 337.13 nm and 357.7 nm. Both correspond to the 2nd positive system of N_2 ($C^3\Pi_u$-$B^3\Pi_g$). No intensity peaks are noted at wavelengths characteristic of O_2 or O_3 molecules. Laroussi et.al. [15] noted a similar result, using DBD plasma in volume configuration. Since the spectral signature shows no noticeable wavelengths below 290 nm, it is unlikely that shortwave UV radiation (200–300 nm) plays a major role in surface DBD plasma sterilization.

From Figure 8 above, it is seen that the peak intensities occur in the wavelength range 300–500 nm. Figure 9 below is an expanded version of Figure 8, focusing on the spectral signature in the range 330–350 nm.

In Figure 9(A) above, the spectral data sampled at 30s, 60s, 90s and 120s during the 2-minute interval is shown. All four plots show similar peaks i.e. Plasma generated using a clean FR4 device for a 2-minute time interval, shows similar intensity values at all times. However, it is evident in Figure 9(B) that the intensity value at 30s is less than that at 120s. As time progresses, it is also observed that the amount of liquid bacterial sample on the inoculated device decreases and intensity increases. This dependence of spectroscopic intensity on amount of liquid bacterial sample is discussed in detail later on.

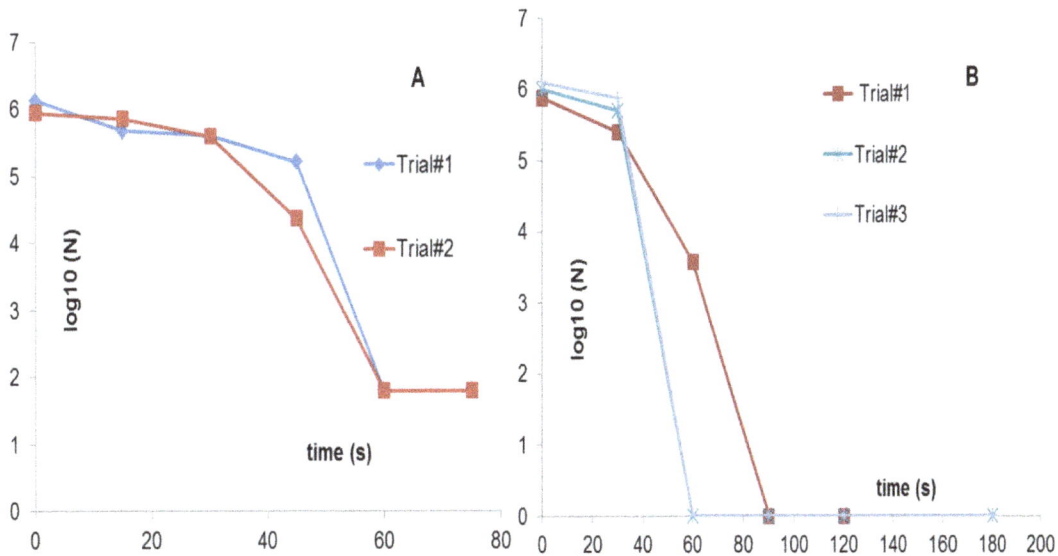

Figure 5. Sterilization plots obtained using *S. cerevisiae* (yeast) with (A) sawtooth electrode (B) comb-like electrode. Earlier sterilization trials were conducted using the sawtooth configuration. However, with extended usage, the sawtooth configuration began posing a problem, which is why a new comb-like electrode was designed and implemented.

Ozone Studies

Ozone is an effective bactericidal agent and may play a role in the sterilization process. It is also a respiratory irritant that must be controlled to protect the device operator. Thus, it is important to understand how much ozone is produced and how fast it dissipates. With such a high amount of ozone produced, it is highly important that we understand factors such as the rate of production/dissipation of ozone, its dependence on different dielectrics as well as necessary precautions to be taken for safe operation of these devices.

In order to do this, a 2B Tech ® Ozone meter was used to measure the ozone levels while the plasma devices (both clean and inoculated) were powered for 1 minute. The plasma device and ozone meter were set up in an acrylic enclosure to allow accurate measurement of the concentration of ozone. Enclosure sizes of varying volumes were tested in order to understand how the volume of the enclosure affected the concentrations of the emitted ozone and its subsequent diffusion and breakdown. The volume of the smallest enclosure was 840 in³. The ozone probe was placed 6.5″ above the chamber floor and ~1″ away from the device. The ratio of the volumes of enclosures #2,3,4 w.r.t to the smallest enclosure (#1) was 2:4:32. Figure 10 below shows this dependence. The X-axis denotes volume of the enclosure (in³) while the Y-axis denotes ozone levels (ppm).

From Figure 10, the following observations are made. The clean FR4 device generates the maximum amount of ozone, followed by the clean SC device (~28% less). Both the inoculated FR4 and SC device generate considerably lesser amounts of ozone than their

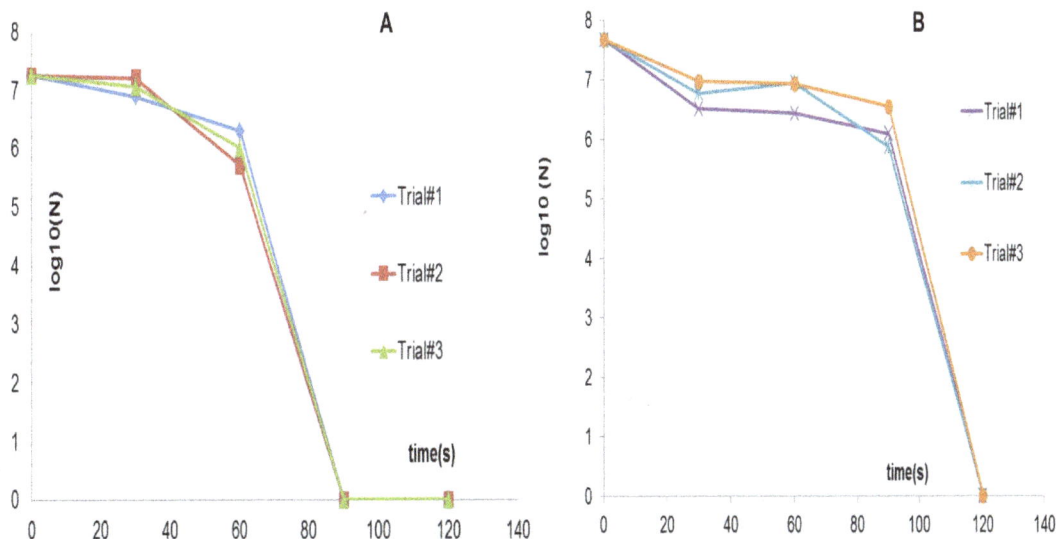

Figure 6. Sterilization plots obtained using *E. coli* as the test pathogen using (A) FR4 dielectric (B) semi-ceramic (SC) dielectric. The former achieves complete sterilization, starting from an initial concentration of 10⁷ cfu in 90s, while the latter achieves the same in 120 s.

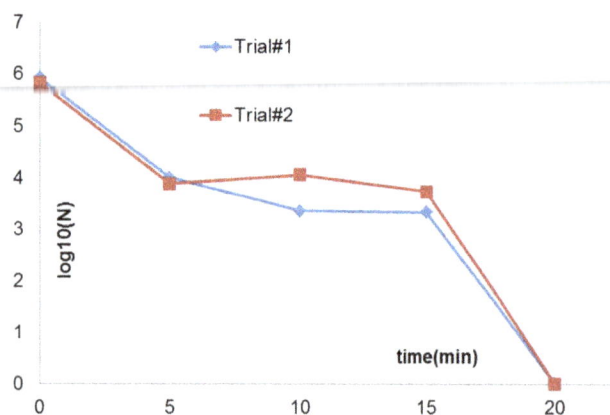

Figure 7. Sterilization plots using *G. stearothermophilus* as the test pathogen. FR4 dielectric was used for these tests. Complete sterilization, starting from an initial concentration of 10^6 cfu was obtained in 20 min.

clean counterparts (~ 70% higher) do. Figure 11 given below depicts this difference in ozone production between a clean and inoculated device. For the sake of simplicity, this difference is shown only for the FR4 dielectric.

Two observations are made from Figure 11 (A) and (B) above. The first observation reinforces that made in Figure 10 i.e. as the volume of the enclosure decreases, the ozone concentration increases. This is because as the volume of the enclosure decreases, the concentration of ozone confined within the enclosure increases. Secondly, the large difference in ozone concentrations between a clean and an inoculated FR4 device is to be noted. In each enclosure, the maximum ozone level noted is about 20%–60% more in the clean case as compared to the inoculated case. This seems to be proportional to the amount of liquid present on the surface of the device. For an inoculated device, while plasma is generated, initially very low levels of ozone are produced. As the liquid evaporates, the amount of ozone produced increases. As

with spectroscopic intensity, this dependence of produced ozone on the amount of liquid sample presents provides a significant insight into the plasma sterilization process and will be discussed later on.

Power Measurements

To better understand the mechanism of plasma sterilization, it was necessary to measure the input power being fed into both clean and inoculated FR4 and semi-ceramic (SC) devices. An experiment was performed in which an inoculated device was powered for 2 minutes. The input power to the device was measured every 15s, using the Agilent ® DSO1004 Oscilloscope and a current probe. Figure 12 below gives this plot of the power varying over time, both for the FR4 as well as semi-ceramic (SC) devices.

In Figure 12, the power varies between 10–12 W for a clean FR4 device. The average power measured over this 2 min interval is 9.67 W. Similarly, for a clean SC device, the power varies between 6–7 W, with an average measured power of 5.8 W over 2 minutes. However, for the inoculated FR4 and SC device, it is observed that the input power follows a steadily increasing trend, starting from ~2 W and gradually increasing to the input power values noted in the case of the clean FR4 or SC devices.

Temperature Measurements

A FLIR A320 ® Infrared camera was used to measure the substrate surface temperature during plasma generation. The infrared camera uses an uncooled micro-bolometer to detect infrared energy (heat) and converts it into an electronic signal, which is then processed to produce a thermal image that can be processed to obtain surface temperature.

In order to obtain the thermographic image of each plasma device, while it was being operated, the plasma device was powered for 2 minutes, during which thermographic images of the plasma device were obtained by the infrared camera at a sampling rate of 0.5 Hz. After turning off the plasma device, the camera continued to record images for another 2 minutes, thus yielding 48 frames. These images were transferred in real-time to a computer, wherein they were subjected to additional data processing.

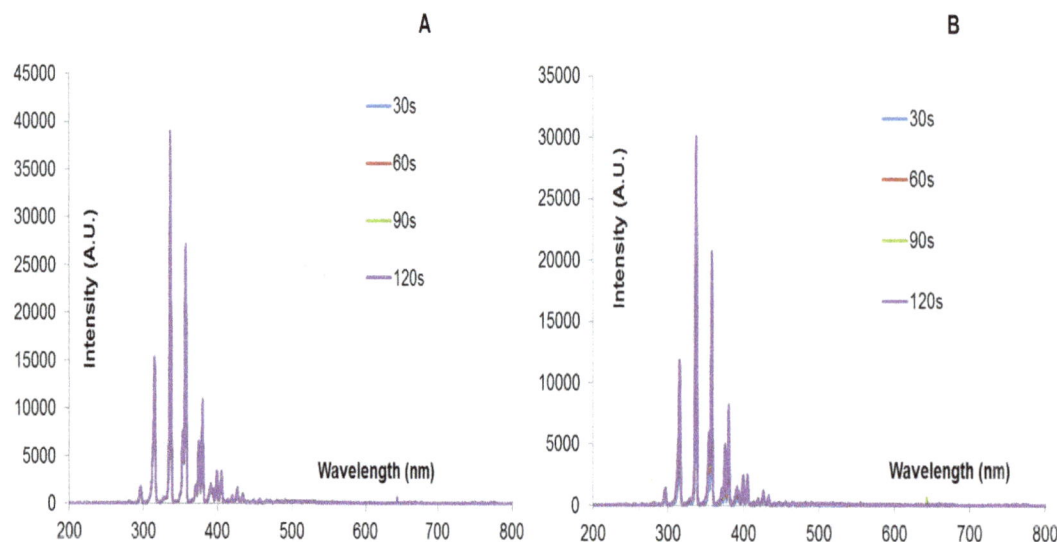

Figure 8. Spectral signature of (A) a clean FR4 device (B) an inoculated FR4 device i.e. a device on which 20 μl of *E. coli* was applied uniformly. Spectral signature was recorded every 10s. Devices were powered for a total of 2 minutes. Only spectra at 30s, 60s, 90s and 120s are shown. Y-axis lists emission intensity in arbitrary unit.

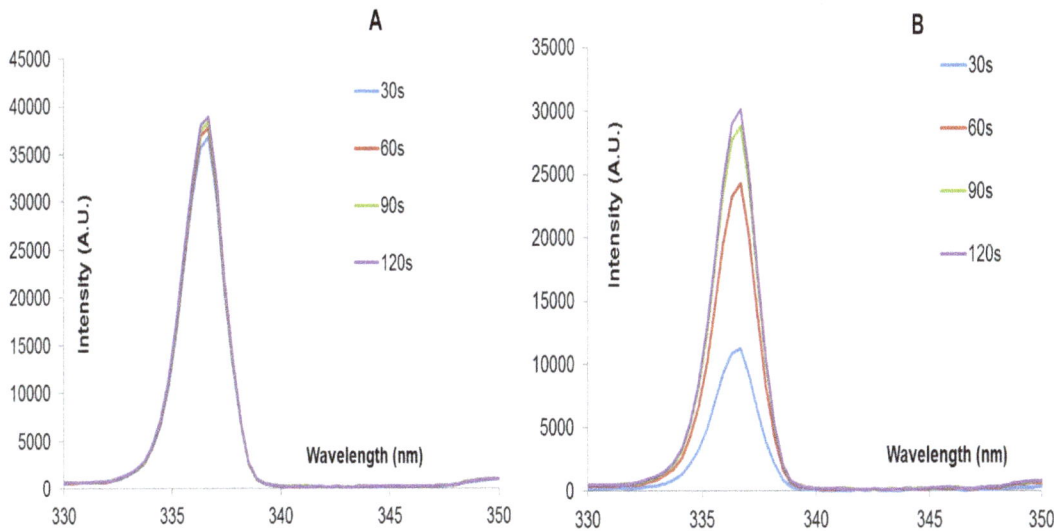

Figure 9. Expanded version of the spectral signature of (A) a clean FR4 device (B) an inoculated FR4 device. This is an expanded image of Figure 8, wherein, the intensity peak in the wavelength range 330–350 nm is depicted to highlight the intensity difference between the clean and inoculated case at different sampling times (30s, 60s, 90s, and 120s).

Figure 13 below shows the comparison of substrate temperature for FR4 and SC dielectrics run clean or inoculated with *E. coli*. In order to compare substrate temperatures, the average temperature over the entire surface area of the substrate was calculated for each frame. This is then plotted against time (s). The average of three sets of data has been plotted in Figure 13. Note that the camera starts recording 5s after the plasma is turned on.

In Figure 13 above, it is evident that during plasma generation the FR4 surface is at a much higher average temperature than the SC surface in both the clean and inoculated cases. However, average surface temperatures range between 30°C–66°C (for FR4) and 20°C–49°C (for SC). Ayan et. al. [20] evaluated the heating

effect of DBD plasma, in which they measured surface temperatures of 310–350 K (36.8°C –76.8°C) and rotational temperatures (gas temperatures) of 340–360 K (66.8°C–86.8°C), using both sinusoidal and microsecond pulsed discharge. Their results indicated that in both types of discharge, while the rotational (gas) temperature was lower, the vibrational temperature was an order of magnitude higher than the rotational temperature, thus probably enhancing chemistry and leading to sterilization. Our measured surface temperatures are cooler than those of Ayan et. al. and thus are less likely to contribute to microbial killing, although temperatures of 56°C are sufficient to denature the 30S ribosomal subunit [21].

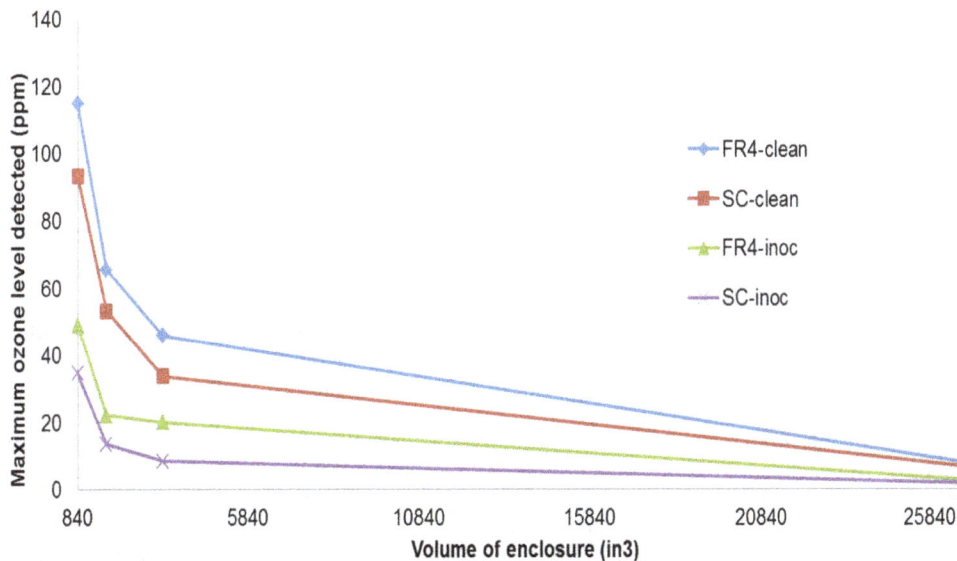

Figure 10. Variation of the maximum ozone levels w.r.t the volume of the acrylic enclosure. Four plots are shown. Each plot shows the maximum ozone level noted in each enclosure for the respective device. "FR4-clean" shows this plot for a clean FR4 device, generating plasma for 1 minute. "FR4-inoc" shows this plot for an inoculated FR4 device, generating plasma for 1 minute. "SC-clean" shows this plot for a clean semi-ceramic (SC) device, generating plasma for 1 minute. "SC-inoc" shows this plot for an inoculated semi-ceramic (SC) device generating plasma for 1 minute.

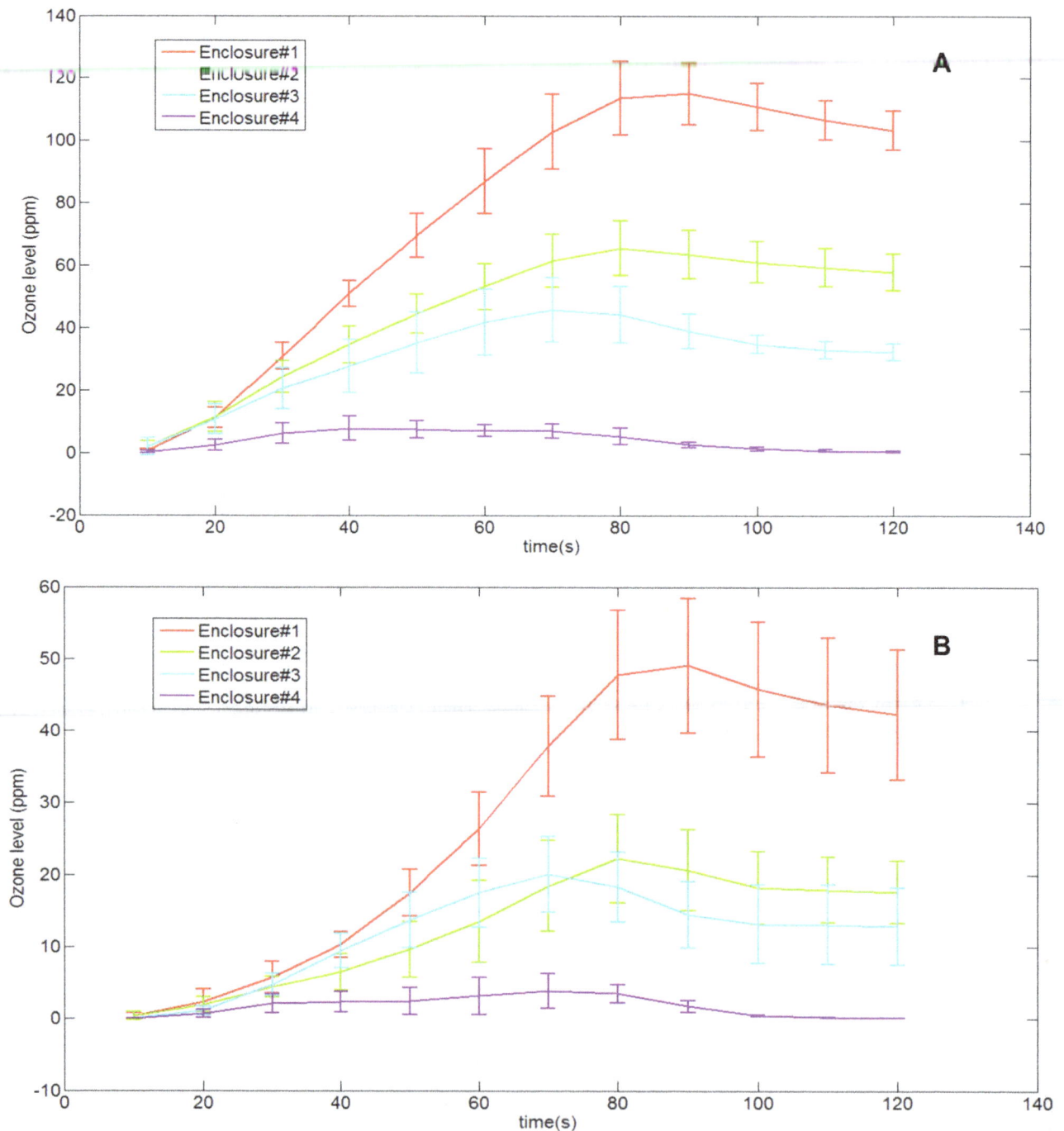

Figure 11. Depicting the trend of ozone production and dissipation for four different acrylic enclosures of different volumes using A) Clean FR4 devices (B) Inoculated FR4 devices. These devices were powered for 1 minute. Ozone data is sampled every 10s.

From Figure 13, it is also observed that while the standard deviation of the temperature measurements in the clean FR4 and SC cases is minimal, it is slightly higher in the case of temperature measurements in the inoculated cases. This standard deviation is greater during the first 60s for inoculated FR4, while for SC, it is greater during 60–105s. The significance of this will be discussed later on while discussing the effect of the liquid on the dielectric surface. It is to be noted that the slightly large variation in standard deviation for inoculated SC, during the latter part of the curve

after plasma is turned off, stems from the fact that for one of the temperature data-sets, the plasma was turned off at $120 \pm 10s$.

Thus, from the results above, it is observed that the sterilization plots (Figures 5, 6, 7) prove that the DBD surface plasma experimental setup can completely sterilize vegetative pathogens in 2–3 minutes, starting from an initial concentration of 10^7 cfu. Plasma treatment using the setup can also lead to complete inactivation in bacterial spores in 20 minutes, starting from an initial concentration of 10^6 cfu. Furthermore, spectroscopic, ozone, power and temperature studies show higher spectroscopic

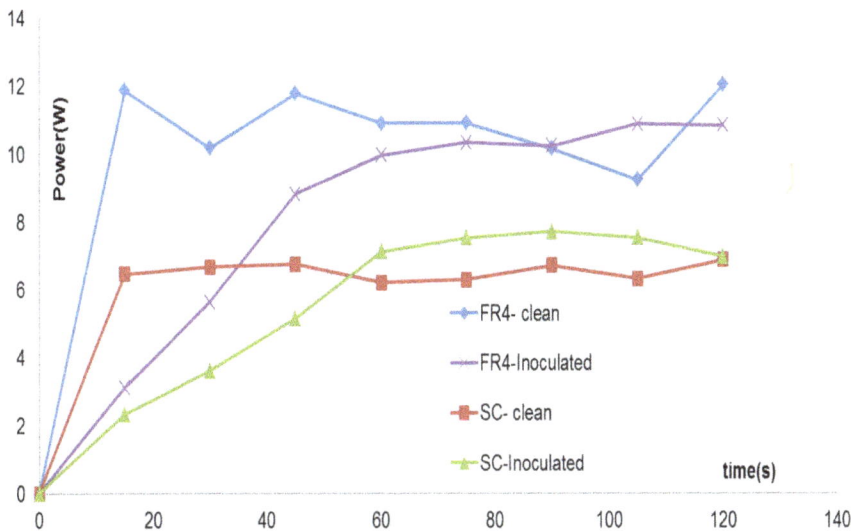

Figure 12. Variation of the input power over time for clean and inoculated FR4 and SC devices. Input power was sampled every 15s over a 2 minute interval for all four cases. Devices were powered during the entire 2-minute interval.

intensities, greater ozone levels, higher input power and higher surface temperatures in the case of the FR4 dielectric as compared to the semi-ceramic (SC) dielectric. It is also noted that these parameters differ widely between a clean and inoculated case for both dielectrics. The significance of these results is discussed in the next section.

Discussion

The current paper examines sterilization using a DBD surface plasma generator in room air. Sterilization trials using different test pathogens were conducted. Killing times of 2 minutes or less were noted for vegetative cells (*E. coli* and *S. cerevisiae*). *G. stearothermophilus* spores required 20 minutes. These times are faster than previous reports involving volume plasma. Hence, we have been able to obtain fast sterilization using a simple

experimental setup. Current efforts are being targeted at making this setup portable and scaling it up in size to sterilize larger surfaces.

The purpose of this paper was to study two factors affecting the process of plasma sterilization. One was to analyze the effect of different dielectric substrates on the process of plasma sterilization. Two dielectric substrates were used. One was standard FR4, while the other was a semi- ceramic laminate. The dielectric constant (k = 3.8–4.5 mean = 4.15) for the former is not a tightly controlled variable, while the latter has a dielectric constant of k = 3.00±0.05. Thus, the dielectric constant of FR4 is an average of 28% higher than the SC.

The dielectric constant of a material is the ratio of amount of electrical energy stored in a material by an applied voltage, relative to that stored in vacuum. Any of the devices described in this paper can be considered as a parallel plate capacitor, using a

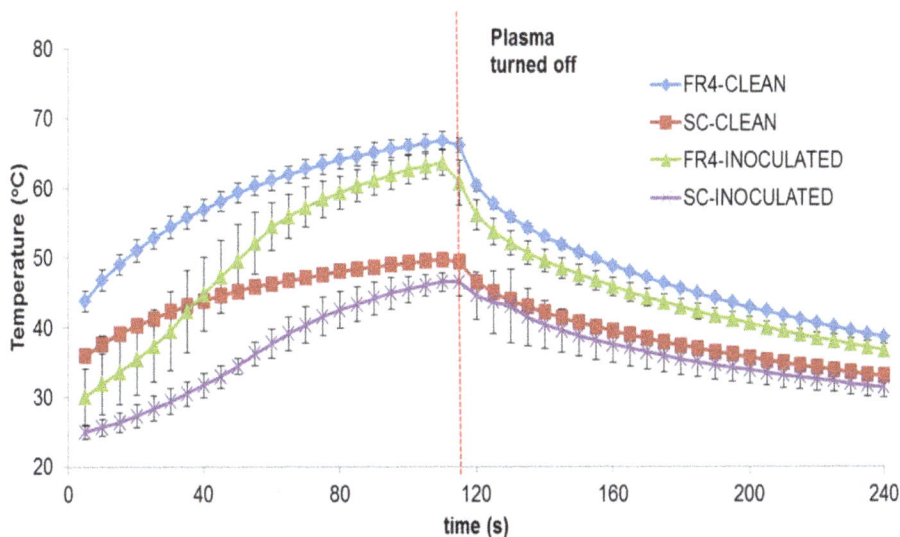

Figure 13. Comparison of surface temperatures during plasma generation for clean and inoculated FR4 and SC devices, measured using an infrared camera. Devices were powered for 2 minutes.

dielectric material of dielectric constant 'k'. Then, the capacitance of such a system is

$$C = \frac{k\varepsilon_0 A}{d}$$

where ε_0 = absolute permittivity of air, A = surface area of the top surface of the device &'d' = the thickness of the dielectric layer. The energy stored in a parallel plate capacitor is $U = \frac{1}{2}CV^2$.

Hence, for the devices considered in this paper, energy stored in the device is directly proportional to the capacitance of the system, which in turn is directly proportional to the dielectric constant of the material. Hence the FR4 device, which has a higher dielectric constant than the SC device, has more energy stored in the dielectric layer, which explains the results noted in Figure 6 i.e. complete sterilization is achieved faster (t = 90s) for FR4 as compared to semi-ceramic (SC) dielectric (t = 120s).

This difference is also supported by spectroscopic studies, which show that while plasma generated in both cases show similar spectra, the emitted intensities differ in both cases i.e. emitted intensity is higher by about 20–30% in the case of FR4. For the sake of simplicity, the spectral signatures for the semi-ceramic (SC) devices were not shown in this paper, as the peaks did not differ from those for FR4, just the intensity. Similarly, as shown in Figure 10 and 11, measured ozone levels show that the ozone levels produced in the case of FR4 devices (both clean and inoculated) are higher (25%–28%) than those produced in the case of semi-ceramic devices (both clean and inoculated). Additionally, Figure 13 shows that the maximum difference in average surface temperature during plasma generation between the FR4 and SC devices is ~18°C. Such a slim difference in average surface temperature leads us to believe that temperature is not the differentiating factor between sterilization times for FR4 and SC. The FR4 and SC plasma devices, when compared, differ only in that they are constructed out of different dielectric materials. This would imply that the difference in sterilization times noted between FR4 and SC plasma devices is due to the difference in dielectric constants. Thus, FR4 proves to be a much more efficient dielectric surface for plasma sterilization.

However, this advantage is offset by the disadvantage posed by the faster degradation of the FR4 dielectric. Preliminary scanning electron microscopy (SEM) studies indicated that after about 30 minutes of plasma generation, the FR4 dielectric starts degrading. This is shown below in Figure 14. One way to explain this might be because FR4, which has a higher dielectric constant, requires a higher input power for plasma generation, which in turn leads to higher surface heating, thus leading to faster breakdown of the dielectric surface in the case of FR4. A more detailed investigation of the surface degradation of both FR4 and SC dielectrics is ongoing and will be reported later.

A note on the operation of DBD plasma devices and the high levels of ozone emission has to be made. During these experiments, the ozone levels within the laboratory did not rise to unsafe levels, but local ozone levels directly above the device did exceed safe levels(as seen in the enclosed chambers). Safe permissible levels for ozone are 0.1 ppm, as per the Occupational Safety and Health Administration (OSHA), 0.1 ppm, as per the National Institute of occupational safety and health (NIOSH) and 0.05 ppm, as per the Food and Drug Administration for indoor medical devices (FDA). To prevent unsafe ozone exposure, experiments were conducted in an acrylic enclosure that was vented away from the operator at the end of the experiment. Charcoal was used as an adsorbent either glued to a wire mesh cage placed over the test bench or in respirators.

The other factor analyzed in this study was the effect of having liquid present on the surface of the plasma device. The evaporation of the liquid *E. coli* sample deposited upon the device surface follows a pattern. Initially the bacterial sample deposited covers the entire electrode surface area, and plasma is visible only around the edges of the electrode. As time progresses, the sample begins to evaporate around the outer edges of the electrode. Gradually, this evaporation begins to spread to other parts of the electrode, until eventually plasma covers the entire electrode surface area. This usually occurs at around t = 90s for the FR4 dielectric and just before t = 120s for the semi-ceramic (SC) dielectric. A steep drop in pathogen concentration is also noted precisely at these time points (Figure 6).

Spectroscopic, ozone, power and temperature data uniformly show that plasma is repressed while visible liquid is present on the test devices. The spectral peaks (Figure 8, 9) are noted at the same wavelengths at each time point; however, their intensities increase as the liquid evaporates. Similarly, it is observed that as the liquid sample evaporates, rate of production of ozone increases (Figure 10, 11). Additionally, temperature data (Figure 13) demonstrates that the standard deviation of temperature measurements in inoculated cases is especially large during the first 60s (for FR4) and during 60–105s (for SC). Visibly, it is observed that the liquid starts evaporating rapidly during these exact time intervals for both dielectrics, which is why a large variation in surface temperature can be observed. Once the liquid is completely evaporated, during the last 30s and 5–10s for FR4 and SC respectively, very little standard deviation is observed.

When the same number of organisms was deposited in a 40 µL volume instead of the standard 20 µL inoculation volume, the "passive phase" wherein there is little or no loss of viability (Figure 6) was extended by about 30s here. Thus, the rapid drop in *E. coli* concentration occurs at t≥120s, as opposed to 90s in the case of the lower inoculation volume (20 µl). This is shown below in Figure 15.

Hence, the point at which all the liquid covering the electrode evaporates and plasma covers the entire electrode surface area is the point at which there is a rise in surface temperature, input power, emitted ozone levels and spectroscopic intensity. This is also the point where the steep drop in pathogen concentration occurs, thus indicating that there is a threshold time-point at which complete sterilization occurs. Hence in our case, liquid seems to inhibit plasma generation and killing and this should be taken into account when designing surface sterilization systems.

One way to explain this liquid dependence is in terms of capacitance. As per the 'parallel-plate capacitor' theory discussed earlier, if the FR4/SC plasma device is considered as a capacitor of capacitance (C_1), then the liquid layer spread uniformly on top of the device can be considered as a second capacitor of capacitance (C_2), connected in 'series' with C_1. Thus the combined capacitance of this system would be

$$C = \frac{C_1 C_2}{C_1 + C_2} = \frac{C_2}{1 + \frac{C_2}{C_1}}$$

The impedance Z_1 of the liquid layer varies inversely with the amount of liquid present on the surface of the device i.e. as the liquid evaporates, impedance Z_1 decreases. Since Z_1 is inversely proportional to capacitance, C_1 ($Z = \frac{1}{j\omega C}$), it follows that as Z_1 decreases, C_1 increases. Following this, as C_1 increases, the capacitance of the overall system (C) increases and thus, the energy stored in the system ($U = \frac{1}{2}CV^2$) increases, proving that the

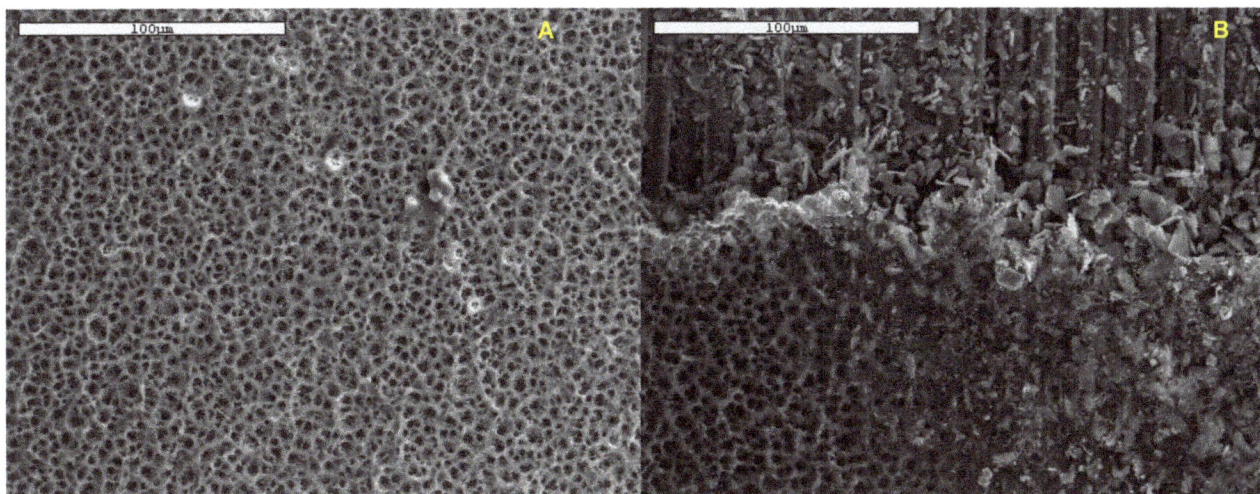

Figure 14. Preliminary SEM studies depicting the appearance of the dielectric substrate in (A) a fresh unused plasma device (B) a plasma device that has been powered continuously for 30 minutes. The devices shown have been imaged at a magnification of 500×. Comparing (A) and (B), it is evident that while (A) shows a fresh dielectric surface, (B) shows a degraded dielectric surface, wherein it appears that the top-layer seems to have eroded away, thus displaying the underlying fibers. A more detailed study of this degradation is ongoing and will be reported later.

amount of the liquid on the surface of the plasma device is actually detrimental to the performance of the plasma device as a sterilizer. This is mirrored in Figure 15 i.e. more the inoculation volume, more the amount of liquid covering the electrode surface, less the input power absorbed and hence more the sterilization time.

Oehmigen et.al. [22] reported experiments wherein they examined the role of acidification in influencing antimicrobial activity due to DBD plasma exposure. They concluded that plasma treatment of non-buffered liquids by indirect surface DBD resulted in acidification and thus, inactivation of suspended bacteria. When they tested the same theory with buffered solutions, they noted that pH decrease was avoided and therefore, antimicrobial plasma activity was reduced. It was suggested that reactive species from the plasma generation are the cause of liquid acidification and bactericidal activity. Along similar lines in our study, plasma devices inoculated with 20 μl of *E. coli* and plasma treated for $\Delta t = 30, 60, 90, 120$s were placed in sterile bags and thoroughly rinsed with 1 ml of Type 1 (ultrapure) Milli-Q® water. For each sample, the pH of the corresponding volume of water was measured using an Accumet® AB 15 pH meter (accuracy of ±

0.01). The process was repeated for both FR4 and SC dielectric devices. Before measuring the pH, the meter was standardized using pH buffer solution. The pH of LB broth used to make the *E. coli* sample was measured as 7.16 and that of the *E. coli* sample itself was measured to be 6.77. The variation of pH is given below in Figure 16.

Figure 16 above indicates that the reduction of pH is greater in the case of FR4 as compared to SC. However unlike the drastic reduction in pH values noted by Oehmigen et.al. [22], there is not a strong pH change in our results (both FR4 and SC). Thus, it is most likely that acidification plays some role but not a major one in bacterial cell death. Note that the pH value does not vary much, except during the last 30s (for FR4) and not at all for SC. This is again indicative of the effect of the liquid on the dielectric surface. Since the liquid bacterial sample deposited on the dielectric substrate does not evaporate until the very end of the sterilization time interval (for both FR4 and SC), the pH does not change very much until the very end. This confirms that the liquid deposited on the dielectric substrate inhibits plasma generation and hampers the sterilization process.

In conclusion, this paper describes the usage of a DBD surface plasma generator using air as the working gas to implement sterilization. Complete sterilization, starting with an initial concentration of 10^6–10^7 cfu, is achieved within 90s to 120s (for vegetative pathogens) and within 20 minutes (for spores). FR4 is more efficient in this aspect, as compared to SC. The intensity of spectral peaks, amount of ozone generated, the absorbed input power and the surface temperature during plasma generation are all higher in the case of FR4. However, preliminary SEM studies also indicate a faster degradation of the FR4 dielectric. Thus, a trade-off may be required between faster sterilization times and durability of the plasma devices.

Spectroscopic studies show that the spectral pattern characteristic of the DBD plasma generated in this setup shows intensity peaks at wavelengths characteristic of the 2nd positive system of N_2. FR4 and SC plasma devices show intensity peaks at same wavelengths, although they differ in intensity values shown at each wavelength. Future studies will include investigating whether this

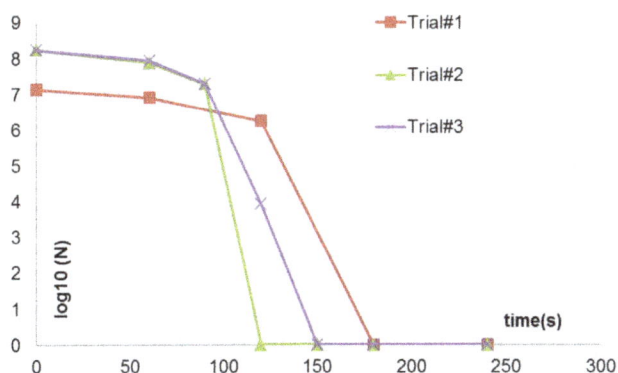

Figure 15. Sterilization plots for inoculation volume = 40 μl of *E. coli*.

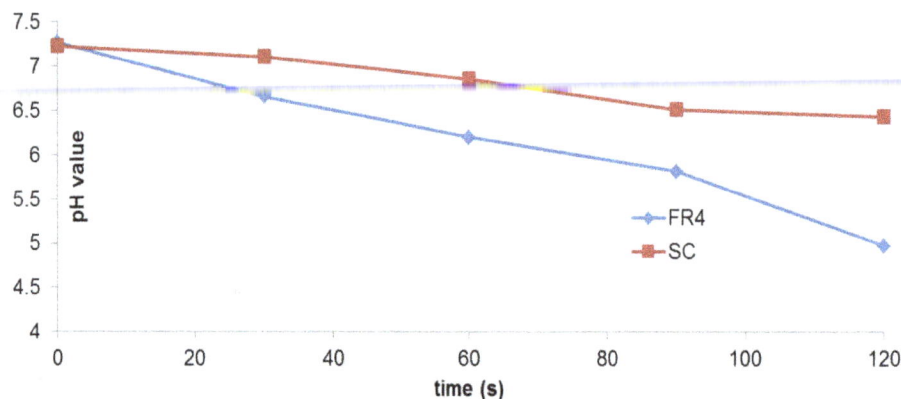

Figure 16. Plot of pH values, obtained by rinsing devices with Millipore water after plasma generation and measuring the pH value of this water in each case. Both FR4 and SC dielectrics are compared. pH values do not change as much in the case of SC, as compared to the case of FR4.

difference in intensity value is integral to the difference in sterilization times when using FR4 and SC. Similarly, ozone studies show that a clean FR4 device produces more ozone than a clean semi-ceramic (SC) device. Additionally an inoculated FR4 or SC device produces less ozone than the respective clean devices. Temperature studies show that the surface temperatures reached during plasma generation are in the range of 30°C–66°C (for FR4) and 20°C–49°C (for SC). pH studies indicate a slight reduction in pH value due to plasma generation, which coupled with temperature studies implies that while temperature and acidification may play a role in DBD plasma sterilization, these are not the dominant roles.

Plasma generation and sterilization are also inhibited by liquid on the electrode, as evidenced by spectroscopic, ozone, temperature and absorbed power measurements in a clean case as compared to an inoculated case. Future studies will include the investigation of sterilization times needed when plasma devices are inoculated and allowed to dry. However, it is clear that the experimental setup will have to be designed, keeping this liquid dependence in mind. Thus, our work shows that DBD surface plasma generators hold great promise for rapid and economical sterilization.

Acknowledgments

Authors would like to thank Dr. David Hahn, Mechanical & Aerospace Engineering (MAE), University of Florida, for sharing his insight and knowledge in setting up the spectroscopic studies. Special thanks are also due to Pengfei Zhao of the APRG, MAE, UF for his help in fine-tuning the spectroscopic equipment and setup. Last but not the least, our gratitude is also due to Raul A. Chinga of APRG, Dept. of Electrical Engineering, UF, who was responsible for designing and implementing a more compact version of the plasma generation setup.

Author Contributions

Conceived and designed the experiments: SR JJ. Performed the experiments: NM. Analyzed the data: NM JJ SR. Wrote the paper: NM JJ SR.

References

1. Conrads H, Schmidt M (2000) Plasma generation and plasma sources. Plasma Sources Sci Technol 9: 441–454.
2. Kogelschatz U (2003) Dielectric barrier Discharges: Their History, Discharge Physics and Industrial Applications. Plasma Chemistry and Plasma Processing 23(1): 1–46.
3. Moisan M, Barbeau J, Moreau S, Pelletier J, Tabrizian M, et al. (2001) Low-temperature sterilization using gas plasmas: a review of the experiments and an analysis of the inactivation mechanism. International Journal of Pharmaceutics 226(1): 1–21.
4. Laroussi M (2005) Low Temperature Plasma-based Sterilization: Overview and State- of- the-Art. Plasma Processes and Polymers 2(5): 391–400.
5. Gallagher M, Vaze N, Gangoli S, Vasilets VN, Gutsol AF, et al. (2007) Rapid inactivation of airborne bacteria using atmospheric pressure dielectric barrier grating discharge. Plasma Science, IEEE Transactions on 35(5): 1501–1510.
6. Menashi WP (1968) U S Patent No. 3383163. Washington, DC: U S Patent and Trademark Office.
7. Hury S, Vidal DR, Desor F, Pelletier J, Lagarde T (1998) A parametric study of the destruction efficiency of bacillus spores in low pressure oxygen-based plasmas. Lett Appl Microbiol 26(6): 417–421.
8. Lerouge S, Wertheimer MR, Marchand R, Tabrizian M, Yahia LH (1999) Effect of gas composition on spore mortality and etching during low-pressure plasma sterilization. J Biomed Mater Res 51(1): 128–135.
9. Moreau S, Moisan M, Tabrizian M, Barbeau J, Pelletier J, et al. (2000) Using the flowing afterglow of a plasma to inactivate B. subtilis spores: Influence of the operating conditions. J Appl Phy 88(2): 1166–1174.
10. Ying J, Chunsheng R, Zhilong X, Dezhen W, Younian W, et al. (2006) Comparison of yeast inactivation treated in He, Air and N2 DBD Plasma. Plasma Science and Technology 8(6): 720.
11. Sladek REJ, Filoche SK, Sissons CH, Stoffels E (2007) Treatment of Streptococcus mutans biofilms with a nonthermal atmospheric plasma. Lett Appl Microbiol 45(3): 318–323.
12. Kalghatgi S, Kelly CM, Cerchar E, Torabi B, Alekseev O, et al. (2011) Effects of non-thermal plasma on mammalian cells. PLoS One 6(1): e16270.
13. Joshi SG, Cooper M, Yost A, Paff M, Ercan UK, et al. (2011) Nonthermal dielectric barrier discharge plasma-induced inactivation involves oxidative DNA damage and membrane lipid peroxidation in Escherichia coli. Antimicrobial agents and chemotherapy 55(3): 1053–1062.
14. Lerouge S, Fozza AC, Wertheimer MR, Marchand R, Yahia LH (2000) Sterilization by low-pressure plasma: the role of vacuum-ultraviolet radiation. Plasmas and Polymers 5(1): 31–46.
15. Laroussi M, Leipold F (2004) Evaluation of the roles of reactive species, heat and UV radiation in the inactivation of bacterial cells by air plasmas at atmospheric pressure. International Journal of Mass Spectrometry 233(1): 81–86.
16. Dobrynin D, Fridman G, Friedman G, Fridman A (2009) Physical and biological mechanisms of direct plasma interaction with living tissue. New Journal of Physics 11(11): 115020.
17. Yu H, Xiu ZL, Ren CS, Zhang JL, Wang DZ, et al. (2005) Inactivation of yeast by dielectric barrier discharge (DBD) Plasma in helium at atmospheric pressure. Plasma Science, IEEE Transactions on 33(4): 1405–1409.
18. Mastanaiah N, Wang CC, Johnson JA, Roy S (2011) A computational diagnostic tool for understanding plasma sterilization. In 49th AIAA Aerospace Sciences Meeting and Exhibit, AIAA Paper.
19. Pearse R, Gaydon A (1976) The identification of molecular spectra. London: Chapman and Hall.
20. Ayan H, Fridman G, Staack D, Gutsol AF, Vasilets VN, et al. (2009) Heating effect of dielectric barrier discharges for direct medical treatment. Plasma Science, IEEE Transactions on 37(1): 113–120.

21. Mackey BM, Miles CA, Parsons SE, Seymour DA (1991) Thermal denaturation of whole cells and cell components of Escherichia coli examined by differential scanning calorimetry. J Gen Microbiol 137(10): 2361–2374.

22. Oehmigen K, Hahnel M, Brandenburg R, Wilke C, Weltmann KD, et al. (2010) The role of acidification for antimicrobial activity of atmospheric pressure plasma in liquids. Plasma Processes and Polymers 7(3–4): 250–257.

12

The Chlorate-Iodine-Nitrous Acid Clock Reaction

Rafaela T. P. Sant'Anna, Roberto B. Faria[*]

Instituto de Química, Universidade Federal do Rio de Janeiro, Rio de Janeiro, RJ, Brazil

Abstract

A new clock reaction based on chlorate, iodine and nitrous acid is presented. The induction period of this new clock reaction decreases when the initial concentrations of chlorate, nitrous acid and perchloric acid increase, but it is independent on the initial iodine concentration. The proposed mechanism is based on the LLKE autocatalytic mechanism for the chlorite-iodide reaction and the initial reaction between chlorate and nitrous acid to produce nitrate and chlorite. This new clock reaction opens the possibility for a new family of oscillating reactions containing chlorate or nitrous acid, which in both cases has not been observed until now.

Editor: Bing Xu, Brandeis University, United States of America

Funding: RTPS received a scholarship from Coordenação de Aperfeiçoamento de Pessoal de Nível Superior - CAPES; RTPS received a scholarship from Conselho Nacional de Desenvolvimento Científico e Tecnológico - CNPq; RBF was funded by Conselho Nacional de Desenvolvimento Científico e Tecnológico - CNPq, #303988/2009-6 and #308497/2013-9; this research was funded by Fundação Carlos Chagas Filho de Amparo à Pesquisa do Estado do Rio de Janeiro-FAPERJ, #E29-111749/2011. The funders had no role in study design, data collection and analysis, decision to publish, or preparation of the manuscript.

Competing Interests: The authors have declared that no competing interests exist.

* Email: faria@iq.ufrj.br

Introduction

The discovery of a clock reaction is a very special and rare nonlinear event because it is necessary an autocatalytic sequence of reactions and a short range of reagent concentrations. The first clock reaction containing chlorate (the chlorate-iodine clock reaction), was discovered by our group when studying the kinetic of the chlorate-iodine reaction [1]. This clock reaction is indeed a photochemically induced clock reaction because it needs UV light stimulation to occur [2]. Because UV light can generate small amounts of ozone from dissolved oxygen in water, we investigated the possibility that an ozone solution could substitute for light and trigger the chlorate-iodine clock reaction. In fact, we were able to show that the ozone-iodine-chlorate clock reaction exists and the initial step is the reaction between ozone and iodide [3]. In this same work [3] we have also shown that not all oxidant species, as for example H_2O_2, were able to react in such way to produce the autocatalysis which results in the clock reaction behavior.

In the present work we show that nitrous acid can substitute for ozone to produce a new clock reaction which was fully characterized. This new chlorate-iodine-nitrous acid clock reaction opens a new opportunity to find an oscillating system containing chlorate or nitrous acid species, in both cases an unprecedented finding. The possibility to find out an oscillating reaction based on this new clock reaction is greater than in the case of the ozone-iodine-chlorate clock reaction [3] because a nitrous acid solution is much easier to work than the unstable ozone solution.

Materials and Methods

All reagents were used as received: sodium chlorate (Fluka), perchloric acid (VETEC), resublimed iodine (VETEC), sodium nitrite (Carlo Erba). The solutions were made using conductivity water (18 MΩ) from a Milli-Q Plus system. All experiments were conducted at $25 \pm 0.1°C$ using Hellma Suprasil quartz cuvettes inserted in a jacketed cuvette holder connected to an Etica electronic thermostatic bath.

The clock reaction was followed at 460 nm, which is the maximum of the iodine band, using an Agilent 8453 spectrophotometer in the kinetic mode. Experimental data were acquired at fixed frequency (cycle time) and the data collection time for each point (integration time) was equal to 0.5 s. In these experiments only the tungsten lamp was turned on (deuterium lamp was turned off) to prevent incidence of ultraviolet light on the sample.

All experiments were done at least in triplicate. The experimental curves presented in the figures are typical ones for each set of reagents concentrations. The clock time of these curves does not deviate more than 5% of any other using the same experimental conditions. Similarly, the absorbance values do not spread more than 2% when repeating the same experiment.

The concentration of the iodine ($\lambda_{max} = 460$ nm; $\varepsilon = 740$ L mol^{-1} cm^{-1}) [4], and nitrous acid ($\lambda_{max} = 358$ nm; $\varepsilon = 52$ L mol^{-1} cm^{-1}) [5] solutions were measured by spectrophotometry using the indicated bands and molar absorptivity. The nitrous acid solutions were always freshly produced by dissolution of sodium nitrite in perchloric acid aqueous solution to produce the desired nitrous acid and H^+ concentration.

A semi-implicit Runge-Kutta numerical integration method [6], codified in Turbo Pascal language, was used to produce the simulation results employing the proposed mechanism.

Results and Discussion

Figure 1 shows that without nitrous acid, no significant change occur at 460 nm during 500 s. However, in the presence of nitrous acid an autocatalytic reaction occurs after a small induction period, which is a typical format for a clock reaction.

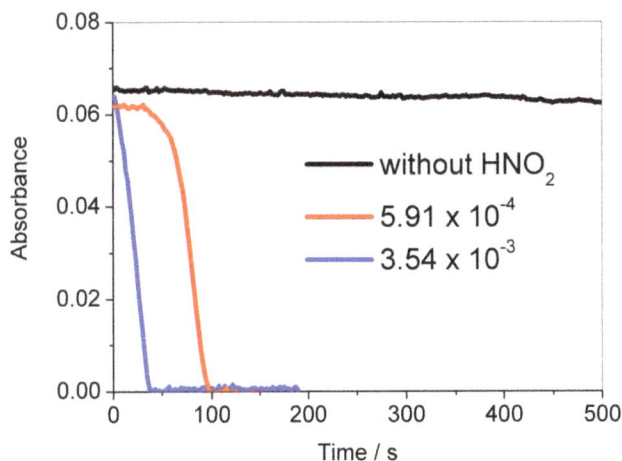

Figure 1. The addition of nitrous acid produces the chlorate-iodine-nitrous acid clock reaction. Following the absorbance at 460 nm for the systems chlorate-iodine-perchloric acid (without addition of HNO_2) and chlorate-iodine-perchloric acid-nitrous acid (HNO_2 concentrations indicated in the figure, mol L^{-1}), both in the absence of ultraviolet light. Other initial concentrations: $[HClO_4] = 0.948$ mol L^{-1}; $[NaClO_3] = 0.0251$ mol L^{-1}; $[I_2] = 8.80 \times 10^{-5}$ mol L^{-1}. Experimental data point were measured at every 2 s.

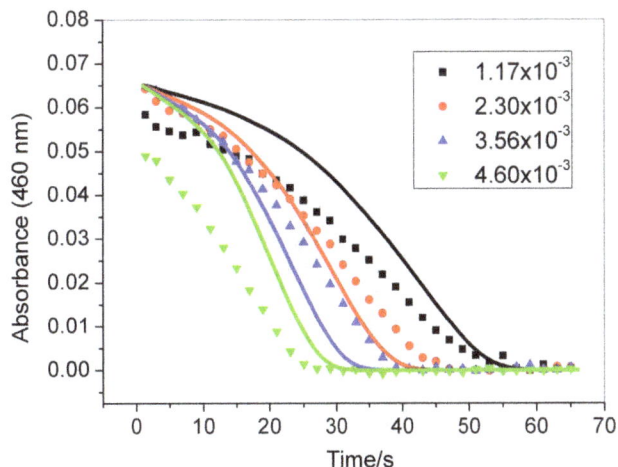

Figure 3. Effect of nitrous acid concentration (indicated in the figure – mol L^{-1}) in the clock time. Other initial concentrations: $[HClO_4] = 0.948$ mol L^{-1}; $[NaClO_3]$ 0.0251 mol L^{-1}; $[I_2] = 8.80 \times 10^{-5}$ mol L^{-1}. Experimental (symbols); simulation (continuous lines). Experimental data point were measured at every 2 s.

Figures 2, 3, and 4 show that the induction period of this chlorate-iodine-nitrous acid clock reaction decreases as the initial concentrations of chlorate, nitrous acid, and perchloric acid are increased. However, the induction period does not change when the initial iodine concentration is modified (Figure 5).

To explain the observed clock behavior we propose the mechanism presented in Table 1, which is based on Reactions 1 to 16, proposed by Lengyel *et al.* [7] to provide the autocatalytic pathway necessary for the clock behavior. Despite of some criticism [8], this mechanism reproduces the chlorite-iodide clock behavior and remains the best available model for this system. To

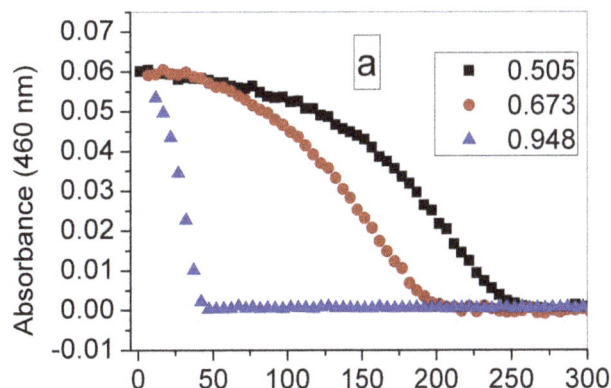

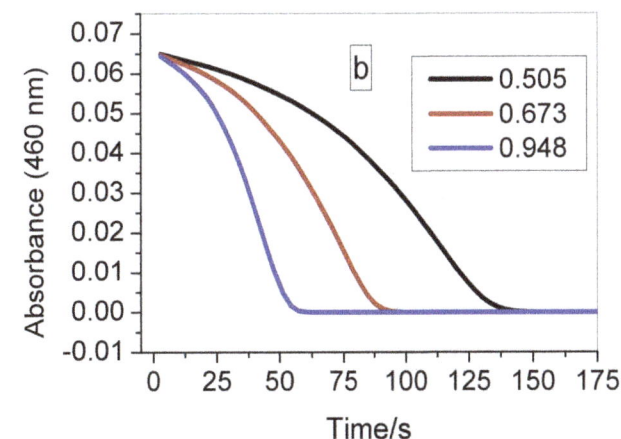

Figure 2. Effect of chlorate concentration (indicated in the figure – mol L^{-1}) in the clock time. Other initial concentrations: $[HClO_4] = 0.948$ mol L^{-1}; $[I_2] = 8.80 \times 10^{-5}$ mol L^{-1}; $[HNO_2] = 1.15 \times 10^{-3}$ mol L^{-1}. Experimental (symbols); simulation (continuous lines). Experimental data point were measured at every 5 s.

Figure 4. Effect of perchloric acid concentration (indicated in the figure – mol L^{-1}) in the clock time. Other initial concentrations: $[I_2] = 8.80 \times 10^{-5}$ mol L^{-1}; $[NaClO_3] = 0.0251$ mol L^{-1}; $[HNO_2] = 1.15 \times 10^{-3}$ mol L^{-1}. (a) Experimental (symbols); (b) Simulation (continuous lines). Experimental data points were measured at every 5 s.

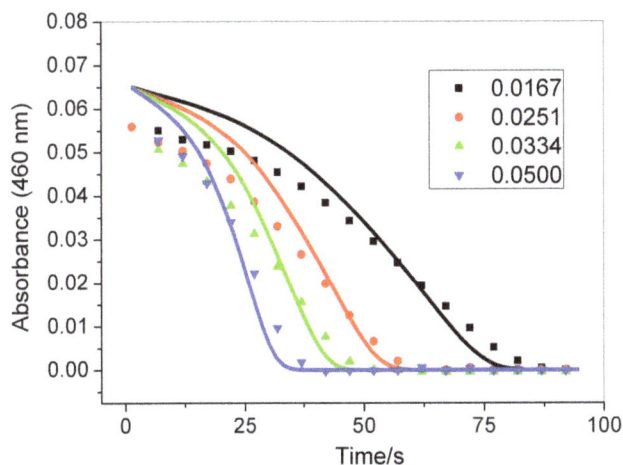

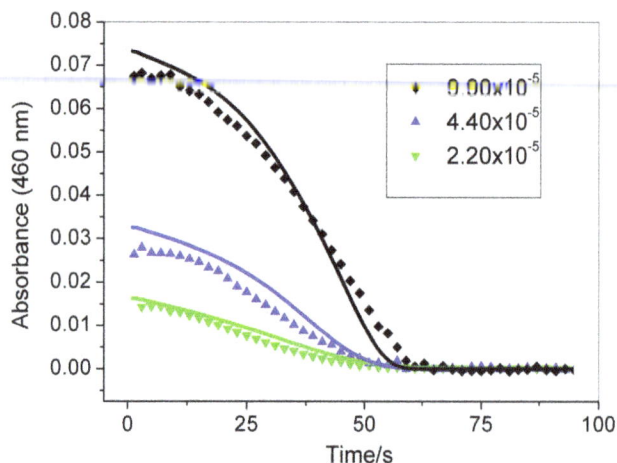

Figure 5. Effect of iodine concentration (indicated in the figure – mol L^{-1}) in the clock time. Other initial concentrations: $[HClO_4] = 0.948$ mol L^{-1}; $[NaClO_3] = 0.0251$ mol L^{-1}; $[HNO_2] = 1.15 \times 10^{-3}$ mol L^{-1}. Experimental (symbols); simulation (continuous lines). Experimental data point were measured at every 2 s.

experimental results only after reduce this rate constant to 1.8×10^{-2} L^3 mol^{-3} s^{-1} (the order of the reaction was changed also; see next paragraph for explanation). Based on the proposed mechanism for the chlorate-chloride reaction [10], it can be considered that the presence of chloride is able to turn chlorate into more reactive species like ClOClO. This may justify the use of a lower rate constant in our system that does not contain HCl.

The qualitative agreement for the acid concentration effect, as shown in Figure 4, was obtained only after the inclusion of one additional H^+ in the rate laws for Reactions (4), (17) and (18). In the case of Reaction (4), the additional H^+ intend to adjust the ratios $[ClO_2^-]/[HClO_2]$ and $[HOI]/[H_2OI^+]$ which must be affected by changes in $[H^+]$ as indicated by Jowsa et al. [8]. This additional H^+ in the rate law for Reation (4) has been used before in the model for the ozone-iodine-chlorate clock reaction [3]. For Reaction (18), the additional H^+ in the rate law can be justified by the same argument, considering the protonation of ClO_3^- and HIO_2, and has also been used in the model for the ozone-iodine-chlorate clock reaction [3]. In the case of Reaction (17) it was very critical to introduce a square dependence on H^+ to obtain the qualitative agreement shown in Figure 4. Again, as in our system we do not use HCl, it is possible that the rate law for this reaction has a greater dependence on H^+ than observed by Emeish [9] in the presence of an excess of chloride.

As can be seen from the simulated results in Figures 2, 3, 4, and 5, the proposed mechanism reproduces the experimental behavior with good agreement for the chlorate, nitrous acid, and iodine effect. In the case of the acid concentration effect, to better show the qualitative agreement between the model and experimental results these are shown using different time scales in Figure 4. The comparison between experimental and simulation results presented in Figure 4 shows that the model reproduces the clock time for $[H^+] = 0,948$ M and produces a short clock time for other acid

this set of reactions we added the reactions of chlorate with nitrous acid and with HIO_2, Reactions (17) and (18), respectively.

Reaction (18) has been used in our model for the chlorate-iodine clock reaction [1] with $k = 7$ L mol^{-1} s^{-1}, and in the model for the ozone-iodine-chlorate clock reaction [3] with the same rate constant used here ($k = 20$ L mol^{-1} s^{-1}).

Reaction (17) has been studied by Emeish [9] which found the rate law $k[ClO_3^-][HNO_2][H^+]$, $k = 2.42$ L^2 mol^{-2} s^{-1}, at 25°C, using HCl to adjust pH. The use of this rate constant in our model produced a very fast clock event. We were able to simulate the

Table 1. Mechanism for the chlorate-iodine-nitrous acid clock reaction.

Number[a]	Reaction	Rate law[b]
1*	$I_2 + H_2O \leftrightarrow HOI + I^- + H^+$	1.98×10^{-3} [I_2]/[H^+]-3.67×10^9 [HOI][I$^-$]
2*	$I_2 + H_2O \leftrightarrow H_2OI^+ + I^-$	5.52×10^{-2} [I_2]-3.48×10^9 [H_2OI^+][I$^-$]
3*	$HClO_2 + I^- + H^+ \rightarrow HOI + HOCl$	7.8 [$HClO_2$][I$^-$]
4**	$HClO_2 + HOI \rightarrow HIO_2 + HOCl$	6.9×10^7 [$HClO_2$][HOI][H^+]
5*	$HClO_2 + HIO_2 \rightarrow IO_3^- + HOCl + H^+$	1.0×10^6 [$HClO_2$][HIO_2]
6*	$HOCl + I^- \rightarrow HOI + Cl^-$	4.3×10^8 [HOCl][I$^-$]
7*	$HOCl + HIO_2 \rightarrow IO_3^- + Cl^- + 2 H^+$	1.5×10^3 [HOCl][HIO_2]
8*	$HIO_2 + I^- + H^+ \leftrightarrow 2 HOI$	1×10^9 [HIO_2][I$^-$][H^+] $- 22$ [HOI]^2
9*	$2 HIO_2 \rightarrow IO_3^- + HOI + H+$	25 [HIO_2]^2
10*	$HIO_2 + H_2OI^+ \rightarrow IO_3^- + I^- + 3 H^+$	110 [HIO_2][H_2OI^+]
11*	$Cl_2 + H_2O \leftrightarrow HOCl + Cl^- + H^+$	22 [Cl_2] $- 2.2 \times 10^4$ [HOCl][Cl$^-$][H^+]
12*	$Cl_2 + I_2 + 2 H_2O \rightarrow 2 HOI + 2 Cl^- + 2 H^+$	1.5×10^5 [Cl_2][I_2]
13*	$Cl_2 + HOI + H_2O \rightarrow HIO_2 + 2 Cl^- + 2 H^+$	1.0×10^6 [Cl_2][HOI]
14*	$HClO_2 \leftrightarrow ClO_2^- + H^+$	2×10^8 [$HClO_2$]-1×10^{10} [ClO_2^-][H^+]
15*	$HOI + H^+ \leftrightarrow H_2OI^+$	1×10^{10} [HOI][H^+]-3.4×10^8 [H_2OI^+]
16*	$I_2 + I^- \leftrightarrow I_3^-$	5.6×10^9 [I_2][I$^-$]-7.5×10^6 [I_3^-]
17	$ClO_3^- + HNO_2 \rightarrow ClO_2^- + NO_3^- + H^+$	1.8×10^{-2} [ClO_3^-][HNO_2][H^+]^2
18	$ClO_3^- + HIO_2 \rightarrow IO_3^- + HClO_2$	20 [ClO_3^-][HIO_2][H^+]

$^{a)}$ * Reactions taken form Lengyel et al. [7]. ** Modified from the Lengyel et al. [7] including the [H^+] effect in the rate law.
$^{b)}$ Rate constants' units are s^{-1}, L mol^{-1} s^{-1}, L^2 mol^{-2} s^{-1}, and L^3 mol^{-3} s^{-1} for first-, second-, third-, and forth-order processes, respectively.

concentrations, presenting a good qualitative agreement in this case.

It is important to notice that the chlorate-iodine-ozone [3] and chlorate-iodine-nitrous acid clock reactions have very different starting reactions. In the first the iodide is oxidized by ozone to HOI which reacts with chlorate producing chlorite which is converted to $HClO_2$. In the present work the first step is the chlorate reaction with nitrous acid that produces nitrate and chlorite that goes to $HClO_2$. The $HClO_2$ then reacts with several iodine species (I-, HOI, and HIO_2) starting the complex autocatalytic set of reactions. In other words, despite the fact that we have substituted ozone for nitrous acid, this species is oxidized by chlorate while ozone is reduced by iodide. It means that this new chlorate-iodine-nitrous acid clock reaction can not be considered a simple and obvious variation of the chlorate-iodine-light clock reaction [1,2] or of the chlorate-iodine-ozone clock reaction [3].

These results above show that a new clock reaction involving chlorate, iodine and nitrous acid has been discovered and was fully characterized. It is the third clock reaction involving chlorate and it is an important step toward the discovery of an oscillating reaction involving chlorate, because the nitrous acid aqueous solution is much easier to handle than the ozone aqueous solution employed in the ozone-iodine-chlorate system [3]. Using this clock reaction and the cross-shape diagram method [11] should allow to find the experimental conditions to observe the first oscillating reaction involving chlorate.

Acknowledgments

We thank Conselho Nacional de Desenvolvimento Científico e Tecnoló-gico-CNPq and Fundação Carlos Chagas Filho de Amparo à Pesquisa do Estado do Rio de Janeiro-FAPERJ for funding this research. We also thank Dr. Istvan Lengyel for the use of his Turbo Pascal code used in the simulations by numerical integration.

Author Contributions

Conceived and designed the experiments: RTPS RBF. Performed the experiments: RTPS. Analyzed the data: RTPS RBF. Contributed reagents/materials/analysis tools: RBF. Wrote the paper: RTPS RBF.

References

1. Oliveira AP, Faria RB (2005) The chlorate-iodine clock reaction. J Am Chem Soc 127: 18022–18023.

2. Galajda M, Lente G, Fábián I (2007) Photochemically induced autocatalysis in the chlorate ion-iodine system. J Am Chem Soc 129: 7738–7739.

3. Sant'Anna RTP, Monteiro EV, Pereira JRT, Faria RB (2013) The ozone-chlorate-iodine clock reaction. Plos One 8: e83706.

4. Rábai G, Beck MT (1987) Kinetics and mechanism of the autocatalytic reaction between iodine and chlorite ion. Inorg Chem 26: 1195–1199.

5. Gomes MG, Borges SSS, Lopes LGF, Franco DW (1993) UV-visible spectrum of nitrous acid in solution: pKa determination and analytical applications. Anal Chim Acta 282: 81–85.

6. Kaps P, Rentrop P (1979) Generalized runge-kutta methods of order four with stepsize control for stiff ordinary differential equations. Numer Math 33: 55–68.

7. Lengyel I, Li J, Kustin K, Epstein IR (1996) Rate constants for reactions between iodine- and chlorine-containing species: A detailed mechanism of the chlorine dioxide/chlorite-iodide reaction. J Am Chem Soc 118: 3708–3719.

8. Jowza M, Sattar S, Olsen RJ (2005) Modeling chlorite-iodide reaction dynamics using a chlorine dioxide-iodide reaction mechanism. J Phys Chem A 109: 1873–1878.

9. Emeish SS (1980) Solvent effect on the rate of N(III)/Cl(V) reaction. Can J Chem 58: 902–905.

10. Sant'Anna RTP, Santos CMP, Silva GP, Ferreira RJR, Côrtes CES, et al. (2012) Kinetics and mechanism of chlorate-chloride reaction. J Braz Chem Soc 23: 1543–1550.

11. Boissonade J, De Kepper P (1980) Transitions from bistability to limit cycle oscillations. Theoretical analysis and experimental evidence in a open chemical system. J Phys Chem 84: 501–506.

13

S-Nitroso-Proteome in Poplar Leaves in Response to Acute Ozone Stress

Elisa Vanzo[1], Andrea Ghirardo[1], Juliane Merl-Pham[2], Christian Lindermayr[3], Werner Heller[3], Stefanie M. Hauck[2], Jörg Durner[3], Jörg-Peter Schnitzler[1]*

1 Research Unit Environmental Simulation, Institute for Biochemical Plant Pathology, Helmholtz Zentrum München, Neuherberg, Germany, **2** Research Unit Protein Science, Helmholtz Zentrum München, Neuherberg, Germany, **3** Institute for Biochemical Plant Pathology, Helmholtz Zentrum München, Neuherberg, Germany

Abstract

Protein S-nitrosylation, the covalent binding of nitric oxide (NO) to protein cysteine residues, is one of the main mechanisms of NO signaling in plant and animal cells. Using a combination of the biotin switch assay and label-free LC-MS/MS analysis, we revealed the S-nitroso-proteome of the woody model plant *Populus* x *canescens*. Under normal conditions, constitutively S-nitrosylated proteins in poplar leaves and calli comprise all aspects of primary and secondary metabolism. Acute ozone fumigation was applied to elicit ROS-mediated changes of the S-nitroso-proteome. This treatment changed the total nitrite and nitrosothiol contents of poplar leaves and affected the homeostasis of 32 S-nitrosylated proteins. Multivariate data analysis revealed that ozone exposure negatively affected the S-nitrosylation status of leaf proteins: 23 proteins were de-nitrosylated and 9 proteins had increased S-nitrosylation content compared to the control. Phenylalanine ammonia-lyase 2 (log2[ozone/control] = −3.6) and caffeic acid O-methyltransferase (−3.4), key enzymes catalyzing important steps in the phenylpropanoid and subsequent lignin biosynthetic pathways, respectively, were de-nitrosylated upon ozone stress. Measuring the *in vivo* and *in vitro* phenylalanine ammonia-lyase activity indicated that the increase of the phenylalanine ammonia-lyase activity in response to acute ozone is partly regulated by de-nitrosylation, which might favor a higher metabolic flux through the phenylpropanoid pathway within minutes after ozone exposure.

Editor: Keqiang Wu, National Taiwan University, Taiwan

Funding: The authors have no support or funding to report.

Competing Interests: The authors have declared that no competing interests exist.

* Email: jp.schnitzler@helmholtz-muenchen.de

Introduction

Nitric oxide (NO), a volatile nitrogen compound, is widely recognized as a signaling molecule in various organisms. In plants, NO is involved in the regulation of various developmental processes, such as root growth, seed germination, flowering, pollen tube re-orientation and stomatal closure [1–6]. Furthermore, NO plays important roles in plant responses to biotic and abiotic stresses [7,8].

Different modes of NO signaling have been reported to mediate this abundance of function in plants. Most NO signaling is accomplished through the posttranslational modification (PTM) of target proteins, such as (i) the nitration of protein tyrosine moieties, (ii) binding to metal centers or (iii) the nitrosylation of cysteine residues [9]. S-nitrosylation, the reversible attachment of a NO moiety to thiol groups of selected cysteine residues functions as the most important PTM in the context of NO signaling. S-nitrosylation can impact protein functionality, stability and cellular localization [10]. The S-nitrosylation of enzymes regulates their activity either negatively or positively. Several detailed analyses of the S-nitrosylation of specific proteins have used NO donors (i.e., S-nitrosoglutathione (GSNO)) to promote S-nitrosylation *in vitro*. In most cases, the NO donor treatment seems to reduce the enzyme activity in plants [11–15]. Reports showing that the

protein S-nitrosylation of specific cysteine residues promotes enzyme activity are scarce in plant science, and only a few hints regarding an activity-enhancing effect have been reported for animals [16–18]. Although de-nitrosylation represents a less-described aspect of NO signaling, the process of de-nitrosylation is also a strictly regulated event involving two recently proposed enzyme systems: (i) the thioredoxin system, which comprises thioredoxin, thioredoxin reductase and NADPH, and (ii) the glutathione/GSNO reductase system [19]. Some proteins are constitutively S-nitrosylated, and de-nitrosylation has been observed after stimulation, leading to the activation of enzyme activity or *vice versa* [20,21]. S-nitrosylation and de-nitrosylation together generate the S-nitroso-proteome of a cell, a dynamic and rapidly changing regulatory network, especially under stress conditions.

More than a decade ago, Jaffrey and colleagues introduced the biotin switch assay [22], which facilitates the identification of S-nitrosylated proteins. By utilizing both the biotin switch assay and mass spectrometry (MS), hundreds of putative S-nitrosylation targets have been identified. In pioneering studies in plants, S-nitrosylation was enforced using NO donors on protein extracts [11,23,24]. More recent research has focused on identifying endogenously S-nitrosylated proteins in unstressed plants [25] and S-nitrosylation patterns in plants that are exposed to different stresses [15,26–28]. Comparative analysis of the S-nitroso-proteome under control and stress conditions is an important tool to provide information about the biological relevance of NO signaling upon various stress conditions. To date, no information is available regarding ozone-induced changes in the S-nitroso-proteome of plants. Although several studies describe the impact of acute ozone exposure on total proteomes of rice, soybean, wheat, and poplar [29–32], the issue of redox-linked protein modifications upon ozone has only been examined in two studies [33,34]. Ozone exerts bi-functional effects on earth. While stratospheric ozone protects life from harmful ultraviolet radiation, tropospheric ozone is an air pollutant that can induce oxidative stress and cell death in plants [35,36] causing considerable agricultural crop losses and damage in forest trees [37,38]. Temporary exposure to ozone at a high-level, termed acute exposure induces changes in gene expression and protein activities often within minutes after the onset of the fumigation [39]. It causes the formation of various reactive oxygen species (ROS) in plant tissues, mainly superoxide anion, hydrogen peroxide (H_2O_2) and hydroxyl radicals, which can induce cell-death lesions in ozone-sensitive plants. This rapid accumulation of ROS upon acute ozone treatment resembles the oxidative burst after plant-pathogen interactions [40,41]. Concomitant with the oxidative burst upon acute ozone fumigation, an accumulation of NO is observed [42–44]. It is thought that, in response to ozone, NO and ROS work synergistically to promote a defense response in plants, mimicking the hypersensitive response (HR) that occurs as a result of incompatible plant-pathogen interactions [41,45]. Therefore, the fine-tuning of the NO/ROS balance is needed [46].

Poplar is the model system for woody plants as its relatively small genome was the first to be sequenced [47]. Also, poplar is suitable for genetic transformation and can be propagated vegetatively, facilitating large-scale production of clones [48]. As demand for renewable bioenergy is increasing, poplar, a fast-growing pioneer tree, is receiving great amounts of attention due to its suitability for heat and power generation [49,50]. This demand implies more future scientific and economical interest in using poplar for analyzing stress responses in woody plant species. Moreover, while knowledge of S-nitrosylated proteins in herbaceous model plants is broad [11,26,28], there is limited information regarding woody plants [51].

In the present study we aimed to identify S-nitrosylated proteins in grey poplar (*Populus* x *canescens*) to improve understanding of the initial steps in plants' ozone response. By performing a biotin switch assay in conjunction with quantitative (label-free) LC-MS/MS analysis, we tried to answer the following questions: Firstly, are there proteins that are constitutively S-nitrosylated in the green tissue (callus and leaf) of poplar? Secondly, does acute oxidative stress (ozone fumigation) induce quantitative and/or qualitative alterations in the pattern of S-nitrosylated proteins in the leaves? And finally, if there are changes upon oxidative stress, can we deduce a regulation scheme that may explain the physiological relevance of S-nitrosylation signaling during plant ozone response?

Here we report a list of constitutively S-nitrosylated proteins in grey poplar under unstressed conditions. The list comprises many proteins not reported in the context of S-nitrosylation so far. Quantitative (label-free) analysis revealed significant and rapid changes in the S-nitroso-proteome of poplar undergoing acute ozone exposure.

Materials and Methods

Plant material, growth conditions and ozone treatment

For the generation of callus tissue, leaf and stem explants of grey poplar (*Populus* x *canescens* INRA clone 7171-B4; syn. *Populus tremula* x *Populus alba* (Aiton.) Smith) were cultured in the dark on callus induction medium for three weeks at 20°C [52]. Explants were transferred to shoot induction medium and were maintained for up to ten weeks under moderate light (16/8 h photoperiod at 125 μmol photons m^{-2} s^{-1}). Fully developed calli (Figure S1) were frozen at −80°C. All media are described elsewhere [53].

The ozone experiments were performed in three independent runs with *P.* x *canescens* plants. Poplar plants were multiplied by micropropagation on half-concentrated MS medium as described elsewhere [53]. After 8 weeks rooted shoots were transferred to soil substrate (50% v/v Fruhstorfer Einheitserde, 50% v/v silica sand (particle size 1–3 mm)) and grown under a plastic lid to maintain high humidity, to which the plants got used, during sterile culture. Plantlets were grown under the following conditions: 27°C/24°C (day/night) and a photoperiod of 16 h with approximately 100 μmol photons m^{-2} s^{-1} during the light period. Plantlets were adapted to ambient humidity conditions by carefully opening the lid after two weeks. After acclimatization, the plants were planted into 2.2 l pots (25% v/v Fruhstorfer Einheitserde, 25% v/v silica sand (particle size 1–3 mm), 50% v/v perlite) and were transferred to the greenhouse. Before ozone fumigation, the plants were grown in the greenhouse for eight weeks in May and June 2012 until they had attained a height of 60–70 cm and had produced 20–22 leaves. No supplemental lighting was provided. Fertilization was performed with Triabon (Compo, Münster, Germany) and Osmocote (Scotts Miracle-Gro, Marysville, USA) (1:1, v/v; 10 g per liter of soil).

Plant responses to ozone are closely linked to the effective dose taken up by the plant via the stomata [54]. Thus, flux-based indices that take into account ozone deposition into the leaf are considered a more reliable indicator of potential ozone damage than exposure time and air ozone concentration [55,56]. To determine the ozone uptake in grey poplar leaves by a given ozone concentration, we quantified the cumulative ozone dose in our experiments. Poplar plants were enclosed in a cuvette made of glass and Teflon (170 l volume, PPFD 250 μmol m^{-2} s^{-1}, air temperature 25°C±1°C, and flux 11.5 l min^{-1}). A fan ensured homogeneous mixing of the chamber air to remove boundary layer resistance at the plant surfaces. CO_2 assimilation, transpiration and foliar ozone flux (nmol m^{-2} s^{-1}) were monitored as differences between cuvette inlet and outlet by infrared-absorption (Fischer-Rosemount Binos 100 4P, Hasselroth, Germany), and chemoluminescence (O341 M, Ansyco Karlsruhe, Germany), respectively (Figure S2). When net CO_2 assimilation was stable, ozone fumigation (800 ppb) was applied for 1 hour. Ozone destruction on inner surfaces of the cuvette and non-stomatal adsorption on outer plant surfaces were taken into account by measuring empty cuvettes (including covered pots with soil) or darkening the plant in the cuvette, respectively. Application of 800 ppb ozone in the inlet air resulted finally in a cumulative uptake of ozone of 110±19 μmol m^{-2} (n = 6±SE). This ozone 'dose' is in the same range as previously described in the context of acute ozone treatments on plants (130–200 μmol ozone m^{-2} [57] and approx. 370 μmol ozone m^{-2} [58]).

Before fumigation started, all plants (control (C) and ozone (O) plants) were allowed to acclimatize to the cuvette conditions for 2 hours (PPFD 250 μmol m^{-2} s^{-1}, air temperature 25°C±1°C). O plants were then exposed to 800 ppb ozone for 1 hour, while control plants (C) were further kept in the control cuvette under ambient air (ozone-free) for 1 hour. For proteomics and biochemical analysis, fully developed leaves (number 9 and 10 from the apex; n = 3 plants for C and O, each) were frozen in liquid nitrogen immediately after exposure. C and O samples were subjected to the biotin switch assay (for detection of S-nitrosylated proteins), overall proteomic analysis (for normalization of S-nitrosylated protein abundance) and *in vivo* PAL activity measurement. Measurements of the *in vitro* phenylalanine ammonia-lyase (PAL) activity were performed on samples taken one month later from grey poplar plants that were grown in the greenhouse for 12 weeks in June, July and August 2012 (leaf number 9 and 10 from the apex).

Detection of endogenously S-nitrosylated proteins in poplar green tissue by modified biotin switch assay

The detection of *in vivo* S-nitrosylated proteins was performed *via* a modified biotin switch assay [22]. One of the most important modifications was the extraction of proteins in the presence of N-ethylmaleimide (NEM), which 'freezes' the S-nitrosylation pattern present at the moment of extraction by blocking all accessible thiol residues. A second change concerned the reduction step. Ascorbate, which serves as specific cysteine-NO reducing agent, was replaced by sinapic acid (SIN) due to its greater degree of specificity [59,60]. In brief, frozen leaf powder (either callus tissue or leaf tissue from control or ozone-treated plants) was mixed with HENT buffer (100 mM HEPES-NaOH pH 7.4, 10 mM EDTA, 0.1 mM Neocuproine, 1% (v/v) Triton X-100) in a mixing ratio of leaf powder:buffer 1:5 (w/v). The HENT buffer contained 30 mM NEM and protease inhibitor cocktail tablets (Complete, Roche, Grenzach-Wyhlen, Germany). The homogenate was mixed on a shaker for 30 s, incubated on ice for 15 min and centrifuged twice (14 000 g for 10 min). The protein concentration of the supernatant was adjusted to 1 µg/µl using the HENT buffer. For the blocking step, four-times the volume (v/v) of HENS (225 mM HEPES-NaOH pH 7.2, 0.9 mM EDTA, 0.1 mM Neocuproine, 2.5% (w/v) SDS) was freshly prepared, and 30 mM NEM was added to the protein extracts; the samples were incubated at 37°C for 30 min. Excess NEM was removed by precipitation with ice-cold acetone, and the protein pellet was re-suspended in 0.5 ml HENS buffer (without NEM) per milligram of protein in the starting sample. Biotinylation was achieved by adding biotin-HPDP and SIN (1 mM and 3 mM final concentrations, respectively) with further incubation at room temperature for 1 hour in the dark. The controls for false-positive signals (FP) were treated with SIN in the presence of NEM for 25 min at 37°C before the biotinylation step. After biotinylation, the proteins were precipitated with acetone and subjected to Western blot analyses and/or affinity purification of biotinylated proteins by NeutrAvidin agarose as described elsewhere [11]. For Western blot analyses, protein pellets were re-suspended in sample buffer and were separated by non-reducing SDS-PAGE on 12% polyacrylamide gels followed by immunoblotting [61]. After immunoblotting, membranes were stained with PonceauS to control protein loadings (Figure S3). After blocking membranes with 1% (w/v) nonfat milk powder and 1% (w/v) bovine serum albumin, the blots were incubated with the anti-biotin mouse monoclonal antibody conjugated with alkaline phosphatase (Sigma-Aldrich, St. Louis, USA; final dilution 1:10 000) overnight at 4°C. The protein bands were visualized using 5-bromo-4-chloro-3-indolyl phosphate and nitro blue tetrazolium. For affinity purification of biotinylated proteins, the precipitated proteins were re-suspended in HENS buffer (100 µl per mg of protein in the starting sample) and 2 volumes of neutralization buffer (20 mM HEPES, pH 7.7, 100 mM NaCl, 1 mM EDTA, and 0.5% (v/v) Triton X-100).

Biotinylated proteins were incubated for 1 hour at room temperature with the NeutrAvidin-agarose (30 µl per mg of protein). The agarose-matrix was washed extensively with 20 volumes of washing buffer (600 mM NaCl in neutralization buffer) and bound proteins were eluted with 100 mM β-mercaptoethanol in elution buffer (20 mM HEPES, pH 7.7, 100 mM NaCl, 1 mM EDTA) and precipitated with ice-cold acetone.

In-solution digest of S-nitrosylated proteins after NeutrAvidin affinity purification

The pellets from the acetone-precipitation (section before) were dissolved in 30 µl of 50 mM ammonium bicarbonate (AmBic). For protein reduction, 2 µl of 100 mM DTT was added and incubated for 15 min at 60°C. After cooling to room temperature, the free cysteine residues were alkylated by adding 2 µl of freshly prepared 300 mM iodoacetamide (IAA) solution for 30 min in the dark. A tryptic digest was performed overnight at 37°C using 0.5 µg of trypsin (Promega, Mannheim, Germany) per sample. To stop the digest, the sample was acidified using trifluoroacetic acid (TFA) and then stored at −20°C.

Preparation of whole-cell extracts (WCE) for overall proteomic analyses

For normalization of the protein abundance of S-nitrosylated proteins, we determined the total protein abundances by preparing WCE of C and O samples. Fifty mg of frozen leaf powder was mixed with 1 ml HENT buffer (for buffer composition see above) containing a protease inhibitor cocktail tablet and incubated on ice for 10 min. After centrifugation for 10 min (14 000 g), the protein extract was passed through a Sephadex G-25 column (GE Healthcare, München, Germany) using HEN (without Triton X-100) as a buffer. The protein concentration was determined by the Bradford assay with bovine serum albumin (BSA) as a standard.

Filter aided proteome preparation (FASP) digest of proteins from WCEs

Of each of three WCEs of C and O samples, an aliquot containing 10 µg of protein was digested using a modified FASP procedure [62]. In brief, the proteins were reduced and alkylated using DTT and IAA and then centrifuged through a 30 kDa cut-off filter device (PALL, Port Washington, USA), washed thrice with UA buffer (8 M urea in 0.1 M Tris/HCl pH 8.5) and twice with 50 mM AmBic. The proteins were digested for 2 hours at room temperature using 1 µg Lys-C (Wako Chemicals, Neuss, Germany) and for 16 hours at 37°C using 2 µg trypsin (Promega, Mannheim, Germany). The peptides were collected by centrifugation (10 min at 14 000 g), and the samples were acidified with 0.5% TFA and stored at −20°C.

Mass spectrometry

Digested samples (after affinity purification or from WCE) were thawed and centrifuged (14 000 g) for 5 min at 4°C. The LC-MS/MS analysis was performed as previously described [63]. Every sample was automatically injected and loaded onto the trap column at a flow rate of 30 µl min^{-1} in 5% buffer B (98% acetonitrile (ACN)/0.1% formic acid (FA) in HPLC-grade water) and 95% buffer A (2% ACN/0.1% FA in HPLC-grade water). After 5 min, the peptides were eluted from the trap column and separated on the analytical column by a 170 min gradient from 5 to 31% of buffer B at 300 nl min^{-1} flow rate followed by a short gradient from 31 to 95% buffer B for 5 min. Between each sample, the gradient was set back to 5% buffer B and left to equilibrate for 20 min. From the MS pre-scan, the 10 most abundant peptide ions were fragmented in the linear ion trap if they showed an intensity of at least 200 counts and if they were at least +2 charged. During fragmentation a high-resolution (6×10^4 full-width half maximum) MS spectrum was acquired in the Orbitrap (Thermo Fischer Scientific, Bremen, Germany) with a mass range from 200 to 1 500 Da.

Label-free analysis using Progenesis LC-MS

The acquired spectra were loaded to the Progenesis LC-MS software (v2.5, Nonlinear Dynamics Ltd, Newcastle upon Tyne, UK) for label-free quantification and analyzed as previously described [63,64]. Features of only one charge or more than eight charges were excluded. The raw abundances of the remaining features were normalized to allow for the correction of factors resulting from experimental variation. Rank 1–3 MS/MS spectra were exported as a MASCOT generic file and used for peptide identification with MASCOT (v2.2 and 2.3.02, Matrix Science, London, UK) in the *Populus trichocarpa* protein database (v4; 17 236 452 residues; 45 036 sequences). The search parameters were 10 ppm peptide mass and 0.6 Da MS/MS tolerance, one missed cleavage allowed.

For the identification and quantification of S-nitroso-proteins, N-ethylmaleinimidation and carbamidomethylation were set as variable modifications, as well as methionine oxidation. A MASCOT-integrated decoy database search calculated a false discovery rate (FDR) of 0.17% using a MASCOT ion score cut-off of 30 and a significance threshold of $P < 0.01$.

For the identification and quantification of total proteins in the WCEs of leaves, carbamidomethylation was set as a fixed modification, and methionine oxidation and deamination of asparagine/glutamine as variable modification. A MASCOT-integrated decoy database search calculated a FDR of <1%. The MASCOT Percolator algorithm was used to distinguish between correct and incorrect spectrum identification [65], with a maximum q value of 0.01. The peptides with a minimum percolator score of 15 were used further.

For each dataset, the peptide assignments were re-imported into the Progenesis LC-MS software. After summing up the abundances of all of the peptides that were allocated to each protein, the identification and quantification results were exported and are given in Table S1.

Statistics

The differences in the S-nitroso-proteome between control and ozone-treated samples were analyzed as previously described [66] using Principal Component Analysis (PCA) and Orthogonal Partial Least Square regression (OPLS) statistical methods from the software packages 'SIMCA-P' (v13.0.0.0, Umetrics, Umeå, Sweden). The results were validated by 'full cross validation' [67] using a 95% confidence level.

Before PCA and OPLS, S-nitroso-protein intensities were normalized to ensure that quantitative differences in S-nitrosylation between the C and O samples were not due to different amounts of the respective proteins. The normalization process and the calculated log2-fold changes (ozone/control) are given in Table S1C. In brief, the abundance of each S-nitrosylated protein (yellow, C1, C2, etc.) was normalized to the corresponding (averaged) protein abundance in the whole-cell extracts (WCE) of the C and O leaves (i.e. green, Avg_C_total, Avg_O_total). For the very lowly abundant proteins, which were not detected in the WCEs (25 proteins out of 172 proteins, highlighted in red), the total protein abundance was calculated based on the fraction of S-nitroso-protein content over the total protein content in the WCE detected protein, i.e multiplying by 100 and dividing by the

median (i.e. 2.2) of percentages between all S-nitroso-proteins and the respective measured total proteins. PCA was performed on normalized, summed S-nitroso-protein intensities (centered and scaled with 1 SD^{-1}) used as X-variables. X-data were pre-processed by logarithmic (base 10) transformation. Three independent biological replicates were used for each C and O treatment and their respective FP samples were included as controls of the biotin switch assay. Therefore, the size of the analyzed matrix was 172-by-8.

OPLS was employed to understand how the pattern of S-nitrosylated proteins changed upon ozone treatment and to discover which S-nitroso-proteins are affected by ozone exposure. OPLS was performed as PCA and by giving as Y-variable the value of 0 to C samples and value of 1 to O samples. S-nitroso-proteins showing Variable of Importance for the Projection (VIP) greater than 1 and uncertainty bars of jack-knifing method [68] smaller than the respective VIP value were defined as discriminant proteins that can separate O from C samples. Additionally, discriminant proteins were tested for significance difference ($P < 0.05$) between C and O samples independently from multivariate data analysis using Student's t-test and applying a FDR of 5% according to the Benjamini Hochberg modified correction (MATLAB R2011b; MathWorks, Natick, USA) [69,70]. PCA and OPLS were also performed on non-normalized abundances, giving similar results.

Determination of nitrite and nitrosothiol content

The quantification of nitrite and nitrosothiols (SNO) in poplar leaf tissue was performed using Sievers' Nitric Oxide Analyzer (NOA 280i, GE Water & Process Technologies, Ratingen, Germany). The instrument is based on ozone-dependent chemiluminescence: in the reaction vessel, nitrite and SNO are reduced to NO, which reacts with ozone to yield light photons. A total of 200 µg frozen plant powder from leaf tissue (C and O) was mixed with 600 µl extraction buffer (1×PBS pH 7.7, 10 mM NEM, 2.5 mM EDTA) and centrifuged twice (14 000 g for 20 min). For nitrite and SNO determination, 100 µl and 200 µl of the plant extracts, respectively, were injected into the reaction vessel of the NOA 280i containing the reducing agent iodine, iodide and acetic acid at room temperature. To measure the SNO content, the plant extracts were pre-treated in a mixing ratio of 9:1 with 5% sulfanilamide (w/v, in 1 M HCL) to chemically remove nitrite. The peak area integration and quantification of nitrite and SNO contents were performed with Sievers NO Analysis Software (v3.2) using nitrite standards.

Determination of PAL activity

The assay for the measurement of the PAL activity was modified from Heide et al. [71]. The WCEs from C and O leaves were used as the starting material. To measure the effect of S-nitrosylation and de-nitrosylation, the WCEs were pre-treated with either GSNO (500 µM) or SIN (3 mM), respectively, for 1 hour at room temperature in the dark. GSNO is a widely used NO donor leading to enhanced S-nitrosylation levels of proteins [11,13,72]. The inhibitory effect of GSNO treatment on the enzyme activity has been shown for several proteins [10]. SIN is used as de-nitrosating agent in the biotin switch assay [59]. The PAL activity assay began by mixing 150 µl of pre-treated WCE, 150 µl of 0.2 M sodium borate buffer (pH 8.9) and 100 µl L-phenylalanine (in sodium borate buffer). The reaction was performed at 37°C for 2 hours and was stopped by adding 100 µl of 6 N HCl. The controls were run without adding L-phenylalanine or WCE. The reaction product *trans*-cinnamic acid was extracted with 500 µl ethyl acetate, stirred for 15 min and then centrifuged for 2 min at 14 000 *g*. The ethyl acetate layer was transferred to a new tube and evaporated. The residue was dissolved in 150 μl methanol and 100 μl double distilled water and analyzed by HPLC using the Beckman Gold 7.11 HPLC system (Beckman Coulter, Krefeld, Germany) with a Bischoff ProntoSIL Spherisorb ODS2 Type NC separation column (5 μm– 250 mm×4.6 mm) (Bischoff Chromatography, Leonberg, Germany) at a flow rate of 1 ml min^{-1}. The sample injection volume was 10 μl, solvent A was water and 5% ammonium formate in formic acid (mixed at a ratio of 98:2, respectively) and solvent B methanol, double distilled water and ammonium formate in FA (mixed at a ratio of 88.2:9.8:2, respectively). The separation program was isocratic with 60% solvent B for 2 min, linear gradient to 100% solvent B for 5 min, 100% solvent B for 4 min, linear gradient to 60% solvent B for 1 min and isocratic with 60% solvent B for 3 min. The absorbance at 273 nm was used for detection. The specific enzyme activity was calculated using the total protein concentration of WCEs.

Results

Detection of *in vivo* S-nitrosylated proteins in green tissues of poplar

We aimed to survey proteins that are constitutively S-nitrosylated in poplar callus and leaf tissues. Therefore, protein extracts from grey poplar callus and leaves from control (C) and ozone-treated (O) plants were subjected to the biotin switch assay and Western blot analysis. Several bands were detected in callus and leaf extracts compared to controls (FP), indicating the presence of S-nitrosylated proteins *in vivo* (Figure S3). The most prominent band in the Western Blot of the leaf tissue was the ribulose-1,5-bisphosphate carboxylase/oxygenase (RuBisCO) large subunit (~50 kDa), a well-known target of S-nitrosylation in plants [11,24–26]. In the callus tissue, which was only slightly green (Figure S1), the amount of RuBisCO is decreased compared to that in the leaf samples, as indicated by the weaker Ponceau S-stained protein band on the gel (Figure S3). No bands were visualized on the Western blot of False Positive (FP) samples, the controls of biotin switch assay.

LC-MS/MS analysis of NeutrAvidin affinity chromatography-purified proteins allowed the identification (MASCOT ion scores >30 and protein identifications based on at least two unique peptides) of a total of 172 S-nitrosylated proteins (Table S1A). About one-third of the proteins (63 proteins) were common in all samples (callus, leaf C and leaf O), whereas 11 of the remaining 109 proteins were exclusively detected in ozone-treated leaves, 6 in the callus and 4 in control leaves (Figure 1A). To functionally categorize the proteins, a MapManBIN search (http://ppdb.tc.cornell.edu/dbsearch/searchacc.aspx) was performed using the accession numbers of the respective Arabidopsis orthologs (Figure 1B, Table S2). The 172 S-nitrosylated proteins were clustered in 9 main functional categories according to MapMan-BIN. 'Photosynthesis' represented the main group (26% of the total number of proteins), comprising the photosynthetic light reactions, Calvin cycle, photorespiration and enzymes of the tetrapyrrole synthesis pathway. 'Amino acid + Protein' represented the next frequent group (23%) containing enzymes of the amino acid metabolism and the protein synthesis, degradation and folding process. 18% of the proteins were associated to 'Primary metabolism'. These are amongst others aconitase (EC 4.2.1.3) and malate dehydrogenase (EC 1.1.1.37) of the citric acid cycle; ribulose-5-phosphate-3-epimerase (EC 5.1.3.1), transketolase (EC 2.2.1.1), and transaldolase (EC 2.2.1.2) of the pentose phosphate pathway and eight glycolytic enzymes, with phosphofructokinase

(EC 2.7.1.11) and pyruvate kinase (EC 2.7.1.40) regulating two committing steps in glycolysis. Ribulose-5-phosphate-3-epimerase of the pentose phosphate pathway was identified for the first time as target of S-nitrosylation in plants. Proteins assigned to 'Redox + Signaling', 'Secondary Metabolism' and 'Structural function' represented 7%, 4% and 4%, respectively. A total of 6% of the proteins could not be functionally annotated or are unknown proteins. Categories containing less than four proteins were summarized in 'Other' (9%). Our MS analysis revealed 43 not previously described candidates for S-nitrosylation in plants, including six enzymes of the polyphenol biosynthetic pathways: phenylalanine ammonia-lyase (EC 4.3.1.24; PAL2 and PAL3), chalcone synthase (EC 2.3.1.74; CHS), naringenin 3-dioxygenase (EC 1.14.11.9; F3H), caffeic acid 3-O-methyltransferase (EC 2.1.1.68; COMT) and polyphenol oxidase (EC 1.14.18.1; PPO) (see Table S1A, highlighted in bold).

Acute ozone stress evokes rapid changes in the S-nitroso-proteome of poplar

Acute ozone fumigation has been shown to induce transient NO production in Arabidopsis and tobacco plants [40,42,43,73]. Using a chemiluminescence-based assay, we measured the nitrosothiol (SNO) and nitrite contents in control and ozone-treated poplar leaves. After 1 hour of ozone exposure the nitrite content in O leaves was 4-fold higher and the SNO content was 3.5-fold higher, although only the nitrite content was found to be statistically different ($P<0.05$, Students *t*-test) (Figure 2).

To gain insight into the change of the S-nitrosylation signaling during ozone stress, we performed a quantitative proteomic analysis (label-free LC-MS/MS). Principal component analysis (PCA) revealed a clear ozone-induced change in the S-nitrosylated

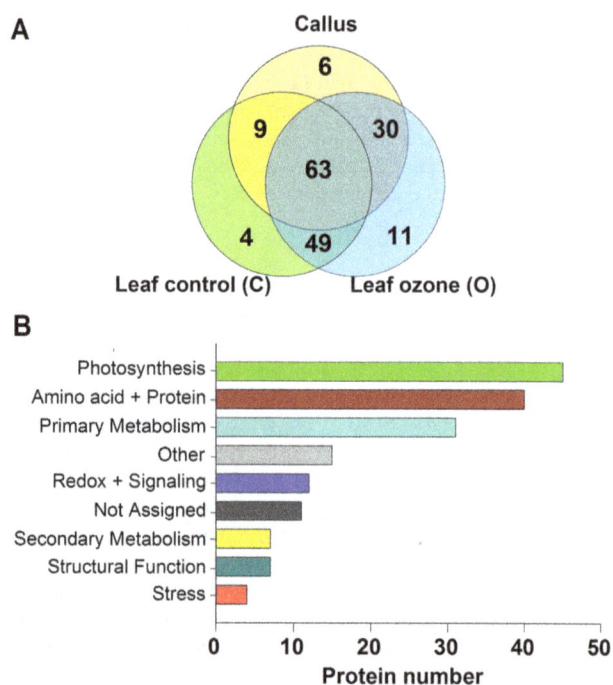

Figure 1. Identification of endogenously S-nitrosylated proteins in grey poplar. (A) VENN diagram visualizing the number of identified S-nitrosylated proteins by LC-MS/MS in callus and leaf tissue (C and O) of poplar. (B) Functional categorization of the 172 identified S-nitrosylated proteins in poplar callus and leaf samples according to MapManBINs (http://ppdb.tc.cornell.edu/dbsearch/searchacc.aspx).

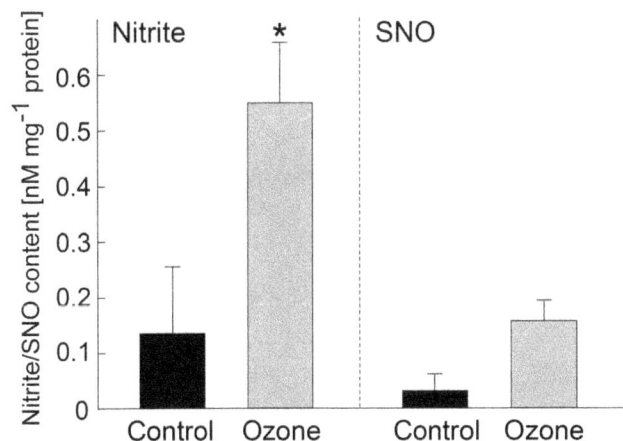

Figure 2. Nitrite and nitrosothiol (SNO) contents in grey poplar leaves upon acute ozone exposure (grey bars) compared to controls (black bars). Poplar plants were fumigated with ozone and harvested immediately after ozone treatment. 200 µg homogenized plant material was mixed with 600 µl PBS buffer containing NEM and EDTA. For nitrite and SNO determination, 100 µl and 200 µl of the plant extracts were, respectively, injected into the reaction vessel of Sievers NO analyzer containing the reducing agent tri-iodide. To measure SNO, the plant extracts were pre-treated with sulfanilamide to chemically remove nitrite. Values represent means ± SE of 3 biological replicates measured in three independent analyses. *$P < 0.05$ (Student's t-test).

protein profile of C and O leaves, as indicated by the separation of C from O samples along the second principal component (PC2) (Figure S4). The first component (PC1) accounts for most of the explained variance (57%) and shows large differences between samples containing S-nitrosylated proteins and their false-positive controls (FP-samples in Figure S4), indicating a reliable detection of S-nitrosylated proteins using the biotin switch assay and LC-MS/MS. OPLS was further employed for studying the S-nitroso-protein patterns of C and O samples in greater detail, and for evaluating to what extent the S-nitrosylated proteins became up- and down-regulated upon ozone exposure. The O samples were separated from the C samples by 43% explained variance of the PC1 (Figure 3A). The proteins that statistically discriminate between C and O samples are summarized in Table 1 together with the corresponding log2-fold changes and the additional statistical analysis (t-test). The analysis revealed that the content of 27 and 13 S-nitrosylated proteins were decreased and increased upon ozone fumigation, respectively. Among those 40 negatively and positively correlated proteins to ozone treatment, 32 proteins significantly changed at $P < 0.05$ (t-test; FDR 5%).

Specifically, O samples were negatively correlated to proteins of the secondary metabolism (see yellow squares in Figure 3B). Three enzymes of the secondary metabolism became de-nitrosylated in O samples as seen by the negative fold changes. The S-nitrosylation level of COMT ($\log2(O/C) = -3.4$) and PAL2 (-3.6) was significantly lower in the O samples ($P < 0.05$) whereas log fold changes < -1 but no statistical significance was found for PPO ($\log2 = -2.0$; t-test, $P = 0.064$). Another group of proteins affected by ozone was related to photosynthesis (Figure 3B, green triangles), mostly comprising proteins of the light-harvesting complex (chlorophyll a/b-binding proteins). Moreover, a significant decrease in S-nitrosylation was observed in the following: cyclophylin-type protein (-1.3), sucrose synthase (-3.0; EC 2.4.1.13), malate dehydrogenase (-1.3; EC 1.1.1.37), glyceraldehyde-3-phosphate dehydrogenase (-0.6; EC 1.2.1.59), stress-related EP3 chitinase (-4.7; EC 3.2.1.14), thaumatin (pathogen-

esis-related protein; -3.8), alpha-N-arabinofuranosidase (-1.0; EC 3.2.1.55), monodehydroascorbate reductase (-0.7; EC 1.6.5.4), multifunctional chaperone (14-3-3 family; -0.3), GDSL-like lipase/acylhydrolase (-1.1), ascorbate peroxidase (-0.9; EC 1.11.1.11) and two aldo/keto reductase family proteins (-0.8; -1.0).

The proteins which positively correlated to O samples in the OPLS are functionally categorized into 'Amino acid+Proteins', 'Photosynthesis', 'Primary Metabolism', 'Redox+Signaling', 'Structural Function' and 'Other' (Table 1, Figure 4). Calmodulin-like protein 6a and peroxiredoxin 5 show a significant increase in S-nitrosylation after ozone fumigation compared to the C samples ($\log2 = +2.8$ and $+0.9$, respectively) as well as aconitase ($+1.0$), chaperonin precursor ($+1.6$), porphobilinogen deaminase ($+2.6$; EC 2.5.1.61), glycine cleavage system protein H precursor ($+1.4$), ribulose-phosphate 3-epimerase ($+1.0$; EC 5.1.3.1), glutamate synthase ($+1.3$; EC 1.4.7.1) and tubulin alpha chain ($+1.1$). Taken together, one hour of ozone fumigation dramatically changed the S-nitrosylation status of leaf proteins, inducing rapid de-nitrosylation and S-nitrosylation events in 32 proteins.

Acute ozone stress and de-nitrosylation increase PAL enzyme activities

To demonstrate the effect of S-nitrosylation on an enzyme activity, we chose to analyze PAL, which is the first regulatory enzyme in the phenylpropanoid pathway of plants. It has been shown that short pulses of ozone (duration <10 hours) increase both PAL activity and PAL gene expression [44,74,75]. Here, we measured a 5-fold increase in PAL activity compared to that in untreated controls after 1 hour of ozone treatment (Figure 5A).

The *Populus trichocarpa* protein database comprises five isoenzymes of PAL (PAL1 to PAL5; Figure S5). Applying the biotin-switch assay, PAL2 and PAL3 were identified as candidates for S-nitrosylation (Table S1A) and PAL2 was revealed as significantly de-nitrosylated after ozone exposure (Table 1). To determine if de-nitrosylation affected the overall PAL enzyme activity, we measured the PAL activities of leaf extracts *in vitro* after incubation with sinapic acid (SIN) or GSNO (Figure 5B). SIN caused a significant ($P < 0.01$, paired t-test) increase in PAL activities, whereas GSNO did not, indicating that de-nitrosylation positively regulates PAL activity. The relative increase of PAL activity was approx. 25% (Figure 5B; insert).

The alignment of the amino acid sequences of the five PAL isoenzymes of *Populus trichocarpa* with PAL sequences from Arabidopsis, tobacco, parsley and pea revealed highly conserved cysteine residues (Figure S5). Of these cysteine residues, two were predicted by the GPS-SNO software to be S-nitrosylation sites in *Populus trichocarpa* PAL2 (Cys557 and Cys691) [76]. As shown in Figure 5C, Cys557 of PAL2 is located within the shield domain (highlighted in green). This domain narrows the substrate access to a small tunnel and is highly mobile [77]. Also in the shield-domain, a phosphorylation site was detected in the PAL of French bean (Thr545) [78]. This amino acid is also highly conserved in the amino acid sequence of poplar PAL (Figure S5). Phosphorylation at Thr545 (at the transition between the core domain and shield domain) might induce conformational changes in the flexible loop of the shield domain of each subunit (functional PAL is a tetramer of identical subunits), thereby changing the accessibility of the substrate to the active site of neighboring units.

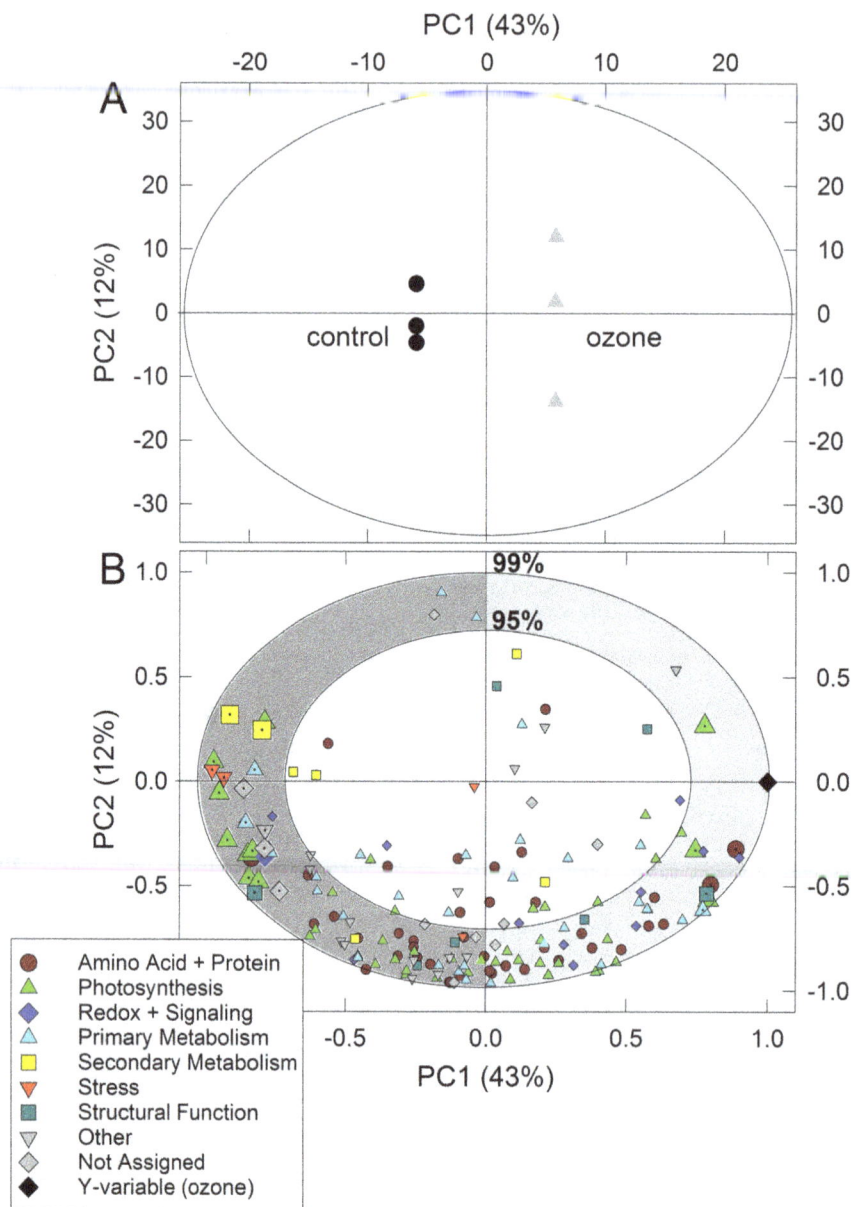

Figure 3. Two-dimensional (A) score and (B) scaled and centered loading plots of orthogonal partial least square regression (OPLS) of S-nitrosylated proteins (normalized by whole-cell extract) in poplar leaf samples determined using biotin switch assay and LC-MS/MS. The explained variance (in percentage) and the number of principal components (PC) are reported in the x- and y-axes in both (A) and (B) plots. The ellipse in (A) indicates the tolerance based on Hotelling's T^2 with the significance level of 0.05. The outer and inner ellipses in (B) indicate 100% and 75% explained variance, respectively. (A) control = black circles, ozone-treated = grey triangles; (B) each functional group of proteins (according to Figure 1C) is indicated with different symbols, zoomed symbols with a dot represent the significantly different proteins between C and O plants tested independently with Student's t-test (P<0.05 applying a FDR of 5%). Symbol legend: dark red circles = Amino acid metabolism and Protein synthesis, folding and degradation; blue diamonds = Redox and Signaling; cyan triangles-up = Primary metabolism; yellow square = Secondary metabolism; red triangles-down = Stress; dark green squares = Structural function; grey triangles-down = other; grey squares = not assigned or not identified. The black diamond indicates the Y-variable, i.e. ozone-treatment.

Discussion

Homeostatic level of S-nitrosylated proteins in poplar green tissues

This study represents the investigation of protein S-nitrosylation patterns in the woody model plant grey poplar. By performing a biotin switch assay optimized for poplar in conjunction with the quantitative LC-MS/MS analysis (label-free), we identified 172 proteins that are S-nitrosylated *in vivo* in callus tissue and fully developed poplar leaves under unstressed conditions and that appear after a short pulse of oxidative stress. The primary outcome is the observation that key enzymes of the phenylpropanoid and subsequent flavonoid and lignin biosynthetic pathways (PAL2, PAL3, CHS, COMT, F3H and PPO) are possible targets of S-nitrosylation. The polyphenol biosynthesis metabolism generates an enormous collection of secondary metabolites that are involved in plant development and stability (i.e., lignin, suberin, and condensed tannins) [79], plant defense responses against abiotic,

Table 1. Variable Importance for the Projection (VIP) for each discriminant protein that separates ozone fumigated poplar leaves (O) from control leaves (C) in the OPLS model.

Functional category	Accession	VIP score	SE	Description	Function (MapManBIN)	Log2 (O/C)
Decrease in S-nitrosylation after ozone						
Amino acid+ Protein	POPTR_0017s10920.1	1.721	0.797	Cyclophylin-type protein	Protein/folding	−1.3*
	POPTR_0021s00380.1	1.256	1.202	T-protein of glycine decarboxylase complex	Amino acid metabolism/degradation	−0.6
Photosynthesis	POPTR_0001s13800.1	1.997	0.513	Chlorophyll a/b-binding protein 2 precursor	Photosynthesis/lightreaction	−2.5**
	POPTR_0015s07340.1	1.955	0.591	Chlorophyll a/b-binding protein 4	Photosynthesis/lightreaction	−2.3**
	POPTR_0005s26080.1	1.896	0.510	Chlorophyll a/b-binding protein 2	Photosynthesis/lightreaction	−1.6**
	POPTR_0001s11600.1	1.745	0.702	Photosystem I reaction center subunit III	Photosynthesis/lightreaction	−1.4*
	POPTR_0019s09140.1	1.738	0.827	Chlorophyll a/b-binding protein CP26	Photosynthesis/lightreaction	−1.5*
	POPTR_0011s02770.1	1.710	0.998	Chlorophyll a/b-binding protein 2	Photosynthesis/lightreaction	−3.7*
	POPTR_0010s22790.1	1.669	0.952	Chlorophyll a/b-binding protein, putative	Photosynthesis/lightreaction	−3.5*
	POPTR_0008s15100.1	1.627	1.103	Photosystem I 20 kD protein	Photosynthesis/lightreaction	−0.5*
	POPTR_0008s06720.1	1.288	1.213	Chlorophyll a/b-binding protein	Photosynthesis/lightreaction	−0.8
Primary Metabolism	POPTR_0018s07380.1	1.760	0.874	Sucrose synthase	Major CHO metabolism/degradation	−3.0*
	POPTR_0018s09380.1	1.697	1.007	Malate dehydrogenase	TCA	−1.3*
	POPTR_0012s09570.1	1.593	1.091	Glyceraldehyde-3-phosphate dehydrogenase	Glycolysis	−0.6*
	POPTR_0008s05640.1	1.233	1.229	Triosephosphate isomerase, cytosolic	Glycolysis	−0.7
Secondary Metabolism	POPTR_0012s00670.1	1.877	0.817	Caffeic acid 3-O-methyltransferase	Secondary metabolism/ phenylpropanoids	−3.4**
	POPTR_0008s03810.1	1.644	1.107	Phenylalanine ammonia-lyase 2	Secondary metabolism/ phenylpropanoids	−3.6*
	POPTR_0001s39630.1	1.415	1.222	Polyphenol oxidase	Secondary metabolism	−2.0
Stress	POPTR_0019s12370.1	2.010	0.496	EP3 chitinase	Stress/biotic	−4.7**
	POPTR_0018s10490.1	1.922	0.775	Thaumatin, pathogenesis-related protein	Stress/abiotic	−3.8**
Structural Function	POPTR_0016s02620.1	1.696	0.805	Alpha-N-arabinofuranosidase	Cell wall	−1.0*
Redox + Signaling	POPTR_0006s11570.1	1.774	1.004	Monodehydroascorbate reductase	Redox/ascorbate and glutathione	−0.7*
	POPTR_0004s10120.1	1.570	1.172	Multifunctional chaperone (14-3-3 family)	Signaling	−0.3*
Other	POPTR_0018s09580.1	1.622	1.146	GDSL-like lipase/acylhydrolase	Misc/GDSL-motif lipase	−1.1*
Not Assigned	POPTR_0005s17350.1	1.777	0.978	Ascorbate peroxidase, putative	Not assigned	−0.9*
	POPTR_0005s22860.1	1.627	1.106	Aldo/keto reductase family protein	Not assigned	−0.8*
	POPTR_0002s05640.1	1.518	1.148	Aldo/keto reductase family protein	Not assigned	−1.0*
Increase in S-nitrosylation after ozone						
Amino acid+ Proteins	POPTR_0002s12130.1	1.824	0.889	Aconitase	Protein/degradation	1.0**
	POPTR_0003s20870.1	1.640	1.017	Chaperonin precursor	Protein/folding	1.6*
Photosynthesis	POPTR_0007s07680.1	1.623	1.062	Porphobilinogen deaminase	Tetrapyrrole synthesis	2.6*
	POPTR_0003s08760.1	1.595	1.050	Glycine cleavage system protein H precursor	Photosynthesis/photorespiration	1.4*
	POPTR_0015s07330.1	1.528	1.066	Ribulose-phosphate 3-epimerase	Photosynthesis/calvincycle	1.0*
	POPTR_0014s11580.1	1.427	1.300	CP12 domain-containing protein	Photosynthesis/calvincycle	1.9
Primary Metabolism	POPTR_0006s03660.1	1.583	1.151	Glutamate synthase, ferredoxin-dependent	N-metabolism	1.3*

Table 1. Cont.

Functional category	Accession	VIP score	SE	Description	Function (MapManBIN)	Log2 (O/C)
	POPTR_0016s03630.1	1.434	1.275	Glutamate synthase, ferredoxin-dependent	N-metabolism	1.9
Redox+ Signaling	POPTR_0001s22980.1	1.848	0.869	Calmodulin-like protein 6a	Signaling/calcium	2.9**
	POPTR_0013s10250.1	1.586	1.077	Peroxiredoxin 5, putative	Redox	0.8*
	POPTR_0002s08260.1	1.416	1.110	Protein disulfide-isomerase, precursor	Redox	0.7
Structural Function	POPTR_0001s29670.1	1.611	1.055	Tubulin alpha chain	Cell/organisation	1.1*
Other	POPTR_0001s23310.1	1.386	1.226	Glycosyl hydrolase family protein	Misc/gluco-, galacto- and mannosidases	1.0

Proteins showing VIP >1 and uncertainty bars of jack-knifing method (SE) < than the respective VIP value were defined as discriminant proteins. Additionally, discriminant proteins were tested for significance difference ($P<0.05$) between C and O plants independently from multivariate data analysis using t-test (* = $P<0.05$; ** = $P<0.01$) applying a FDR of 5%. Log fold changes (ozone/control) were calculated from normalized protein abundances (Table S1C). Annotation and functional categorization was obtained from Phytozome and PPDB (http://www.phytozome.net/; http://ppdb.tc.cornell.edu/dbsearch/searchacc.aspx).

and biotic environmental constraints (i.e., stilbenes, flavonoids, and anthocyanins) [80–83] and signaling (salicylic acid, and benzoic acids) [84].

Strikingly, many enzymes belonging to the plant primary carbohydrate metabolism (citric acid cycle, glycolysis, pentose phosphate pathway) were identified as targets of S-nitrosylation. S-nitrosylation of the glycolytic enzyme GAPDH results in a reduction of enzyme activity in animals and plants and is a well-known example of how S-nitrosylation negatively affects the enzyme function [12,85,86]. It has been proposed that under oxidative stress, the glycolytic pathway is reduced and glucose equivalents are redirected into the pentose phosphate pathway [87–89], which is essential for maintaining the cytoplasmic NADPH concentration as a base for the anti-oxidative defense systems [90,91].

Our proteomic analysis in poplar revealed that photosynthesis is another important cellular process which is regulated by protein S-nitrosylation, supporting observations from previous studies [11,15,24,92,93]. We observed an increase in the number of S-nitrosylated proteins related to photosynthetic processes when

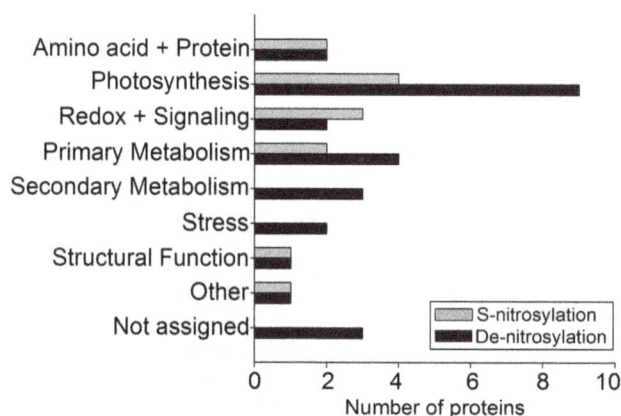

Figure 4. Number of proteins with increased (grey bars) or decreased (dark grey bars) level of S-nitrosylation in O samples of grey poplar according to Table 1. The proteins were grouped based on their function.

comparing green, un-differentiated calli with fully developed poplar leaves. Various members of the photosynthetic light reaction (e.g., oxygen-evolving complex, proteins of the light-harvesting complex, plastocyanin and thylakoid formation protein (THF)), as well as enzymes of the Calvin cycle (i.e. RuBisCO, RuBisCO activase, phosphoglycerate kinase and aldolase), are S-nitrosylated under steady-state conditions. The THF 1 protein, a new candidate in the context of S-nitrosylation, is a re-modeling factor of Photosystem II (PSII)-Light Harvesting Complex II and is involved in the repair cycle of PSII upon photo damage in Arabidopsis [94]. In addition to the regulation of photochemical aspects, four enzymes of the tetrapyrrole biosynthetic pathway are S-nitrosylated in poplar: glutamate 1-semialdehyde aminotransferase (EC 5.4.3.8), porphobilinogen synthase (syn. aminolevulinate dehydratase; EC 4.2.1.24), porphobilinogen deaminase (syn. hydroxymethylbilane synthase; EC 2.5.1.61) and coproporphyrinogen III oxidase (EC 1.3.3.3). These enzymes catalyze the 3rd, 4th, 5th, and 8th steps of the plastidic porphyrin biosynthesis, respectively [95], providing the backbones for chlorophyll and heme molecules. The identification of porphobilinogen synthase and glutamate 1-semialdehyde aminotransferase as targets of S-nitrosylation is described for the first time. Taken together, the identification of enzymes involved in glycolysis, the pentose phosphate pathway and chlorophyll biosynthesis as targets of S-nitrosylation in plants raises the general question as to how S-nitrosylation is globally involved in the fine tuning of the photosynthetic activity, pigment turnover, Calvin cycle processes and subsequent channeling of the photosynthetic metabolites that are related to the modifications of the glycolytic and pentose phosphate pathway proteins.

Acute ozone stress alters the homeostasis of S-nitrosylated proteins in poplar

To provide information regarding the biological relevance of S-nitrosylation, the comparative analysis of the S-nitroso-proteome under control and stress conditions is an important tool. At the present, no information is available regarding ozone-induced changes in the S-nitroso-proteome of plants and this is the first study that comprehensively describes it. Here, we applied a short, strong ozone pulse as a model to trigger ROS and NO formation

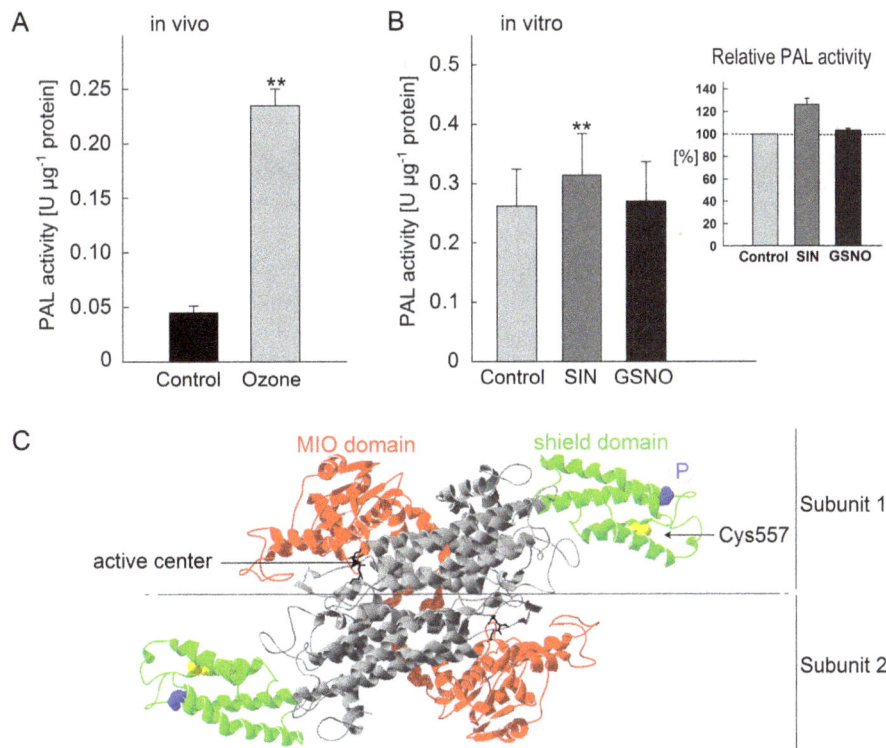

Figure 5. Effect of SIN, GSNO and ozone on the activity of PAL in grey poplar leaves. (A) The PAL activity was analyzed at the end of ozone fumigation or in control samples at the same time point. The values represent the means ± SE of three experiments. **P<0.01 (Student's t-test for unpaired samples). (B) Effect of SIN (removal of NO groups) and GSNO (NO-donor) on in vitro PAL activity. 12-week-old plants grown under greenhouse conditions were used as leaf material. Before measuring the enzyme activity, the leaf extracts were pre-incubated for 1 hour with 3 mM SIN or 500 µM GSNO. Each value represents the mean of nine replicates ± SE and significant differences are given with ** for P<0.01 (Student's t-test for paired samples). The relative PAL activity was calculated by comparing each replicate in the SIN and GSNO group to the corresponding value in the control group. The PAL activity in the control was taken as 100%. (C) Structural model of PAL2 from *Populus trichocarpa*. The three-dimensional structure was modeled according the crystal structure of *Petroselinum crispum* (PDB code: 1w27B) using SWISS-MODEL [125,126]. For clarity, a dimer of the actual tetrameric PAL2 is shown. The flexible shield domain is highlighted in green and contains the cysteine predicted to be S-nitrosylated in PAL2 (Cys557; yellow) and the postulated phosphorylation site at Thr549 (blue). The MIO domain (red) contains the catalytic active center (black).

in order to analyze the fast responses in S-nitrosylation/de-nitrosylation.

The cumulative ozone uptake in the present experiment was 110 ± 19 µmol m^{-2}, a value of comparable magnitude also applied in earlier studies [57,58]. We observed no immediate decrease in net CO_2 assimilation and transpiration rates upon ozone treatment, demonstrating that grey poplar tolerates high acute ozone doses, an observation already mentioned before [58,96].

Short-term acute ozone fumigation is often used to mimic the HR [35,40] and therefore the present shifts in the S-nitrosylation pattern might be transferable to early events in leaf pathogenesis. The accumulation of NO and nitrite is a common feature of short-term and chronic ozone fumigation [42,43,73]. We observed a rapid nitrite increase and a slight increase in the nitrosothiol (SNO) content in response to the short-term ozone treatment (Figure 2). Increased nitrite content is often observed upon abiotic stresses and is linked to S-nitrosylation events [97]. It has been shown that nitrite induces S-nitrosylation and the subsequent inactivation of the protease caspase-3 [98]. Nitrite, a reservoir for NO, can be reduced back to NO via non-enzyme-dependent reactions [99,100] or enzymatically by nitrate reductase [101]. Higher levels of SNOs comprise GSNO, a low-molecular weight SNO. GSNO is considered an intracellular NO donor in trans-nitrosylation reactions [102,103].

Ozone fumigation highly modulated the S-nitroso-proteome. Overall, the S-nitrosylation pattern of 28 proteins was significantly affected. The fact that we observed both S-nitrosylation (one-third of total) and de-nitrosylation (two-thirds of total) events upon ozone stress seems to be contradictory to the accumulation of nitrite and SNO that we measured in response to ozone. However, two other studies [15,51] also show this phenomenon: despite an increase in SNO in *Brassica juncea* plants undergoing low temperature stress, Abat and Deswal [15] identified nine proteins undergoing enhanced S-nitrosylation and eight proteins becoming de-nitrosylated. A similar trend was observed following salt stress [51]. Together with our observations, it becomes clear that the accumulation of NO or SNO in plant cells does not concomitantly induce a global response leading to general S-nitrosylation of target proteins. The NO production/turnover and the NO targets for S-nitrosylation may be spatially or temporally separated. The kinetics and the localization of the NO production in leaves as measured by NO-specific fluorophores demonstrated that the NO production begins within the chloroplasts and subsequently propagates into the nucleus and the cytosol [104,105]. The stress-dependent activation of de-nitrosylases could provide another explanation for de-nitrosylation events in the presence of enhanced NO and nitrite concentrations. In the case of phosphorylation, the phosphorylation status of proteins depends on the activities of kinases and phosphatases. Similarly, the extent of S-nitrosylation in any protein will depend on the rates of S-nitrosylation and de-nitrosylation [6]. Recent observations classify GSNO reductase and the thioredoxin/thioredoxin reductase system as denitrosylases [19,106,107]. These enzymes might themselves be regulated by NO [108–111]. Moreover, de-nitrosylation is not exclusively enzymatically driven. Non-enzymatic de-nitrosylation can occur due to alterations in the cellular redox environment (pH and pO_2 shifts) [112].

De-nitrosylation regulates PAL activity in response to acute ozone

S-nitrosylation events can alter the outcome of a signaling pathway by switching on/off target proteins. The regulation of enzyme activity by S-nitrosylation was demonstrated for several plant proteins [10]. Strikingly, S-nitrosylation is often mentioned in conjunction with an inhibition of the enzyme activity, whereas de-nitrosylation can promote the enzyme activity [13–15,28,113–116]. Our quantitative proteomic analysis revealed three enzymes that are involved in different phenolic pathways as being de-nitrosylated in response to ozone fumigation (COMT, PAL2, PPO). PAL is the first and committing enzyme of the phenylpropanoid pathway, which directs the carbon flow from the shikimate pathway to the various branches of the general phenylpropanoid metabolism. The observed increase in the overall PAL activity upon ozone exposure (5-fold) is higher than the increase in enzyme activity that we observed from incubation with SIN, promoting de-nitrosylation (approximately 25%). This difference might be explained by other PTMs, such as de-phosphorylation events, after ozone exposure. Indeed, a phosphorylation site has been identified in the PAL of French bean (*Phaseolus vulgaris* L.). Phosphorylation at this site (Thr545) was associated with a decrease in the V_{max} and an increased turnover of the PAL subunits [78,117]. Our multiple sequence alignment of the five PAL isoenzymes of *Populus trichocarpa* showed that the phosphorylation site of the French bean PAL corresponds with highly conserved threonine residues in poplar. Analogous to phosphorylation, S-nitrosylation/de-nitrosylation may also modify the catalytic activity of PAL by changing the substrate access to the active center as indicated by the observed increase of the overall *in vitro* PAL activity following incubation with SIN. Future studies with heterologous expressed PAL proteins (i.e. PcPAL2 and PcPAL3) site-directly modified at the Cys557 probably will clarify this initial observation. Interestingly, NO fumigation or NO donor treatment is often associated with a rapid increase in the activity of PAL concomitant with an accumulation of phenolic and flavonoid compounds [118–120]. In our work, COMT became significantly de-nitrosylated upon ozone fumigation. COMT is an essential enzyme that is involved in monolignol biosynthesis, influencing the lignin content and lignin structure in poplar [121]. In poplar, the transcription of COMT, and hence the lignin content, can be induced by ozone, leading to an enhanced ozone tolerance due to the antioxidative properties of lignin [122].

As de-nitrosylation of enzymes is often observed with an up-regulation of enzyme activity [19] we hypothesize that de-nitrosylation represents another PTM to regulate enzyme activities, contributing to a fast up-regulation of the metabolic flux through the phenylpropanoid pathway within minutes upon onset of the stress event producing the biosynthetic precursor of all kinds of defense metabolites in ozone-stressed poplars. At longer time scales (hours to days) accumulation of polyphenols would be achieved by an up-regulation of *de-novo* synthesis [44,74,82,123,124]. This hypothesis, however, needs to be proven by future detailed biochemical analysis of the respective enzymes.

Supporting Information

Figure S1   Pictures of *Populus* x *canescens* callus tissue. Leaf (A) and stem (B) explants of poplar were cultured in darkness on callus induction medium for 3 weeks at 20°C. (C) Fully developed calli were transferred to shoot induction medium and maintained under moderate light (16/8 h photoperiod, PPFD 125 µmol photons m^{-2} s^{-1}).

Figure S2   Experimental setup to test the uptake of ozone into the poplar leaves. (A) Glass cuvette with enclosed poplar plant. (B) Ozone concentrations measured at the inlet (empty circles) and outlet (filled circles) of the empty cuvette and when the plant was enclosed. (C) Response of poplar assimilation (filled squares) and transpiration rates (empty squares) to acute, short-term ozone exposure. (D) Foliar ozone flux during light and dark periods. Grey shaded box indicates when the light was turned off.

Figure S3   Western blot showing *in vivo* S-nitrosylated proteins in callus, leaf tissue (C) and leaves subjected to ozone (O), including controls for false-positives (FP). Callus and leaf extracts underwent the biotin switch assay, were separated by SDS-PAGE and were blotted onto nitrocellulose membrane. Biotinylated (= S-nitrosylated) proteins were detected by an anti-biotin antibody. The lower part shows the PonceauS-stained membrane for loading control. The biotin switch assay was repeated three times with similar results.

Figure S4   Two-dimensional (A) score and (B) scaled and centered loading plots of principal component analysis (PCA) of S-nitrosylated proteins in poplar leaf samples identified using biotin switch assay and LC-MS/MS. The explained variance (in percentage) and the number of principal components (PC) are reported in x- and y-axes in both (A) and (B) plots. Ellipse in (A) indicates the tolerance based on Hotelling's T2 with significance level of 0.05. The outer and inner ellipses in (B) indicate 100% and 50% explained variance, respectively. A, control = black circles; ozone-treated = grey triangles; false positive (FP) control = black circle white-filled; FP ozone-treated = grey triangle white-filled; B, each functional group of proteins is indicated with different symbols, zoomed symbols with a dot represent the significantly different proteins between C and O plants tested independently with Student's t-test (P<0.05 applying a FDR of 5%). Symbol legend: dark red circles = Amino acid metabolism and Protein synthesis, folding and degradation; blue diamonds = Redox and Signaling; cyan triangles-up = Primary metabolism; yellow squares = Secondary metabolism; red triangles-down = Stress; dark green squares = Structural function; grey triangles-down = other; grey squares = not assigned or not identified.

Figure S5   Multiple alignment of PAL protein sequences from different species. The alignment was performed with COBALT tool from NCBI. The five isoenzymes of *Populus trichocarpa* PAL (ACC63888.1, EEE89380, ACC63887.1, EEF04645, and XP_002315308) were aligned with PAL1 from parsley (*Petroselinum crispum*: CAA68938.1), Arabidopsis (*Arabidopsis thaliana*: AEC09341.1), tobacco (*Nicotiana tabacum*: BAA22963.1) and pea (*Pisum sativum*: Q01861.1). All of the cysteine residues are highlighted in yellow. Cysteine residue predicted to be targets of S-nitrosylation by GPS-SNO software [76] are highlighted in red. The active center of the PAL is defined by the Ala-Ser-Gly tripeptide (framed in green). Red letters indicate highly conserved positions (identical amino acid in all aligned species) and blue letters indicate less conserved ones.

Table S1 Full data set for protein identification of (A) S-nitroso-proteins, (B) total proteins from whole cell extracts (WCE) and (C) corresponding protein abundances after label-free quantification and normalization process. (A) Identified S-nitrosylated proteins of grey poplar callus and leaf tissue (control (C), ozone-fumigated (O)). Only proteins with at least two unique peptides were used for further analyses (black) (grey = proteins with < than 2 unique peptides used for quantification). Enzymes of the polyphenol metabolism are highlighted in bold. (B) Identified total proteins of grey poplar whole cell extracts (WCE) from C and O samples. MASCOT search of acquired spectra conveyed a total of 1,461 detectable proteins in C and O samples with 3 replicates in each. Annotations of proteins are only given for those proteins detected as S-nitrosylated in Table S1A (black). (C) Normalization of protein abundances from S-nitroso-proteins with the averages of the protein abundances from total proteins. The abundance of each S-nitrosylated protein was normalized to the corresponding, averaged abundance in the whole-cell extracts (WCE) of the C and O leaves. For the very low abundant proteins, which were not detected in the WCE, the S-nitroso-protein was normalized to a "calculated total protein" (marked in red). The "calculated total protein" was the S-nitroso-protein abundance multiplied by 100 and divided by the median (i.e. 2.2) of percentages between all S-nitroso-protein and the respective measured total proteins. Log2 fold changes were calculated from the means of the normalized C and O replicates.

Table S2 Functional categorization of S-nitrosylated proteins according to MapManBINs (http://ppdb.tc.cornell.edu/dbsearch/mapman.aspx). BINs (major functional categories) and subcategories (subBINs) are given with the corresponding number of proteins.

Acknowledgments

We greatly acknowledge the technical assistance of S. Stich (HMGU-BIOP, Neuherberg, Germany) for the PAL analysis and H. Lang (HMGU-EUS, Neuherberg, Germany) for the ozone fumigation.

Author Contributions

Conceived and designed the experiments: EV JPS CL JD. Performed the experiments: EV AG JM. Analyzed the data: EV AG JM WH SMH. Contributed to the writing of the manuscript: EV AG JM CL WH SMH JD JPS.

References

1. Prado AM, Porterfield DM, Feijo JA (2004) Nitric oxide is involved in growth regulation and re-orientation of pollen tubes. Development 131: 2707–2714.
2. Beligni MV, Lamattina L (2000) Nitric oxide stimulates seed germination and de-etiolation, and inhibits hypocotyl elongation, three light-inducible responses in plants. Planta 210: 215–221.
3. He Y, Tang RH, Hao Y, Stevens RD, Cook CW, et al. (2004) Nitric oxide represses the Arabidopsis floral transition. Science 305: 1968–1971.
4. Neill SJ, Desikan R, Clarke A, Hancock JT (2002) Nitric oxide is a novel component of abscisic acid signaling in stomatal guard cells. Plant Physiology 128: 13–16.
5. Pagnussat GC, Simontacchi M, Puntarulo S, Lamattina L (2002) Nitric oxide is required for root organogenesis. Plant Physiology 129: 954–956.
6. Lane P, Hao G, Gross SS (2001) S-nitrosylation is emerging as a specific and fundamental posttranslational protein modification: head-to-head comparison with O-phosphorylation. Sci STKE 2001: re1.
7. Corpas FJ, Leterrier M, Valderrama R, Airaki M, Chaki M, et al. (2011) Nitric oxide imbalance provokes a nitrosative response in plants under abiotic stress. Plant Sci 181: 604–611.
8. Yu MD, Yun BW, Spoel SH, Loake GJ (2012) A sleigh ride through the SNO: regulation of plant immune function by protein S-nitrosylation. Current Opinion in Plant Biology 15: 424–430.
9. Gow AJ, Farkouh CR, Munson DA, Posencheg MA, Ischiropoulos H (2004) Biological significance of nitric oxide-mediated protein modifications. American Journal of Physiology - Lung Cellular and Molecular Physiology 287: L262–268.
10. Astier J, Kulik A, Koen E, Besson-Bard A, Bourque S, et al. (2012) Protein S-nitrosylation: what's going on in plants? Free Radical Biology & Medicine 53: 1101–1110.
11. Lindermayr C, Saalbach G, Durner J (2005) Proteomic identification of S-nitrosylated proteins in Arabidopsis. Plant Physiology 137: 921–930.
12. Padgett CM, Whorton AR (1995) S-nitrosoglutathione reversibly inhibits GAPDH by S-nitrosylation. American Journal of Physiology 269: C739–749.
13. Lindermayr C, Saalbach G, Bahnweg G, Durner J (2006) Differential inhibition of Arabidopsis methionine adenosyltransferases by protein S-nitrosylation. Journal of Biological Chemistry 281: 4285–4291.
14. Romero-Puertas MC, Laxa M, Matte A, Zaninotto F, Finkemeier I, et al. (2007) S-nitrosylation of peroxiredoxin II E promotes peroxynitrite-mediated tyrosine nitration. Plant Cell 19: 4120–4130.
15. Abat JK, Deswal R (2009) Differential modulation of S-nitrosoproteome of Brassica juncea by low temperature: change in S-nitrosylation of Rubisco is responsible for the inactivation of its carboxylase activity. Proteomics 9: 4368–4380.
16. Qu J, Nakamura T, Cao G, Holland EA, McKercher SR, et al. (2011) S-Nitrosylation activates Cdk5 and contributes to synaptic spine loss induced by beta-amyloid peptide. Proceedings of the National Academy of Sciences of the United States of America 108: 14330–14335.
17. Liu DH, Yuan FG, Hu SQ, Diao F, Wu YP, et al. (2013) Endogenous nitric oxide induces activation of apoptosis signal-regulating kinase 1 via S-nitrosylation in rat hippocampus during cerebral ischemia-reperfusion. Neuroscience 229: 36–48.
18. Hu SQ, Ye JS, Zong YY, Sun CC, Liu DH, et al. (2012) S-nitrosylation of mixed lineage kinase 3 contributes to its activation after cerebral ischemia. Journal of Biological Chemistry 287: 2364–2377.
19. Benhar M, Forrester MT, Stamler JS (2009) Protein denitrosylation: enzymatic mechanisms and cellular functions. Nature Reviews Molecular Cell Biology 10: 721–732.
20. Mannick JB, Hausladen A, Liu L, Hess DT, Zeng M, et al. (1999) Fas-induced caspase denitrosylation. Science 284: 651–654.
21. Erwin PA, Lin AJ, Golan DE, Michel T (2005) Receptor-regulated dynamic S-nitrosylation of endothelial nitric-oxide synthase in vascular endothelial cells. Journal of Biological Chemistry 280: 19888–19894.
22. Jaffrey SR, Snyder SH (2001) The biotin switch method for the detection of S-nitrosylated proteins. Sci STKE 2001: pl1.
23. Serpa V, Vernal J, Lamattina L, Grotewold E, Cassia R, et al. (2007) Inhibition of AtMYB2 DNA-binding by nitric oxide involves cysteine S-nitrosylation. Biochemical and Biophysical Research Communications 361: 1048–1053.
24. Abat JK, Mattoo AK, Deswal R (2008) S-nitrosylated proteins of a medicinal CAM plant Kalanchoe pinnata-ribulose-1,5-bisphosphate carboxylase/oxygenase activity targeted for inhibition. FEBS Journal 275: 2862–2872.
25. Fares A, Rossignol M, Peltier JB (2011) Proteomics investigation of endogenous S-nitrosylation in Arabidopsis. Biochemical and Biophysical Research Communications 416: 331–336.
26. Romero-Puertas MC, Campostrini N, Matte A, Righetti PG, Perazzolli M, et al. (2008) Proteomic analysis of S-nitrosylated proteins in Arabidopsis thaliana undergoing hypersensitive response. Proteomics 8: 1459–1469.
27. Chaki M, Fernandez-Ocana AM, Valderrama R, Carreras A, Esteban FJ, et al. (2009) Involvement of reactive nitrogen and oxygen species (RNS and ROS) in sunflower-mildew interaction. Plant and Cell Physiology 50: 265–279.
28. Ortega-Galisteo AP, Rodriguez-Serrano M, Pazmino DM, Gupta DK, Sandalio LM, et al. (2012) S-Nitrosylated proteins in pea (Pisum sativum L.) leaf peroxisomes: changes under abiotic stress. Journal of Experimental Botany 63: 2089–2103.
29. Agrawal GK, Rakwal R, Yonekura M, Kubo A, Saji H (2002) Proteome analysis of differentially displayed proteins as a tool for investigating ozone stress in rice (Oryza sativa L.) seedlings. Proteomics 2: 947–959.
30. Cho K, Shibato J, Agrawal GK, Jung YH, Kubo A, et al. (2008) Integrated transcriptomics, proteomics, and metabolomics analyses to survey ozone responses in the leaves of rice seedling. Journal of Proteome Research 7: 2980–2998.
31. Bohler S, Bagard M, Oufir M, Planchon S, Hoffmann L, et al. (2007) A DIGE analysis of developing poplar leaves subjected to ozone reveals major changes in carbon metabolism. Proteomics 7: 1584–1599.

32. Renaut J, Bohler S, Hausman JF, Hoffmann L, Sergeant K, et al. (2009) The impact of atmospheric composition on plants: a case study of ozone and poplar. Mass Spectrom Rev 28: 495–516.

33. Galant A, Koester RP, Ainsworth EA, Hicks LM, Jez JM (2012) From climate change to molecular response: redox proteomics of ozone-induced responses in soybean. New Phytologist 194: 220–229.

34. Qiu QS, Huber JL, Booker FL, Jain V, Leakey ADB, et al. (2008) Increased protein carbonylation in leaves of Arabidopsis and soybean in response to elevated [CO(2)]. Photosynthesis Research 97: 155–166.

35. Kangasjarvi J, Jaspers P, Kollist H (2005) Signalling and cell death in ozone-exposed plants. Plant Cell and Environment 28: 1021–1036.

36. Langebartels C, Wohlgemuth H, Kschieschan S, Grun S, Sandermann H (2002) Oxidative burst and cell death in ozone-exposed plants. Plant Physiology and Biochemistry 40: 567–575.

37. Fares S, Vargas R, Detto M, Goldstein AH, Karlik J, et al. (2013) Tropospheric ozone reduces carbon assimilation in trees: estimates from analysis of continuous flux measurements. Global Change Biology 19: 2427–2443.

38. Ainsworth EA, Rogers A, Leakey ADB (2008) Targets for crop biotechnology in a future high-CO_2 and high-O_3 world. Plant Physiology 147: 13–19.

39. Vainonen JP, Kangasjarvi J (2014) Plant signalling in acute ozone exposure. Plant Cell and Environment.

40. Rao MV, Davis KR (2001) The physiology of ozone induced cell death. Planta 213: 682–690.

41. Lamb C, Dixon RA (1997) The Oxidative Burst in Plant Disease Resistance. Annual Review of Plant Physiology and Plant Molecular Biology 48: 251–275.

42. Ahlfors R, Brosche M, Kollist H, Kangasjarvi J (2009) Nitric oxide modulates ozone-induced cell death, hormone biosynthesis and gene expression in Arabidopsis thaliana. Plant Journal 58: 1–12.

43. Mahalingam R, Jambunathan N, Gunjan SK, Faustin E, Weng H, et al. (2006) Analysis of oxidative signalling induced by ozone in Arabidopsis thaliana. Plant Cell and Environment 29: 1357–1371.

44. Pasqualini S, Piccioni C, Reale L, Ederli L, Della Torre G, et al. (2003) Ozone-induced cell death in tobacco cultivar Bel W3 plants. The role of programmed cell death in lesion formation. Plant Physiology 133: 1122–1134.

45. Durner J, Wendehenne D, Klessig DF (1998) Defense gene induction in tobacco by nitric oxide, cyclic GMP, and cyclic ADP-ribose. Proceedings of the National Academy of Sciences of the United States of America 95: 10328–10333.

46. Wang Y, Lin A, Loake GJ, Chu C (2013) H_2O_2-induced leaf cell death and the crosstalk of reactive nitric/oxygen species. Journal of Integrative Plant Biology 55: 202–208.

47. Tuskan GA, Difazio S, Jansson S, Bohlmann J, Grigoriev I, et al. (2006) The genome of black cottonwood, Populus trichocarpa (Torr. & Gray). Science 313: 1596–1604.

48. Jansson S, Douglas CJ (2007) Populus: A model system for plant biology. Annual Review of Plant Biology 58: 435–458.

49. Aylott MJ, Casella E, Tubby I, Street NR, Smith P, et al. (2008) Yield and spatial supply of bioenergy poplar and willow short-rotation coppice in the UK. New Phytologist 178: 358–370.

50. Laureysens I, De Temmerman L, Hastir T, Van Gysel M, Ceulemans R (2005) Clonal variation in heavy metal accumulation and biomass production in a poplar coppice culture. II. Vertical distribution and phytoextraction potential. Environmental Pollution 133: 541–551.

51. Tanou G, Job C, Rajjou L, Arc E, Belghazi M, et al. (2009) Proteomics reveals the overlapping roles of hydrogen peroxide and nitric oxide in the acclimation of citrus plants to salinity. Plant Journal 60: 795–804.

52. Behnke K, Ehlting B, Teuber M, Bauerfeind M, Louis S, et al. (2007) Transgenic, non-isoprene emitting poplars don't like it hot. Plant Journal 51: 485–499.

53. Leple J, Brasileiro A, Michel M, Delmotte F, Jouanin L (1992) Transgenic poplars: expression of chimeric genes using four different constructs. Plant Cell Reports 11: 137–141.

54. Emberson LD, Wieser G, Ashmore MR (2000) Modelling of stomatal conductance and ozone flux of Norway spruce: comparison with field data. Environmental Pollution 109: 393–402.

55. Beauchamp J, Wisthaler A, Hansel A, Kleist E, Miebach M, et al. (2005) Ozone induced emissions of biogenic VOC from tobacco: relationships between ozone uptake and emission of LOX products. Plant Cell and Environment 28: 1334–1343.

56. Fares S, Goldstein A, Loreto F (2010) Determinants of ozone fluxes and metrics for ozone risk assessment in plants. Journal of Experimental Botany 61: 629–633.

57. Pasqualini S, Antonielli M, Ederli L, Piccioni C, Loreto F (2002) Ozone uptake and its effect on photosynthetic parameters of two tobacco cultivars with contrasting ozone sensitivity. Plant Physiology and Biochemistry 40: 599–603.

58. Behnke K, Kleist E, Uerlings R, Wildt J, Rennenberg H, et al. (2009) RNAi-mediated suppression of isoprene biosynthesis in hybrid poplar impacts ozone tolerance. Tree Physiology 29: 725–736.

59. Kallakunta VM, Staruch A, Mutus B (2010) Sinapinic acid can replace ascorbate in the biotin switch assay. Biochimica et Biophysica Acta 1800: 23–30.

60. Huang B, Chen C (2006) An ascorbate-dependent artifact that interferes with the interpretation of the biotin switch assay. Free Radical Biology & Medicine 41: 562–567.

61. Laemmli UK (1970) Cleavage of structural proteins during the assembly of the head of bacteriophage T4. Nature 227: 680–685.

62. Wisniewski JR, Zougman A, Nagaraj N, Mann M (2009) Universal sample preparation method for proteome analysis. Nat Methods 6: 359–362.

63. Hauck SM, Dietter J, Kramer RL, Hofmaier F, Zipplies JK, et al. (2010) Deciphering membrane-associated molecular processes in target tissue of autoimmune uveitis by label-free quantitative mass spectrometry. Molecular & Cellular Proteomics 9: 2292–2305.

64. Merl J, Ueffing M, Hauck SM, von Toerne C (2012) Direct comparison of MS-based label-free and SILAC quantitative proteome profiling strategies in primary retinal Muller cells. Proteomics 12: 1902–1911.

65. Brosch M, Yu L, Hubbard T, Choudhary J (2009) Accurate and sensitive peptide identification with Mascot Percolator. Journal of Proteome Research 8: 3176–3181.

66. Velikova V, Ghirardo A, Vanzo E, Merl J, Hauck SM, et al. (2014) The Genetic Manipulation of Isoprene Emissions in Poplar Plants Remodels the Chloroplast Proteome. Journal of Proteome Research.

67. Eriksson L, Johansson E, Kettaneh-Wold N, Wold S (2006) Multi- and Megavariate Data Analysis. Part I: Basic Principles and Applications. Umeå, Sweden: Umetrics Academy.

68. Efron B, Gong G (1983) A leisurely look at the bootstrap, the jack-knife, and cross-validation. American Statistician 37: 36–48.

69. Benjamini Y, Hochberg Y (1995) Controlling the false discovery rate: a practical and powerful approach to multiple testing. Journal of the Royal Statistical Society Series B (Methodological) 57: 289–300.

70. Benjamini Y, Krieger AM, Yekutieli D (2006) Adaptive linear step-up procedures that control the false discovery rate. Biometrika 93: 491–507.

71. Heide L, Nishioka N, Fukui H, Tabata M (1989) Enzymatic regulation of shikonin biosynthesis in Lithospermum erythrorhizon cell cultures. Phytochemistry 28: 1873–1877.

72. Hess DT, Matsumoto A, Kim SO, Marshall HE, Stamler JS (2005) Protein S-nitrosylation: purview and parameters. Nature Reviews Molecular Cell Biology 6: 150–166.

73. Ederli L, Morettini R, Borgogni A, Wasternack C, Miersch O, et al. (2006) Interaction between nitric oxide and ethylene in the induction of alternative oxidase in ozone-treated tobacco plants. Plant Physiology 142: 595–608.

74. Eckey-Kaltenbach H, Ernst D, Heller W, Sandermann H, Jr. (1994) Biochemical Plant Responses to Ozone (IV. Cross-Induction of Defensive Pathways in Parsley (Petroselinum crispum L.) Plants). Plant Physiology 104: 67–74.

75. Francini A, Nali C, Pellegrini E, Lorenzini G (2008) Characterization and isolation of some genes of the shikimate pathway in sensitive and resistant Centaurea jacea plants after ozone exposure. Environmental Pollution 151: 272–279.

76. Xue Y, Liu Z, Gao X, Jin C, Wen L, et al. (2010) GPS-SNO: computational prediction of protein S-nitrosylation sites with a modified GPS algorithm. PLoS One 5: e11290.

77. Ritter H, Schulz GE (2004) Structural basis for the entrance into the phenylpropanoid metabolism catalyzed by phenylalanine ammonia-lyase. Plant Cell 16: 3426–3436.

78. Allwood EG, Davies DR, Gerrish C, Ellis BE, Bolwell GP (1999) Phosphorylation of phenylalanine ammonia-lyase: evidence for a novel protein kinase and identification of the phosphorylated residue. FEBS Letters 457: 47–52.

79. Davin LB, Jourdes M, Patten AM, Kim KW, Vassao DG, et al. (2008) Dissection of lignin macromolecular configuration and assembly: comparison to related biochemical processes in allyl/propenyl phenol and lignan biosynthesis. Natural Product Reports 25: 1015–1090.

80. Jeandet P, Delaunois B, Conreux A, Donnez D, Nuzzo V, et al. (2010) Biosynthesis, metabolism, molecular engineering, and biological functions of stilbene phytoalexins in plants. Biofactors 36: 331–341.

81. Agati G, Biricolti S, Guidi L, Ferrini F, Fini A, et al. (2011) The biosynthesis of flavonoids is enhanced similarly by UV radiation and root zone salinity in L. vulgare leaves. Journal of Plant Physiology 168: 204–212.

82. Rosemann D, Heller W, Sandermann H (1991) Biochemical Plant Responses to Ozone: II. Induction of Stilbene Biosynthesis in Scots Pine (Pinus sylvestris L.) Seedlings. Plant Physiology 97: 1280–1286.

83. Yu CK, Springob K, Schmidt J, Nicholson RL, Chu IK, et al. (2005) A stilbene synthase gene (SbSTS1) is involved in host and nonhost defense responses in sorghum. Plant Physiology 138: 393–401.

84. Dempsey DA, Vlot AC, Wildermuth MC, Klessig DF (2011) Salicylic Acid biosynthesis and metabolism. Arabidopsis Book 9: e0156.

85. Mohr S, Hallak H, de Boitte A, Lapetina EG, Brune B (1999) Nitric oxide-induced S-glutathionylation and inactivation of glyceraldehyde-3-phosphate dehydrogenase. Journal of Biological Chemistry 274: 9427–9430.

86. Holtgrefe S, Gohlke J, Starmann J, Druce S, Klocke S, et al. (2008) Regulation of plant cytosolic glyceraldehyde 3-phosphate dehydrogenase isoforms by thiol modifications. Physiologia Plantarum 133: 211–228.

87. Ralser M, Wamelink MM, Kowald A, Gerisch B, Heeren G, et al. (2007) Dynamic rerouting of the carbohydrate flux is key to counteracting oxidative stress. J Biol 6: 10.

88. Janero DR, Hreniuk D, Sharif HM (1994) Hydroperoxide-induced oxidative stress impairs heart muscle cell carbohydrate metabolism. American Journal of Physiology 266: C179–188.

89. Godon C, Lagniel G, Lee J, Buhler JM, Kieffer S, et al. (1998) The H_2O_2 stimulon in *Saccharomyces cerevisiae*. Journal of Biological Chemistry 273: 22480–22489.

90. Nogae I, Johnston M (1990) Isolation and characterization of the *ZWF1* gene of *Saccharomyces cerevisiae*, encoding glucose-6-phosphate dehydrogenase. Gene 96: 161–169.

91. Pollak N, Dolle C, Ziegler M (2007) The power to reduce: pyridine nucleotides-small molecules with a multitude of functions. Biochemical Journal 402: 205–218.

92. Tanou G, Job C, Belghazi M, Molassiotis A, Diamantidis G, et al. (2010) Proteomic signatures uncover hydrogen peroxide and nitric oxide cross-talk signaling network in citrus plants. Journal of Proteome Research 9: 5994–6006.

93. Marcus Y, Altman-Gueta H, Finkler A, Gurevitz M (2003) Dual role of cysteine 172 in redox regulation of ribulose 1,5-bisphosphate carboxylase/oxygenase activity and degradation. Journal of Bacteriology 185: 1509–1517.

94. Huang W, Chen Q, Zhu Y, Hu F, Zhang L, et al. (2013) *Arabidopsis Thylakoid Formation 1* Is a Critical Regulator for Dynamics of PSII-LHCII Complexes in Leaf Senescence and Excess Light. Molecular Plant 6: 1673–1691.

95. Richter AS, Grimm B (2013) Thiol-based redox control of enzymes involved in the tetrapyrrole biosynthesis pathway in plants. Frontiers in Plant Science 4: 371.

96. Strohm M, Eiblmeier M, Langebartels C, Jouanin L, Polle A, et al. (1999) Responses of transgenic poplar (*Populus tremula×P-alba*) overexpressing glutathione synthetase or glutathione reductase to acute ozone stress: visible injury and leaf gas exchange. Journal of Experimental Botany 50: 365–374.

97. Ziogas V, Tanou G, Filippou P, Diamantidis G, Vasilakakis M, et al. (2013) Nitrosative responses in citrus plants exposed to six abiotic stress conditions. Plant Physiology Biochem 68: 118–126.

98. Lai YC, Pan KT, Chang GF, Hsu CH, Khoo KH, et al. (2011) Nitrite-mediated S-nitrosation of caspase-3 prevents hypoxia-induced endothelial barrier dysfunction. Circulation Research 109: 1375–1386.

99. Samouilov A, Kuppusamy P, Zweier JL (1998) Evaluation of the magnitude and rate of nitric oxide production from nitrite in biological systems. Archives of Biochemistry and Biophysics 357: 1–7.

100. Zweier JL, Wang P, Samouilov A, Kuppusamy P (1995) Enzyme-independent formation of nitric oxide in biological tissues. Natural Medicines 1: 804–809.

101. Rockel P, Strube F, Rockel A, Wildt J, Kaiser WM (2002) Regulation of nitric oxide (NO) production by plant nitrate reductase *in vivo* and *in vitro*. Journal of Experimental Botany 53: 103–110.

102. Meyer DJ, Kramer H, Ozer N, Coles B, Ketterer B (1994) Kinetics and equilibria of S-nitrosothiol-thiol exchange between glutathione, cysteine, penicillamines and serum albumin. FEBS Letters 345: 177–180.

103. Hogg N (1999) The kinetics of S-transnitrosation-a reversible second-order reaction. Analytical Biochemistry 272: 257–262.

104. Wendehenne D, Pugin A, Klessig DF, Durner J (2001) Nitric oxide: comparative synthesis and signaling in animal and plant cells. Trends in Plant Science 6: 177–183.

105. Gould KS, Lamotte O, Klinguer A, Pugin A, Wendehenne D (2003) Nitric oxide production in tobacco leaf cells: a generalized stress response? Plant Cell and Environment 26: 1851–1862.

106. Liu L, Hausladen A, Zeng M, Que L, Heitman J, et al. (2001) A metabolic enzyme for S-nitrosothiol conserved from bacteria to humans. Nature 410: 490–494.

107. Liu L, Yan Y, Zeng M, Zhang J, Hanes MA, et al. (2004) Essential roles of S-nitrosothiols in vascular homeostasis and endotoxic shock. Cell 116: 617–628.

108. Andoh T, Chiueh CC, Chock PB (2003) Cyclic GMP-dependent protein kinase regulates the expression of thioredoxin and thioredoxin peroxidase-1 during hormesis in response to oxidative stress-induced apoptosis. Journal of Biological Chemistry 278: 885–890.

109. Schulze PC, Liu H, Choe E, Yoshioka J, Shalev A, et al. (2006) Nitric oxide-dependent suppression of thioredoxin-interacting protein expression enhances thioredoxin activity. Arteriosclerosis Thrombosis and Vascular Biology 26: 2666–2672.

110. Barglow KT, Knutson CG, Wishnok JS, Tannenbaum SR, Marletta MA (2011) Site-specific and redox-controlled S-nitrosation of thioredoxin. Proceedings of the National Academy of Sciences of the United States of America 108: E600–606.

111. Sengupta R, Ryter SW, Zuckerbraun BS, Tzeng E, Billiar TR, et al. (2007) Thioredoxin catalyzes the denitrosation of low-molecular mass and protein S-nitrosothiols. Biochemistry 46: 8472–8483.

112. Singh RJ, Hogg N, Joseph J, Kalyanaraman B (1995) Photosensitized decomposition of S-nitrosothiols and 2-methyl-2-nitrosopropane. Possible use for site-directed nitric oxide production. FEBS Letters 360: 47–51.

113. Belenghi B, Romero-Puertas MC, Vercammen D, Brackenier A, Inze D, et al. (2007) Metacaspase activity of *Arabidopsis thaliana* is regulated by S-nitrosylation of a critical cysteine residue. Journal of Biological Chemistry 282: 1352–1358.

114. Kato H, Takemoto D, Kawakita K (2013) Proteomic analysis of S-nitrosylated proteins in potato plant. Physiologia Plantarum 148: 371–386.

115. Palmieri MC, Lindermayr C, Bauwe H, Steinhauser C, Durner J (2010) Regulation of plant glycine decarboxylase by s-nitrosylation and glutathionylation. Plant Physiology 152: 1514–1528.

116. Yun BW, Feechan A, Yin M, Saidi NB, Le Bihan T, et al. (2011) S-nitrosylation of NADPH oxidase regulates cell death in plant immunity. Nature 478: 264–268.

117. Allwood EG, Davies DR, Gerrish C, Bolwell GP (2002) Regulation of CDPKs, including identification of PAL kinase, in biotically stressed cells of French bean. Plant Molecular Biology 49: 533–544.

118. Dong J, Zhang M, Lu L, Sun L, Xu M (2012) Nitric oxide fumigation stimulates flavonoid and phenolic accumulation and enhances antioxidant activity of mushroom. Food Chemistry 135: 1220–1225.

119. Kovacik J, Klejdus B, Backor M (2009) Nitric oxide signals ROS scavenger-mediated enhancement of PAL activity in nitrogen-deficient *Matricaria chamomilla* roots: side effects of scavengers. Free Radical Biology & Medicine 46: 1686–1693.

120. Xiao WH, Cheng JS, Yuan YJ (2009) Spatial-temporal distribution of nitric oxide involved in regulation of phenylalanine ammonialyase activation and Taxol production in immobilized *Taxus cuspidata* cells. Journal of Biotechnology 139: 222–228.

121. Jouanin L, Goujon T, de Nadai V, Martin MT, Mila I, et al. (2000) Lignification in transgenic poplars with extremely reduced caffeic acid O-methyltransferase activity. Plant Physiology 123: 1363–1374.

122. Cabane M, Pireaux JC, Leger E, Weber E, Dizengremel P, et al. (2004) Condensed lignins are synthesized in poplar leaves exposed to ozone. Plant Physiology 134: 586–594.

123. Heller W, Rosemann D, Osswald WF, Benz B, Schonwitz R, et al. (1990) Biochemical response of Norway spruce (*Picea abies* (L.) Karst.) towards 14-month exposure to ozone and acid mist: part I–Effects on polyphenol and monoterpene metabolism. Environmental Pollution 64: 353–366.

124. Sharma YK, Davis KR (1994) Ozone-Induced Expression of Stress-Related Genes in Arabidopsis thaliana. Plant Physiology 105: 1089–1096.

125. Guex N, Peitsch MC (1997) SWISS-MODEL and the Swiss-PdbViewer: an environment for comparative protein modeling. Electrophoresis 18: 2714–2723.

126. Arnold K, Bordoli L, Kopp J, Schwede T (2006) The SWISS-MODEL workspace: a web-based environment for protein structure homology modelling. Bioinformatics 22: 195–201.

14

Do Author-Suggested Reviewers Rate Submissions More Favorably than Editor-Suggested Reviewers? A Study on *Atmospheric Chemistry and Physics*

Lutz Bornmann[1]*, Hans-Dieter Daniel[2,3]

1 Office of Research Analysis and Foresight, Max Planck Society, Munich, Germany, **2** Professorship for Social Psychology and Research on Higher Education, ETH Zurich, Zurich, Switzerland, **3** Evaluation Office, University of Zurich, Zurich, Switzerland

Abstract

Background: Ratings in journal peer review can be affected by sources of bias. The bias variable investigated here was the information on whether authors had suggested a possible reviewer for their manuscript, and whether the editor had taken up that suggestion or had chosen a reviewer that had not been suggested by the authors. Studies have shown that author-suggested reviewers rate manuscripts more favorably than editor-suggested reviewers do.

Methodology/Principal Findings: Reviewers' ratings on three evaluation criteria and the reviewers' final publication recommendations were available for 552 manuscripts (in total 1145 reviews) that were submitted to *Atmospheric Chemistry and Physics*, an interactive open access journal using public peer review (authors' and reviewers' comments are publicly exchanged). Public peer review is supposed to bring a new openness to the reviewing process that will enhance its objectivity. In the statistical analysis the quality of a manuscript was controlled for to prevent favorable reviewers' ratings from being attributable to quality instead of to the bias variable.

Conclusions/Significance: Our results agree with those from other studies that editor-suggested reviewers rated manuscripts between 30% and 42% less favorably than author-suggested reviewers. Against this backdrop journal editors should consider either doing without the use of author-suggested reviewers or, if they are used, bringing in more than one editor-suggested reviewer for the review process (so that the review by author-suggested reviewers can be put in perspective).

Editor: Pedro Antonio Valdes-Sosa, Cuban Neuroscience Center, Cuba

Funding: The study was funded by the Max Planck Society. However, the funder had no role in study design, data collection and analysis, decision to publish, or preparation of the manuscript.

Competing Interests: The authors have declared that no competing interests exist.

* E-mail: bornmann@gv.mpg.de

Introduction

In the research on journal peer review, there are said to be biases, if – independently of the quality of submitted manuscripts – attributes of the reviewers (such as the nomination of a reviewer by the author or the editor) are correlated statistically with the reviewers' ratings [1]. Arkes [2] defines bias "as any systematic effect on ratings unrelated to the true quality of the object being rated. Thus, bias consists of effects that reduce the validity of ratings through contamination, but not random error" (p. 378). According to Jayasinghe [3] "a random error is an 'unexplained' error whereas systematic bias such as leniency/ harshness of reviewers ... can be explained or statistically controlled" (p. 35).

Reviewers for a manuscript can be selected by editors (1) on the basis of their personal knowledge and familiarity from past experience, (2) from a database of previous reviewers cross-referenced by name and specialty, (3) from references listed in the manuscript, and (4) based on suggestions made by the authors of the manuscript [4]. For Tonks [5], an assistant editor at the *British Medical Journal* (BMJ), the selection of author-suggested reviewers

(R_a) "could improve the quality of peer review in two important ways. Firstly, authors are often better placed than editors to know whom to approach for a considered, balanced, and credible opinion in their field of research. The best reviewers are not those with the most experience or eminence and may be unknown to anyone outside the subject. This is a particular problem for editors of general journals, who review manuscripts from a wide range of disciplines. Secondly, nominated reviewers will enrich the BMJs database, keeping us in touch with young active researchers and giving us a broader population of reviewers."

According to the "Ethical Guidelines for Publication in Journals and Reviews" of the European Association for Chemical and Molecular Sciences [6], editors have the responsibility "to consider the use of an author's suggested reviewers for his/her submitted manuscript, but to ensure that the suggestions do not lead to a positive bias." R_a may be biased in favor of the authors [7]. The danger with R_a is that "they can be the authors' best friends" [8] (p. 15). It is feared that through the use of R_a in addition to editor-suggested reviewers (R_e) (meaning reviewers selected by the editor not on the basis of a suggestion by the author), the one (R_a) rates a manuscript systematically more leniently than the other (R_e). (We assume this leniency effect, although an R_e is not necessarily unknown to the authors.)

A number of studies of different journals showed that this fear is justified. A study by Schroter, Tite, Hutchings, and Black [9] on the peer review process at 10 biomedical journals found that R_a "tended to make more favorable recommendations for publication" (p. 314) than R_e [10]. Similar findings were reported by Scharschmidt, Deamicis, Bacchetti, and Held [11] for the *Journal of Clinical Investigation*, Earnshaw and Farndon [12] for the *British Journal of Surgery*, Goldsmith, Blalock, Bobkova, and Hall [13] for the *Journal of Investigative Dermatology*, Wager, Parkin, and Tamber [14] for medical journals in the BMC (BioMed Central) series, Rivara, Cummings, Ringold, Bergman, Joffe, and Christakis [15] for a pediatric journal, and Bornmann and Daniel [16] for *Angewandte Chemie International Edition* (AC-IE). In addition, Jayasinghe, Marsh and Bond [17] found similar results in the area of grant peer review.

In this study we aim to test whether there is a potential source of bias in the manuscript reviewing in public peer review at an interactive open access journal, *Atmospheric Chemistry and Physics* (ACP), through the use of R_a and R_e. Using modern information technology, in particular the Internet, the ACP and other interactive open access journals have now become established in science that work with a "new" system of public peer review [18,19]. Compared to the traditional system, the new system of peer review in an electronic environment is seen to have the following advantages, among others: (1) submitted manuscripts are immediately published as "discussion papers" on the journal's website, (2) reviewers' comments on the quality of the content of the manuscript and authors' replies to the reviewers' critical comments are publicly exchanged, and (3) reviewers' arguments are publicly heard, and, if comments are openly signed, reviewers can also claim authorship for their contributions [20].

Even if all studies so far have found that R_a rate manuscripts systematically more favorably than R_e, it would be expected that public peer review at ACP does not show this effect. (With the exception of Wager, Parkin, and Tamber [14], the aforementioned studies conducted up to now examined traditional peer review.) Public peer review is supposed to bring a new openness to the reviewing process that will enhance its objectivity [21]. Publishing reviews is supposed to lead to reviewers using argumentation and judging solely on the basis of scientific criteria, so that the reviewer's ratings will not be influenced by potential sources of bias. We investigated the extent to which this expectation can be confirmed, taking the example of ACP.

Methods

Manuscript review at ACP

ACP was launched in September 2001. It is produced and published by the European Geosciences Union (EGU) (http://www.egu.eu) and Copernicus Publications (http://publications.copernicus.org/). ACP is freely accessible via the Internet (www.atmos-chem-phys.org). It has the second highest annual Journal Impact Factor (JIF) (provided by Thomson Reuters, Philadelphia, PA, USA) in the category "Meteorology & Atmospheric Sciences" (at 4.881 in the 2009 Journal Citation Reports, Science Edition). ACP has a two-stage publication process [20,22] that is described on the ACP website as follows: In the first stage, manuscripts that pass a rapid pre-screening process (access review) are immediately published as "discussion papers" on the journal's website (by doing this, they are published in *Atmospheric Chemistry and Physics Discussions*, ACPD). These discussion papers are then made available for "interactive public discussion," during which the comments of reviewers (usually, reviewers that already conducted the access review), additional comments by other interested members of the scientific community, and the authors' replies are published alongside the discussion paper. The reviewers can be R_a or R_e.

During the discussion phase, the designated reviewers are asked to answer to the following questions according to the ACP's principal evaluation criteria (see http://www.atmospheric-chemistry-and-physics.net/review/ms_evaluation_criteria.html, from which the following information is taken): (1) scientific significance ("Does the manuscript represent a substantial contribution to scientific progress within the scope of ACP (substantial new concepts, ideas, methods, or data?"), (2) scientific quality ("Are the scientific approach and applied methods valid? Are the results discussed in an appropriate and balanced way (consideration of related work, including appropriate references)?"), and (3) presentation quality ("Are the scientific results and conclusions presented in a clear, concise, and well-structured way (number and quality of figures/tables, appropriate use of English language)?"). The response categories for the three questions are: (1) excellent, (2) good, (3) fair, and (4) poor. In addition to the principal evaluation criteria, the reviewers are asked to give a final publication recommendation: "Do you recommend acceptance of the manuscript?" Here, the response categories are: (1) yes, without alterations, (2) yes, after minor alterations, (3) yes, after major alterations, and (4) no. Besides giving the formal ratings to the four questions, the reviewers also have the opportunity to write a commentary.

The ratings are submitted in parallel to the commentaries, but they are not open, because they are meant to support the editorial decision rather than the scientific discussion. This policy was introduced in 2001. According to the experiences and the philosophy of ACP's chief-executive editor Ulrich Pöschl, prescribed publication of formal ratings is likely to do more harm than good (e.g., initiation/escalation of unnecessary controversies). Most other journals pursuing public peer review do not prescribe publication of formal ratings either, and some of them explicitly instruct reviewers not to include formal ratings in their public comments (see, e.g., http://adv-model-earth-syst.org/index.php/JAMES/about/faq). At ACP, the editors leave it up to the reviewers if they want to include ratings in their public comments, and sometimes they do (~30%). With increasing acceptance and spread of public review it may become beneficial and appropriate to prescribe publication of formal ratings. For now, however, the ACP editors prefer a mix of open commentaries and non-public ratings for the discussion phase.

After the end of the discussion phase every author has the opportunity to submit a revised manuscript taking into account the reviewers' comments and the comments of interested members of the scientific community. Based on the revised manuscript and in view of the access peer review and interactive public discussion, the editor accepts or rejects the revised manuscript for publication in ACP. For this publication decision, further external reviewers may be asked to review the revision, if needed. In general, an editor accepts a manuscript for publication in ACP, if – similar to the "clear-cut" rule of the journal AC-IE [23] – all reviewers rate the manuscript favorably (see here http://www.atmospheric-chemistry-and-physics.net/review/ms_evaluation_criteria.html).

Database for the present study

For the investigation of peer review at ACP we had data for 1111 manuscripts that went through the complete ACP selection process in the years 2001 to 2006 [24,25,26]. Of the 1111 manuscripts, 1032 (93%) manuscripts were published as discussion papers; 79 (7%) were rejected during access review for publication as discussion papers. Reviewers' ratings on the evaluation criteria and reviewers' final publication recommendations, made during the discussion phase of the reviewing process, were available for 552 (55%) of the 1008 manuscripts. This reduction in number is due to the fact that the ratings have been stored electronically by the publisher only since 2004. Of the 552 manuscripts, 16% ($n = 87$) have one review, 64% ($n = 356$) have two, 17% ($n = 92$) have three, 3% ($n = 15$) have four, and 2 manuscripts have five independent reviews. Of the total 1145 reviews, 304 (27%) were by R_a and 841 (73%) by R_e.

Of the 1111 manuscripts submitted between 2001 and 2006, 958 (86%) were published in ACPD and ACP, 74 (7%) were published in ACPD but not in ACP (here, the editor rejected the revised manuscript), and 79 (7%) were published neither in ACPD nor in ACP (these manuscripts were rejected during the access review). The search for the fate of the manuscripts that were not published in ACP ($n = 153$) revealed that 38 (25%) were published as contributions in other journals. No publication information was found for 115 (75%) manuscripts, whereby 70 of the 115 manuscripts (61%) were published in ACPD. The 38 manuscripts that were published as contributions in other journals were published in 25 different journals within a time period of five years (that is, between 2005 and 2009). Six manuscripts were published in the *Journal of Geophysical Research*; three manuscripts were published in *Geophysical Research Letters*. The other 23 journals published one or two of these manuscripts each [25].

Statistical procedures

Normally, when examining the association of a bias variable and reviewers' ratings it is impossible to establish unambiguously whether a particular group of manuscripts receives more favorable reviewers' ratings due to this variable, or if the more favorable ratings are simply a consequence of the manuscripts' scientific quality [27]. For this reason, the statistical analysis should control for the scientific quality of a manuscript [28]. Smart and Waldfogel [29] call this approach "a clean test for the existence of discrimination" (p. 5), which in this study was realized through different statistical methods in two independent analysis steps.

To test whether R_a rate more leniently than R_e, we used what is called a within-manuscript analysis as a first step. This analysis approach was proposed by Jayasinghe, Marsh, and Bond [30] for grant peer review research. They analyzed reviewers' gender as a potential source of bias in the Australian Research Council (Canberra) peer review and conducted "a within-proposal analysis based on those proposals with at least one male external reviewer and at least one female external reviewer" (p. 353). Some years later Wager, Parkin, and Tamber [14] investigated in the area of journal peer review "pairs of reviews from 100 consecutive submissions to medical journals in the BMC series (with one author-nominated and one editor-chosen reviewer and a final decision)."

At ACP between 2004 and 2006 135 of a total of 552 manuscripts (25%) were reviewed by a pair of R_a and R_e. Differences in the ratings by the two reviewers of these manuscripts (related paired samples of R_a and R_e) were investigated using the marginal homogeneity test [31], which generalizes the McNemar test from binary response to multinomial response. The method developed in the present release of StatXact [32] applies to ordered response. As the ACP data for the marginal homogeneity test are sparse, exact p-values were calculated.

As in the within-manuscript analysis only 135 of the 552 manuscripts could be included, an ordinal regression model

(ORM) was computed as a second step to analyze ratings of R_a and R_e. Using ORM, the association between several independent variables (here: suggestion of a reviewer and citations as an indicator for scientific quality) and an ordinal-scaled dependent variable (here: the reviewers' ratings) can be determined: "As with the binary regression model, the ORM is nonlinear, and the magnitude of the change in the outcome probability for a given change in one of the independent variables depends on the levels of all the independent variables" [33] (p. 183). For the analysis, the ACP data is a dataset where the assumption of independence between individual ratings of the reviewers may not hold, as the reviews are nested within manuscripts. In order to take the dependencies between individual ratings into account in the estimation of the ORMs, we used the "cluster" option in Stata [34]. Specifying this option leads to robust standard errors in the sense that the estimates provide correct standard errors in the presence of the effects of clustered data [33]. "The performance of the cluster-robust estimator is good with 50 or more clusters, or fewer if the clusters are large and balanced" [35] (p. 514). In this study we have 552 unbalanced clusters (manuscripts with one to five reviewers).

By fitting an ordinary ORM with robust standard errors for clustered data instead of fitting a variance components model (a multilevel model for ordinal responses), we were treating the within-cluster dependence as a "nuisance" and not as a phenomenon that we were interested in [36]. A Wald test by Brant [37] was performed to test the parallel regression assumption for each independent variable considered in the ORM [38]. As the test provides evidence that the assumption was violated for the variable "number of citations for a manuscript," the variable was entered into the regression analysis as a log-transformed variable.

Out of a lack of other operationalizable indicators, it is common in research evaluation to use citation counts as an indicator for scientific quality. According to van Raan [39] citations provide "a good to even very good quantitative impression of at least one important aspect of quality, namely international impact" (p. 404). According to Lindsey [40] citations are "our most *reliable* convenient measure of quality in science – a measure that will continue to be widely used" (p. 201). In the present study we retrieved citation counts for manuscripts accepted by ACP or rejected and published elsewhere for a fixed time window of three years after the publication year. "Fixed citation windows are a standard method in bibliometric analysis, in order to give equal time spans for citation to articles published in different years, or at different times in the same year" [41] (p. 243). The citation analyses for the present study were conducted based on Chemical Abstracts (CA) (Chemical Abstracts Services, Columbus, Ohio, USA). CA is a comprehensive database of publicly disclosed research in chemistry and related sciences (see http://www.cas.org/).

As the citation counts were captured ex post – that is, after the editors' publication decisions (at ACP or another journal) – they are included in the regression models only as control variables. This means that in the analysis the interest was *not* the correlation between citation counts and reviewers' ratings but instead the correlation between the bias variable and ratings, when manuscript impact is *statistically controlled*. In statistical bias analysis this procedure is called the control variable approach [42].

Results

Table 1 shows the minimum, maximum, mean, standard deviation, and median of the ratings by R_a and R_e on the scientific significance, scientific quality, and presentation quality of a

Table 1. Minimum (min), maximum (max), mean, standard deviation (sd), and median of ratings by R_a and R_e on the scientific significance, scientific quality, and presentation quality of a manuscript and final publication recommendations.

Reviewer group	Scientific significance*	Scientific quality*	Presentation quality*	Final publication recommendation$
R_a				
n	304.00	304.00	304.00	304.00
min	1.00	1.00	1.00	1.00
max	4.00	4.00	4.00	4.00
mean	1.89	2.00	2.02	2.27
sd	0.66	0.69	0.69	0.55
median	2.00	2.00	2.00	2.00
R_e				
n	841.00	841.00	841.00	841.00
min	1.00	1.00	1.00	1.00
max	4.00	4.00	4.00	4.00
mean	2.07	2.25	2.20	2.40
sd	0.71	0.76	0.73	0.63
median	2.00	2.00	2.00	2.00
Total				
n	1145.00	1145.00	1145.00	1145.00
min	1.00	1.00	1.00	1.00
max	4.00	4.00	4.00	4.00
mean	2.02	2.19	2.15	2.37
sd	0.70	0.75	0.72	0.61
median	2.00	2.00	2.00	2.00

Notes.
*Response categories: (1) excellent, (2) good, (3) fair, and (4) poor.
$ Response categories: (1) yes, without alterations, (2) yes, after minor alterations, (3) yes, after major alterations, (4) no.

manuscript and the final publication recommendation. Whereas the arithmetic average ratings by R_e are more negative on all evaluation criteria and for the final publication recommendation than the ratings by R_a, the median ratings of the two groups do not differ on either evaluation criteria or final publication recommendation. The median ratings for the two reviewers groups are always 2. The results shown in Table 1 are not really meaningful, as they do not refer to differences between R_a and R_e on one and the same manuscript.

Table 2 presents the results of the within-manuscript analysis. For each evaluation criterion and for the final publication recommendation the table shows the difference between the ratings of reviewers for those manuscripts (n = 135) that were each reviewed by an R_a and an R_e. The table shows the number of those manuscripts (row percents) for which the ratings by R_a and R_e did not differ (column: "no difference"), the rating by R_a was more positive than the rating by R_e (column: "R_a is more positive than R_e"), and the rating by R_e was more positive than the rating

Table 2. Differences between the ratings by R_a and R_e on three evaluation criteria and on the reviewers' final publication recommendations for those manuscripts that were each reviewed by both an R_a and an R_e (n = 135).

Evaluation criteria and final publication recommendation	Difference between R_a and R_e (row percent):			Marginal homogeneity test (χ^2)
	No difference	R_a is more positive than R_e	R_e is more positive than R_a	
Scientific significance§	53	29	18	1.75
Scientific quality§	47	30	23	1.54
Presentation quality§	49	31	20	1.53
Final publication recommendation$	67	22	11	1.87*

Notes.
§Response categories: (1) excellent, (2) good, (3) fair, and (4) poor.
$ Response categories: (1) yes, without alterations, (2) yes, after minor alterations, (3) yes, after major alterations, (4) no.
*p<.05 (marginal homogeneity test for ordered data; one-side p-value, difference occurs in the hypothesized direction).

by R_a (column: "R_e is more positive than R_a"). As the distribution of the percentage values for all evaluation criteria and for the final publication recommendation show, there are clearly more manuscripts rated more favorably by R_a than by R_e than there are manuscripts rated more favorably by R_e than by R_a. For instance, 22% of the final publication recommendations made by R_a are more positive than those made by R_e. There are more positive recommendations by R_e than by R_a for only 11% of the manuscripts (there is no difference between the recommendations by the two reviewer groups for 67% of the manuscripts). Hence, overall for this group of manuscripts R_a rated more favorably than R_e more frequently than vice versa. Using the marginal homogeneity test, we examined whether the ratings by R_a and R_e also differed statistically significantly. As the results of the test in Table 2 show, the difference is statistically significant only for the final publication recommendation. The differences between the ratings on the evaluation criteria are non-significant.

The differing results of the marginal homogeneity test could indicate that with the same ratings on all evaluation criteria, R_a tend to make a more positive final publication recommendation than R_e. To test this hypothesis, in a further analysis we selected those manuscripts among the 135 manuscripts reviewed by both

R_a and R_e that were rated the same on all evaluation criteria by both reviewers. This was the case for 18% of the manuscripts (n = 24). Table 3 shows the reviewers' ratings on the evaluation criteria and their final publication recommendations for the 24 manuscripts. Whereas the final publication recommendations by both reviewers were the same for 21 manuscripts, for 3 manuscripts the final publication recommendations by R_a were more favorable than the recommendations by R_e. No manuscript received a more favorable final publication recommendation by R_e than by R_a.

In closing, we tested differences between the ratings by R_a and R_e using ORMs. An ORM was computed for each evaluation criterion and the final publication recommendation. Table 4 presents a description of the dependent and independent variables that were included in the total of four ORMs. The independent variables are "Author-suggested reviewer" (R_a or R_e) and the log-transformed citation counts. Table 5 shows the results of the ORMs. For all ORMs the variable "Author-suggested reviewer" has a statistically significant effect in the expected direction: If the review is by R_a, the ratings on all criteria as well as the final publication recommendation are statistically significantly more favorable than the ratings, if the review is by R_e – independently of

Table 3. Final publication recommendation by R_a and R_e for those manuscripts, for which an R_a and an R_e gave identical ratings on three evaluation criteria (n = 24).

Evaluation criteria§			Final publication recommendation$	
Scientific significance	Scientific quality	Presentation quality	R_e	R_a
1	1	1	Minor	**Accept**
1	1	1	Minor	Minor
1	1	2	Minor	Minor
1	2	2	Minor	Minor
2	1	1	Minor	Minor
2	2	2	Major	Minor
2	2	2	Minor	Minor
2	2	2	Minor	Minor
2	2	2	Minor	Minor
2	2	2	Minor	Minor
2	2	2	Minor	Minor
2	2	2	Minor	Minor
2	2	2	Minor	Minor
2	2	2	Minor	Minor
2	2	2	Minor	Minor
2	2	2	Major	**Minor**
2	2	3	Minor	Minor
2	2	3	Minor	Minor
2	3	3	Major	**Minor**
2	3	3	Major	Major
2	3	3	Major	Major
3	2	3	Minor	Minor
3	3	2	Major	Major
3	3	3	Major	Major

In the table, three final publication recommendations where R_a made a more favorable recommendation than R_e are shown in **bold**.
Notes.
§Response categories: (1) excellent, (2) good, (3) fair, and (4) poor.
$ Response categories: (Accept) yes, without alterations, (Minor) yes, after minor alterations, (Major) yes, after major alterations.

Table 4. Description of the dependent and independent variables included in the ORM.

Variable	Range of values	Arithmetic mean
Dependent variable		
Scientific significance	1→4	2.02
Scientific quality	1→4	2.19
Presentation quality	1→4	2.15
Final publication recommendation	1→4	2.37
Independent variables		
Author-suggested reviewer (1 = R_a, 0 = R_e)	0→1	0.27
Citation counts for the first three years after the publication year (measured ex post, log-transformed)	0→4.56	1.80

the quality of the reviewed manuscript (measured ex-post using citation counts). To be able to assess the size of the effect of the variable "Author-suggested reviewer" on the ratings, after the ORMs we computed percent changes in expected ratings for a unit increase (from rating by R_e to rating by R_a) [33]. As the results in Table 5 show, in reviews by R_e ratings can be expected that are between 30% and 42% less favorable than the ratings by R_a.

Discussion

Compared to most of the studies on potential sources of bias in the manuscript reviewing process published up to now, the present study used an optimized strategy with two independent analysis steps. In both steps there was a control for the scientific impact of the research reported in a manuscript in order to be able to determine – *independently of their quality* – whether manuscripts that were reviewed by R_a are reviewed more favorably than manuscripts that were reviewed by R_e. The results of this study are therefore more solid than the results of most of the studies published up to now that did not control for the scientific impact of manuscripts in the evaluation.

In a first step of analysis, we used a within-manuscript approach. Even though this analysis revealed a statistically significant difference between the reviews by R_a and R_e only with regard to the final publication recommendation (and not for the evaluation criteria), there is a tendency in the dataset towards more manuscripts that R_a rate more favorably than R_e than the opposite case. In addition, with the same ratings on the evaluation criteria, R_a tends towards a more positive than a more negative final publication recommendation than R_e. In a second step of analysis, an ORM was computed. This analysis showed that both for the evaluation criteria and the final publication recommendations, more positive ratings can be expected by R_a than by R_e. All in all, the results for the journal ACP agree with the results of other studies (see the introduction section) and indicate that the bias variable "Author-suggested reviewer" has an effect on the reviewing process.

Table 5. Results of the ORM predicting reviewers' ratings for three evaluation criteria and the final publication recommendation.

Independent variable	Scientific significance§	Scientific quality§	Presentation quality§	Final publication recommendation$
Fixed part: Maximum likelihood estimates				
Citation counts for the first three years after the	−0.434***	−0.434***	−0.278***	−0.368***
publication year (measured ex post, log-transformed)	(−5.81)	(−7.04)	(−4.04)	(−4.86)
Author-suggested reviewer (1 = R_a, 0 = R_e)	−0.408**	−0.552***	−0.406**	−0.350*
	(−3.08)	(−4.30)	(−3.10)	(−2.57)
Fixed part: Thresholds				
K_1	−2.337***	−2.706***	−2.228***	−4.383***
	(−12.93)	(−17.38)	(−13.41)	(−17.16)
K_2	0.567***	0.00570	0.292*	−0.125
	(3.68)	(0.05)	(2.02)	(−0.82)
K_3	2.740***	2.142***	3.026***	2.385***
	(11.78)	(12.60)	(12.84)	(11.72)
$n_{reviews}$	1145	1145	1145	1145
$n_{manuscripts}$	552	552	552	552
Reviews per	min = 1	min = 1	min = 1	min = 1
manuscript	mean = 2.1	mean = 2.1	mean = 2.1	mean = 2.1
(cluster)	max = 5	max = 5	max = 5	max = 5
Change in expected rating for a unit increase in "Author-suggested reviewer"	-34%	-42%	-33%	-30%

Notes.
*p<0.05,
**p<0.01,
***p<0.001
§Response categories: (1) excellent, (2) good, (3) fair, and (4) poor.
$ Response categories: (1) yes, without alterations, (2) yes, after minor alterations, (3) yes, after major alterations, (4) no.
t statistics in parentheses.

However, even though the results of the study indicate that there are differences between the ratings by R_a and R_e, the results should be seen as only an *indication* of a potential source of bias in the ACP peer review process and *not* as proof of favoritism of certain manuscripts by R_a. Strictly speaking, solid findings on the existence of biases in peer review processes can be produced only by experimental studies in which the research objects (such as manuscripts) are randomly assigned to a treatment and control group (such as R_a and R_e) [43]. As a study of that kind would influence the review process, there is a risk of infringing the rules of good scientific practice, as pointed out by critical commentaries on the study published by Peters and Ceci [44] (see *Behavioral and Brain Sciences*, 1982, pp. 196–246, and *Behavioral and Brain Sciences*, 1985, pp. 743–747). In that study manuscripts with fictitious author names and institutional affiliations were submitted to journals for publication.

Regardless of what the results of experimental studies of that kind would be, we can probably assume that there can be no peer review system without the influence of potential sources of bias. Scientists, too, are only human: "Philosophers and sociologists agree that the notion of a truly objective disinterested 'seeker after truth' is incompatible with the realities of social existence. We all have personal interests and institutional values that we are bound to promote in our scientific work … It will surely defend objectivity as an ideal, impossible to realize completely in practice but always to be respected and desired" [45] (p. 754). To obtain an indication of the systematic influence of sources of bias in a peer review process, in research evaluation it is proposed that the process of peer reviewing should be studied continuously and that any evidence of bias in the process should be brought to the attention of the editor for correction and modification of the process [46,47]. Hojat, Gonnella, and Caelleigh [48] demanded "that the journal editors conduct periodic internal and external evaluations of their journals' peer review process and outcomes" (p. 75) to assure the integrity of the process. In the most comprehensive review of research on biases in peer review, Godlee and Dickersin [49] also concluded that "journals should continue to take steps to minimize the scope for unacceptable biases, and researchers should continue to look for them" (p. 112).

If indications of the effect of sources of bias are found in a peer review process, Thorngate, Dawes, and Foddy [50] recommend the following measures "to fix the problem … One possible

solution is to replace biased judges with neutral ones. Another is to train and to motivate offending judges to mend their judgmental ways. A third is to add more judges in hopes that their biases will counterbalance each other and produce a neutral group consensus. Each is worthy of brief consideration" (p. 55). This study showed, in agreement with all other studies, for the bias variable investigated that independently of the quality of a manuscript, better ratings can be expected from R_a than from R_e. Many journals use precautions to avoid biased review from R_a, e.g., by stipulating that reviewers do not work in the same institution, have never published with them, etc. If reviewers have a disqualifying conflict they should excuse themselves or not be used. However, personal relationships are harder to quantify than financial links so they are often overlooked. Journal editors should therefore consider, if R_a are used, bringing in more than one R_e for the review process so that the review by R_a can be put in perspective.

Acknowledgments

We would like to thank Dr. Hanna Joos (at the Institute for Atmospheric and Climate Science of ETH Zurich, Switzerland) and Dr. Hanna Herich (at EMPA, a research institution within the ETH Domain) for the investigation of the manuscripts rejected by *Atmospheric Chemistry and Physics* and published elsewhere. We thank Dr. Ulrich Pöschl, Chief Executive Editor of *Atmospheric Chemistry and Physics*, the Editorial Board of *Atmospheric Chemistry and Physics*, and Copernicus Publications (Göttingen, Germany) for permission to conduct the evaluation of the selection process of the journal, and thank the members of Copernicus Systems + Technology (Berlin, Germany) for their generous technical support during the carrying out of the study. We also thank Dr. Werner Marx and Dr. Hermann Schier of the Central Information Service for the institutes of the Chemical Physical Technical (CPT) Section of the Max Planck Society (located at the Max Planck Institute for Solid State Research in Stuttgart, Germany) for conducting the citation search for citations of the accepted and rejected (but published elsewhere) manuscripts in the literature database Chemical Abstracts. The authors wish to express their gratitude to Liz Wager for her helpful comments.

Author Contributions

Conceived and designed the experiments: LB. Performed the experiments: LB. Analyzed the data: LB. Contributed reagents/materials/analysis tools: LB. Wrote the paper: LB HDD.

References

1. Weller AC (2002) Editorial peer review: its strengths and weaknesses. Medford, NJ, USA: Information Today, Inc.
2. Arkes HR (2003) The nonuse of psychological research at two federal agencies. Psychological Science 14: 1–6.
3. Jayasinghe UW (2003) Peer review in the assessment and funding of research by the Australian Research Council. Greater Western Sydney, Australia: University of Western Sydney.
4. Lee K, Boyd E, Bero L (2004) A look inside the black box: a description of the editorial process at three leading biomedical journals; Ottawa, Canada.
5. Tonks A (1995) Reviewers chosen by authors. British Medical Journal 311: 210.
6. European Association for Chemical and Molecular Sciences (2006) Ethical guidelines for publication in journals and reviews. Brussels, Belgium: European Association for Chemical and Molecular Sciences (EuCheMS).
7. Perlman D, Dean E (1987) The wisdom of Salomon: avoiding bias in the publication review process. In: Jackson DN, Rushton J, eds. Scientific excellence Origins and assessment. London, UK: Sage. pp 204–221.
8. Anon (2007) Gatekeepers of science. Interview with Peter Stern, editor at *Science* magazine. BIF Futura 22: 14–18.
9. Schroter S, Tite L, Hutchings A, Black N (2006) Differences in review quality and recommendations for publication between peer reviewers suggested by authors or by editors. JAMA 295: 314–317.
10. Grimm D (2005) Suggesting or excluding reviewers can help get your paper published. Science 309: 1974.
11. Scharschmidt BF, Deamicis A, Bacchetti P, Held MJ (1994) Chance, concurrence, and clustering - analysis of reviewers recommendations on 1,000 submissions to the *Journal of Clinical Investigation*. Journal of Clinical Investigation 93: 1877–1880.
12. Earnshaw JJ, Farndon JR (2000) A comparison of reports from referees chosen by authors or journal editors in the peer review process. Annals of the Royal College of Surgeons of England 82: 133–135.
13. Goldsmith LA, Blalock E, Bobkova H, Hall RP (2005) Effect of authors' suggestions concerning reviewers on manuscript acceptance. In: Rennie D, Godlee F, Flanagin A, Smith J, eds. 5th International Congress on Peer Review and Biomedical Publication: Chicago, IL, USA.
14. Wager E, Parkin E, Tamber P (2006) Are reviewers suggested by authors as good as those chosen by editors? Results of a rater-blinded, retrospective study. BMC Medicine 4: 13.
15. Rivara FP, Cummings P, Ringold S, Bergman AB, Joffe A, et al. (2007) A comparison of reviewers selected by editors and reviewers suggested by authors. Journal of Pediatrics 151.
16. Bornmann L, Daniel H-D (2009) Reviewer and editor biases in journal peer review: an investigation of manuscript refereeing at *Angewandte Chemie International Edition*. Research Evaluation 18: 262–272.
17. Jayasinghe UW, Marsh HW, Bond N (2003) A multilevel cross-classified modelling approach to peer review of grant proposals: the effects of assessor and researcher attributes on assessor ratings. Journal of the Royal Statistical Society Series a-Statistics in Society 166: 279–300.
18. Bailey CW (2005) Open Access Bibliography. Washington, DC, USA: Association of Research Libraries.
19. Pöschl U (2010) Interactive open access publishing and peer review: the effectiveness and perspectives of transparency and self-regulation in scientific communication and evaluation. Liber Quarterly 19: 293–314.
20. Koop T, Pöschl U (2006) Systems: an open, two-stage peer-review journal. The editors of *Atmospheric Chemistry and Physics* explain their journal's approach.

Retrieved 26 June 2006, from http://www.nature.com/nature/peerreview/debate/nature04988.html.

21. Bingham CM, Higgins G, Coleman R, Van Der Weyden MB (1998) *The Medical Journal of Australia* Internet peer-review study. Lancet 352: 441–445.

22. Pöschl U (2004) Interactive journal concept for improved scientific publishing and quality assurance. Learned Publishing 17: 105–113.

23. Bornmann L, Daniel H-D (2009) The luck of the referee draw: the effect of exchanging reviews. Learned Publishing 22: 117–125.

24. Bornmann L, Daniel H-D (2010) Reliability of reviewers' ratings at an interactive open access journal using public peer review: a case study on *Atmospheric Chemistry and Physics*. Learned Publishing 23: 124–131.

25. Bornmann L, Marx W, Schier H, Thor A, Daniel H-D (2010) From black box to white box at open access journals: predictive validity of manuscript reviewing and editorial decisions at *Atmospheric Chemistry and Physics*. Research Evaluation 19: 81–156.

26. Bornmann L, Neuhaus C, Daniel H-D (in press) The effect of a two-stage publication process on the Journal Impact Factor: a case study on the interactive open access journal *Atmospheric Chemistry and Physics*. Scientometrics.

27. Budden AE, Aarssen L, Koricheva J, Leimu R, Lortie CJ, et al. (2008) Response to Whittaker: challenges in testing for gender bias. TRENDS in Ecology & Evolution 23: 480–481.

28. Laband DN, Piette MJ (1994) Favoritism versus search for good papers: empirical evidence regarding the behavior of journal editors. Journal of Political Economy 102: 194–203.

29. Smart S, Waldfogel J (1996) A citation-based test for discrimination at economics and finance journals. NBER working Paper, No 5460 Cambridge, MA, USA: National Bureau of Economic Research.

30. Jayasinghe UW, Marsh HW, Bond N (2001) Peer review in the funding of research in higher education: the Australian experience. Educational Evaluation and Policy Analysis 23: 343–346.

31. Agresti A (2002) Categorical data analysis. Hoboken, NJ, USA: John Wiley & Sons, Inc.

32. Cytel Software Corporation (2010) StatXact: version 9. Cambridge, MA, USA: Cytel Software Corporation.

33. Long JS, Freese J (2006) Regression models for categorical dependent variables using Stata. College Station, TX, USA: Stata Press, Stata Corporation.

34. StataCorp (2009) Stata statistical software: release 11. College Station, TX, USA: Stata Corporation.

35. Nichols A (2007) Causal inference with observational data. Stata Journal 7: 507–541.

36. Rabe-Hesketh S, Skrondal A (2008) Multilevel and longitudinal modeling using Stata. College Station, TX, USA: Stata Press.

37. Brant R (1990) Assessing proportionality in the proportional odds model for ordinal logistic regression. Biometrics 46: 1171–1178.

38. Long JS (1997) Regression models for categorical and limited dependent variables. Thousand Oaks, California, USA: Sage.

39. van Raan AFJ (1996) Advanced bibliometric methods as quantitative core of peer review based evaluation and foresight exercises. Scientometrics 36: 397–420.

40. Lindsey D (1989) Using citation counts as a measure of quality in science. Measuring what's measurable rather than what's valid. Scientometrics 15: 189–203.

41. Craig ID, Plume AM, McVeigh ME, Pringle J, Amin M (2007) Do open access articles have greater citation impact? A critical review of the literature. Journal of Informetrics 1: 239–248.

42. Cole S, Fiorentine R (1991) Discrimination against women in science: the confusion of outcome with process. In: Zuckerman H, Cole JR, Bruer JT, eds. The outer circle Women in the scientific community. London, UK: W W Norton & Company. pp 205–226.

43. Shadish WR, Cook TD, Campbell DT (2002) Experimental and quasi-experimental designs for generalized causal inference. Boston, MA, USA: Houghton Mifflin Company.

44. Peters DP, Ceci SJ (1982) Peer-review practices of psychological journals - the fate of accepted, published articles, submitted again. Behavioral and Brain Sciences 5: 187–195.

45. Ziman J (1996) Is science losing its objectivity? Nature 382: 751–754.

46. Geisler E (2001) The mires of research evaluation. The Scientist 15: 39.

47. Bornmann L, Mutz R, Daniel H-D (2008) How to detect indications of potential sources of bias in peer review: a generalized latent variable modeling approach exemplified by a gender study. Journal of Informetrics 2: 280–287.

48. Hojat M, Gonnella JS, Caelleigh AS (2003) Impartial judgment by the "gatekeepers" of science: fallibility and accountability in the peer review process. Advances in Health Sciences Education 8: 75–96.

49. Godlee F, Dickersin K (2003) Bias, subjectivity, chance, and conflict of interest. In: Godlee F, Jefferson J, eds. Peer review in health sciences, 2nd ed London, UK: BMJ Publishing Group. pp 91–117.

50. Thorngate W, Dawes RM, Foddy M (2009) Judging merit. New York, NY, USA: Psychology Press.

15

Can We Identify Non-Stationary Dynamics of Trial-to-Trial Variability?

Emili Balaguer-Ballester[1,2]*, Alejandro Tabas-Diaz[1], Marcin Budka[1]

1 Faculty of Science and Technology, Bournemouth University, United Kingdom, **2** Bernstein Center for Computational Neuroscience, Medical Faculty Mannheim and Heidelberg University, Mannheim, Germany

Abstract

Identifying sources of the apparent variability in non-stationary scenarios is a fundamental problem in many biological data analysis settings. For instance, neurophysiological responses to the same task often vary from each repetition of the same experiment (trial) to the next. The origin and functional role of this observed variability is one of the fundamental questions in neuroscience. The nature of such trial-to-trial dynamics however remains largely elusive to current data analysis approaches. A range of strategies have been proposed in modalities such as electro-encephalography but gaining a fundamental insight into latent sources of trial-to-trial variability in neural recordings is still a major challenge. In this paper, we present a proof-of-concept study to the analysis of trial-to-trial variability dynamics founded on non-autonomous dynamical systems. At this initial stage, we evaluate the capacity of a simple statistic based on the behaviour of trajectories in classification settings, the trajectory coherence, in order to identify trial-to-trial dynamics. First, we derive the conditions leading to observable changes in datasets generated by a compact dynamical system (the Duffing equation). This canonical system plays the role of a ubiquitous model of non-stationary supervised classification problems. Second, we estimate the coherence of class-trajectories in empirically reconstructed space of system states. We show how this analysis can discern variations attributable to non-autonomous deterministic processes from stochastic fluctuations. The analyses are benchmarked using simulated and two different real datasets which have been shown to exhibit attractor dynamics. As an illustrative example, we focused on the analysis of the rat's frontal cortex ensemble dynamics during a decision-making task. Results suggest that, in line with recent hypotheses, rather than internal noise, it is the deterministic trend which most likely underlies the observed trial-to-trial variability. Thus, the empirical tool developed within this study potentially allows us to infer the source of variability in *in-vivo* neural recordings.

Editor: Dante R. Chialvo, National Research & Technology Council, Argentina

Funding: The author ATD has been funded by The Graduate School PhD studentship scheme, Bournemouth University. This paper has been funded by the BU Open Access Publication Fund. The research leading to these results has received partial funding from the European Union 7th Framework Programme (FP7/2007–2013) under grant agreement 251617. The funders had no role in study design, data collection and analysis, decision to publish, or preparation of the manuscript.

Competing Interests: The authors have declared that no competing interests exist.

* E-mail: eb-ballester@bournemouth.ac.uk

Introduction

Non-stationary time series are very common in physical and biological systems. Thus, approaches to the analysis of time series in dynamic scenarios have been developed in a wide range of areas such as geophysics (e.g. [1,2] and references therein), econometrics [3] or human neurophysiology [4] to name just a few. For instance, electroencephalographic responses (EEG) often appear non-stationary; therefore it is crucial to extract invariant, stationary components of the signal for performing reliable analyses [2,4].

More generally, responses of the brain to the same stimulus typically vary across multiple instances of the same experiment (trials) [5–12]. The origin of the trial-to-trial variability is currently one of the most actively debated topics in neuroscience. Trial-to-trial variability has been observed in multiple modalities of neural recordings [5,7,13–17] and it has been studied using a variety of techniques ranging from multivariate statistics to information-theoretic approaches (e.g. [7,18–20]). However, despite the large

number of studies over recent decades, the dynamical substrate of such observed variability is largely unknown [5,13].

Understanding the main causes of trial variability in neural recordings is a major challenge for current data analysis techniques. Often such variability is attributed to the irregular responses in cortical neurons (due to the probabilistic nature of synaptic transmission; see e.g. [5,21–24]), but other potential causes are the chaotic dynamics of complex neural networks [25–27] or the lack of specificity in top-down brain dynamics [13]. Thus it is important to design new data analysis methods in order to discern whether observed variability is essentially driven by stochastic or by deterministic processes.

Data analysis methods for non-stationary environments are a very active research direction in machine learning and computational statistics. Attention has typically been focused on change detection (e.g. [28–34]) and on designing strategies yielding to competitive predictions in dynamic settings e.g. in areas such as streaming data mining [29,35,36], on-line dimensionality reduction [37], metalearning [38] or Gaussian Processes [39] to name a few. Recent studies identified invariant subspaces, allowing the design of robust models specifically for each stationary data segment [4,6]. Nevertheless, a common assumption in such approaches is that stationarity is preserved in short segments of the time series (for instance [6]). In this setting, the source of non-stationarity is typically attributed to a "temporal drift" in the statistical moments of likelihood distributions $P(x|C;t)$, generating x patterns of each class C [6,36].

In this proof-of-concept initial study we propose a different angle for the analysis of multivariate recordings based on non-autonomous dynamical systems. The challenge is to discern whether the observed trial-to-trial variability in recordings is caused by deterministic dynamics or by stochastic fluctuations. Towards this goal, we first analysed a compact low-order nonlinear dynamical system with random initial conditions. As the simplest possible model exhibiting two attractors, we used the Duffing equation[40–43], a ubiquitous model arising in many physics and engineering areas such as nonlinear electrical circuits, optics (e.g. [44,45] and references therein), quantum field theory [45,46] or the study of chaotic oscillatory behaviour [43]. Similar but less parsimonious multi-stable canonical systems have been recently used for modelling how biological systems effectively operate in non-stationary environments, such as human alpha rhythms underlying EEG recordings [47]. Smooth variations of the high-order perturbation term typically enable such class of models to express a wide dynamic repertoire [47], as is the case in the compact system that we show in this work.

We also propose a simple measure of classifier performance based on the coherent behavior of trajectories with respect to class-boundaries and analyse its response depending on the source of non-stationarity. Time series driven by non-autonomous (time-varying) dynamics show an abrupt variation in the trajectory coherence statistic which is not present in randomly generated data, as commonly assumed in current approaches [29]. Thus, this statistic acts as an immediate signature of a significant variation in the underlying dynamics. Our analyses enable us, for instance, to inform models on the necessity of updating their parameters towards maintaining a competitive performance in non-stationary conditions.

The analysis is then extended to multivariate classification problems in real datasets exhibiting non-stationary dynamics, consisting of atmospheric pollutants and neural recordings time series. As an illustrative example, we focused on multi-unit recordings in rodent frontal cortex networks in behaving animals during the performance of a difficult task [48,49]. Recently, it has been proposed that behavioural trial-to-trial variability could be the result of the imprecision of top-down processes involved in the performance of cognitively demanding tasks [13,50], while variability in cell-to-cell responses – the commonly accepted source of the observed variance [21–23,51] – may play a secondary role [13]. Thus, as an illustrative example, we focus on multi-unit recordings in rodent frontal cortex networks. Equipped with the analyses presented here, we suggest that a deterministic trend plays a major role in the observed trial-to-trial variability during decision making.

Results

The following section introduces intuitively the canonical system used in the study (the Duffing family) and frames it in the context of a supervised machine learning task – classification. This system plays the role of a ubiquitous model for understanding complex classification problems from a nonlinear dynamics angle. Results lead to a proposition in Text S1 and to a general conjecture, which we have benchmarked in real non-stationary datasets discussed in

Text S2 and Figure S2. In the last section, these approaches are applied to neural recordings.

Canonical model of binary classification in non-stationary settings

Our first aim is to infer the conditions in which arbitrarily small perturbations in parameters of underlying dynamics can be discriminated from random fluctuations. The first step is to model a non-stationary two-class classification problem.

The simplest, yet ubiquitous ordinary dynamical system capable of a range of attracting dynamics is the Duffing nonlinear equation, encompassing first order and cubic nonlinearities (the perturbation term) as well as an external force:

$$\ddot{x}(t) + \delta\dot{x}(t) - \beta x(t) + \alpha x^3(t) = \Omega \cdot \cos(wt), \qquad (1)$$

or equivalently,

$$y = \dot{x}(t); \quad \dot{y}(t) = -\delta y(t) + \beta x(t) - \alpha x^3(t) + \Omega \cdot \cos(wt),$$

where α, β and $\delta \in \Re$ are model parameters. This dissipative autonomous system generates a wide range of attracting phenomena such as bi-stability, periodic orbits and fractal attractors. Thus, it has provided a useful paradigm during recent decades for the study of nonlinear oscillations and chaotic dynamical systems [45]. Despite its simplicity, exact solutions of this system are generally not known, although they have been the focus of many studies during recent decades [41–43,45], thus numerical simulations are needed.

For a range of parameter values ($\delta \geq 8\beta$; $\beta,\alpha > 0$; $\Omega = 0$) the system has a simple behaviour: a saddle point at $x = 0$ and two sinks at the symmetric equilibrium points $x_1 = -\sqrt{\frac{\beta}{\alpha}}$, $x_2 = \sqrt{\frac{\beta}{\alpha}}$ (Figure 1A; see also Methods).

A nonlinear two-class classification problem is then naturally defined: Figures 1A and 2B show the basin of attraction of the two sinks, constructed by generating random initial conditions from a fixed, two-dimensional Gaussian distribution centred at the origin (standard dev. 4), which are then subjected to the flow indicated in Equation 1.

Blue and red dots show fixed points towards which trajectories converge. Trajectories belong to the class C_1 (red) if they are attracted to the left sink or to the class C_2 (blue) if they converge to the right sink. Figure 2B shows a more detailed display of the basins of attraction of the sinks (using 10^4 random initial conditions). Groups of class C_1 trajectories are interleaved with groups of C_2 trajectories in the phase space; hence basins of attraction furnish the spiral structure shown in Figure 2B.

Such simple dynamics typically breaks down with changes of β,δ,Ω parameters (e.g. it undergoes supercritical pitchfork bifurcation and periodic orbits appear for $\delta \simeq 0$, Figure 1B; a chaotic attractor emerges for a range of Ω values, Figure 1C [45]), yielding to abrupt variations in posterior probabilities of class-membership $P(C|x,y;t)$ (see insets in the figures and Methods for details).

This setting has parallels with the so-called "concept shift" in data mining literature [38,52] and is not of interest here as detection of abrupt changes is often successfully addressed by standard change detection approaches (e.g. [29,36]). Thus, such kind of relatively obvious non-stationary changes, typically induced by bifurcations are not considered in this work.

In contrast, and crucially, here we are only interested in inferring very subtle variations in the underlying system dynamics

A

B

C

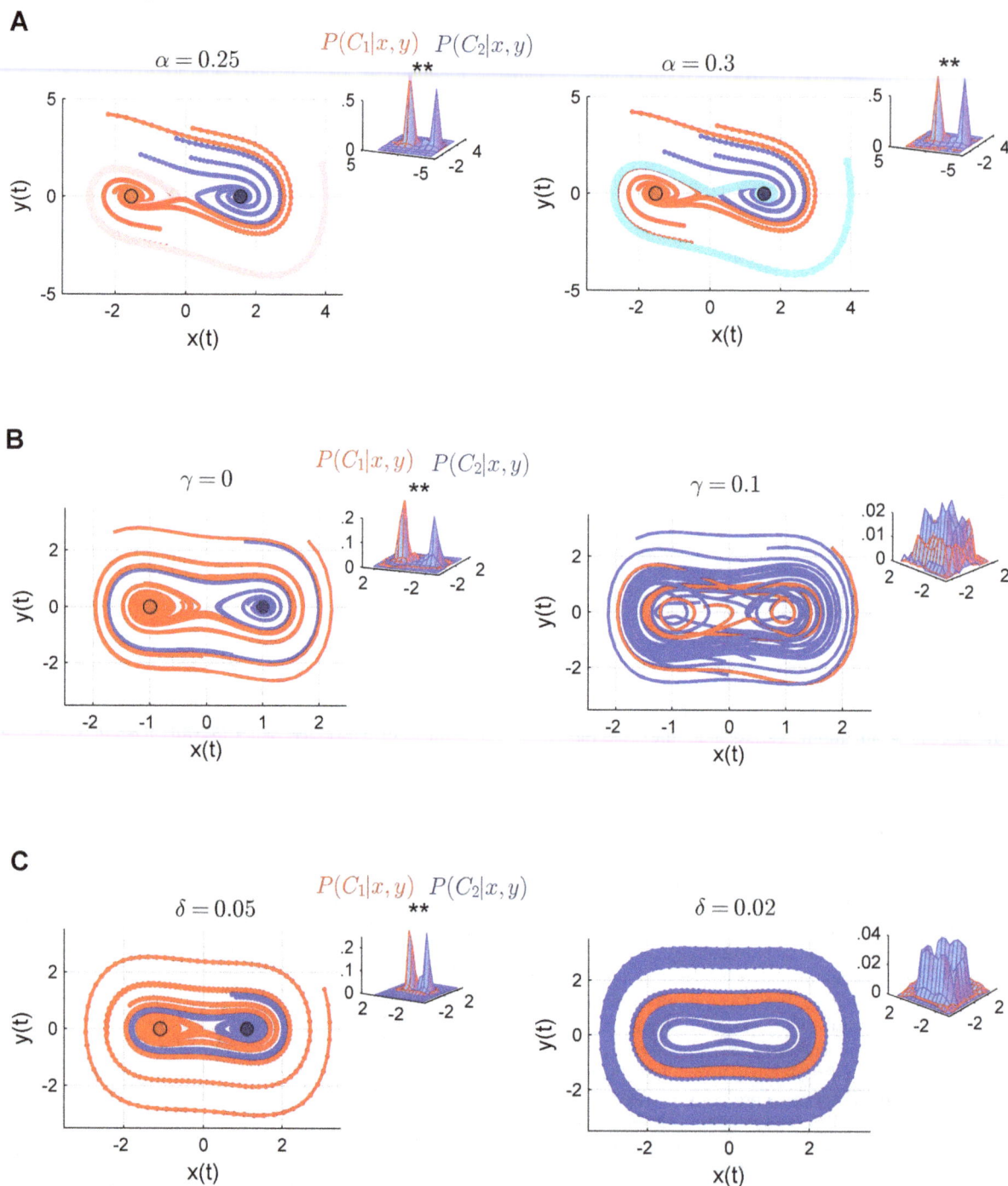

Figure 1. Duffing non-linear oscillator (Equation 1, see parameter values in Methods). (A) A small perturbation leading to a subtle drift in the relative distance between fixed points. Each subplot shows 10 trajectories (i.e. 10 different initial conditions randomly drawn, see text). Light red (left) and blue (right) lines indicate an example of a trajectory that changes its class (i.e. it is attracted to the opposite sink) after the small perturbation induced. Insets show class-posterior probabilities of each phase space vector belonging to the basin of attraction of one of the two sinks (see Methods for details). Two stars (**) indicate significant differences between means in the x-axis at $p < 0.001$; which remain after a subtle variation in the $\Delta\alpha$ of the perturbation parameter of the Duffing system. (B) and (C): Perturbation in other parameters induces bifurcations leading to chaotic oscillations (B) or global limit cycles (C) e.g. [42]. As in plot A, inset shows the class-posteriors, which are severely transformed after such parameter variations.

which are not evident from standard statistical analysis. To this end, we modify slightly the relative distance of the attractors while the dynamics are essentially unchanged by inducing a small perturbation in α, which can be approximated to $\alpha(t_0) \rightarrow$

$\alpha(t_1) = \alpha(t_0) + \Delta\alpha$ on a first-order level (all other parameters are fixed). As the fixed points become closer to each other i.e. the α parameter increases (Figure 1A, inset) distribution modes significantly differ (multivariate analysis, Wilks' $\Lambda = 0.35$,

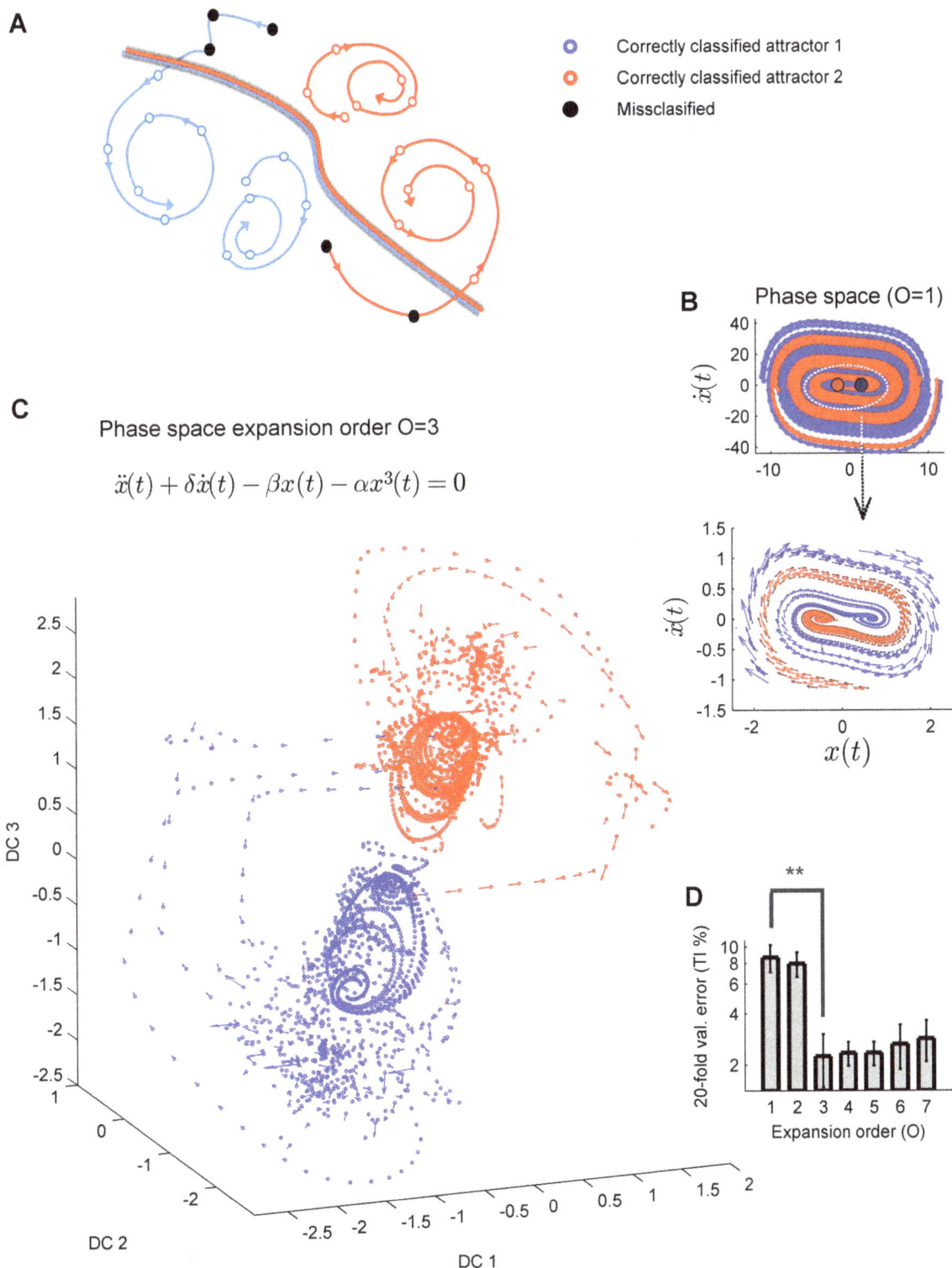

Figure 2. Trajectory behaviour in Duffing systems. (A) Schema illustrating convergent trajectories with respect to attracting state boundaries (see also Figure S1). (B) Phase space flow (using 10^4 initial conditions). (C) Projection into the three maximally discriminating directions (gram-schmidt ortonomalized) of an expanded space of order three. (D) This optimally regularized discriminant was used to compute the 20-fold cross validation of the trajectory incoherence index (TI) i.e. those different from any of the trajectories shown in plot (A) across initial conditions. The expansion order 3 yields to a maximal out-of-sample convergence; highly significant with respect to the phase space ($O = 1$) shown in plot B ($p < 0.001$, see main text).

$\chi^2(2) = 1.58 \, 10^3$, p<0.001). However, some of those trajectories crossing the vicinity of the centre fixed point $(0,0)$ are attracted to the opposite sink i.e. they belong to a different class (Figure 1A).

Thus, intuitively, we expect that a classifier which models the two posteriors with negligible error at $t = t_0$, will fail to predict the true class of such trajectories at $t = t_1$ after a subtle drift on the α

parameter. This arbitrary accurate classifier at $t=0$ is blind to such a subtle, yet fundamental change in the latent dynamics.

Is there a way to discriminate deterministic variations from changes of probabilistic nature? The following sections show how trajectories which changed the attractor in non-stationary settings allow us to discern the source of the observed data variability.

Reconstructing attractor dynamics

The analysis starts by devising an optimal classifier for the autonomous (stationary) system shown in Equation 1. The two basins of attractions (the regions of the space in which trajectories ultimately converge towards the corresponding attractor) are not separable in the original phase space (Figure 2B). Thus, an optimally expanded space was used to compute boundaries between classes with a minimum generalization error (the space with the lowest dimensionality allowing us to reach a Bayes-optimal error, see Methods and Figure 2).

Multinomial expansions of a phase spaces are also suitable spaces i.e. the trajectory flow will consistently converge to the corresponding attractor as in the original phase space, while the basin of attraction tends to be linearly separable [48,53,54]. Here we used embedding spaces of different dimensionality spanned by high-order interactions up to a $o^{th} order$ of the original dimensions (see Methods).

In general, distances in such high dimensional spaces cannot be feasibly computed due to a range of problems collectively referred to as the "curse of the dimensionality" in the machine learning literature [55], and especially the distance concentration phenomenon [56]. Thus it is in general not possible to analyse trajectory dynamics directly in large embedding spaces. Nevertheless, a classifier allows us to estimate relative positions of input vectors with respect to the class boundaries (Figure 2). By tracking the predicted label of the l vectors encompassing a single trajectory, we can access and assess the behaviour of the *class*-trajectory in the state space.

In simple terms, a class-trajectory initiated at $x(t_0) = (x(t_0),y(t_0))$ is considered as convergent into a specific volume of the space if *all* its vectors from a certain time $t>>t_0$ are correctly classified (empirically, it will suffice in this simulation with the last $l/4$ trajectory vectors) i.e. they are assigned to the closer attractor (see schema in Figures 2A and S1A). For instance, trajectories shown in Figure 2A are examples of convergent trajectories, because they either cycle within or finish in the region of the space delimited by its class i.e. its basin of attraction.

We can thus define a natural statistic for time series, the lack of coherence of class-trajectories (trajectory incoherence, TI), as the fraction of complete trajectories which are not convergent. In other words TI is the percentage of trajectories which are *not* of the type of trajectories shown in Figure 2A (see Methods for a more precise definition and Text S1).

TI is thus a quantitative index of trajectory behaviour in non-accessible, high-dimensional state spaces (not to be confused with the exponential divergence of nearby trajectories given by the maximum lyapunov exponent, used as a signature of chaos, for instance [57,58]). In Figure 2 we estimated TI by cross-validating a regularized Fisher discriminant (*kernelized* for effectively operating in high dimensional state spaces as detailed in Methods [59,60]). Not surprisingly, an embedding space of third order, precisely the nonlinear order in Equation 1, is the most suitable to capture the attractor dynamics i.e. with the lowest TI. In the light of this simple index, we next studied the behaviour of trajectories in time-varying scenarios.

Detection of latent non-stationary trends

The analysis continues with a parsimonious simulation of a multi-stage data acquisition setting in noise. We induce a temporal dependency on the perturbation term of the Duffing model (Equation 1),

$$\ddot{x}(t) + \delta\dot{x}(t) - \beta x(t) + \alpha(t)x^3(t) = 0; \qquad (2)$$

which now has a simple non-autonomous dynamics. We must stress that we are interested here in subtle i.e. non-statistically detectable (on a single-trial basis) variations in the relative position of the attractors in the phase space; which essentially preserve their dynamics (unlike more abrupt non-stationary changes, e.g. Figures 1C and D) therefore bifurcations are typically excluded from this analysis. This subtle non-stationarity is induced by arbitrarily small perturbations in the parameter α, thus, it will suffice to analyse the behavior of TI for a first order expansion of $\alpha(t)$ in equation 2. An analysis of the perturbation effect in the system dynamics can be found in Text S1.

Figure 3A shows a few randomly generated trajectories, see also schema in Figure S1A. As stated previously, when α linearly increases the distance between attractors decreases and some trajectories crossing $x=0$ will be potentially attracted to the opposite spiral (see also Text S1). For instance, after six trials in Figure 3A a single trajectory changes the attractor, while no significant change in the statistical moments will be observed, as discussed below.

A simulation of this setting is shown in Figures 3B–D and S1. As expected, the error monotonically increases while distance between fixed point decreases. Critically, there are no statistical differences in the classification error from one trial to the next (two-tailed pairwise t-tests, $t(598) < 1.26, p > 0.21$, normality accepted according to Lilliefors tests, $p > 0.05$). Other standard classification accuracy measures (Wilk's Lambda, higher order statistics such as Jensen-Shanon divergence between posteriors or certainty measures [61]) showed similar insensitivity to those subtle changes (Figures S1B and S2C).

In this simulation, CE does not increase significantly with respect to the first trial before trial number 6 i.e the comparison of trial 1 versus trial 6 is the first to achieve significance ($t(598)=1.02, p=0.012$, Figure 3B). Thus, when information on the classification performance in previous trials is not accessible, statistics will fail to detect such an event on a single-trial basis. This historical information is often not available.

Class-trajectory coherence statistic (TI), in contrast, allows the detection of such critical change on a trial-by-trial basis. The fraction of misclassified trajectories progressively increases with respect to the previous trial and reaches trial-to-trial significance on trial 6 ($t(598)=1.97, p=0.048$) precisely when CE is significant with respect to the reference trial. Therefore TI immediately alerts on the loss of generalization capability of the classification model, unlike the classification error and related statistics (Figure 3C, thick triangle markers). Consistently, the Priestley-Subba-Rao test (PSR) of non-stationarity shown in Figure S1C (see Methods, [32–34]) is non-significant for all trial-to-trial pairwise comparisons of x and y time series until trial 5 (non-parametric MannWhitney $U(4998) < 6.3 \times 10^6, p > 0.503$); while it reaches trial-to-trial significance precisely on trial 6 (Figure S1C, MannWhitney $U(4998)=6.4 \times 10^6, p=0.0478$; normality rejected according to Lilliefors test, $p<0.01$) fully in line with TI results.

Note also that initial conditions were randomly drawn from a normal distribution spanning up to four standard deviations, suggesting that TI is robust to high levels of this input noise at

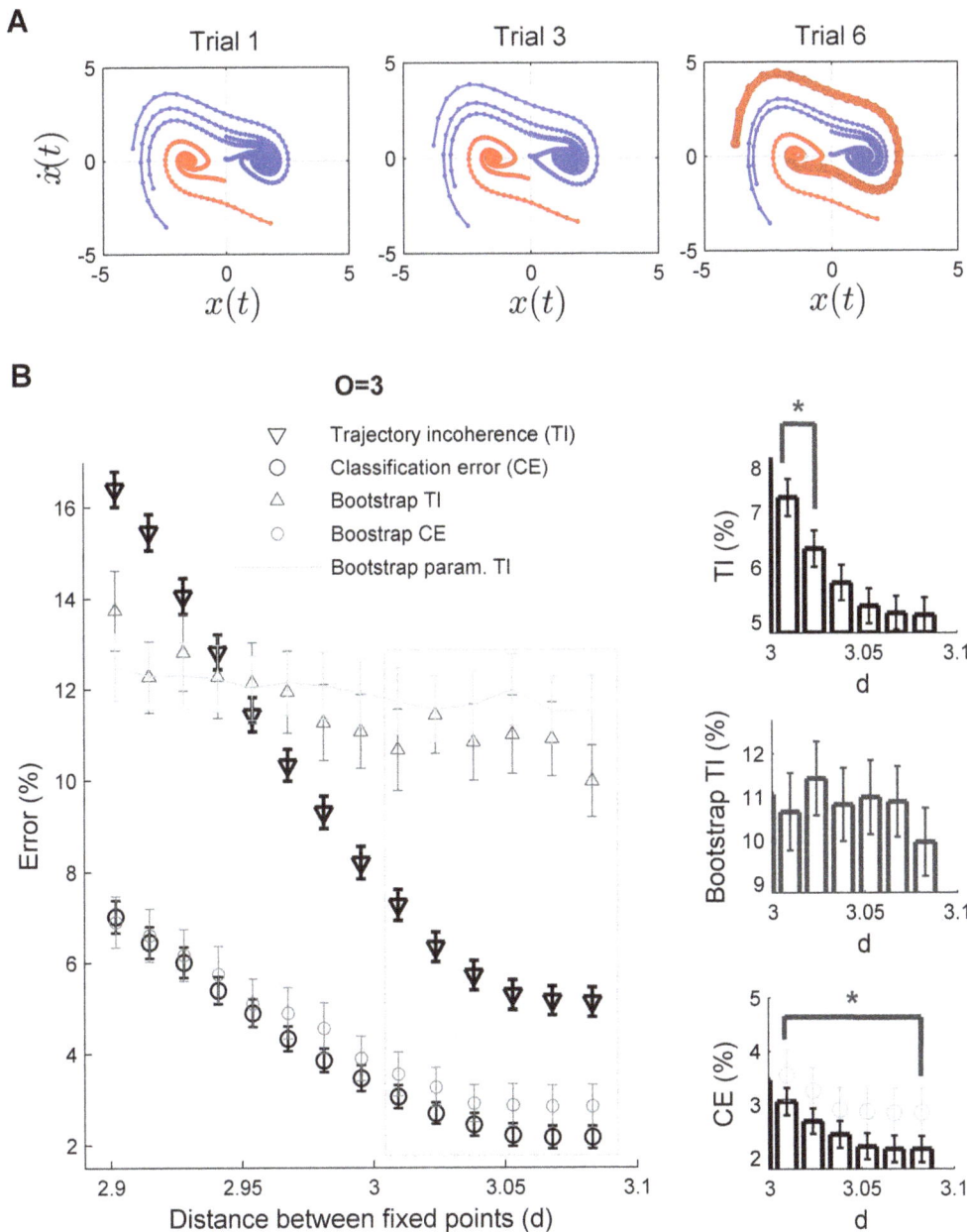

Figure 3. Non-autonomous drift in a non-linear dynamical system (unforced Duffing oscillator). (c.f. Figure S1). (A) Example of a linear variation in the perturbation term α (see also Equation 1). As fixed points approach each other, few trajectories change the basin of attraction and thus the class-membership. (B) Optimally regularized kernel-fisher discriminant in a third order expanded space was used to compute the classification error (CE) and trajectory incoherence (TI) as the distance between fixed point varies (shown mean values of 1000 initial conditions for each trial, error bars are SEM). The discriminant subspace is computed for the first trial and then fixed and applied to subsequent trials (note that only validation results from trials 2–14 are shown in the figure). Insets show amplified versions. Both CE (bottom inset) and TI (top inset) increase over trials, but TI enables us to detect, on a single trial basis, when a significant change occurs. When the temporal contingency within each trajectory is disrupted (bootstrap data, middle inset) TI is no longer sensitive to trial-to-trial variations, indicating the absence of a deterministic trend driving the observed dynamics. When bootstraps are generated by randomly sampling the increment of α (from a uniform distribution of the same range), no trend in TI is observed either (thin grey line), as expected. These results are fully in line with statistical analyses shown in Figures S1B and S1C.

99.9% confidence. However, and significantly, this is only the case if the underlying source of non-stationarity is deterministic. Figure 3B also shows bootstrap data, constructed by shuffling vectors $\mathbf{x}(t)$ within each trajectory, while class-associations are maintained. Thus, CE is not altered, but the temporal flow within trajectories breaks down. In this setting, there is no guarantee that trajectories are attracted to any volume and thus TI should not vary significantly (Figure 3B, grey triangle markers), suggesting

that multi-stable deterministic dynamics does not play a major role in the observed data. Likewise no trend in TI is observed either when trajectories are preserved, but the perturbation term varies randomly from trial to trial; in other words when the autonomous duffing system is deterministic but its non-autonomous dynamics is stochastic (grey line in Figure 3B), as envisaged.

These results have been illustrated for the Duffing family, but this analysis potentially has a wider scope of application.

As a simple, intuitive example, consider an autonomous (static) dynamical system parameterized by α equipped with i.i.d. random initial conditions; this system generates an observable dataset of n trajectories of length l patterns each. Consider also an accurate classifier in a Bayes sense for such stationary dataset. Then, a small parameter perturbation such as the ones illustrated in Figures 1 and 3 will have a completely different effect on CE and TI. Since at least one trajectory of length l will converge to a different attractor (see also Text S1 Lemma 1), the change on TI is at least $\frac{1}{n}$.

$$\Delta TI = TI(\alpha + \Delta\alpha) \geq \frac{1}{n}; \qquad (3)$$

By definition of TI only the last $\tilde{l} \leq l$ vectors from a trajectory of length l will be misclassified, thus:

$$\Delta TI \geq \frac{l}{n \cdot l} \geq \frac{\tilde{l}}{n \cdot l}; \qquad (4)$$

As the classification error is the fraction of misclassified vectors $CE = \frac{\tilde{l}}{n \cdot l}$, trivially, the following relation holds:

$$\Delta TI \geq \Delta CE; \qquad (5)$$

This is precisely the result shown in Figure 3C i.e. TI increases more abruptly than the classification error.

In contrast, if we consider an identical dataset in which *all* $n \times l$ patterns (not just the initial conditions) have been i.i.d randomly drawn i.e. where there are no coherent trajectories, a change in the parameters of the generative distribution does not guarantee Equation 5 bound. Thus, TI would not be sensitive to any changes and CE would be a more appropriate estimate in this i.i.d. data. This effect is shown in Figure 3C, where the order or vectors within trajectories has been randomly altered before the system undergoes a parameter drift (Bootstrap TI in Figure 3C).

The approach devised here could be thus applied to multi-stable scenarios, where a "snapshot" of attracting dynamics is observed in each trial. As a real data example, we applied analyses in a well-known, multivariate time series where attractors subtly drift over time, discussed in detail in Text S2. The dataset consists of hourly concentrations of ozone, meteorological variables and other atmospheric pollutants (Text S2). Ozone time series are well known to exhibit daily periodicity which is modulated by a subtle seasonal trend [62,63]; thus they will serve to benchmark further simulation results before the analysis of neural data in the next section.

This first illustrative analysis is shown in Figure S2. Precisely as in the dynamical system simulations, a signature of non-autonomous dynamics is indicated by an abrupt increase in TI not accompanied by a sudden change in CE, suggesting a deterministic trend in the observed trial-to-trial variability (see details in Figure S2 and Text S2).

In summary, results obtained for the Duffing family of dynamical systems are potentially extendible to more general settings, exhibiting a repertoire of attracting dynamics in noise. The next section shows another example of application of our approach, the investigation of trial-to-trial variability in *in vivo* recordings.

Trial to trial variability in neural ensembles

Neuronal responses to the same task often differ from trial to trial, particularly when recorded in higher cognitive areas [5]. The origin and functional role of this variability has recently attracted a lot of attention in neuroscience [5,13,64], and has been analysed using a variety of statistical and information-theoretic approaches (e.g. [6,7,18–20,65]).

The analysis developed in this work enable us to infer whether the observed trial-to-trial variability is essentially driven by stochastic processes as typically assumed in previous studies. We focus on a cognitively demanding task to investigate the trial-to-trial dynamics of neural ensemble recordings in rodent frontal cortex. Figure 4A shows an example of a memory-guided decision-making radial arm-maze experiment (e.g. [48,49]). In a nutshell, the animal visits a series of baited arms during the training phase (termed choice epochs) in order to consume the reward (termed reward epochs), followed by a delay phase in which no task is performed (omitted in the Figure). Subsequently, during the test phase, the rat visits different arms to obtain the reward again. Activity of a neural ensemble was recorded in a rat frontal cortex during several consecutive trials (Methods). We next defined a classification problem where classes correspond to short (± 1 sec.) temporal periods surrounding choices and reward epochs during training and test periods, respectively (the rest of the firing rate vectors are not considered in the analysis). For more details on the task, see Methods and [49]). Figure 4B shows the projection into the three maximally discriminating dimensions of the optimally expanded space. In this case the reconstruction started with a delay-coordinate map before the nonlinear expansion map [53,54,66] as a previous step for disambiguating the trajectory flow (see [48,67]). As in Figure 2, arrows indicate the flow field of neural population states; which moves quickly between different task phases, suggesting the presence of attracting states. Attracting dynamics of neural ensembles have previously been found in different areas such as the olfactory bulb of insects, rodent hippocampus [68–71] and in prefrontal cortex [48].

However, Figure 4C also shows responses from trial to trial subtly differ: there is an apparent clockwise rotation of the task-epoch trajectories suggesting a consistent temporal drift, which may be the cause of such non-stationarity. The approach developed here helps to discern whether the origin of such shift can be solely attributed to stochastic fluctuations.

A sufficient condition of non-autonomous dynamics is a sharp increase in TI index just at the trial when the classification error significantly increases (with respect to any previous trial); as devised in the previous section. This is precisely the result of the analysis shown in Figure 4C, where TI abruptly changes on the third trial (Mann-Whitney $U(5) = 18, p = 0.046$; normality rejected according to Lilliefors test, $p < 0.05$). As in Figure 3, this trial-to-trial variation is non-significant for CE by large margins ($p > 0.15$ for any trial-to-trial test comparison) while the comparison of trials 1 and 3 CE reaches significance (Mann-Whitney-U, $p = 0.0079$)

In order to ensure further the significance of these analyses, bootstraps were constructed by shuffling the firing rate vectors within trajectories while preserving the trials order [48]. According to previous section results, ΔTI should no longer be informative, as shown in Figure 3C. This prediction is again fully in line with results reported in Figure 4C.

Overall, Figure 4 shows that during the performance of this cognitively demanding task, the process underlying trial-to-trial variability in frontal cortex ensemble recordings is essentially non-autonomous. The aim of this single example is only to illustrate the capacity of the proposed approach. However, this striking result suggests that intrinsic, random fluctuations may not be the only

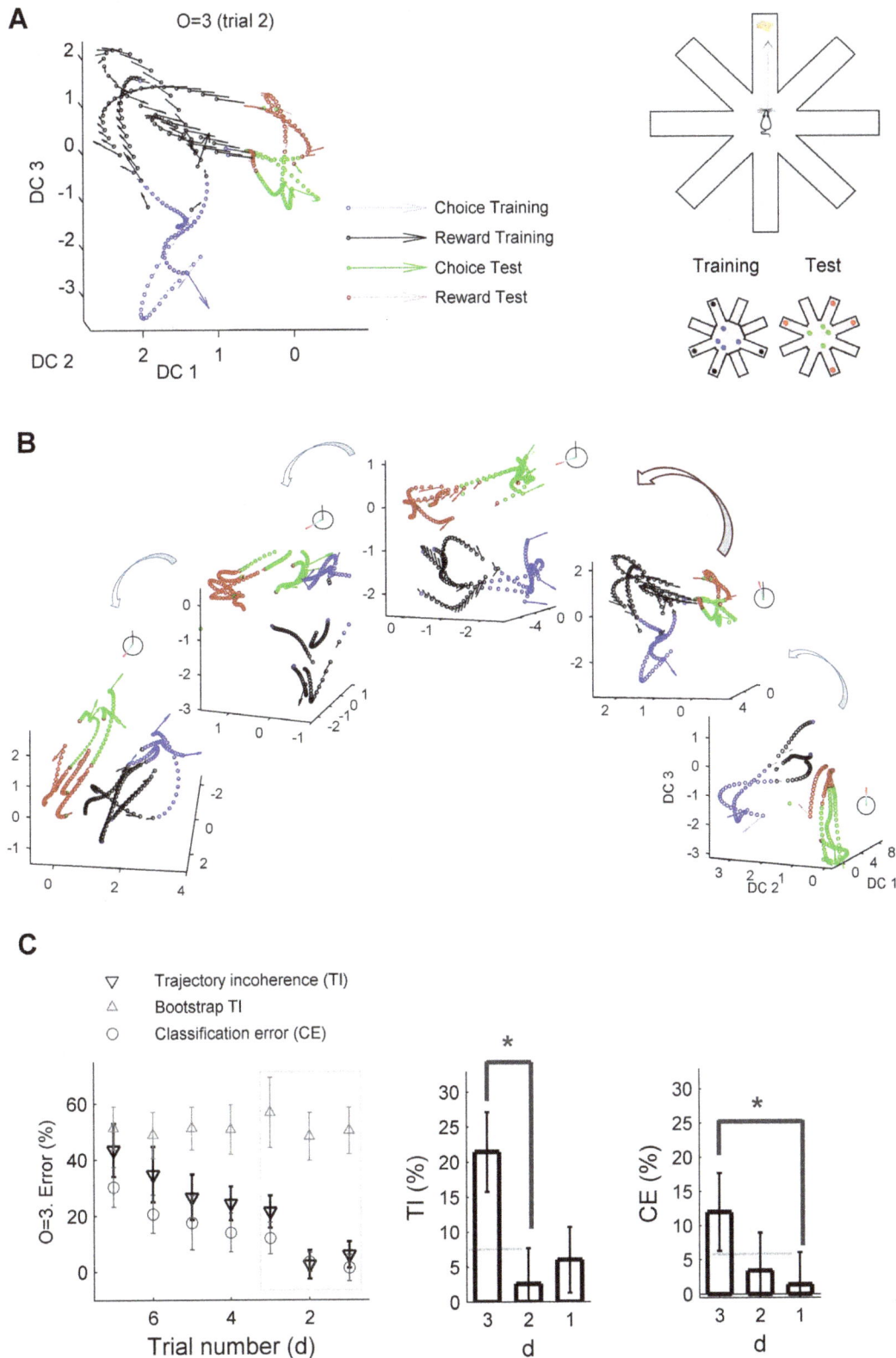

Figure 4. In vivo neural ensemble recordings in rat frontal cortex. (A) Example of a delay-coordinate map expanded to a third order state space; see Methods and [48]) projected onto the three maximally discriminating dimensions (ortonormalized). Different colours correspond to different stages of the task (a radial arm-maze, inset left). (B) A clockwise rotation of the task-stage states from trial to trial seems to take place, suggesting a deterministic drift in the putatively attracting sets associated with task epochs. (C) Non-stationary drift in ensemble recordings. Analyses on an expanded space of third order where optimised for the first trial, the maximally discriminant subspace is fixed and then used to compute CE and TI in the next trials. As in the theoretical model (Figures 1–3) and in the real data example (Figure S2), TI increases faster than CE. Again consistently with previous results, when temporal order of vectors is shuffled, TI is not sensitive to trial-to-trial shifts in dynamics.

cause of the observed variability in ensemble recordings, as commonly assumed in neural modelling [5].

Discussion

In this proof-of-concept study we devised a sufficient condition to identify when a multivariate dataset has undergone changes in its parameters' dynamics from trial to trial. The proposed statistic, class-trajectory coherence (or lack thereof) is an easily accessible value, sensitive to subtle departures of deterministic nature in multi-attracting dynamics subject to input noise. This analysis is particularly advantageous when statistical moments do not significantly vary from trial to trial and thus a significant trend cannot be statistically proven on a trial-to-trial basis by standard testing.

The fraction of non-coherent trajectories is a sufficient statistics i.e. if data is independently drawn, both trajectory and classification errors would behave similarly, indicating that deterministic hypothesis cannot be accepted. The importance of this study hence also stems from the fact that i.i.d. data generation is still the typical assumption in current data mining approaches for non-stationary problems [36]. For i.i.d. data, the classification error or derived measures are appropriate empirical estimators of the "true error" (the asymptotic risk, a well-known result in statistical learning [72]) and trajectory analyses are not necessary. A number of tests for non-stationary time series have been proposed in the statistical literature based e.g. on fourier analyses [33,34] or more recently on wavelet spectrum analyses (for instance see [32]); such tests are also powerful tools when the sampling size is significant (unlike the *in vivo* ensemble recordings analysed here).

As an example of a real-world application, we used two well-known and completely different datasets where attracting dynamics was observed. The main focus of our analysis was on in-vivo neural ensemble recordings, where trial-to-trial variability is often observed. The origin of trial-to-trial variability in neural recordings is a fundamental question in neuroscience, touching the essentials of our understanding of neural computations. Among the many possible causes, it has been traditionally accepted that the intrinsic irregularity of spike probability is the origin of most of the observed trial-to-trial variance, mainly due to probabilistic nature of synaptic transmission [5]. Thus, very recently, efforts have been applied to devising suitable methods for the analyses of non-stationary spike trains[6,65]. In a similar spirit, recent models have sought to infer time-varying statistics of synaptic conductances from membrane recordings (e.g. [73–75]).

However, there are no empirical demonstrations of whether internal, random fluctuations always drive the observed trial-to-trial variance in neural recordings. The hypothesis stating that the observed trial-to-trial variably has a stochastic, internal origin has recently been debated [5]. For instance, Beck and colleagues [13] proposed that spike irregularity is often a minor contributor to the unexplained variance, while suboptimal inference (the imprecision associated with deterministic approximations in complex computations) may be the dominant component of behavioural variability in difficult tasks. Thus, most of the variability may be originated rather by complex or chaotic deterministic processes [13], whose parameters can be top-down modulated by active attention (e.g. [50,76]) or by stimulus expectancy [18].

The analyses performed within this study are in line with this hypothesis: we have observed that trial-to-trial variability processing in frontal cortex has a deterministic component. Nevertheless, in this work we show only a limited dataset as an illustrative example because our focus here is rather methodological (an exhaustive analysis on ensemble recordings is not in the scope of this preliminary study).

Our initial analyses are also potentially relevant in the context of biophysical modelling. It has recently been proposed that structured stochastic fluctuations have a highly beneficial function by enhancing the dynamical repertoire of multi-attractor landscape of deterministic networks shaped by anatomical structures in cortex [15,64]. In contrast, in other contemporary models, the richness of observed activity pattern dynamics is provided by purely deterministic, transient dynamical objects. Such *heteroclinic channels* [77,78] are not attractor states, but still retain the neural activity trajectories only for a limited amount of time, even without the intervention of stochastic variability. The class-trajectory coherence statistic presented here would help to validate empirically or disconfirm these two theories.

In a wider scope, understanding the dynamics underlying non-stationary recordings is a ubiquitous problem of computational biology and data analysis. Contemporary machine learning approaches focus on designing algorithms capable of operating in non-stationary settings (e.g. [30,36,37,52]). In this context, the results of this study suggest that trajectory coherence can be used to detect when a classifier needs updating on a single trial basis. This is a critical advantage of our method as with sufficiently smooth drift, an arbitrarily large number of historical results may otherwise be required, which is often computationally impractical in real life settings (e.g. in data streams or online settings [30]) and sometimes not even experimentally accessible.

In summary, in this opening work, we have provided simulated and real challenging scenarios where standard statistics are unable to identify a deterministic trend on a trial-by-trial basis. Analyses developed in this study help to circumvent drawbacks of existing data analysis tools in order to potentially enable a deeper insight into the dynamic sources of the observed trial-to-trial variability in neural recordings.

Materials and Methods

Analyses

Compact non-autonomous dynamical system. The unforced Duffing oscillator for $\delta \geq 8\beta, \beta > 0, \Omega = 0$, as indicated in the Results section, has a simple behaviour consisting of three fixed points (two spiral sinks and a centre). Trivially, the linearized system matrix,

$$J = \begin{pmatrix} 0 & 1 \\ \beta - 3x^2 & -\delta \end{pmatrix};$$ (6)

has eigenvalues $-\frac{\delta}{2} \pm \frac{\sqrt{\delta^2 + 4\beta}}{2}$ for $x = 0$ and $-\frac{\delta}{2} \pm \frac{\sqrt{\delta^2 - 8\beta}}{2}$ for the two attractors $x_{1,2}$ (e.g. [42]). The basic set of parameters used in static simulations (Figures 1–2) were $\alpha = 0.25, \beta = 0.6, \delta = 0.5, \Omega = 0$ (Figure 1A, left plot, Figure 2B) [42]. In Figure 1, only the parameter specified in the plot title is varied, while the rest of parameters are held constant.

A discrete trajectory of class C_1 (c.f. C_2) of length l is defined as

$$T(t_0) = (\mathbf{x}(t_0), \mathbf{x}(t_1), ... \mathbf{x}(t_l));$$ (7)

where $\mathbf{x}(t) = (x,y)$, the initial condition $\mathbf{x}(t_0)$ belongs to the basin of attraction of the positive attractor (blue, class C_2; c.f. red, class C_1) i.e. the continuous counterpart of such discrete trajectory asymptotically converges to the two fixed points $(x_1, 0)$ (c.f. $(x_2, 0)$).

Can We Identify Non-Stationary Dynamics of Trial-to-Trial Variability?

149

In Figures 1 and 2, a class C_1-trajectory is a set of $l = 100$ consecutive patterns with a random initial condition $\mathbf{x}(t=0)$ i.i.d. drawn from $\aleph(0,3)$ such that $\|x(t_l) - x_1\| < \|x(t_l) - x_2\|$ (c.f. $\|x(t_l) - x_1\| > \|x(t_l) - x_2\|$ for class C_2).

Posterior probability distributions shown in Figure 1 ($P(C_1|x,y;t)$ and $P(C_2|x,y;t)$) are computed by tiling the phase space in equal rectangular bins; the limits of the grid are defined by the maximum and minimum values of x and y axes in each simulation. The histogram of classes (i.e. of the corresponding attractors of phase space vectors) is then computed and normalized, yielding to posteriors estimates.

The model used in this work is the simplest dynamical system that can implement a binary classification problem (as defined herein). Although exact solutions are generally unknown, approximations can be established (e.g. Text S1, [45]) enabling us further insights into the system dynamics. A more detailed study of the behaviour of the non-autonomous Duffing oscillator can be found in Text S1.

Reconstruction of attractor dynamics. Kernel algorithms (e.g. [31,59,79]) were used to solve the non-autonomous classification problem in a phase space where basins of attractions are separable. Recently, embedding delay-coordinate maps were combined with nonlinear expanded spaces to reconstruct neural activity trajectories [48]. A polynomial expansion of a phase space is a potentially valid reconstruction of attractor dynamics in moderate noise conditions (for instance [53]) and a well-know reproducing-kernel Hilbert space [59]. Thus, an expanded space of dimension $\frac{p+2!}{2p!}$ is devised here by including high-order interactions up to a pth order of the phase space variables. The dot product of two feature vectors is the inhomogeneous polynomial kernel of a Mercer type [59,80],

$$k(t,t') = \Phi\Phi^T = (1 + \mathbf{x}(t)\mathbf{x}(t'))^p - 1; \tag{8}$$

A regularised kernel Fisher discriminant was then 20-fold cross-validated (Figure 2C, D) in blocks of 10^5 patterns (1,000 trajectories of 100 patterns each on this test set). Optimal regularization penalties, specific of each expanded space, were previously established on an independent (validation) dataset leading to the minimum TI index; see details of this process in [48,60]. Normality is preserved in the discriminant subspace (Lilliefords non-parametric test, $p < 0.05$) as expected from the Central Limit Theorem [55,59,81], leading to a negligible cross validation error for the optimal expanded space (see Figure 2D).

Figure 2A shows an intuitive schema on the class-trajectory coherence index (TI). To be more precise, consider an autonomous dynamical system parameterized by p coefficients α in a dynamical regime corresponding to multiple attracting sets:

$$\dot{\mathbf{x}}(t) = A(\mathbf{x}(t), \alpha); \tag{9}$$

where $\mathbf{x} = (x, \dot{x})$ is a d-dimensional phase space and A is a nonlinear differential operator.

This system, equipped with i.i.d. initial random conditions, defines a natural classification problem. The system generates an observable dataset D of size $n \times l$ patterns (n discrete trajectories of length l). In this context, $f(\mathbf{x}(t))$ is an arbitrary classifier such that the "true" (asymptotic) risk [6,59,72] $e(\alpha)$ given that the pattern $\mathbf{x}$ belongs to class C_i

$$e(\alpha) = P(c(\mathbf{x}) \neq C_i, \alpha); \tag{10}$$

is minimum. The empirical estimator of the true error is the classification error CE shown in the figures. Taking into account the definition of class-trajectory (Equation 7), we term $f(T)$ as the predicted class for each point in the trajectory

$$f(T) = (f(x(t_0)), f(x(t_1)), ..., f(x(t_l))); \tag{11}$$

Thus, a divergent or incoherent class-trajectory is the one in which *all* vectors from a certain t_i are incorrectly classified. In other words, considering trajectory of class C i.e. in which all points of the trajectory belong to this class, a divergent class-trajectory verifies

$$f(x(t_i)) \neq C \; \forall t > t_i; \tag{12}$$

For simplicity, we will indicate the last condition as $f(T) \neq C$. The true trajectory error is then

$$e_T(\alpha) = P(f(T) \neq C, \alpha); \tag{13}$$

The lack of trajectory coherence index, TI, shown in figures is the empirical estimator of the true trajectory error e_T.

Analysis of the non-autonomous system. Endowed with the definition of TI, we can infer the conditions for a classifier to be no longer optimal when the system undergoes gradual non-stationary drift. In short, Text S1 analyses show how an arbitrarily small parameter perturbation $\Delta\alpha$ causes at least one trajectory to change its basin of attraction i.e. its class as was demonstrated empirically in Figures 1–3. In Figure 3 α increases by 10% after each time step. The dataset size is the same as in the previous sections (1000 randomly generated initial conditions i.i.d. normally drawn, zero mean and $\sigma = 4$).

As suggested in this section, $\Delta TI \geq \Delta CE$ cannot be established in general: for i.i.d. data from a generative distribution Q, the change induced in the distribution parameters $Q(\alpha + \Delta\alpha)$ does not necessarily entail a change in TI. For instance, given $1 < \tilde{l} < l$ misclassified i.i.d. patterns, the log-likelihood that they belong to the same trajectory is typically very small, and thus we cannot expect a different behaviour of TI and CE statistics (Figure 3C, TI bootstrap; see also Figures 4 and S2 bootstrap data).

The classical Priestley and Subba Rao (PSR) test of non-stationarity (Figure S1C) was used to analyse the simulated dataset shown in Figure 3, because it typically requires large sample sizes for a robust estimation (e.g. [32,34]). The simplest version of the test consists of analysing the logarithmic of the time-varying spectrum,

$$X(t,w) = log(f(x(t))); \quad Y(t,w) = log(f(y(t))); \tag{14}$$

where f is an estimator of the fourier spectrum and w is the frequency. The logarithm typically stabilizes the variance and thus enables us to assume a linear model for $Y(t,w)$, $X(t,w)$ with constant covariance. Differences between non-stationary means in segments of Y, X are then analysed using standard statistical testing [32,34] as shown in Results section and in Figure S1C.

Data acquisition

Behavioural task and electrophysiological recordings. Electrophysiology and preprocessing. The animal recorded was treated in accordance with the ethical guidelines set forth by the Canadian Council for Animal Care. Procedures have been approved by the Animal Care and Biosafety Committee of the University of British Columbia (UBC) and conform to the UBC policy 41 regarding research and teaching involving animals. For a detailed description of the surgical and probe making procedures see [48,49]. In brief, electrophysiological data was recorded via a 24 single-wire tungsten array implanted into the ACC of the behaving rodent; recordings were sampled at 30 kHz, band-pass filtered from 600–6000 Hz. Spike channels were then amplified, sorted and classified offline using the Spikesort 3D unsupervised clustering software (Neuralyx; Bozeman, MT, USA) as explained in [49].

Spike trains from the 24 simultaneously recorded units were convolved with Gaussian functions to obtain statistically reliable estimates of spike densities. The value of the optimal bandwidth for each neuron (variance of the gaussian kernel) was optimized using a multivariate kernel density estimation approach as described in [82] (see also [83]). Spike density estimates were then binned at 100 ms, so that 95% of bins contained 1-0 spikes.

Behaviour. Behavioural data were captured via a video camera (Cohu, Poway, CA), recorded in Noldus Ethovision (Noldus, Leesburg, VA) and also stored for off-line analysis. The rat was trained on the delayed spatial win shift run on an eight arm radial arm maze where all arms where initially baited. Each trial consisted of a training, test phase (separated by one minute delay not considered in this study). During the training phase, four of eight arms where opened to enable acquisition of a sugar reward (Noyes, Lancaster, NH). After the delay, all eight arms were opened during the test phase and errors were scored as re-entries into previously visited arms (Figure 4A). This task was performed ten times (trials). The animal scored no error during this task in any of the trials.

In this study we focused on four periods with different cognitive demands, namely reward epochs (dark gray and red dots) during the training or test phases, respectively and correct choice epochs during training and test phases (blue and green, respectively). Reward epochs were defined as the ± 1 s periods around the point in which the animals nose reached the sugar pellet; similarly choice epochs were defined as 1 s periods around each arm entry (see [48]).

Standard statistical testing, atmospheric pollution supplemental dataset and software. Statistical test details can be found in the corresponding sections. Nonparametric tests were used based on conservatively designed bootstrap data (200 replications used for two-sided comparisons at $p = 0.01$, [81]) as explained in the corresponding text sections and figure captions.

Analyses presented in this work are also benchmarked with an additional illustrative dataset where the presence of attracting states is well-known. Data used in this research belongs to the Department of Agriculture, Generalitat Valenciana (Regional Government), Valencia, Spain; and it was recorded in a rural area of particular agricultural interest. Data consists of hourly concentrations of ozone, NO, NO_2 and hourly recordings of meteorological variables for over a two month period. Ozone concentration is known to exhibit regular daily oscillations yet subtle seasonal variations [62,63] and thus this data is an ideal testbed for the TI index. Details of this dataset and analyses performed can be found in Text S2 and Figure S2.

Software for analysing trajectory dynamics is freely available under the terms of the GNU licence as Software S1. Updates of this software are available at http://www.bccn-heidelberg-mannheim.de and http://www.researchgate.net/profile/EmiliBalaguer-Ballester/ websites.

Supporting Information

Figure S1 Non-autonomous drift in the duffing dynamical system (cont. from Figure 3). (A) Schema illustrating convergent trajectories with respect to attracting state boundaries in the reference set (top left), in the prediction (validation) set after a deterministic drift preserving the initial conditions (top right) and when those initial conditions are randomly drawn (bottom); the later setting is related to the analyses shown in Figure 3. As illustrated in the figure, the behavior of CE and TI indexes is remarkably different. (B) The left axis shows the Jensen-Shannon divergence between predicted posteriors provided by the discriminant analysis (same dataset as in Figure 3). As in Figure 3 analyses, regularized kernel-fisher discriminant in a third order expanded space was optimized for the first trial and applied to the subsequent trials. As the distance between fixed point varies, like in CE, the Jensen-Shannon divergence increases approximately monotonically in a logarithmic shape, thus it is not sensitive to any change in dynamics (two-tailed t-tests, $t(598) < 0.49, p > 0.63$, normality accepted at $p = 0.05$ according to Lilliefors test). The right axes show the Wilks Λ value, which behaves in similar way to CE and Jensen-Shannon divergences. All trial-to-trial comparisons are again non-significant $(t(598) > -1.1, p > 0.28$, normality accepted at $p = 0.05$). Moreover, the first significant result is achieved in the pairwise comparison form trial 1 to trial 6 $(t(598) > -2.7, p = 0.007)$, fully in line with CE results shown in Figure 3. (C) Priestley-Subba-Rao test (PSR) of non-stationarity [32–34](see main text and Methods). Again fully in line with TI results (Figure 3) only the pairwise comparison from trial 5 to trial 6 reaches significance (MannWhitney $U(4998) = 6.4 \times 10^6$, $p = 0.0478$; normality rejected according to Lilliefors test, $p < 0.01$).

Figure S2 Example of the analysis of a non-stationary dataset. (A) Hourly ozone (O_3) ground concentration, nitric oxides (NO_2, NO) temperature and relative humidity during a summer week. Ozone is an atmospheric pollutant synthesised primarily from NO_2 (red line in the plot) by the catalysis of solar radiation. Ozone levels are divided into three ranges (low, moderate and high). (B) An optimally regularized discriminant defined in an expanded phase space of third order is used to map precursors and atmospheric variables to O_3 classes. As in Figure 3, the discriminant subspace is computed for the first trial (i.e. the first week of data) and then used to compute CE and TI on the next trials. In week 6, an abrupt increase of TI is not accompanied by a trial-to trial change in CE, suggesting a deterministic origin of the observed non-stationary in hourly ozone concentrations. Lowest plot shows the certainty in the classification (see Text S2).

Software S1 Demo trajectories reconstruction toolbox; pls revise this cite in the text and EM.

Text S1 Local trajectory analyses in a Duffing system.

Text S2 Illustrative dataset in a non-stationary environment.

Acknowledgments

Authors thank Daniel Durstewitz for his invaluable input. We also thank Chris Lapish and Jeremy Seamans for their generous support.

Author Contributions

Conceived and designed the experiments: EBB ATD MB. Performed the experiments: EBB ATD. Analyzed the data: EBB. Contributed reagents/materials/analysis tools: EBB ATD MB. Wrote the paper: EBB MB ATD.

References

1. Mann M (2004) On smoothing potentially non-stationary climate time series. Geophys Res Lett 31: L07214.
2. Haraa S, Kawaharaa Y, Washioa T, von Bnau P, Tokunagac T, et al. (2012) Separation of stationary and non-stationary sources with a generalized eigenvalue problem. Neural Networks 33: 7–20.
3. Csorgo H, Horvarth L (2009) Nonparametric methods for change point problems, Kluwer Academic Pub., volume 7. pp. 403–425.
4. von Bunau P, Meinecke F, Kiraly F, Robert-Muller K (2009) Finding stationary subspaces in multivariate time series. Phys Rev Lett 103: 214101.
5. Masquelier T (2013) Neural variability, or lack thereof. Frontiers in Comput Neurosci 7: 1–7.
6. Quiroga Lombard C, Hass J, Durstewitz D (2013) A method for stationarity-segmentation of spike train data with application to the pearson cross-correlation. J Neurophysiol 110: 562–572.
7. Churchland M, Abbot L (2012) Two layers of neural variability. Nat Neurosci 15: 1472–1474.
8. Zohary E, Shadlen M, Newsome W (1994) Correlated neuronal discharge rate and its implications for psychophysical performance. Nature 370: 140–143.
9. Whitsel B, Schreiner R, Essick G (1977) Analysis of variability in somatosensory cortical neuron discharge. J Neurophysiol 40: 589–607.
10. Werner G, Mountcastle V (1963) Variability of central neural activity in a sensory system, and its implications for central reection of sensory events. J Neurophysiol 29: 958–977.
11. Durstewitz D, Vittoz N, Floresco S, Seamans J (2010) Abrupt transitions between prefrontal neural ensemble states accompany behavioral transitions during rule learning. Neuron 66: 438–448.
12. Churchland M, Afshar A, Shenoy K (2006) A central source of movement variability. Neuron 52: 1085–1096.
13. Beck J, Ma WJ, Pitkow X, Latham PE, Pouget A (2012) Not noisy, just wrong: The role of suboptimal inference in behavioral variability. Neuron 74: 33–39.
14. Toups J, Fellous JM, Thomas P, Sejnowski T, Tiesinga P (2012) Multiple spike time patterns occur at bifurcation points of membrane potential dynamics. PLoS Comput Biol 8: e1002615.
15. Deco V Gand Jirsa (2012) Ongoing cortical activity at rest: criticality, multistability and ghost attractors. J Neurosci 32: 3366–3375.
16. Braun J, Mattia M (2010) Attractors and noise: twin drivers of decisions and multistability. Neuroimage 52: 740–751.
17. Bernal-Casas D, Balaguer-Ballester E, Gerchen M, Iglesias S, Walter H, et al. (2013) Multi-site reproducibility of prefrontal-hippocampal connectivity estimates by stochastic dynamic causal models. Neuroimage 82: 555–563.
18. Churchland M, Yu B, Cunningham J, Sugrue L, Cohen M, et al. (2010) Stimulus onset quenches neural variability: a widespread cortical phenomenon. Nat Neurosci 13: 369–378.
19. Hyman J, Ma L, Balaguer-Ballester E, Durstewitz D, Seamans J (2012) Contextual encoding by ensembles of medial prefrontal cortex neurons. PNAS 109: 5086–5091.
20. Scaglione A, Moxon K, Aguilar J, Foffani G (2011) Trial-to-trial variability in the responses of neurons carries information about stimulus location in the rat whisker thalamus. PNAS 108: 14956–61.
21. Deneve S, Latham P, Pouget A (2001) Efficient computation and cue integration with noisy population codes. Nat Neurosci 4: 826–831.
22. Stiefel K, Englitz B, Sejnowski T (2013) Origin of intrinsic irregular firing in cortical interneurons. PNAS 110: 7886–91.
23. Faisal A, Selen L, Wolpert D (2008) Noise in the nervous system. Nat Rev Neurosci 66: 292–303.
24. Moreno-Bote R, Knill D, Pouget A (2011) Bayesian sampling in visual perception. PNAS 108: 12491–12496.
25. Sussillo D, Abbott L (2009) Generating coherent patterns of activity from chaotic neural networks. Neuron 63: 544–557.
26. Litwin-Kumar A, Doiron B (2012) Slow dynamics and high variability in balanced cortical networks with clustered connections. Nat Neurosci 15: 1498–1505.
27. Renart A, de la Rocha J, Bartho P, Hollender L, Parga N, et al. (2010) The asynchronous state in cortical circuits. Science 327: 587–590.
28. Blythe DAJ, von Bunau P, Meinecke FC, Robert-Muller K (2012) Feature extraction for changepoint detection using stationary subspace analysis. IEEE Trans Neural Networks 23: 631–643.
29. Kuncheva L (2013) Change detection in streaming multivariate data using likelihood detectors. IEEE Transactions on Knowledge and Data Engineering 25: 1175–1180.
30. Zliobaite I, Bifet A, Pfahringer B, Holmes G (2013) Active learning with drifting streaming data. IEEE Transactions on Neural Networks and Learning Systems In press.
31. Volpi M, Tuia D, Camps-Valls G, Kanevski M (2012) Unsupervised change detection with kernels. IEEE Geosciences and Remote Sensing Letters 9: 1026–1030.
32. Nason G (2013) A test for second-order stationarity and approximate confidence intervals for localized autocovariances for locally stationary time series. J R Statist Soc B 75: 879904.
33. Chen J, Hu N (2014) A frequency domain test for detecting nonstationary time series. Computational Statistics and Data Analysis In press.
34. Priestley M, Subba T (1969) A test for non-stationarity of time-series. J R Statist Soc B 31: 140149.
35. Bouchachia A (2011) Incremental learning with multi-level adaptation. Neurocomputing 74: 1785–1799.
36. Sayed-Mouchaweh M, Lughofer E, editors(2012) Learning in Non-Stationary Environments. Springer.
37. Honeine P (2012) Online kernel principal component analysis: A reduced-order model. Journal of Machine Learning Research 34: 1814–1826.
38. Brazdil P, Gama J, Soares C (2009) Meta-Learning: Applications to Data Mining. Springer.
39. Robinson J, Hartemink A (2010) Learning non-stationary dynamic bayesian networks. Journal of Machine Learning Research 11: 3647–3680.
40. Du J, Cui M (2010) Solving the forced duffing equation with integral boundary conditions in the reproducing kernel space. International Journal of Computer Mathematics 87: 2088–2100.
41. Sabarathinama S, Thamilmaran K, Borkowski L, Perlikowski P, Brzeski P, et al (2013) 10: 3098–3017.
42. Wiggins S (2013) Introduction to applied nonlinear dynamical systems and chaos. Springer.
43. Holmes P, Whitley D (1983) On the attracting set for duffing's equation. Physica D 7: 111–123.
44. Jiang ZP (2002) Advanced feedback control of the chaotic duffinng equation. IEEE Trans Circuits Syst 49: 244–249.
45. Feng Z, Chen G, Hsu S (2006) A qualitative study of the damped duffing equation and applications. American Institute of Mathematical Sciences 6: 1097–1112.
46. Ha J, Nakagiri S (2004) Identification problems for the damped klein-gordon equation. Math Anal Appl 289: 77–89.
47. Freyer F, Roberts J, Ritter P, Breakspear M (2012) A canonical model of multistability and scaleinvariance in biological systems. PLoS Comput Biol 8: e1002634.
48. Balaguer-Ballester E, Lapish C, Seamans J, Durstewitz D (2011) Attracting dynamics of frontal cortex ensembles during memory guided decision making. PLoS Computational Biology 7: e1002057
49. Lapish C, Durstewitz D, Chandler L, Seamans J (2008) Successful choice behavior is associated with distinct and coherent network states in anterior cingulate cortex. Proc Natl Acad Sci USA 105: 12010–12015 (*first two authors contributed equally).
50. Balaguer-Ballester E, Clark N, Coath M, Krumbholz K, Denham S (2009) Understanding pitch perception as a hierarchical process with top-down modulation. PLoS Comput Biol 5: e1000301.
51. Stein R, Gossen E, Jones KE (2005) Neuronal variability: noise or part of the signal? Nat Rev Neurosci 6: 389–397.
52. Gama J, Sebastio R, Rodrigues P (2013) On evaluating stream learning algorithms. Machine Learning 90: 317–346.
53. Sauer T, Yorke J, Casdagli M (1992) Embedology. J Stat Phys 65: 579–616.
54. Provenzale A, Smith L, R V, Murante G (1992) Distinguishing between low-dimensional dynamics and randomness in measured time series. Physica D 5: 28–31.
55. Bishop C (2007) Pattern recognition and machine learning. Springer.
56. Budka M, Gabrys B (2011) Electrostatic field framework for supervised and semi-supervised learning from incomplete data. Natural Computing 10: 921–945.
57. Kantz H, Schreiber T (2004) Nonlinear time series analysis. Cambridge University Press.
58. Balaguer-Ballester E, Soria E, Palomares A, Martn-Guerrero J (2008) Predicting service request in support centres based on nonlinear dynamics, arma modelling and neural networks. Expert Sys with App 34: 665–672.
59. Scholkopf B, Smola A (2002) Learning with kernels. MIT Press.
60. Saadi K, Talbot N, Cawley G (2007) Optimally regularised kernel fisher discriminant classification. Neural Networks 20: 832–841.

61. Schapire R, Freund Y, Bartlett P, Sun Lee W (1998) Boosting the margin: A new explanation for the effectiveness of voting methods. The annals of statistics 26: 1651–1686.

62. Gomez-Sanchis J, Martin-Guerrero J, Soria-Olivas E, Vila-Frances J, Carrasco J, et al. (2006) Neural networks for analysing the relevance of input variables in the prediction of tropospheric ozone concentration. Atmospheric Environment 40: 6173–6180.

63. Balaguer-Ballester E, Camps-Valls G, Carrasco-Rodriguez JL, E S, del Valle-Tascon S (2002) Effective one-day ahead prediction of hourly surface ozone concentrations in eastern spain using linear models and neural networks. Ecological Modelling 156: 27–41.

64. Deco G, Jirsa V, McIntosh A (2011) Emerging concepts for the dynamical organization of restingstate activity in the brain. Nat Rev Neurosci 12: 43–56.

65. Staude B, Grun S, Rotter S (2010) Higher-order correlations in non-stationary parallel spike trains: statistical modeling and inference. Front Comput Neurosci 16: doi:10.3389/fncom.2010.00016

66. Balaguer-Ballester E, Coath M, Denham S (2007) A model of perceptual segregation based on clustering the time series of the simulated auditory nerve firing probability. Biol Cybern 97: 479–491.

67. Durstewitz D, Balaguer-Ballester E (2010) Statistical approaches for reconstructing neuro-cognitive dynamics from high-dimensional neural recordings. Neuroforum 1: 89–98.

68. Niessing J, Friedrich R (2010) Olfactory pattern classification by discrete neuronal network states. Nature 465: 47–54.

69. Wills T, Lever C, Cacucci F, Burgess N, O'Keefe J (2005) Attractor dynamics in the hippocampal representation of the local environment. Science 308: 873–876.

70. Mazor O, Laurent G (2005) Transient dynamics versus fixed points in odor representations by locust antennal lobe projection neurons. Neuron 48: 661–673.

71. Bathellier B, Buhl D, Accolla R, Carleton A (2008) Dynamic ensemble odor coding in the mamalian olfactory bulb: Sensory information at different time scales. Neuron 57: 586–598.

72. Vapnik V (1998) Statistical learning theory. Wiley-Interscience.

73. Milad L, Ping Z, Srikanta S, Taro T (2013) Inferring trial-to-trial excitatory and inhibitory synaptic inputs from membrane potential using gaussian mixture kalman filtering. Frontiers in Comput Neurosci 7: 00109.

74. Paninski L, Vidne M, DePasquale B, Ferreira DG (2012) Inferring synaptic inputs given a noisy voltage trace via sequential monte carlo methods. J Comput Neurosci 33: 1–19.

75. Kobayashi R, Shinomoto S, Lansky P (2012) Estimation of time-dependent input from neuronal membrane potential. Neural Comput 23: 3070–3093.

76. Ledberg A, Montagnini A, Coppola R, Bressler S (2012) Reduced variability of ongoing and evoked cortical activity leads to improved behavioral performance. PLoS ONE 7: e43166.

77. Rabinovich M, Huerta R, Laurent G (2008) Transient dynamics for neural processing. Science 321: 48–50.

78. Rabinovich M, Varona P (2011) Robust transient dynamics and brain functions. Front Comput Neurosci 6: doi:10.3389/fncom.2011.00024

79. Ben-Hur A, Ong C, Sonnenburg S, Scholkopf B, Ratsch G (2008) Support vector machines and kernels for computational biology. PLoS Comput Biol 4: e1000173.

80. Smola A, vri Z, Williamson R (2001) Regularization with dot-product kernels, MA: MIT Press, volume 7. pp. 308–314.

81. Hastie T, Tibshirani R, Friedman J (2009) The elements of statistical learning. Springer.

82. Duong T, Hazelton M (2005) Cross-validation bandwidth matrices for multivariate kernel density estimation. Scand J Statist 32: 485–506.

83. Omi T, Shinomoto S (2011) Optimizing time histograms for non-poissonian spike trains. Neural Computation 23: 3125–3144.

16

Role of the Adiponectin Binding Protein, T-Cadherin (cdh13), in Pulmonary Responses to Subacute Ozone

David I. Kasahara[1], Alison S. Williams[1], Leandro A. Benedito[1], Barbara Ranscht[2], Lester Kobzik[1], Christopher Hug[3], Stephanie A. Shore[1]*

1 Department of Environmental Health, Harvard School of Public Health (HSPH), Boston, Massachusetts, United States of America, **2** Department of Neurosciences, University of California San Diego, San Diego, California, United States of America, **3** Division of Pulmonary Medicine, Children's Hospital Boston, Harvard Medical School (HMS), Boston, Massachusetts, United States of America

Abstract

Adiponectin, an adipose derived hormone with pleiotropic functions, binds to several proteins, including T-cadherin. We have previously reported that adiponectin deficient ($Adipo^{-/-}$) mice have increased IL-17A-dependent neutrophil accumulation in their lungs after subacute exposure to ozone (0.3 ppm for 72 hrs). The purpose of this study was to determine whether this anti-inflammatory effect of adiponectin required adiponectin binding to T-cadherin. Wildtype, $Adipo^{-/-}$, T-cadherin deficient ($T\text{-}cad^{-/-}$), and bideficient ($Adipo^{-/-}/T\text{-}cad^{-/-}$) mice were exposed to subacute ozone or air. Compared to wildtype mice, ozone-induced increases in pulmonary IL-17A mRNA expression were augmented in $T\text{-}cad^{-/-}$ and $Adipo^{-/-}$ mice. Compared to $T\text{-}cad^{-/-}$ mice, there was no further increase in IL-17A in $Adipo^{-/-}/T\text{-}cad^{-/-}$ mice, indicating that adiponectin binding to T-cadherin is required for suppression of ozone-induced IL-17A expression. Similar results were obtained for pulmonary mRNA expression of saa3, an acute phase protein capable of inducing IL-17A expression. Comparison of lung histological sections across genotypes also indicated that adiponectin attenuation of ozone-induced inflammatory lesions at bronchiolar branch points required T-cadherin. BAL neutrophils and G-CSF were augmented in $T\text{-}cad^{-/-}$ mice and further augmented in $Adipo^{-/-}/T\text{-}cad^{-/-}$ mice. Taken together with previous observations indicating that augmentation of these moieties in ozone exposed $Adipo^{-/-}$ mice is partially IL-17A dependent, the results indicate that effects of T-cadherin deficiency on BAL neutrophils and G-CSF are likely secondary to changes in IL-17A, but that adiponectin also acts via T-cadherin independent pathways. Our results indicate that T-cadherin is required for the ability of adiponectin to suppress some but not all aspects of ozone-induced pulmonary inflammation.

Editor: Christian Taube, Leiden University Medical Center, The Netherlands

Funding: This study was supported by the U.S. National Institute of Health [HL-084044, ES-013307, and ES-00002]. The funders had no role in study design, data collection and analysis, decision to publish, or preparation of the manuscript.

Competing Interests: The authors have declared that no competing interests exist.

* E-mail: sshore@hsph.harvard.edu

Introduction

Ozone (O_3) is an environmental pollutant generated by chemical reactions of automobile emissions (NO and hydrocarbons) with sunlight. O_3 acts as oxidizing agent on cell membranes and on proteins and lipids in the lung and airway lining fluid, leading to epithelial injury and an inflammatory response that includes induction of acute phase cytokines and chemokines, and neutrophil influx [1,2,3].

Adiponectin, an adipose-derived hormone that decreases in obesity [4], has important anti-inflammatory effects. For example, adiponectin treatment decreases endotoxin-induced pro-inflammatory cytokine expression and augments anti-inflammatory IL-10 expression in monocytes and macrophages [5,6]. Exogenous administration of adiponectin also decreases allergic airways inflammation in mice [7]. In addition, we have previously reported that compared to wildtype (WT) mice, adiponectin deficient ($Adipo^{-/-}$) mice exposed to subacute O_3 (0.3 ppm for 24 to 72 h) have increased neutrophilic inflammation, and increased pulmonary expression of certain cytokines and chemokines, including IL-17A and G-CSF [8].

Several adiponectin binding proteins have been cloned including AdipoR1, AdipoR2, and T-cadherin (T-cad) [9,10], all of which are expressed in the lungs [11,12]. T-cad (cdh13 or H-cadherin) is a 95 kd glycoprotein which differs from other cadherin proteins by lacking both transmembrane and cytoplasmatic domains. Instead, T-cadherin is anchored, mainly on the apical surface of cells [13], via a glycosylphosphatidylinositol (GPI) linkage [14]. Importantly, T-cadherin primarily binds the hexameric and high molecular weight isoforms of adiponectin [10]. These are also the isoforms that dominate in the lung lining fluid [15]. In the heart, T-cadherin appears to mediate the beneficial effects of adiponectin. Following pressure overload, mice deficient in Tcadherin ($T\text{-}cad^{-/-}$ mice), exhibit increased cardiac hypertrophy compared to WT mice [16], similar to $Adipo^{-/-}$ mice. Similarly, the size of infarctions in hearts of mice subjected to ischemia-reperfusion is greater in $T\text{-}cad^{-/-}$ than WT mice [16]. Furthermore, the ameliorative effects of adiponectin in these models are not observed in $T\text{-}cad^{-/-}$ mice [16].

The purpose of this study was to examine the hypothesis that T-cadherin is required for the anti-inflammatory effects of adiponectin that limit the pulmonary inflammation induced by subacute

O_3. To address this hypothesis, we assessed pulmonary inflammation in T-$cad^{-/-}$ mice and their WT controls exposed to either air or O_3 (0.3 ppm) for 72 hours. For comparison we also examined $Adipo^{-/-}$ mice and their WT controls. T-cadherin functions not only as an adiponectin binding protein, but also as a cell-cell adhesion molecule that can impact cell polarization, migration, adhesion, and survival [14]. Hence, effects of T-cadherin deficiency on responses to O_3 may be the result of the cell-adhesion rather than the adiponectin-binding properties of T-cadherin. To address this issue, we also examined mice deficient in both T-cadherin and adiponectin ($Adipo^{-/-}/T$-$cad^{-/-}$ mice). We reasoned if effects of T-cadherin deficiency were a reflection of adipnectin binding to T-cadherin, then we would not see any difference between T-cadherin deficient mice and mice deficient in both adiponectin and T-cadherin.

Methods

Animals

This study was approved by the Harvard Medical Area Standing Committee on Animals under protocol number 03078 and carried out in accordance with the recommendations in the Guide for the Care and Use of Laboratory Animals from the National Institute of Health. All efforts were made to minimize suffering. $Adipo^{-/-}$ and T-$cad^{-/-}$ mice were obtained from Dr. Matsuzaka (Osaka, Japan). T-$cad^{-/-}$ mice [17] and $Adipo^{-/-}$ were bred together to obtain $Adipo^{-/-}/T$-$cad^{-/-}$ mice as previously described [18]. Others have reported small but potentially significant genetic differences in C57BL/6 mice from different vendors [19]. Differences in the microbiome between C57BL/6 mice from Jackson Laboratories and Taconic Farms have also been reported [20]. Hence, as controls for the $Adipo^{-/-}$ mice, we used C57BL/6 from Jackson Laboratories (WT-jax) because this was the genetic background for the $Adipo^{-/-}$ mice. As controls for the T-$cad^{-/-}$ and $Adipo^{-/-}/T$-$cad^{-/-}$ mice, we used C57BL/6 mice from Taconic Farms (WT-tac) because this was the background for the T-$cad^{-/-}$ mice. Note that we generated the $Adipo^{-/-}/T$-$cad^{-/-}$ mice by backcrossing offspring from $Adipo^{-/-} \times T$-$cad^{-/-}$ matings onto T-$cad^{-/-}$ mice [18].

Protocol

Age and gender matched $Adipo^{-/-}$, T-$cad^{-/-}$, $Adipo^{-/-} T$-$cad^{-/-}$, and control mice were exposed to either air or O_3 (0.3 ppm, 72 hours) as previously described [8]. Our experience is that neutrophil recruitment plateaus after 48 h of exposure, but significant changes in BAL macrophages do not occur in wildtype mice until 72 h of exposure [8]. Our previous data also indicate that statistically significant changes in BAL macrophages do not occur in wildtype mice until 72 h of exposure [8]. Immediately after the exposure, mice were euthanized by i.p. overdose of sodium pentobarbital. Two cohorts of mice were used. In the first, blood was drawn, the trachea was cannulated to perform bronchoalveolar lavage (BAL), and lungs were stored at $-80°$C for extraction of total RNA. In the second cohort, lungs were fixed for histological assessment of O_3-induced pulmonary lesions [21].

Ozone exposure

Mice were exposed to either O_3 (0.3 ppm) or ambient air for 72 hours as previously described [8]. Briefly, cages without the microisolator cover were placed in a steel and plexiglass chamber, and supplied with a mixture of O_3 and air. O_3 was produced by passing medical grade oxygen through a high voltage ozonizer and bled into the chamber. O_3 concentration in the chamber was controlled by regulating the amount of ambient air flowing into

the chamber. Mice were supplied with normal chow and water *ad libitum*.

Bronchoalveolar lavage and serum

BAL was performed using two instillations of 1 mL of cold PBS. BAL samples were centrifuged, supernatants were assessed for inflammatory cytokines, and BAL cells were resuspended and counted using a hemocytometer. Cytospin was performed for differential cell analysis. Blood was drawn by cardiac puncture to obtain serum.

Cytokines and Chemokines

A panel of 32 cytokines, chemokines and growth factors (eotaxin, G-CSF, GM-CSF, IFNγ, IL-1α, IL-1β, IL-2, IL-3, IL-4, IL-5, IL-6, IL-7, IL-9, IL-10, IL-12p40, IL-12p70, IL-13, IL-15, IL-17, IP-10, KC, LIF, LIX, MCP-1, M-CSF, MIG, MIP-1α, MIP-1β, MIP-2, RANTES, TNF-α, and VEGF) were quantified in BAL by a multiplex assay (Eve Technologies, Alberta, Canada) as previously described [22] and a commercial ELISA was used for quantification of sTNFR1 and adiponectin (R&D Systems, MN).

RNA generation and qRT-PCR

Total lung RNA was extracted using the protocol provided in the RNAeasy kit (Qiagen, MD), and quantified by Nanodrop (ThermoScientific, NJ). cDNA was synthesized from RNA by using a commercial kit [8]. Real time PCR (RT-PCR) was used to assess the mRNA expression of IL-17A, serum amyloid A3 (*saa3*), and *Ki67* (a marker of cell proliferation), by the SYBR green method. Primer sequences are provided in Table 1. Gene expression was normalized to 18S (internal control) followed by analysis by the $\Delta\Delta$CT method.

Histology

Mice were euthanized and lungs inflated to 20 cmH$_2$O with 4% PBS buffered paraformaldehyde (pH = 7.4) overnight. Lungs were washed with PBS and transferred to 70% ethanol, and dehydrated. The left lung was sectioned, embedded in paraffin, and stained with hematoxylin and eosin. The severity of lesions located at terminal bronchioles was assessed by scoring the number of cellular layers below the epithelium as follows: 0 for no lesions, 1: 1–2 cellular layers, 2: for 3 cellular layers, 3: for 4 cellular layers, and 4: for 5 cells or more layers. Total score for each mouse was computed by averaging the scores of all terminal bronchioles present on all sections of left lung.

Statistical Analysis

The significance of differences between groups was assessed using Factorial ANOVA combined with LSD Fisher as post-hoc analysis (Statistica Software, Statsoft, OK). Data that were not normally distributed were log transformed before statistical

Table 1. Primers sequence used for qRT-PCR.

Gene	Sense	Anti-sense
18S	5'-GTAACCCGTTGAACCCCATT-3'	5'-CCATCCAATCGGTAGTAGCG-3'
IL-17A	5'-CCAGGGAGAGCTTCATCTGT-3'	5'-AGGAAGTCCTTGGCCTCAGT-3'
ki67	5'-AGGGTAACTCGTGGAACCAA-3'	5'-GGAGGTGAAAACCACACTGG-3'
saa3	5'-CCTGGGCTGCTAAAGTCATC-3'	5'-CACTCATTGGCAAACTGGTC-3'

analysis. Means and standard errors from log values were retrocalculated with appropriate error propagations. $P < 0.05$ (two tail) was considered significant. Because the controls for the $Adipo^{-/-}$ and the $T\text{-}cad^{-/-}$ and $Adipo^{-/-}/T\text{-}cad^{-/-}$ mice were different, the data from these mice were analyzed separately. All values are expressed as mean±standard mean of error.

Results

BAL and serum adiponectin

Serum adiponectin was substantially higher in $T\text{-}cad^{-/-}$ than WT mice (Fig. 1A), consistent with previous observations [15,17,18]: adiponectin bound to T-cadherin on the endothelium serves as a repository for adiponectin that is delivered back to the circulation in $T\text{-}cad^{-/-}$ mice [16,17,23]. There was no effect of O_3 exposure on serum adiponectin (Fig. 1A). In contrast, O_3 exposure caused a marked increase in BAL adiponectin in $T\text{-}cad^{-/-}$ mice (Fig. 1B), likely as a result of increased transit from the blood into the lung consequent to O_3-induced increases in the permeability of the alveolar/capillary barrier, as previously discussed [8]. BAL adiponectin was slightly lower in air exposed $T\text{-}cad^{-/-}$ versus WT mice and slightly higher in O_3 exposed $T\text{-}cad^{-/-}$ versus WT mice although these changes were not significant (Fig. 1B).

Effect of T-cadherin deficiency on O_3-induced pulmonary inflammation

Compared to air, factorial ANOVA demonstrated that O_3 increased BAL neutrophils, macrophages, and protein (Fig. 2), consistent with previous reports by ourselves and others using this type of O_3 exposure [8,24]. BAL neutrophils and protein were higher in O_3-exposed $Adipo^{-/-}$ versus WT mice, consistent with our previous observations [8]. Note that BAL neutrophils are also significant greater in $Adipo^{-/-}$ versus WT mice after 24 or 48 h of exposure [8]. We also observed significantly more BAL neutrophils in O_3-exposed $T\text{-}cad^{-/-}$ mice versus their corresponding WT controls. BAL neutrophils were also higher in O_3-exposed $Adipo^{-/-}/T\text{-}cad^{-/-}$ than $T\text{-}cad^{-/-}$ mice. No significant change in BAL protein was observed in $T\text{-}cad^{-/-}$ mice exposed to O_3 compared to WT mice under the same exposure ($p = 0.06$), despite a trend towards an increase in $T\text{-}cad^{-/-}$ mice. Interestingly, BAL

protein was significantly greater in O_3-exposed $Adipo^{-/-}/T\text{-}cad^{-/-}$ versus WT mice ($p = 0.005$). There was no significant effect of genotype on ozone induced changes in BAL macrophages (Fig. 2B). Note that it is not possible to directly compare $Adipo^{-/-}/T\text{-}cad^{-/-}$ and $Adipo^{-/-}$ mice due to differences in the background strain (see methods).

To further evaluate the impact of T-cadherin deficiency on O_3 induced inflammation, we performed a multiplex assay of cytokines and chemokines. Of the factors assayed by multiplex, factorial ANOVA indicated a significant effect of O_3 exposure in both cohorts of mice ($WT/Adipo^{-/-}$ and $WT/T\text{-}cad^{-/-}/Adipo^{-/-}/Tcad^{-/-}$) for G-CSF, IL-5, IL-6, LIF, KC, and eotaxin. BAL IL-6 and G-CSF were significantly higher in $Adipo^{-/-}$ versus WT mice exposed to O_3 (Fig. 3A, B), consistent with our previous observations [8]. We also observed significantly higher BAL LIF and IL-5 in O_3-exposed $Adipo^{-/-}$ versus WT mice (Fig. 3C, D). BAL G-CSF was also significantly higher in $T\text{-}cad^{-/-}$ versus WT mice exposed to O_3, and higher still in $Adipo^{-/-}/T\text{-}cad^{-/-}$ versus $T\text{-}cad^{-/-}$ mice (Fig. 3B). Surprisingly, O_3-induced changes in BAL IL-5 were significantly reduced by T-cadherin deficiency, and this change was reversed by combined adiponectin and T-cadherin deficiency. Neither BAL IL-6 nor LIF was significantly affected by T-cadherin deficiency, although there was significantly greater BAL LIF in O_3-exposed $Adipo^{-/-}/T\text{-}cad^{-/-}$ versus WT mice and a similar trend for IL-6. There was no genotype effect for either eotaxin or KC (data not shown). IL-17A was below the limit of detection of the Bioplex assay, but IL-17A mRNA expression was induced by subacute O_3 exposure. O_3-induced increases in IL-17A were significantly greater in $Adipo^{-/-}$ than WT mice (Fig. 3E), consistent with our previous observations [8]. O_3-induced increases in IL-17A were also significantly greater in $T\text{-}cad^{-/-}$ versus WT mice, and there was no further increase in $Adipo^{-/-}/T\text{-}cad^{-/-}$ versus $T\text{-}cad^{-/-}$ mice.

We also measured BAL concentrations of sTNFR1 (soluble TNFα receptor 1), the extracellular domain of TNFR1 (Fig. 3F). sTNFR1 is cleaved from cell surfaces by the enzyme TACE (TNFα converting enzyme). TACE activity is increased by conditions associated with oxidative stress [25]. Compared to air, O_3 exposure resulted in a marked increase in BAL sTNFR1 in all mouse genotypes used in this study (Fig. 3F). BAL sTNFR1 was

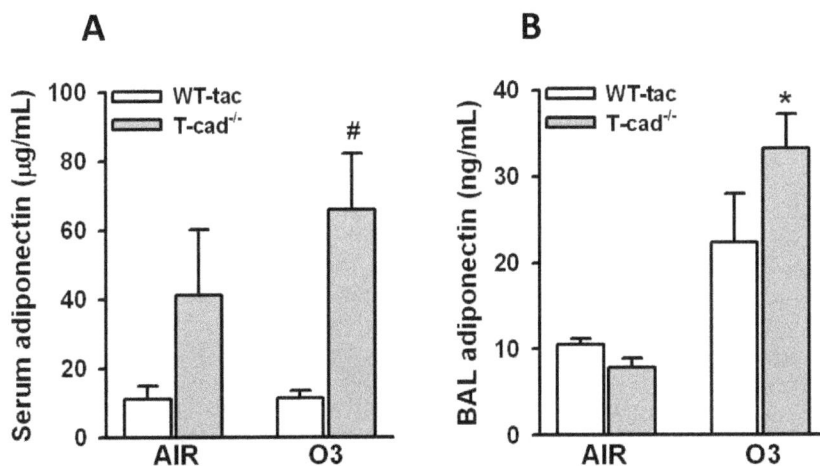

Figure 1. Serum and BAL adiponectin. Total adiponectin was measured by ELISA in serum (A) and bronchoalveolar lavage (BAL) fluid (B) of wildtype (Taconics) and T-cadherin deficient ($T\text{-}cad^{-/-}$) mice exposed to air and ozone (O_3, 0.3 ppm) for 72 h. * p<0.05 versus genotype matched air exposed mice; #p0.05 versus wildtype mice with the same exposure. Results of adiponectin in BAL are expressed as mean ± SEM of data from 5 mice exposed to air per group and 7 ozone exposed mice per group. The number of mice used for measurements of serum adiponectin was 4 mice per group.

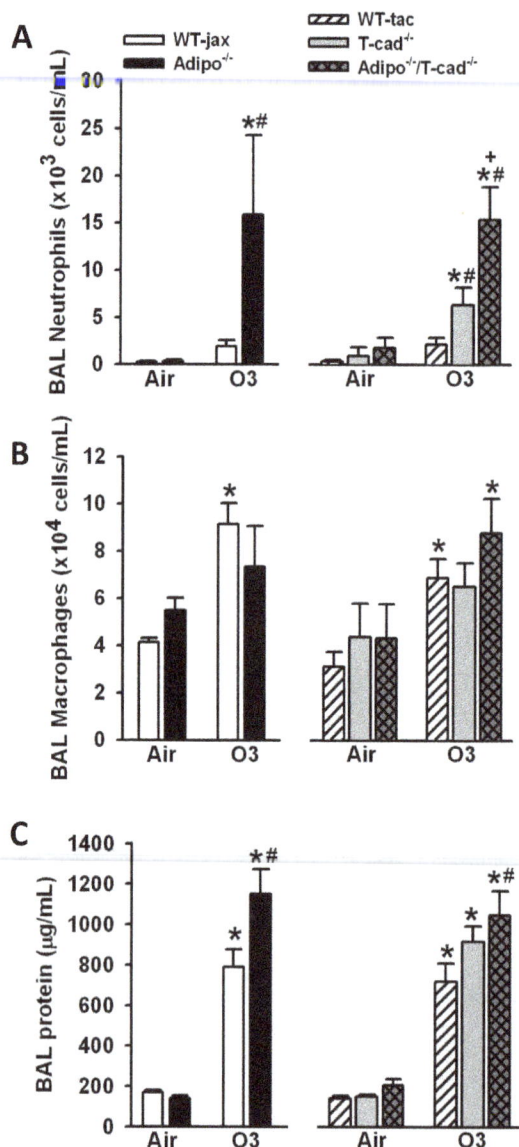

Figure 2. Lung inflammation and injury. BAL neutrophils (A), macrophages (B) and protein (C) in mice exposed to room air or O_3 (0.3 ppm) for 72 hours. *p<0.05 versus air exposed mice of the same genotype, #p<0.05 versus wildtype mice with the same exposure, + p<0.05 versus T-cadherin deficient mice with the same exposure. Results are mean ± SEM of data from 4–7 air exposed mice and 6–10 ozone exposed mice.

These lesions were characterized by focal interstitial expansion by mononuclear cells and reactive hyperplasia of epithelial cells (see Fig. 4B arrow, and a higher magnification of a different airway in Fig. 4C). The severity of the lesions varied, and was quantified as described in Methods. Compared to air, O_3 exposure resulted in a significant increase in terminal bronchiolar lesions (Fig. 4D). O_3-induced lesions were significantly greater in $Adipo^{-/-}$ versus WT mice. O_3-induced lesions were also significantly greater in T-$cad^{-/-}$ versus WT mice, but there was no further augmentation in $Adipo^{-/-}$ / T-$cad^{-/-}$ versus T-$cad^{-/-}$ mice (Fig. 4D).

qRT-PCR

To further evaluate the role of T-cadherin in adiponectin dependent effects on O_3-induced inflammation, we examined the mRNA expression of Ki67 and saa3. We chose to examine Ki67 because it is a well-established marker of cell proliferation [26]. Others have reported BrdU labeling within the terminal bronchiolar epithelium after O_3 exposure [27] consistent with epithelial repair following injury, and the lesions we observed in mice exposed to O_3 included epithelial cell hyperplasia (Fig. 4). We chose to examine saa3 because others have reported that it is increased by O_3 to a greater extent in lungs of other types of O_3-sensitive versus O_3-resistant mice [28,29] qRT-PCR data indicated a robust induction of Ki67 and saa3 by O_3 (Fig. 5). O_3-induced increases in Ki67 were not affected by either adiponectin or T-cadherin deficiency (Fig. 5A). O_3-induced expression of saa3 was higher in $Adipo^{-/-}$ versus WT mice (Fig. 5B). O_3-induced expression of saa3 was also significantly greater in T-$cad^{-/-}$ than WT mice, and there was no further increase in $Adipo^{-/-}$ / T-$cad^{-/-}$ versus T-$cad^{-/-}$ mice.

Discussion

Adiponectin deficiency augments the pulmonary inflammation induced by subacute O_3 exposure in mice [8], indicating an anti-inflammatory role for adiponectin. Results from the current study show that T-cadherin, an adiponectin binding protein, is required for aspects of this anti-inflammatory effect of adiponectin.

We have previously reported that BAL concentrations of adiponectin increase following subacute O_3 exposure and that O_3-induced neutrophilic influx into the lungs is augmented in mice deficient in adiponectin [8]. These results indicate an anti-inflammatory role for adiponectin during subacute ozone exposure. We have also demonstrated that the augmented neutrophilia observed in $Adipo^{-/-}$ mice is the result of increased IL-17A expression and consequent G-CSF production [8]. The goal of this study was to determine whether T-cadherin, an adiponectin binding protein, contributes to these anti-inflammatory effects of adiponectin. Although several other adiponectin binding proteins have been described [9,30], we chose to examine T-cadherin because we have shown, using CHO cells overexpressing T-cadherin, that T-cadherin primarily binds the hexameric and high molecular weight (HMW) isoforms of adiponectin [10], the isoforms most abundant in the lung lining fluid [15]. T-cadherin has also been shown to bind adiponectin in vivo: T-cadherin is prominently expressed on the apical surface of endothelial cells and immunohistochemistry indicates strong binding of adiponectin to these cells in wildtype mice. In contrast, this binding is lost in T-$cad^{-/-}$ mice [17]. In addition, T-cadherin is required for the protective effects of adiponectin against cardiac hypertrophy induced by pressure overload and against cardiac injury induced by ischemia-reperfusion [31]. Our results indicate that T-cadherin deficiency mimics aspects of the effects of adiponectin deficiency. After subacute O_3, T-$cad^{-/-}$ mice, like $Adipo^{-/-}$ mice, had

significantly greater in O_3-exposed $Adipo^{-/-}$ versus WT mice, consistent with previous observations [8]. There was no significant difference in BAL sTNFR1 in O_3-exposed T-$cad^{-/-}$ versus WT mice. However, BAL sTNFR1 was significantly higher in O_3-exposed $Adipo^{-/-}$ / T-$cad^{-/-}$ mice than in either WT or T-$cad^{-/-}$ mice. The results are consistent with the hypothesis that adiponectin deficiency results in increased oxidative stress leading to greater TACE activation, and that T-cadherin is not involved in this effect of adiponectin.

Histology

In O_3-exposed mice, histopathologic examination revealed inflammatory lesions localized at bronchiolar branch points.

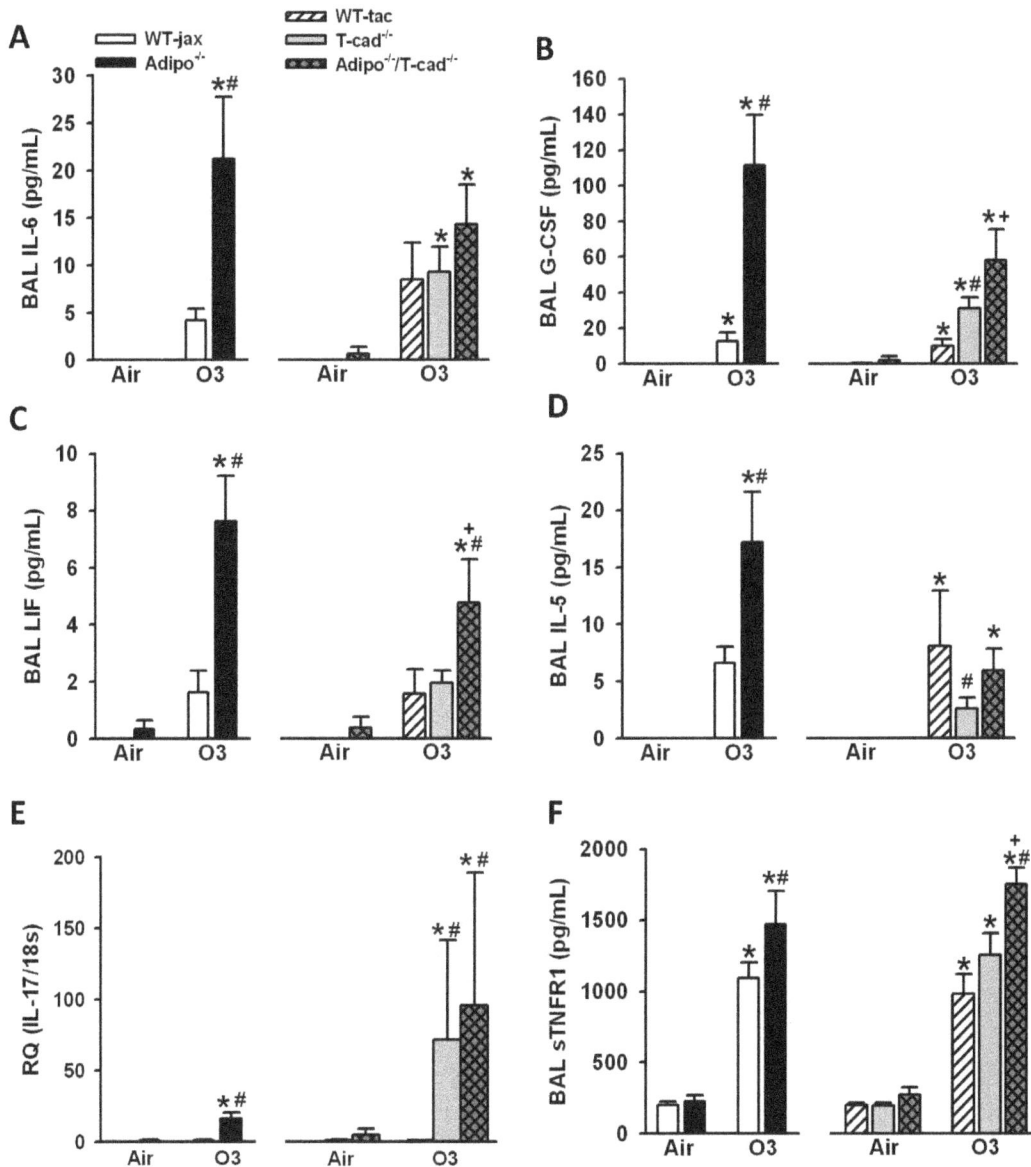

Figure 3. Cytokine and chemokine expression. BAL IL-6 (A), G-CSF (B), LIF (C), IL-5 (D), pulmonary IL-17A mRNA expression (E), and soluble TNFR1 (sTNFR1) (F) in mice exposed to room air or O_3 (0.3 ppm) for 72 hours. *$p<0.05$ versus air exposed mice of the same genotype, #$p<0.05$ versus wildtype mice with the same exposure, + $p<0.05$ versus T-cadherin deficient mice with the same exposure. Results are mean ± SEM of data from 3–5 air exposed mice and 3–7 ozone exposed mice for IL-6, G-CSF, LIF, and IL-5; 4–9 air exposed and 4–9 for ozone exposed mice for IL-17A mRNA; and 4–7 air exposed and 6–10 ozone exposed mice for sTNFR1;.

increased BAL neutrophils and G-CSF, increased pulmonary IL-17A mRNA expression, as well as increased terminal bronchiolar lesions compared to wildtype mice (Fig. 2A,3B,3E, and 4D), suggesting that binding to T-cadherin is required for effects of adiponectin on these outcomes. In contrast, O_3-induced increases in BAL protein, a marker of lung injury, O_3-induced increases in sTNFR1, a marker of oxidative stress, and O_3-induced increases in BAL IL-6 and LIF were augmented in *Adipo*$^{-/-}$ versus wildtype mice, but not in *T-cad*$^{-/-}$ versus wildtype mice (Figs. 2,3). IL-6 is required for recruitment of neutrophils after subacute ozone [2,32]. The role of LIF has not been established, but may be similar to IL-6 since it is a member of the same family of cytokines and shares signal transduction pathways with IL-6 [33]. The results indicate that adiponectin-dependent changes in sTNF1, IL-

6, and LIF involve adiponectin acting through other adiponectin binding proteins such as AdipoR1, AdipoR2, and calreticulin [9,30], or through non receptor mediated effects of adiponectin [34].

In addition to its role as an adiponectin binding protein, T-cadherin has other functions. It acts as a binding protein for lipoproteins [10,35,36,37] and also plays an important role in neuron growth [38] and in polarization and migration of endothelial cells [14]. T-cadherin also reduces surfactant protein D secretion in a pulmonary epithelial cell line [11]. It is possible that loss of such functions, rather than loss of the ability of T-cadherin to bind adiponectin, accounts for the observed effects of T-cadherin deficiency in O_3 exposed mice. To test this hypothesis, we also examined *Adipo*$^{-/-}$/*T-cad*$^{-/-}$ mice. Although we did not

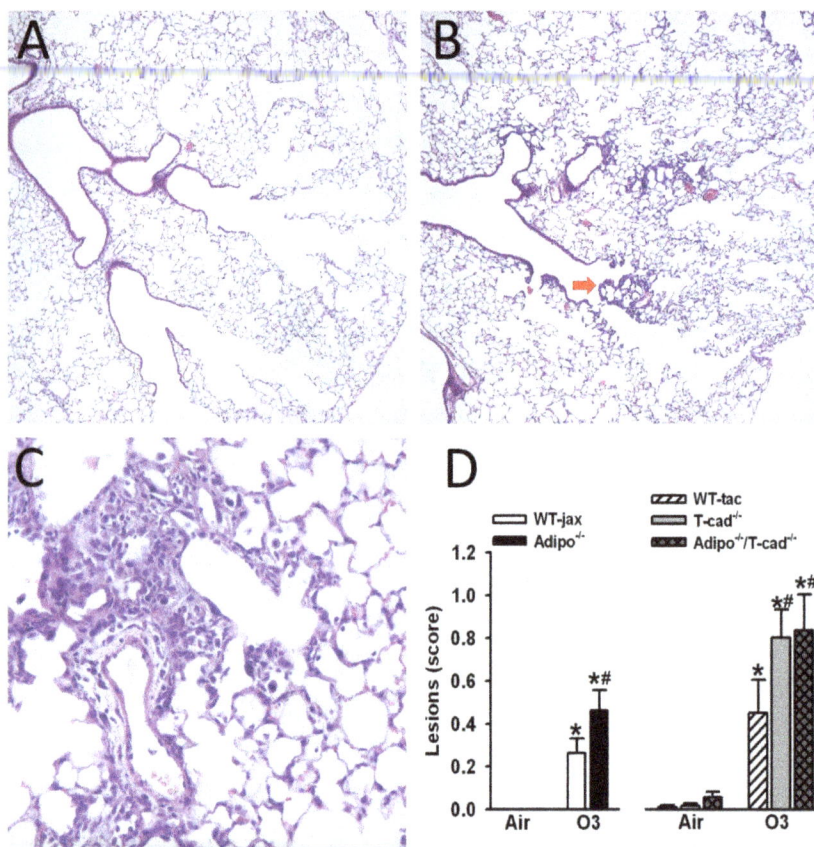

Figure 4. Terminal bronchiolar lesions. H&E stained histological sections of lungs of wildtype mice exposed to air (A) or ozone (B) showing bronchiolar/central acinar lesions in ozone exposed mice (magnification 100×). Arrow in red is pointing a terminal bronchiole with lesion. (C) A detail of these lesions in an ozone exposed mouse (magnification 400×). (D) An average lesion score was calculated for each mouse as described in Methods and these were averaged across mice from each genotype. * $p < 0.05$ versus air exposed mice of the same genotype, # $p < 0.05$ versus wildtype mice with the same exposure. Results are mean ± SEM of data from 6–12 air exposed mice and 8–16 ozone exposed mice.

assess adiponectin binding to T-cadherin in this study, our results suggest that adiponectin binding to T-cadherin is required for the ability of adiponectin to inhibit O_3-induced IL-17A and *saa3* expression, as well as the development of terminal bronchiolar lesions. O_3-induced IL-17A mRNA expression, terminal bronchiolar lesions, and *saa3* expression, were augmented in *T-cad* $^{-/-}$ mice, but no further augmentation was observed in mice with combined adiponectin and T-cadherin deficiency mice

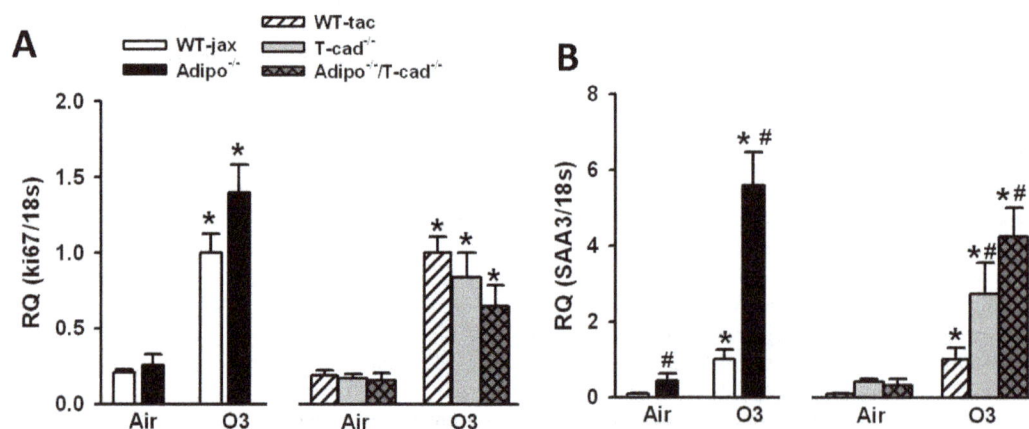

Figure 5. Real time PCR. Pulmonary mRNA expression of ki67 (A) and saa3 (B) in lungs of mice exposed to room air or O_3 (0.3 ppm) for 72 h. Expression was normalized to 18 s and expressed relative to wildtype mice exposed to O_3. *$p < 0.05$ versus air exposed mice of the same genotype, #$p < 0.05$ versus wildtype mice with the same exposure. Results are mean ± SEM of data from 4–7 air exposed mice and 6–10 ozone exposed mice.

(Fig. 3E,4D,5B), indicating that the presence of T-cadherin was necessary for the ability of adiponectin to inhibit these outcomes. In contrast, compared to mice deficient in T-cadherin alone, combined adiponectin and T-cadherin deficiency further augmented BAL neutrophils and G-CSF (Figs. 2A, 3B). Augmented ozone-induced increases in BAL neutrophils and G-CSF are partially dependent on IL-17A [8]. Taken together with the observation that the increased IL-17A expression in $Adipo^{-/-}$ mice requires T-cadherin (Fig. 3E), the results are consistent with the hypothesis that the augmented O_3-induced increases in BAL neutrophils and G-CSF in $T\text{-}cad^{-/-}$ versus wildtype mice derive from increases in IL-17A, whereas the additional effects of combined adiponectin and T-cadherin deficiency are the result of adiponectin acting through non-IL-17A dependent pathways (Figure 6).

In contrast to the increased inflammatory responses observed in $T\text{-}cad^{-/-}$ versus wildtype mice after subacute O_3 exposure reported here, we have previously reported *reduced* airway inflammation in $T\text{-}cad^{-/-}$ versus wildtype mice after allergen sensitization and challenge [18]. In that study, we concluded that T-cadherin does not mediate the effects of adiponectin. Instead, we suggested that the effects of T-cadherin deficiency may result from augmented circulating adiponectin in $T\text{-}cad^{-/-}$ mice (Fig. 1A) acting on other adiponectin binding proteins on circulating or lymphoid tissue Th2 lymphocytes, a key effector cell for allergic airway responses. As discussed above, T-cadherin is highly expressed in endothelial cells [14], where it appears to act as a repository for adiponectin. In the absence of T-cadherin, this adiponectin is delivered back to the blood resulting in increased circulating concentrations [15,16,17,18,23], as observed (Fig. 1A). One explanation for the divergent effects of T-cadherin deficiency in that study [18] versus this one is as follows. As discussed above, T-cadherin does appear to at least partially mediate the effects of adiponectin that reduce inflammation induced by subacute O_3. In contrast, other adiponectin binding proteins appear to mediate the anti-inflammatory effects of adiponectin after allergen challenge [18]. The difference in adiponectin binding proteins employed by adiponectin in the two models is likely related to the cell types involved in the two different types of pulmonary inflammation. Whereas $CD4^+$ lymphocytes are critical for allergic airways inflammation, the response to subacute O_3 mainly involves the innate immune response, particularly epithelial cells, macrophages, and pulmonary $\gamma\delta$ T-cells [8].

In wildtype mice, we observed an increase in BAL concentrations of the Th2 cytokine, IL-5 after subacute O_3 exposure (Fig. 3D). Others have also reported an increase in BAL IL-5 after O_3 exposure in mice, albeit using a different ozone exposure protocol [39]. The role of IL-5 in mediating responses to O_3 has not been established. In contrast to the effects of T-cadherin

deficiency on other O_3-induced changes in the lung, BAL IL-5 was reduced in T-cadherin deficient mice and restored in adiponectin/T-cadherin bideficient mice. This response is similar to the effects of T-cadherin deficiency on Th2 cytokines in allergen sensitized and challenged mice described above [18], suggesting that the source of IL-5 after O_3 may be Th2 cells.

Despite the increased circulating concentrations of adiponectin in $T\text{-}cad^{-/-}$ mice (Fig. 1A), we and others [15,23] have reported reduced BAL adiponectin in naïve $T\text{-}cad^{-/-}$ versus wildtype mice, similar to the results of this study (Fig. 1B, air exposed mice). Such observations indicate that in unchallenged mice, adiponectin is not transported into the lung via simple diffusion: if accumulation of BAL adiponectin relied solely on diffusion, it would be greater in $T\text{-}cad^{-/-}$ versus WT mice, since these mice have greater serum adiponectin (Fig. 1A). Instead, we suggested that T-cadherin may serve to transport adiponectin across the alveolar capillary barrier [15]. Such a hypothesis is consistent with the observation that HMW adiponectin, the adiponectin isoform most readily bound to T-cadherin [10], is also the major isoform present in BAL fluid of naïve mice, whereas the trimeric isoform, which should diffuse most easily, is barely detectable [15,23]. Loss of such a transport function for T-cadherin is unlikely to explain the effects of $T\text{-}cad^{-/-}$ observed in this study since BAL concentrations of adiponectin were actually greater in O_3 exposed $T\text{-}cad^{-/-}$ versus wildtype mice (Fig. 1B). Subacute O_3 exposure results in a marked increase in the permeability of the lungs (Fig. 2C), consistent with lung injury. Based on previous observations indicating that trimeric adiponectin accounted for the majority of the increased BAL adiponectin after ozone [8], we reasoned that in the setting of increased lung permeability, diffusion rather than T-cadherin-mediated transport begins to dominate movement of adiponectin from the blood into the lungs.

Others have demonstrated epithelial injury in mice exposed to O_3 in this manner, especially in the terminal bronchioles and central acinus, as evidenced by increased BrdU incorporation into these cells, likely reflecting cell proliferation after injury [27,40,41]. Consistent with these observations, RT-PCR confirmed increased O_3-induced expression of one of these genes, *Ki67* (Fig. 5A), a common marker of cell proliferation [26]. However, neither adiponectin deficiency nor T-cadherin deficiency had any effect on *Ki67* mRNA expression (Fig. 5A), suggesting that adiponectin does not regulate cell proliferation in this model. In contrast, we observed effects of both adiponectin deficiency and T-cadherin deficiency on the extent of terminal bronchiolar lesions observed after subacute O_3 (Fig. 4). Furthermore, adiponectin/T-cadherin bideficiency did not further augment the effects of T-cadherin deficiency on these lesions (Fig. 4), indicating that the effects of T-cadherin deficiency were the result of loss of adiponectin binding to this receptor. Taken together, the results suggest that the impact of adiponectin on terminal bronchiolar lesions is the result of more inflammatory cell recruitment to these sites of injury rather than more epithelial proliferation.

Saa3 is an acute phase protein capable of recruiting monocytes and macrophages to sites of inflammation, perhaps by forming a complex with the extracellular matrix protein, hyaluronan [42]. O_3-induced increases in *saa3* mRNA expression were substantially (almost 6 fold) greater in $Adipo^{-/-}$ versus WT mice (Fig. 5B). T-cadherin deficiency also resulted in a marked increase in *saa3* expression in O_3-exposed mice, and no further augmentation was observed in mice with combined adiponectin and T-cadherin deficiency (Fig. 5B), indicating that the presence of T-cadherin was necessary for adiponectin to suppress *saa3* mRNA expression, similar to our observations with IL-17A (Fig. 3E). We have previously reported that interstitial macrophages and $\gamma\delta$ T cells

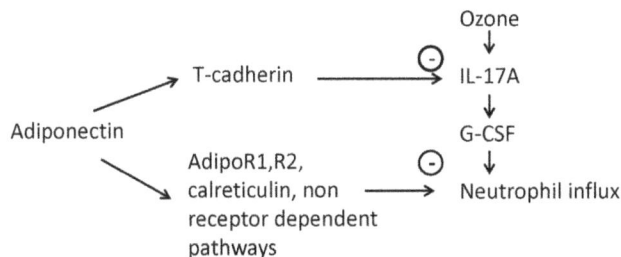

Figure 6. Schematic representation of T-cadherin dependent and independent effects of adiponectin that act to inhibit ozone induced neutrophil influx into the lungs.

are the source of the augmented IL-17A produced in the lungs following subacute O_3 exposure in $Adipo^{-/-}$ mice [8]. $Saa3$ has the capacity to induce IL-17A expression in T cells [43] and it is possible that it also regulates IL-17A expression in lung after subacute O_3 exposure.

One technical issue requires discussion. We exposed mice to O_3 for 72 hours. Others have shown gene specific differences in the kinetics of gene expression following O_3 [28,29]. Hence, it is likely that cytokines and chemokines other than those we identified in Fig. 3 as being impacted by O_3 were induced at earlier times in the exposure and then declined. Moreover, it is possible that the earlier expression of these moieties contributes to responses observed after 72 h exposure. However, we have examined the time course of key outcomes described here that differ in WT versus $Adipo^{-/-}$ mice (BAL neutrophils, IL-17 mRNA expression) and have found that these genotype-related differences exist throughout the 72 h exposure period described here[8].

In summary, our results confirm an anti-inflammatory role for adiponectin in pulmonary responses to subacute O_3 and indicate

that adiponectin binding to T-cadherin is required for aspects of this response, including the induction of IL-17A and consequent recruitment of neutrophils to the lungs.

Acknowledgments

The authors wish to thank Dr. Roderick Bronson at the Harvard Medical School Rodent Histopathology Laboratory for initial assessment of histological sections. The authors would also like to thank Dr. Huiqing Si for her valuable help in breeding mice used in this manuscript.

Author Contributions

Conceived and designed the experiments: DIK SAS. Performed the experiments: DIK ASW LAB LK CH. Analyzed the data: DIK ASW LK SAS. Contributed reagents/materials/analysis tools: BR CH. Wrote the paper: DIK LK SAS. Discussed the manuscript: DIK SAS LK CH BR. Provided transgenic mice: BR CH.

References

1. Cho HY, Zhang LY, Kleeberger SR (2001) Ozone-induced lung inflammation and hyperreactivity are mediated via tumor necrosis factor-alpha receptors. Am J Physiol Lung Cell Mol Physiol 280: L537–546.

2. Johnston RA, Schwartzman IN, Flynt L, Shore SA (2005) Role of interleukin-6 in murine airway responses to ozone. Am J Physiol Lung Cell Mol Physiol 288: L390–397.

3. Zhao Q, Simpson LG, Driscoll KE, Leikauf GD (1998) Chemokine regulation of ozone-induced neutrophil and monocyte inflammation. Am J Physiol 274: L39–46.

4. Arita Y, Kihara S, Ouchi N, Takahashi M, Maeda K, et al. (1999) Paradoxical decrease of an adipose-specific protein, adiponectin, in obesity. Biochem Biophys Res Commun 257: 79–83.

5. Ohashi K, Parker JL, Ouchi N, Higuchi A, Vita JA, et al. (2010) Adiponectin promotes macrophage polarization toward an anti-inflammatory phenotype. J Biol Chem 285: 6153–6160.

6. Wolf AM, Wolf D, Rumpold H, Enrich B, Tilg H (2004) Adiponectin induces the anti-inflammatory cytokines IL-10 and IL-1RA in human leukocytes. Biochem Biophys Res Commun 323: 630–635.

7. Shore SA, Terry RD, Flynt L, Xu A, Hug C (2006) Adiponectin attenuates allergen-induced airway inflammation and hyperresponsiveness in mice. J Allergy Clin Immunol 118: 389–395.

8. Kasahara DI, Kim HY, Williams AS, Verbout NG, Tran J, et al. (2012) Pulmonary inflammation induced by subacute ozone is augmented in adiponectin-deficient mice: role of IL-17A. J Immunol 188: 4558–4567.

9. Yamauchi T, Kamon J, Ito Y, Tsuchida A, Yokomizo T, et al. (2003) Cloning of adiponectin receptors that mediate antidiabetic metabolic effects. Nature 423: 762–769.

10. Hug C, Wang J, Ahmad NS, Bogan JS, Tsao TS, et al. (2004) T-cadherin is a receptor for hexameric and high-molecular-weight forms of Acrp30/adiponectin. Proc Natl Acad Sci U S A 101: 10308–10313.

11. Takeuchi T, Misaki A, Fujita J, Sonobe H, Ohtsuki Y (2001) T-cadherin (CDH13, H-cadherin) expression downregulated surfactant protein D in bronchioloalveolar cells. Virchows Arch 438: 370–375.

12. Nakanishi K, Takeda Y, Tetsumoto S, Iwasaki T, Tsujino K, et al. (2011) Involvement of endothelial apoptosis underlying chronic obstructive pulmonary disease-like phenotype in adiponectin-null mice: implications for therapy. Am J Respir Crit Care Med 183: 1164–1175.

13. Koller E, Ranscht B (1996) Differential targeting of T- and N-cadherin in polarized epithelial cells. J Biol Chem 271: 30061–30067.

14. Philippova M, Joshi MB, Kyriakakis E, Pfaff D, Erne P, et al. (2009) A guide and guard: The many faces of T-cadherin. Cell Signal 21: 1035–1044.

15. Zhu M, Hug C, Kasahara DI, Johnston RA, Williams AS, et al. (2010) Impact of adiponectin deficiency on pulmonary responses to acute ozone exposure in mice. Am J Respir Cell Mol Biol 43: 487–497.

16. Denzel MS, Scimia MC, Zumstein PM, Walsh K, Ruiz-Lozano P, et al. (2010) T-cadherin is critical for adiponectin-mediated cardioprotection in mice. J Clin Invest 120: 4342–4352.

17. Hebbard LW, Garlatti M, Young LJ, Cardiff RD, Oshima RG, et al. (2008) T-cadherin supports angiogenesis and adiponectin association with the vasculature in a mouse mammary tumor model. Cancer Res 68: 1407–1416.

18. Williams AS, Kasahara DI, Verbout NG, Fedulov AV, Zhu M, et al. (2012) Role of the adiponectin binding protein, T-cadherin (cdh13), in allergic airways responses in mice. PLoS One 7: e41088.

19. Bothe GW, Bolivar VJ, Vedder MJ, Geistfeld JG (2004) Genetic and behavioral differences among five inbred mouse strains commonly used in the production of transgenic and knockout mice. Genes Brain Behav 3: 149–157.

20. Ivanov, II, Frutos Rde L, Manel N, Yoshinaga K, Rifkin DB, et al. (2008) Specific microbiota direct the differentiation of IL-17-producing T-helper cells in the mucosa of the small intestine. Cell Host Microbe 4: 337–349.

21. Herbert RA, Hailey JR, Grumbein S, Chou BJ, Sills RC, et al. (1996) Two-year and lifetime toxicity and carcinogenicity studies of ozone in B6C3F1 mice. Toxicol Pathol 24: 539–548.

22. Williams AS, Chen L, Kasahara DI, Si H, Wurmbrand AP, et al. (2012) Obesity and airway responsiveness: Role of TNFR2. Pulm Pharmacol Ther [Epub ahead of print]: PMID 22584291.

23. Konter JM, Parker JL, Baez E, Li SZ, Ranscht B, et al. (2012) Adiponectin Attenuates Lipopolysaccharide-Induced Acute Lung Injury through Suppression of Endothelial Cell Activation. J Immunol 188: 854–863.

24. Kleeberger SR, Levitt RC, Zhang LY (1993) Susceptibility to ozone-induced inflammation. I. Genetic control of the response to subacute exposure. Am J Physiol 264: L15–20.

25. Serino M, Menghini R, Fiorentino L, Amoruso R, Mauriello A, et al. (2007) Mice heterozygous for tumor necrosis factor-alpha converting enzyme are protected from obesity-induced insulin resistance and diabetes. Diabetes 56: 2541–2546.

26. Gerdes J, Schwab U, Lemke H, Stein H (1983) Production of a mouse monoclonal antibody reactive with a human nuclear antigen associated with cell proliferation. Int J Cancer 31: 13–20.

27. Yu M, Pinkerton KE, Witschi H (2002) Short-term exposure to aged and diluted sidestream cigarette smoke enhances ozone-induced lung injury in B6C3F1 mice. Toxicol Sci 65: 99–106.

28. Backus GS, Howden R, Fostel J, Bauer AK, Cho HY, et al. (2010) Protective role of interleukin-10 in ozone-induced pulmonary inflammation. Environ Health Perspect 118: 1721–1727.

29. Bauer AK, Rondini EA, Hummel KA, Degraff LM, Walker C, et al. (2011) Identification of candidate genes downstream of TLR4 signaling after ozone exposure in mice: a role for heat-shock protein 70. Environ Health Perspect 119: 1091–1097.

30. Takemura Y, Ouchi N, Shibata R, Aprahamian T, Kirber MT, et al. (2007) Adiponectin modulates inflammatory reactions via calreticulin receptor-dependent clearance of early apoptotic bodies. J Clin Invest 117: 375–386.

31. Denzel MS, Scimia M-C, Zumstein PM, Walsh K, Ruiz-Lozano P, et al. (2010) T-cadherin is critical for adiponectin-mediated cardioprotection in mice. The Journal of Clinical Investigation 120: 4342–4352.

32. Shore SA, Johnston RA, Schwartzman IN, Chism D, Krishna Murthy GG (2002) Ozone-induced airway hyperresponsiveness is reduced in immature mice. J Appl Physiol 92: 1019–1028.

33. Gearing DP, Comeau MR, Friend DJ, Gimpel SD, Thut CJ, et al. (1992) The IL-6 signal transducer, gp130: an oncostatin M receptor and affinity converter for the LIF receptor. Science 255: 1434–1437.

34. Wang Y, Lam KS, Xu JY, Lu G, Xu LY, et al. (2005) Adiponectin inhibits cell proliferation by interacting with several growth factors in an oligomerization-dependent manner. J Biol Chem 280: 18341–18347.

35. Kuzmenko YS, Kern F, Bochkov VN, Tkachuk VA, Resink TJ (1998) Density- and proliferation status-dependent expression of T-cadherin, a novel lipoprotein-binding glycoprotein: a function in negative regulation of smooth muscle cell growth? FEBS Lett 434: 183–187.

36. Niermann T, Kern F, Erne P, Resink T (2000) The glycosyl phosphatidylinositol anchor of human T-cadherin binds lipoproteins. Biochem Biophys Res Commun 276: 1240–1247.

37. Resink TJ, Kuzmenko YS, Kern F, Stambolsky D, Bochkov VN, et al. (1999) LDL binds to surface-expressed human T-cadherin in transfected HEK293 cells and influences homophilic adhesive interactions. FEBS Lett 463: 29–34.

38. Ranscht B, Dours-Zimmermann MT (1991) T-cadherin, a novel cadherin cell adhesion molecule in the nervous system lacks the conserved cytoplasmic region. Neuron 7: 391–402.

39. Kierstein S, Krytska K, Sharma S, Amrani Y, Salmon M, et al. (2008) Ozone inhalation induces exacerbation of eosinophilic airway inflammation and hyperresponsiveness in allergen-sensitized mice. Allergy 63: 438–446.

40. Mautz WJ, Kleinman MT, Bhalla DK, Phalen RF (2001) Respiratory tract responses to repeated inhalation of an oxidant and acid gas-particle air pollutant mixture. Toxicol Sci 61: 331–341.

41. Kleeberger SR, Levitt RC, Zhang LY, Longphre M, Harkema J, et al. (1997) Linkage analysis of susceptibility to ozone-induced lung inflammation in inbred mice. Nat Genet 17: 475–478.

42. Han CY, Subramanian S, Chan CK, Omer M, Chiba T, et al. (2007) Adipocyte-derived serum amyloid A3 and hyaluronan play a role in monocyte recruitment and adhesion. Diabetes 56: 2260–2273.

43. Ather JL, Ckless K, Martin R, Foley KL, Suratt BT, et al. (2011) Serum amyloid A activates the NLRP3 inflammasome and promotes Th17 allergic asthma in mice. J Immunol 187: 64–73.

17

Molecular Storage of Ozone in a Clathrate Hydrate: An Attempt at Preserving Ozone at High Concentrations

Takahiro Nakajima[1][¤], Taisuke Kudo[1], Ryo Ohmura[1], Satoshi Takeya[2], Yasuhiko H. Mori[1]*

1 Department of Mechanical Engineering, Keio University, Yokohama, Japan, **2** Research Institute of Instrumentation Frontier, National Institute of Advanced Industrial Science and Technology (AIST), Tsukuba, Japan

Abstract

This paper reports an experimental study of the formation of a mixed $O_3 + O_2 + CO_2$ hydrate and its frozen storage under atmospheric pressure, which aimed to establish a hydrate-based technology for preserving ozone (O_3), a chemically unstable substance, for various industrial, medical and consumer uses. By improving the experimental technique that we recently devised for forming an $O_3 + O_2 + CO_2$ hydrate, we succeeded in significantly increasing the fraction of ozone contained in the hydrate. For a hydrate formed at a system pressure of 3.0 MPa, the mass fraction of ozone was initially about 0.9%; and even after a 20-day storage at $-25°C$ and atmospheric pressure, it was still about 0.6%. These results support the prospect of establishing an economical, safe, and easy-to-handle ozone-preservation technology of practical use.

Editor: Yang Gan, Harbin Institute of Technology, China

Funding: This study was supported by a Grant-in-Aid for the Global COE Program for the "Center for Education and Research of Symbiotic, Safe and Secure System Design" from the Ministry of Education, Culture, Sport and Technology, Japan, and by a JKA promotion fund from Keirin racing. The funders had no role in study design, data collection and analysis, decision to publish, or preparation of the manuscript.

Competing Interests: The authors have declared that no competing interests exist.

* E-mail: yhmori@mech.keio.ac.jp

¤ Current address: Technical Institute, Corporate Technology Division, Kawasaki Heavy Industries, Ltd., Akashi City, Japan

Introduction

Ozone (O_3) is known as a powerful oxidant and, due to this nature, it is widely used for, for example, the decontamination of air and water, the sterilization of perishables, the disinfection of medical instruments, and the cleaning or surface-conditioning processes in the semiconductor industry. However, it is neither very easy nor economical to use ozone in consumer applications such as sanitizing foods and drinking water, removing pesticide residues from fruits and vegetables, treating water in aquariums for suppressing bacteria growth, etc. This is because, to artificially generate ozone, we need a high-voltage electric device such as a corona discharger or a cold plasma generator and, once generated, ozone in the gaseous state rapidly decomposes to oxygen (O_2). Thus, it is generally believed that ozone can neither be stored nor transported and must be produced on site. For the limited consumer use of ozone, ozonated water (liquid water in which ozone is physically dissolved) and ozonated ice (water ice holding microbubbles of an ozone-containing gas) are commercially available. However, the ozone concentration in such ozonated water or ice is generally on the order of 1 or 10 ppm even in its fresh state, and rapidly decays with time. The above state of affairs seriously restricts the situations allowing the use of ozone. If we find a convenient means for transporting ozone from its production site to any place where ozone is needed, the utility of ozone will be significantly expanded.

Clathrate hydrates (abbreviated hydrates) are crystalline solid compounds each composed of host water molecules hydrogen-bonded into a structure of interlinked cages. Unless the given pressure is extremely high (typically on the order of gigapascals), each cage contains at most one guest molecule of a substance other than water [1]. That is, the guest molecules in a hydrate are isolated by the cage walls due to van der Waals forces and thereby prevented, in general, from mutual interactions. This indicates that hydrates have a high potential of storing chemically unstable substances, such as ozone, in the form of encaged guest molecules.

The idea of storing ozone in a hydrate was first presented by McTurk and Waller [2,3] in 1964. They reported the formation of an ozone-containing hydrate in an experimental system containing pure ozone, carbon tetrachloride (CCl_4) and water, and, based on their X-ray diffraction measurement, indicated that this hydrate was a double $O_3 + CCl_4$ hydrate in structure II. However, they provided neither any quantitative evaluation of the ozone content nor any experimental evidence for actual ozone preservation in their hydrate.

The pure ozone used by McTurk and Waller [2,3] is not easily available. Besides, it is explosive and very difficult to handle in practice. Carbon tetrachloride is effective as a *help guest* for lowering the pressure required for hydrate formation, though it is toxic and may be unsuitable for some applications. An attempt at forming a hydrate from a dilute ozone-containing gas (a mixture of $\sim5\%$ O_3 and $\sim95\%$ O_2 generated from a commercial ozone generator) in the absence of any help guest was reported by Masaoka et al. [4]. They formed a hydrate at a pressure of 13 MPa and a temperature of $-25°C$, and determined the ozone content of the hydrate to be 2.3 g/L ($\approx 0.2\%$ in mass fraction). They also performed a storage test of the hydrate at the same pressure−temperature conditions as those in the formation process, i.e., 13 MPa and $-25°C$, and observed only a slight decrease in the ozone content of the hydrate during 10-days storage after its formation.

More recently, Muromachi et al. [5] formed a hydrate from a ozone + oxygen gas mixture (~8% in mole fraction of ozone) and carbon tetrachloride or xenon (Xe) at a pressure of 0.25 or 0.35 MPa and a temperature of 0.1°C. They showed that, if cooled to −20°C under an aerated atmospheric-pressure condition, the $O_3 + O_2 + CCl_4$ and $O_3 + O_2 + Xe$ hydrates could preserve ozone for more than 20 days at mass fractions around 0.2% and 0.1%, respectively. Subsequently, Muromachi et al. [6] performed phase-equilibrium measurements for the $O_3 + O_2 + CCl_4$ and $O_3 + O_2 + CH_3CCl_2F$ hydrates.

For practical applications of ozone-containing hydrates, the use of a toxic or very expensive substance as the help guest should be avoided. On the other hand, the hydrates can be desirably formed at moderate pressures and preserved at a moderately cooled atmospheric-pressure condition. In order to satisfy these requirements, Nakajima et al. [7] selected carbon dioxide (CO_2) as the help guest, and formed an $O_3 + O_2 + CO_2$ hydrate from an $O_3 + O_2$ gas mixture (10−12% in O_3 mole fraction) blended with pure CO_2 in a molar ratio of 1:7 at a pressure of 1.9 MPa and a temperature of 0.1°C. They performed preservation tests with the $O_3 + O_2 + CO_2$ hydrate at different storage temperatures from −5°C to −30°C under an aerated atmospheric-pressure condition. The results of these tests showed that, for ozone preservation over 20 days at a mass fraction around 0.1%, the storage temperature should be −25°C or lower. This inferiority in ozone-preserving function of the $O_3 + O_2 + CO_2$ hydrate in comparison with the $O_3 + O_2 + CCl_4$ hydrate [5] is, as demonstrated by a relevant phase-equilibrium study [8], essentially due to the higher phase-equilibrium pressure for the former than the latter at any given temperature and hence inevitable. Despite such a thermodynamic disadvantage of the $O_3 + O_2 + CO_2$ hydrate as compared to the $O_3 + O_2 + CCl_4$ hydrate, the former represents a good compromise between the ozone preservability versus the biological safety and the economy in system operation, and is possibly the best selection as the ozone storage medium for practical use. Moreover, we can expect that the gas mixture released from an $O_3 + O_2 + CO_2$ hydrate will have some synergetic effect of the ozone and carbon dioxide for sterilizing foods [9] and will possibly be more suitable for food-industrial applications than the $O_3 + O_2$ or $O_3 +$ air gas mixtures directly generated from commercial ozone generators.

This study is an extension of the first $O_3 + O_2 + CO_2$ hydrate study by Nakajima et al. [7] discussed above. The major objectives of this study were (a) to generate $O_3 + O_2 + CO_2$ hydrates having higher ozone fractions, and (b) to perform ozone preservation tests with these hydrates in order to examine their practical utility. The study was successful regarding both of these objectives. We confirmed that, with simple modifications of the hydrate-forming procedure, the initial ozone content of a hydrate can be multiplied severalfold as compared to that previously observed [7], and that more than half of such a high ozone content remains after a 20-day hydrate storage under an aerated atmospheric-pressure condition at a temperature of −25°C.

Experimental Section

The general experimental scheme used in this study was the same as that used in our previous study [7] that first dealt with an $O_3 + O_2 + CO_2$ hydrate. However, some core portions of the hydrate-forming apparatus were modified this time in order to increase the water-to-hydrate conversation ratio (i.e., to reduce the fraction of ice in the hydrate + ice solid mixtures for use in ozone

preservation tests) and to allow pressurizing the $O_3 + O_2$ gas mixture released from an ozone generator before mixing it with CO_2 gas. Details of the materials, equipment and procedure used in this study are described below.

Materials

The raw materials used for forming the $O_3 + O_2 + CO_2$ hydrates were oxygen certified to the purity of 99.9% (volume basis) and carbon dioxide certified to the purity of 99.995% (volume basis) by their supplier (Japan Fine Products Corp., Kawasaki, Kanagawa Prefecture, Japan), and water deionized and distilled in our laboratory. Oxygen was used for generating an $O_3 + O_2$ gas mixture (>11% in mole fraction of O_3) with the aid of a dielectric-barrier-discharge-based ozone generator (ED-OGS-HP1, EcoDesign Co., Ltd., Saitama Prefecture, Japan).

Apparatus

The experimental setup used to form the hydrates is schematically illustrated in Fig. 1. By comparing this figure to Fig. S1 in our previous paper [7], one may realize how the setup used this time had been modified from its predecessor. The hydrate-forming reactor [indicated as (l) in Fig. 1] was completely renewed. It was a pan-type stainless-steel vessel with a 65-mm inside diameter and 100-cm^3 inside volume. An impeller-blade stirrer magnetically connected to the drive shaft of an external, variable-speed, ac motor was inserted into this reactor in order to provide its contents with stronger mixing than a magnetic stirrer inside a tall cylindrical reactor used in the previous setup [7] did. As before, the reactor was immersed in a temperature-controlled bath containing an aqueous ethylene-glycol solution.

Another modification in the experimental setup was the installation of two gas-pressurizing chambers, which are indicated as (h) and (i) in Fig. 1. They were stainless-steel cylinders with capacities of 3785 cm^3 and 2250 cm^3, respectively. The smaller chamber (i) was used to store the $O_3 + O_2$ gas mixture supplied from the ozone generator (c) at a pressure up to about 0.3 MPa, while the larger one (h) was initially charged with water. By injecting oxygen gas supplied from the external high-pressure cylinder (a) into the larger chamber, water could be displaced from the larger chamber to the smaller chamber, thereby increasing the pressure of the $O_3 + O_2$ gas mixture to the prescribed level.

Procedure

The procedure of forming an $O_3 + O_2 + CO_2$ hydrate using the renewed setup (Fig. 1) is as described below. First, the reactor (l) was charged with ~30 g of water and immersed in a bath of an aqueous ethylene-glycol solution temperature-controlled at 0.1°C. The reactor, the gas-pressurizing chambers (h) and (i), and the gas-mixing chamber (d) were then flushed at least five times with pure oxygen gas, then evacuated. After confirming that the mole fraction of ozone in the gas mixture released from the ozone generator (c) was in the range of 10−12%, the smaller gas-pressurizing chamber (i) and the gas-mixing chamber (d) were charged with this mixture up to a pressure of 0.3 MPa. Oxygen gas from the high-pressure cylinder (a) was then injected into the larger gas-pressurizing chamber (h) to make the water stored in it flow into the smaller chamber (i) and thereby to make the $O_3 + O_2$ gas mixture flow out of the latter chamber into the gas-mixing chamber (d), until the pressure inside the gas-mixing chamber increased to the prescribed level. When the pressure inside the gas-mixing chamber did not sufficiently increase at this stage, the above serial operations beginning with the charging of the smaller gas-pressurizing chamber with a fresh $O_3 + O_2$ gas mixture was repeated until the pressure was raised to the

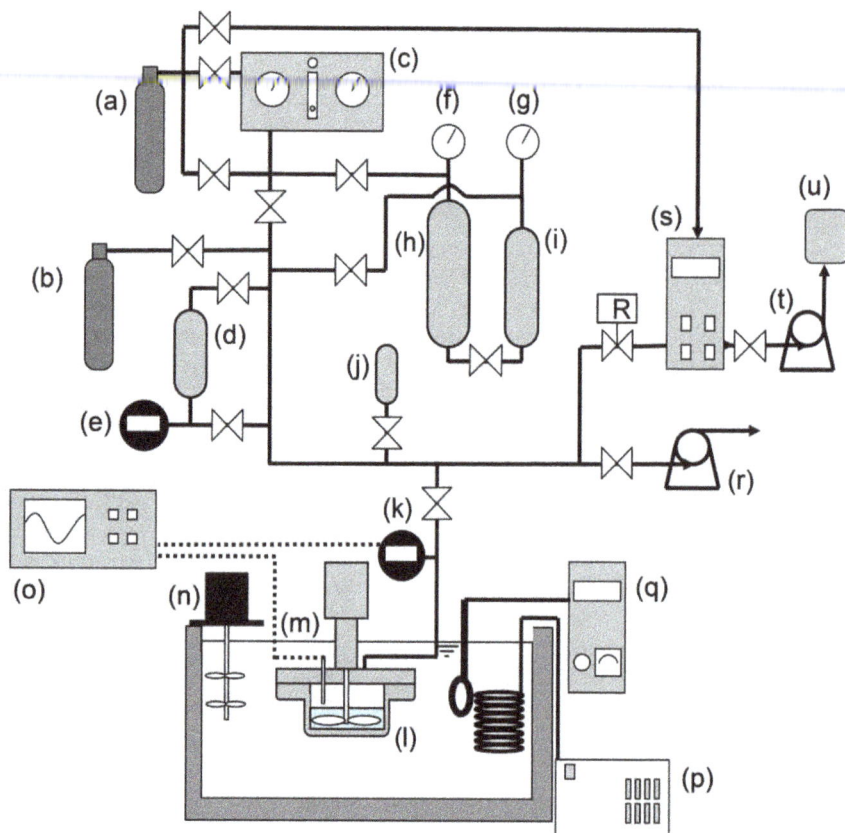

Figure 1. Schematic illustration of the experimental setup for forming the O_3+ O_2+ CO_2 hydrates. This setup consists of (a) an oxygen cylinder, (b) a carbon-dioxide cylinder, (c) an ozone generator, (d) a gas-mixing chamber, (e) a pressure gauge, (f) and (g) pressure gauges, (h) and (i) gas-pressurizing chambers, (j) a gas-sampling chamber, (k) a pressure gauge, (l) a hydrate-forming reactor, (m) a Pt-wire resistance thermometer, (n) a stirrer, (o) a data logger, (p) an immersion cooler, (q) a PID-controlled heater, (r) a vacuum pump, (s) an ozone monitor, (t) a vacuum pump, and (u) an ozone decomposer.

prescribed level. Subsequently, CO_2 gas was supplied to the gas-mixing chamber until the pressure inside increased to 3 or 4 MPa. The O_3+ O_2+ CO_2 gas mixture thus prepared was supplied to the hydrate-forming reactor (l) until the pressure inside the reactor increased to the prescribed level (2.0, 2.5 or 3.0 MPa). A series of intermittent batch operations for forming a hydrate was then started by turning on the stirrer in the reactor. For preventing the system pressure from significantly decreasing from the prescribed level and for minimizing the change in composition of the gas mixture inside the reactor, each batch operation was not allowed to continue for a long period but interrupted by a gas-exchange operation for replacing the residual gas mixture inside the reactor by a fresh gas mixture newly prepared in, and supplied from, the gas-mixing chamber (d). Such a change in the batch and gas-exchange operations was repeated several times until no decrease in system pressure was detected during each batch operation and hence we judged that the hydrate formation had already ceased. The reactor was then cooled to $-15°C$ to freeze the contents of the reactor except for the residual gas mixture. The gas mixture was then discharged into a $50\text{-}cm^3$ gas-sampling chamber (j) for its compositional analysis using a gas chromatograph (Agilent 3000 Micro Gas Chromatograph). After removing the reactor from the bath and dipping it into a liquid-nitrogen pool, the formed hydrate was removed from the reactor and crushed into particles with a $5-7$ mm linear dimension. A small portion ($\sim 1-2$ g in mass) of

the hydrate was sampled for an iodometric measurement for determining its initial ozone content. The rest of the hydrate was stored in a Pyrex test tube, if it was to be used for a subsequent preservation test.

The procedure of the ozone preservation tests performed in this study was completely the same as that employed in our previous studies [5. 7], hence its description is omitted here. The technique used in the PXRD measurement with some hydrate sample was also the same as that described elsewhere [7].

Results and Discussion

Before presenting the results of a series of ozone preservation tests performed in this study, we need to specify the actual contents of what we have called "hydrates" and "ozone fractions" in the preceding sections in relation to the previous hydrate-preservation studies [4,5,7]. Because liquid water used for forming hydrates in any of these studies could not be completely converted to a hydrate, each preservation-test specimen prepared by cooling the formed hydrate, together with residual water, to a test temperature below 0°C must have been a mixture of the hydrate and water ice. Inevitably, the "ozone faction" measured by some macroscopic means (e.g., the iodometric technique [5,7]) is not an intrinsic ozone fraction of the hydrate but an *effective* fraction defined as the ratio of the mass of ozone contained in a given hydrate + ice mixture to that of the mixture itself. This means that an increase in such an effective ozone fraction, x_{O3}, may be achieved either by (i) decreasing the ice fraction in the hydrate + ice mixtures for storing ozone or by (ii) increasing the true ozone fraction in the hydrate. As described in the Experimental Section, we attempted at realizing both of these means in order to significantly increase x_{O3} as compared to its magnitude (~0.1%) observed in the previous study of this series [7]. The former means was realized by intensifying the stirring of the gas/liquid contents in the hydrate-forming rector, while the latter was carried out by varying the system pressure as well as the composition of the feed gas supplied to the reactor during each hydrate-forming operation. Varying the feed-gas composition was made by controlling the ratio of CO_2 addition to the $O_3 + O_2$ mixture generated from an ozone generator at a nearly fixed O_3 fraction (10−12%).

Figure 2 shows a powder X-ray diffraction (PXRD) pattern (measured at 98 K) of an $O_3 + O_2 + CO_2$ hydrate formed from a mixture of $O_3 + O_2$ and CO_2 in a nearly 2:8 molar ratio at a system pressure p of 2.0 MPa and a temperature T of 0.1°C. This pattern indicates that the hydrate sample used here was a mixture of a hydrate in structure I (sI) with the lattice constant of 11.8294(4) Å and water ice in two different crystal forms, i.e., hexagonal ice, Ih, and cubic ice, Ic. The mass fraction of the hydrate was estimated to be 0.89, which was significantly higher than the corresponding estimate (~0.3) for the $O_3 + O_2 + CO_2$ hydrate samples formed in the previous study [7].

We performed hydrate-forming experiments at three different system pressures (2.0, 2.5 and 3.0 MPa) and four different $O_3 + O_2$ versus CO_2 molar ratios (1:9, 2:8, 3:7 and 4:6, each accompanied by slight run-to-run scatter) in the feed gas. The hydrate formed in each experiment was subjected to an iodometric measurement to determine its ozone content, i.e., the initial x_{O3} value for the hydrate which we denote $x_{O3,init}$ hereafter. Figures 3 and 4 show the variations in $x_{O3,init}$ depending on the system pressure p for each hydrate-forming operation and the feed-gas composition, respectively, in which the feed-gas composition is represented by X_{O3}, the mole fraction of ozone in the gas phase in contact with the hydrate at the end of each hydrate-forming operation (consult Tables S1 and S2 and Fig. S1 in Supporting Information S1 for the complete sets of $x_{O3,init}$ and X_{O3} data and

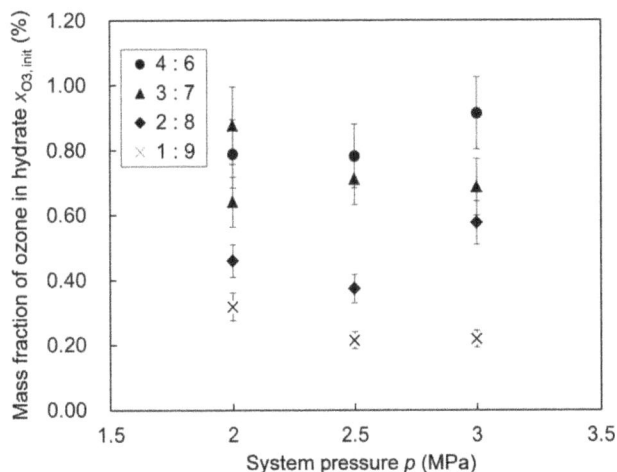

the graphical plots of the $x_{O3,init}$ data). We note that $x_{O3,init}$ shows no systematic dependence on the system pressure p (Fig. 3) but a quasi-linear dependence on X_{O3} (Fig. 4). This fact indicates that $x_{O3,init}$ was primarily controlled by the competitive fractional filling of the hydrate cages by O_3, O_2 and CO_2 molecules and that most of the hydrate cages were occupied by some of these guest molecules even at the lowest system pressure, $p = 2.0$ MPa, prescribed in the present experiments. Our estimation of the cage occupancies by O_3, O_2 and CO_2 molecules is described in Supporting Information S2. Consult Table S3 and Fig. S2 in Supporting Information S2 for the estimated occupancy values.

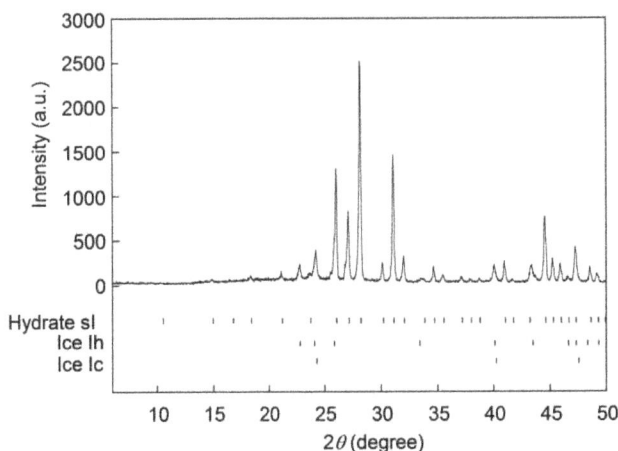

Figure 3. The initial ozone fraction in the formed hydrate versus the system pressure. The legend inserted in the graph indicates the $O_3 + O_2$ versus CO_2 molar ratio in the feed gas used for each operation. Each data point represents the arithmetic mean of the three $x_{O3,init}$ values obtained for the different hydrate samples. The error bar for each data point represents the uncertainty of the ozone-fraction measurement by iodometry.

Figure 2. PXRD profile of an $O_3 + O2 + CO_2$ hydrate at 98 K. The solid curve shows the intensities observed using Cu−Kα radiation. The top row of tick marks represent the calculated peak positions for the structure I hydrate, and the lower two rows represent those for the hexagonal ice Ih and cubic ice Ic, respectively. The hydrate sample (accompanied by ice crystals) used in this PXRD measurement was formed from a mixture of $O_3 + O2$ and CO_2 in a nearly 2:8 molar ratio at the condition of $p = 2.0$ MPa and $T = 0.1$°C.

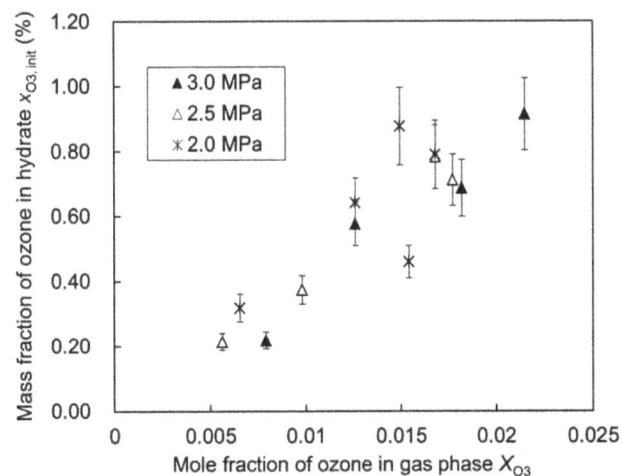

Figure 4. The initial ozone fraction in the formed hydrate versus the gas-phase composition. The mole fraction of ozone, X_{O3}, shown here is for the gas phase inside the reactor when the hydrate formation ceased. The legend inserted in the graph indicates the system pressure p during each hydrate-forming operation. The error bar for each data point represents the uncertainty of the ozone-fraction measurement by iodometry.

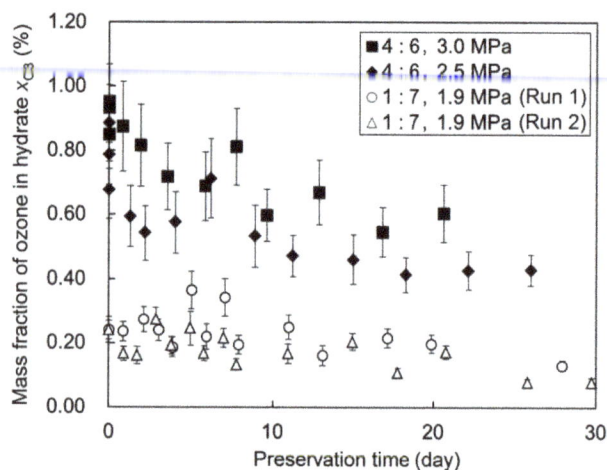

Figure 5. Results of the ozone preservation tests. This graph shows the time evolution of ozone fraction (mass basis) in each $O_3 + O_2 + CO_2$ hydrate stored under an aerated atmospheric-pressure (0.101 MPa) condition temperature-controlled at $-25°C$. The comparison of the ozone preservation test data obtained in this study (marked by closed symbols) and those from a previous study [7] (marked by open symbols) are compared. The legend inserted in the graph indicates the $O_3 + O_2$ versus CO_2 molar ratio in the feed gas and the system pressure p for each hydrate-forming operation. The error bar for each data point represents the uncertainty of the ozone-fraction measurement by iodometry.

For the hydrates formed at the two higher system pressures, $p = 2.5$ and 3.0 MPa, ozone preservation tests were performed following the procedure employed in the previous relevant studies [5,7]. The hydrates were stored in an aerated atmospheric-pressure (0.101 MPa) condition temperature-controlled at $-25°C$. Figure 5 shows the results of these tests, i.e., two x_{O3}-data sets each obtained by continually sampling the stored hydrate for iodometric measurements during a period extending to $20-26$ days after the formation of the hydrate. In addition, Fig. 5 shows, for comparison, two x_{O3}-data sets previously obtained with hydrates formed at a lower system pressure and a lower $(O_3 + O_2)$-to-CO_2 ratio. Obviously, the data obtained in the present preservation tests show much higher x_{O3} values than the previous data through the entire hydrate-storage period. The former exhibited x_{O3} values of $0.4-0.6\%$ at stages of ~20-days storage, which is a few times higher than the x_{O3} values exhibited by the latter at the same stages. The relatively sharp decrease in x_{O3} observed in the present tests, particularly during the initial several days of storage, is presumably ascribable to the higher $O_3 + O_2$ fractions (or the lower CO_2 fractions) in the hydrates used in the tests as compared to the hydrates used in the previous study [7]. Because the thermodynamic equilibrium condition shifts toward a higher pressure or lower temperature with an increase in the $O_3 + O_2$ fraction, or a decrease in the CO_2 fraction, in the hydrate [8], we can reasonably assume that the hydrates used in the present preservation tests suffered higher thermodynamic driving forces

for their dissociation than those used in the previous study [7] under the same storage condition. In addition, the lower ice fraction mixed with the hydrates used in the present tests possibly weakened the ice-barrier effect for suppressing the hydrate dissociation. As recognized in Fig. 5, the present test data show an asymptotic decrease in x_{O3} with time, with an apparent half-life period of $20-25$ days. This period is considered to be long enough to allow the practical use of ozone-containing hydrates for most industrial, medical and consumer needs.

Conclusions

This study demonstrated that ozone can be stored up to a mass fraction of $\sim0.9\%$ in a structure-I hydrate (containing water ice by $\sim10\%$) formed from a ternary (ozone + oxygen + carbon dioxide) gas mixture containing ozone up to a mole fraction of $\sim2\%$. This is the highest record of an ozone fraction in artificially formed hydrates ever reported in the literature. Such an ozone fraction in a formed hydrate may be further increased by increasing the ozone fraction in the feed gas at the cost of the increasing risk of its explosion [10]. The magnitude of the in-hydrate ozone fraction that we achieved in this study seems to be a good compromise between the demand for increasing the ozone content of a formed hydrate and the need for securing operational safety of the hydrate-forming process using an ozone-containing gas mixture as the feed gas.

Besides the magnitude of the ozone fraction in a freshly formed hydrate, the preservability of ozone encaged in the hydrate is of practical importance. The preservation tests performed in this study revealed that the in-hydrate ozone fraction asymptotically decreases with time from its initial value, about $0.8-0.9\%$, but still remains at about $0.4-0.6\%$ after a 20-day hydrate storage in an aerated atmospheric-pressure condition cooled at $-25°C$. This finding strongly suggests the practical utility of mixed ozone + oxygen + carbon dioxide hydrates for the industrial, medical and consumer uses of ozone.

Acknowledgments

We thank Mr. Sanehiro Muromachi, a graduate student at Keio University, for his help in the cage occupancy calculations described in Supporting Information S2.

Author Contributions

Conceived and designed the experiments: TN RO YHM. Performed the experiments: TN TK ST. Analyzed the data: TN TK ST. Wrote the paper: YHM.

References

1. Sloan ED, Koh CA (2008) Clathrate Hydrates of Natural Gases, 3rd ed. Boca Raton: CRC Press. p. 266.
2. McTurk G, Waller JG (1964) Ozone−carbon tetrachloride double hydrate. Nature 202: 1107.
3. Waller JG, McTurk G (1964) Novel ozone inclusion compound. U.K. Patent Specification 961115.

4. Masaoka T, Yamamoto A, Motoi K (2007) Storing method of ozone, method of producing solid material incorporating ozone, food preservation material and food preserving method. Japan Patent Publication 2007-210881. Released online in the Patent and Utility Model Gazette DB of the Japan Patent Office, 23 August 2007. Available: http://www4.ipdl.inpit. go.jp/Tokujitu/tjsogodben.ipdl?N0000 = 115. Accessed 2012 Aug 18.

5. Muromachi S, Ohmura R, Takeya S, Mori YH (2010) Clathrate hydrates for ozone preservation. J Phys Chem B 114: 11430−11435.

6. Muromachi S, Nakajima T, Ohmura R, Mori YH (2011) Phase equilibrium for clathrate hydrates formed from an ozone + oxygen gas mixture coexisting with carbon tetrachloride or 1,1-dichloro-1-fluoroethane. Fluid Phase Equilib 305: 145−151.

7. Nakajima T, Akatsu S, Ohmura R, Takeya S, Mori YH (2011) Molecular storage of ozone in a clathrate hydrate formed from an O_3+ O_2+ CO_2 gas mixture. Angew Chem Int Ed 50: 10340−10343.

8. Muromachi S, Ohmura R, Mori YH (2012) Phase equilibrium for ozone-containing hydrates formed from an (ozone + oxygen) gas mixture coexisting with gaseous carbon dioxide and liquid water. J Chem Thermodyn 49: 1−6.

9. Mitsuda H, Ominami H, Yamamoto A (1990) Synergistic effect of ozone and carbon dioxide gases for sterilizing food. Proc Japan Acad B 66: 68−72.

10. Koike K, Nifuku M, Izumi K, Nakamura S, Fujiwara S, et al. (2005) Explosion properties of highly concentrated ozone gas. J Loss Prev Process Ind 18: 465−468.

18

Long-Term Effects of Liming on Health and Growth of a Masson Pine Stand Damaged by Soil Acidification in Chongqing, China

Zhiyong Li[1,2]*, Yanhui Wang[2], Yuan Liu[3], Hao Guo[4], Tao Li[5], Zhen-Hua Li[2], Guoan Shi[1]

1 College of Agriculture, Henan University of Science and Technology, Luoyang, China, **2** Institute of Forest Ecology, Environment and Protection, Chinese Academy of Forestry, Beijing, China, **3** University of Chinese Academy of Sciences, Beijing, China, **4** Institute of Desertification Studies, Chinese Academy of Forestry, Beijing, China, **5** Hebei University, Baoding, China

Abstract

In the last decades, the Masson pine (*Pinus massoniana*) forests in Chongqing, southwest China, have increasingly declined. Soil acidification was believed to be an important cause. Liming is widely used as a measure to alleviate soil acidification and its damage to trees, but little is known about long-term effects of liming on the health and growth of declining Masson pine forests. Soil chemical properties, health condition (defoliation and discoloration), and growth were evaluated following application of limestone powder (0 (unlimed control), 1, 2, 3, and 4 t ha^{-1}) in an acidified and declining Masson pine stand at Tieshanping (TSP) of Chongqing. Eight years after liming, in the 0–20 cm and 20–40 cm mineral soil layers, soil pH values, exchangeable calcium (Ca) contents, and Ca/Al molar ratios increased, but exchangeable aluminum (Al) levels decreased, and as a result, length densities of living fine roots of Masson pine increased, with increasing dose. Mean crown defoliation of Masson pines (dominant, codominant and subdominant pines, according to Kraft classes 1–3) decreased with increasing dose, and it linearly decreased with length densities of living fine roots. However, Masson pines (Kraft classes 1–3) in all treatments showed no symptoms of discoloration. Mean current-year twig length, twig dry weight, needle number per twig, needle length per twig, and needle dry weight per twig increased with increasing dose. Over 8 years, mean height increment of Masson pines (Kraft classes 1–3) increased from 5.5 m in the control to 5.8, 6.9, 8.3, and 9.5 m in the 1, 2, 3, and 4 t ha^{-1} lime treatments, and their mean DBH (diameter at breast height) increment increased from 3.1 to 3.2, 3.8, 4.9, and 6.2 cm, respectively. The values of all aboveground growth parameters linearly increased with length densities of living fine roots. Our results show that liming improved tree health and growth, and these effects increased with increasing dose.

Editor: Raffaella Balestrini, Institute for Plant Protection (IPP), CNR, Italy

Funding: This study was sponsored by the State Forestry Administration of China (SFA) with the project (201304301–05), the Chinese–Norwegian cooperation project "Forest in south China: an important sink for reactive nitrogen and a regional hotspot for N2O?" (209696/E10), and the Key Laboratory of Forest Ecology and Environment of SFA. The funders had no role in study design, data collection and analysis, decision to publish, or preparation of the manuscript.

Competing Interests: The authors have declared that no competing interests exist.

* E-mail: pphdll@126.com

Introduction

Atmospheric emissions of acidifying compounds resulting from anthropogenic activities since the beginning of the industrial revolution have led to extensive soil acidification in terrestrial ecosystems [1–3]. Soil acidification has been blamed for causing indirect forest damage in Europe, North America, and Asia in the past few decades [4–10]. Some experiments have been implemented with liming in forests in Europe and the northeastern United States. Investigations on forest liming have shown that it enhanced cation exchange capacity (CEC), base saturation (BS), and contents of exchangeable Ca and Mg in the forest floor and the mineral soil, while Al levels and acidity decreased [11–15]. Effects of liming on the growth of tree fine roots were reported to be positive [16–18], although some researchers almost found the contrary, depending on tree species, tree age, site, soil conditions, form and dose of liming, and the time lapse since liming [19–21]. In contrast to expectations, however, liming has not made an overall improvement in aboveground growth of trees; instead, it has often produced either no effect or a detrimental effect on stand

growth [22–26]. But, positive effects of liming on stand growth have also been observed [27–29]. The effect of liming on tree growth is dependent on the soil chemical properties, particularly the deficiency of Ca and Mg [29]. In addition, the reaction of forest stands to liming can be affected by such factors as tree species and age [30].

Masson pine (*Pinus massoniana*) is one of the most widespread forest species across subtropical China and it is highly sensitive to acid deposition [31]. Chongqing, located in the upper reaches of the Yangtze River, is an industrial and commercial municipality in southwest China. Total areas of pure Masson pine forests are in excess of 1.2 million hectares and estimated to account for more than 50% of the total forests in the region [32]. Obviously, this conifer species plays a very important role in protecting the environment of the middle and lower reaches of the Yangtze River. On the other hand, Chongqing ranks amongst the regions with the highest levels of acid deposition in China [33–35]. Due to extensive use of coal with high sulphur (S) contents, basin topography conducive to the accumulation of air pollutants, and low wind velocity, Chongqing has undergone severe deposition of

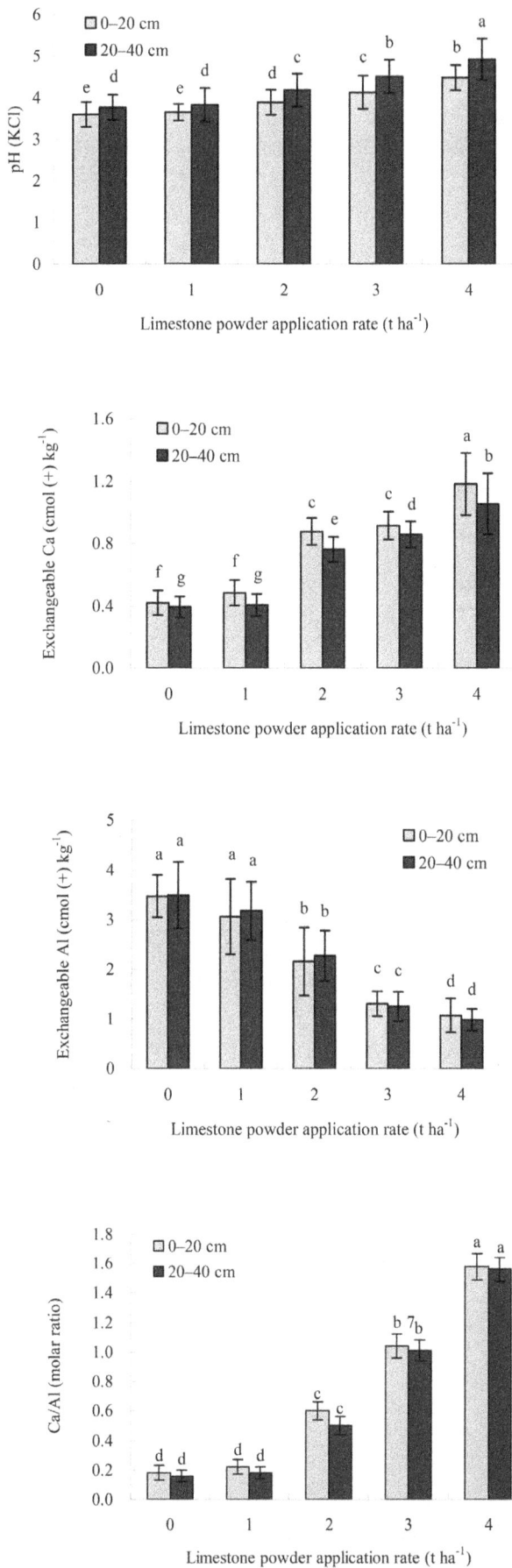

Figure 1. Soil pH values, exchangeable Ca contents, exchangeable Al levels, and Ca/Al molar ratios in the 0–20 cm and 20–40 cm of the Masson pine stand 8 years after liming. Error bars represent the standard deviation (SD) of the mean. Different letters above the error bars indicate significant differences at the 0.05 level (ANOVA and Duncan's multiple range test), n = 3.

acidifying S compounds since the late 1970s, when the Chinese Economic Reform Policy started [1,33,36–38]. More than half of 40 districts and counties within the region was designated as being part of so-called acid rain control areas, i.e. parts of China where pH value of rain water is lower than 4.5 and sulfur deposition exceeds the critical load, and the acid rain problem has been addressed by the national authorities [39]. The soil of Masson pine forests in Chongqing is generally acidic, with low amounts of exchangeable Ca and high amounts of exchangeable Al [33,35,40,41]. Masson pine is highly sensitive to Al [42]. Since the 1980s, large patches of Masson pine forest in some areas have exhibited serious decline symptoms, mainly including fine root death, tip necrosis of needles, thinned crown, reduced needle length, premature needle abscission, dieback of twigs and branches, and reduced growth, which was believed to be caused by soil acidification [4,33,35,43,44].

A common measure to counteract soil acidification is liming of soils [45–47]. In the last decades, however, only a few studies have been conducted to investigate the effects of liming on Masson pine forests growing on acidified soil, mainly with a single lime dose over relatively short time spans. For example, Du and Tian [44] found increased pH values in the 0–25 cm mineral soil on a limed young Masson pine forest 2 years after liming. Huang et al. [48] observed increased Ca contents in current-year needles at limed plots established in a 40-year-old pure Masson pine forest about 1 year after liming with 3 t ha^{-1} limestone powder. Up to now, little is known about the long-term effects of different lime doses on Masson pine. Thus, fifteen permanent plots treated with five doses of limestone powder were set up in a Masson pine stand damaged by soil acidification in Chongqing in 2004, and an investigation was carried out 8 years after liming. The aim of this work was to assess the effects of the five doses on the health and growth of Masson pine.

Study Sites and Methods

Site description

This study was conducted at Tieshanping (TSP) (29°38′N, 106°41′E, 512–579 m a.s.l.). TSP, covering about 1200 ha, is an almost pure Masson pine forest located on a sandstone ridge approximately 25 km northeast from the urban centre of Chongqing (28°10′–32°13′N, 105°17′–110°11′E) in southeast Sichuan Basin, southwest China. Masson pines have been planted since the early 1960s. The study site has a subtropical humid climate, with a mean annual air temperature of 18°C, a mean annual precipitation of 1100 mm that mainly occurs from April to October, and a mean annual relative humidity of 80%. TSP has been experiencing a long-term severe acid deposition, with low mean annual pH (4.0–4.2) and high frequency of acid rain (rainfall of pH<5.6 being about 90.0%) [35,38,49]. High deposition of S has been frequently reported. For example, during 2002 and 2003, throughfall S input was reported to be 16.0 g S m^{-2} a^{-1}, and throughfall N input was measured at 4.0 g N m^{-2} a^{-1} [50]. The forest soil is yellow mountain soil (Haplic Acrisol, WRB) developed on sandstone and prone to acidification. The mean mineral soil depth is between 40 and 80 cm, and the soil was rather homogeneous [33,34]. This acidic soil is representative for

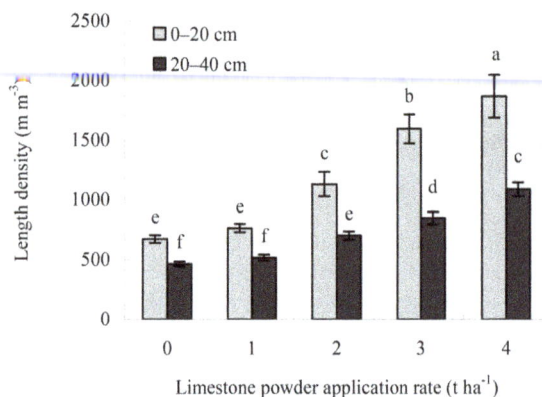

Figure 2. Length densities of living fine roots of Masson pine in the 0–20 cm and 20–40 cm of the Masson pine stand 8 years after liming. Error bars represent the SD of the mean. Different letters above the error bars indicate significant differences at the 0.05 level (ANOVA and Duncan's multiple range test), n = 3.

Chongqing and characterized by low levels of exchangeable Ca and high levels of exchangeable Al [33,35,41]. In the last decades the Masson pine forest ecosystem has declined due to severe soil acidification [33,35].

Plot establishment and liming treatments

In early May 2004, we selected a 26-year-old Masson pine stand, damaged (defoliation >25%) by soil acidification [35] for this limestone powder dose experiment. This stand grew on almost level ground and was composed of evenly-spaced and similarly-sized Masson pines, which were similar in defoliation. Therein, only a few scattered individuals of undergrowth species, e.g. *Schima superba*, *Cunninghamia lanceolata*, *Schoepfia jasminodora*, and *Lithocarpus glaber* grew. The density of Masson pines for this stand was 1192 trees ha^{-1}. In early June 2004, the mean defoliation, discoloration, height, and DBH of Masson pines (dominant, codominant and subdominant pines, Kraft classes 1–3) for this stand were 34%, 1%, 13.3 m, and 15.5 cm, respectively.

The trial was established in this stand as a randomized complete block experiment with five limestone powder (49.5% CaO, particle size <0.25 mm) treatments (0 (unlimed control), 1, 2, 3, and 4 t ha^{-1}) and three replicates. There were total fifteen plots. Each plot measured 10 m×10 m, with about 1 m buffer strips around

each plot. The overall size of the stand within which all plots were installed was about 0.3 ha. Limestone powder was broadcast one time and evenly on the forest floor by hand in early June 2004.

Soil and living fine root sampling and measurements

In mid June 2012, two dominant Masson pines (Kraft class 1) were selected near the centre of each plot for sampling. Humus and soil were collected from the east, south, west, and north directions 1.0 m away from the main stem of each selected tree. Samples of the humus layer and the 0–20 cm, 20–40 cm and 40–60 cm mineral soil layers were taken using a cylindrical auger with an inner diameter of 10 cm and a height of 25 cm. The four samples collected around each selected tree in each plot were pooled to produce one composite sample per layer. In all, there were 120 composite samples representing five treatments and three field replicates. All composite samples were placed in plastic bags and transported to the laboratory, where they were bulked and crumbled manually. Living roots of different diameters of Masson pine were picked out according to the characteristics of appearance, color, elasticity, and odor [35] and placed into 0.2-mm sieves using tweezers. Meanwhile, roots of other plant species and gravel were removed. The living roots of Masson pine were carefully cleaned and washed in slow running water to remove soil from roots, spread out on clean sheets of paper for 5–10 minutes in the shade, kept in sealed bags, and stored at −20°C until later measurements. After being air-dried at room temperature (25°C), all soil samples were ground, passed through a 2-mm sieve, and stored at 4°C in the dark before chemical analysis.

Soil pH was determined in a 1:2.5 slurry of soil:1 M KCl solution using a combination glass electrode, after shaking for 0.5 h and then equilibrating for 0.5 h. The accuracy was 0.01 pH. Exchangeable Ca and Al were extracted with 1 M KCl solution and determined using an atomic absorption spectrometer (AAnalyst 300, Perkin-Elmer, CT, USA). The Ca/Al molar ratios were calculated from exchangeable Ca and Al contents of all humus and soil samples. Lengths of living fine roots (diameter ≤2 mm) were measured using an automated root scanning and analysis system (WinRHIZO, Regent Instruments Inc., Canada). Length densities (m m^{-3}) of living fine roots were calculated from all samples according to the volumes of the humus and soil samples. During sampling and measurements, polypropylene gloves were used to avoid contamination. In this study, we chose to analyze the soil chemical properties and fine root growth of Masson pine in the 0–20 cm and 20–40 cm mineral soil layers where the roots were mainly distributed [33,35].

Table 1. Relationships between the length densities (y, m m^{-3}) of living fine roots and the soil chemical properties (x) in the 0–20 cm and 20–40 cm.

Soil horizon (cm)	x	Unit of x	Regression equation y = ax+b	Correlation coefficient R^2	Degree of freedom df	F-value	P-value
0–20	pH (KCl)	–	y = 1408.5x-4356.6	0.9741	4	150.44	<0.05
	Exchangeable Ca	cmol (+) kg^{-1}	y = 1562.4x-7.207	0.9212	4	46.76	<0.05
	Exchangeable Al	cmol (+) kg^{-1}	y = −485.98x+2284.2	0.9707	4	132.52	<0.05
	Ca/Al (molar ratio)	–	y = 866.51x+579.7	0.9757	4	160.61	<0.05
20–40	pH (KCl)	–	y = 528.91x-1519.6	0.9970	4	1329.33	<0.05
	Exchangeable Ca	cmol (+) kg^{-1}	y = 864.05x+122.69	0.9564	4	87.74	<0.05
	Exchangeable Al	cmol (+) kg^{-1}	y = −220.62x+1217.6	0.9342	4	56.79	<0.05
	Ca/Al (molar ratio)	–	y = 86.627x+456.97	0.9913	4	455.77	<0.05

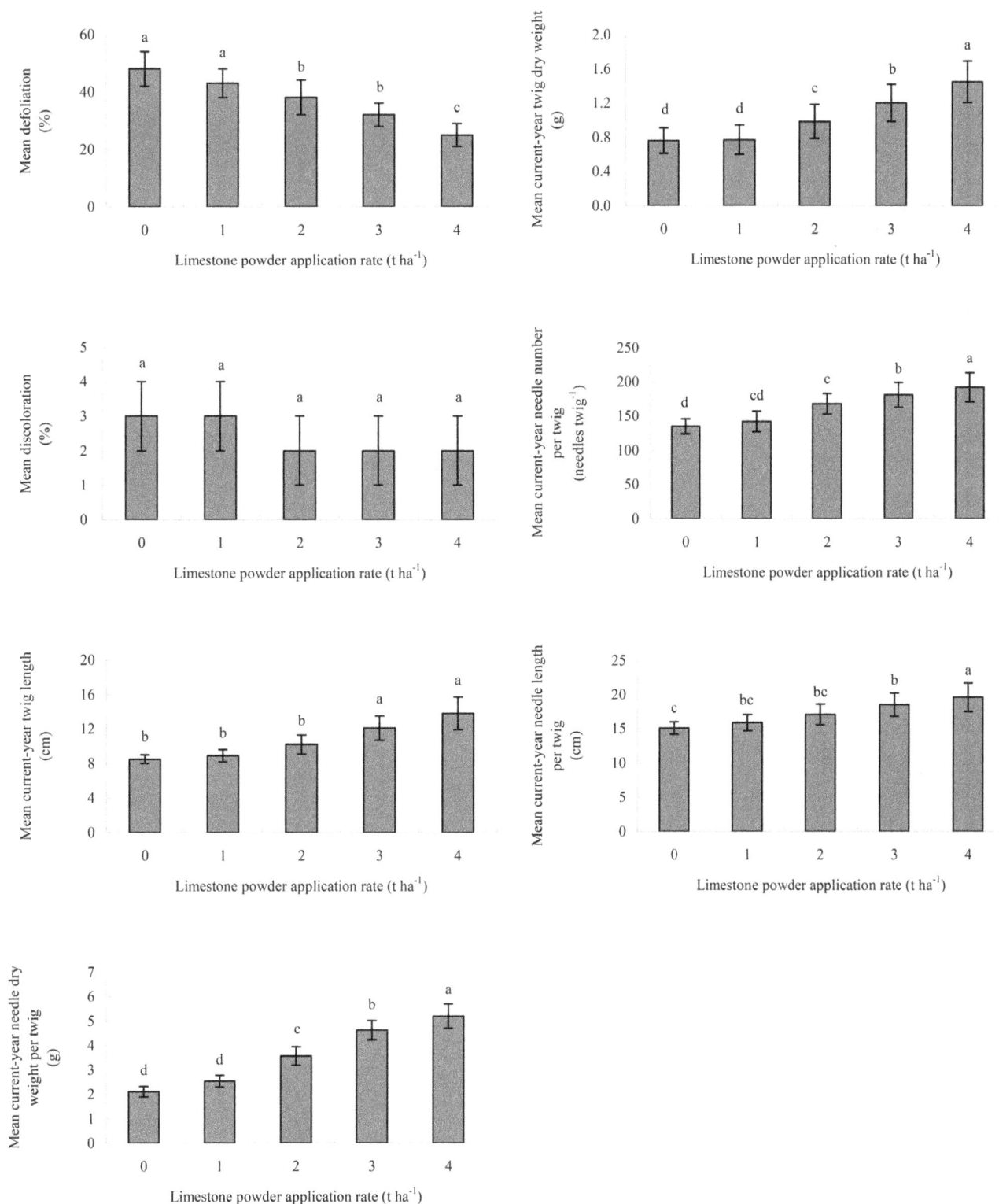

Figure 3. Crown condition of pines (Kraft classes 1–3) in the Masson pine stand 8 years after liming. Error bars represent the SD of the mean. Different letters above the error bars indicate significant differences at the 0.05 level (ANOVA and Duncan's multiple range test), n = 3.

Health and growth monitoring

In early June 2004 and in early June 2012, the assessments of crown health condition (defoliation and discoloration), height, and DBH were performed twice on all dominant, codominant and subdominant Masson pines (Kraft classes 1–3) per plot according to the ICP–Forests manual [51]. Defoliation and discoloration in the assessable crown were estimated based on 5% classes as compared with a local reference tree, which is defined as the healthiest tree that could grow at a particular site, taking into

Table 2. Relationships between the crown condition (y) and the length densities (x, m m^{-3}) of living fine roots in the 0–20 cm and 20–40 cm.

Soil horizon (cm)	y	Unit of y	Regression equation $y=ax+b$	Correlation coefficient R^2	Degree of freedom df	F-value	P-value
0–20	Mean defoliation	%	$y=-0.0172x+57.915$	0.9720	4	138.86	<0.05
	Mean current-year twig length	cm	$y=0.0043x+5.5405$	0.9903	4	408.37	<0.05
	Mean current-year twig dry weight	g	$y=0.0006x+0.354$	0.9845	4	254.06	<0.05
	Mean current-year needle number per twig	Needles twig^{-1}	$y=0.0465x+107.49$	0.9658	4	112.96	<0.05
	Mean current-year needle length per twig	cm	$y=0.0035x+12.984$	0.9892	4	366.37	<0.05
	Mean current-year needle dry weight per twig	g	$y=0.0025x+0.5518$	0.9922	4	508.82	<0.05
20–40	Mean defoliation	%	$y=-0.0351x+62.617$	0.9816	4	213.39	<0.05
	Mean current-year twig length	cm	$y=0.0087x+4.4043$	0.9890	4	359.64	<0.05
	Mean current-year twig dry weight	g	$y=0.0012x+0.2003$	0.9920	4	496.00	<0.05
	Mean current-year needle number per twig	Needles twig^{-1}	$y=0.0936x+95.809$	0.9456	4	69.53	<0.05
	Mean current-year needle length per twig	cm	$y=0.0071x+12.075$	0.9776	4	174.57	<0.05
	Mean current-year needle dry weight per twig	g	$y=0.0051x-0.068$	0.9631	4	104.40	<0.05

Table 3. Relationships between the mean height and DBH increments (y) and the length densities (x, m m^{-3}) of living fine roots in the 0–20 cm and 20–40 cm.

Soil horizon (cm)	y	Unit of y	Regression equation y=ax+b	Correlation coefficient R^2	Degree of freedom df	F-value	P-value
0–20	Mean height increment	m	y=306.23x-998.03	0.9951	4	812.33	<0.05
	Mean DBH increment	cm	y=387.93x-438.03	0.9565	4	87.95	<0.05
20–40	Mean height increment	m	y=150.05x-356.26	0.9893	4	369.83	<0.05
	Mean DBH increment	cm	y=192.68x-92.876	0.9771	4	170.67	<0.05

account factors such as altitude, latitude, tree age, site conditions and social status. The local reference tree has 0% defoliation and 0% discoloration. Pines with defoliation of 0–10%, >10–25%, >25–60%, >60–<100%, and 100% are classified as no defoliation, slight defoliation, moderate defoliation, severe defoliation, and death, respectively. Pine trees with defoliation >25% were classified as "damaged" [52]. Pine trees with discoloration of 0–10%, >10–25%, >25–60%, and >60% are categorized as no discoloration, slight discoloration, moderate discoloration, and severe discoloration, respectively. Tree height was measured with a Haga hypsometer, and DBH using a diameter tape.

At TSP, current-year needles make up the majority of tree foliage. In early June 2012, five current-year twigs with needles were sampled from the upper third of the crown on the south side of each of the two selected trees near the centre of each plot. The five current-year twigs with needles collected from each selected tree per plot were pooled to form a sample. Thirty samples were transported in clean plastic bags to the laboratory, where for each sample we separated needles from twigs, and measured twig length, twig dry weight, needle number per twig, needle length per twig, and needle dry weight per twig. Twig and needle lengths were measured to the nearest millimeter. The samples were ovendried at 60°C for about one week to a constant weight and weighed. Mean twig lengths, mean twig dry weights, mean needle numbers per twig, mean needle lengths per twig, and mean needle dry weights per twig were calculated from all samples.

Statistical analysis

The data were expressed as mean ± standard deviation (SD). They were assessed using one-way analysis of variance (ANOVA) with the SPSS 12.0 software package. The fixed model was used in ANOVA. Multiple comparisons of means among treatments were performed using Duncan test. $P<0.05$ was considered significant.

Results

Soil chemistry

Liming increased the pH, exchangeable Ca, and Ca/Al molar ratio, but decreased the exchangeable Al in the 0–20 cm and 20–40 cm mineral soil layers (Fig. 1). The pH, exchangeable Ca, and Ca/Al molar ratio in the same layers were found to be in the following order of lime treatments: 4 t ha^{-1}>3 t ha^{-1}> 2 t ha^{-1}>1 t ha^{-1}>0 t ha^{-1}, while the exchangeable Al exhibited the reverse pattern. Overall, there were no statistical differences in the soil chemical parameters in the same layers between the 0 and 1 t ha^{-1} lime treatments, whereas significant differences were observed among the 1, 2, 3, and 4 t ha^{-1} lime treatments ($P<0.05$).

Fine root growth

Liming improved the growth of fine roots of Masson pine in the 0–20 cm and 20–40 cm mineral soil layers (Fig. 2). The length densities of living fine roots in the same layers were observed to be in the following order of lime treatments: 4 t ha^{-1}>3 t ha^{-1}> 2 t ha^{-1}>1 t ha^{-1}>0 t ha^{-1}. There were no statistical differences in the length density in the same layers between the 0 and 1 t ha^{-1} lime treatments, whereas significant differences were found among the 1, 2, 3, and 4 t ha^{-1} lime treatments ($P<0.05$). The length densities of living fine roots were significantly positively correlated with the pH values, exchangeable Ca contents, and Ca/Al molar ratios, but significantly negatively with the exchangeable Al levels in both soil layers (Table 1).

Crown condition

Liming decreased the defoliation of Masson pines (Kraft classes 1–3), but increased their current-year twig and needle growth (Fig. 3). The mean defoliation varied with lime treatments, following an order of 0 t ha^{-1}>1 t ha^{-1}>2 t ha^{-1}> 3 t ha^{-1}>4 t ha^{-1}. In contrast, the mean current-year twig length, twig dry weight, needle number per twig, needle length per twig, and needle dry weight per twig showed the reverse behavior. The mean defoliation of Masson pines (Kraft classes 1–3) in the 4 t ha^{-1} lime treatment was 25%, which is the lower limit of the moderate defoliation range (>25%–60%). However, the mean discoloration of Masson pines (Kraft classes 1–3) in all treatments lay within no discoloration range (0–10%). The mean defoliation was significantly negatively correlated with the length densities of living fine roots in both soil layers, but the values of current-year twig and needle growth parameters were significantly positively correlated with them (Table 2).

Height and DBH growth

Liming increased the height and DBH of Masson pines (Kraft classes 1–3) from early June 2004 to early June 2012. Among all treatments, the 4 t ha^{-1} lime treatment was the greatest in terms of both the mean height increment and the mean DBH increment of Masson pines (Kraft classes 1–3), it was followed by, in decreasing order, the 3, 2, 1, and 0 t ha^{-1} lime treatments. The mean height increments of Masson pines (Kraft classes 1–3) were 5.5±0.4, 5.8±0.5, 6.9±0.5, 8.3±0.6, and 9.5±0.6 m in the 0, 1, 2, 3, and 4 t ha^{-1} lime treatments, and their mean DBH increments 3.1±0.2, 3.2±0.2, 3.8±0.3, 4.9±0.4, and 6.2±0.4 cm, respectively. There were no statistical differences in both parameters between the 0 and 1 t ha^{-1} lime treatments, but significant differences were detected among the 1, 2, 3, and 4 t ha^{-1} lime treatments ($P<0.05$). The mean height increment and mean DBH increment were significantly positively correlated with the length densities of living fine roots in the 0–20 cm and 20–40 cm mineral soil layers (Table 3).

Discussion

Liming in early June 2004 increased the soil pH and exchangeable Ca and decreased the exchangeable Al to a 40 cm depth on the Masson pine stand; this effect increased with increasing dose. It is worth noting that in 0–20 cm and 20–40 cm in the 4 t ha^{-1} lime treatment the soil pH values increased to 4.48 and 4.92, respectively (Fig. 1), which are very near to or in the optimum pH (KCl) range (4.50–6.00) for the growth of Masson pine [53]. As a result, amelioration of the decline of the Masson pine stand has been observed 8 years after liming; this effect also increased with increasing dose. From the viewpoint of forest health restoration, the dose of 4 t ha^{-1} limestone powder is suggested, since this dose induced the best health and growth of Masson pine. More long-term observations are needed to fully assess the effects of liming on the health and growth of the Masson pine stand.

Fine roots as uptake organs to sustain aboveground health and growth of forests are susceptible to changes in the soil environment [54]. Due to soil acidification, toxic Al is released, which adversely impacts the growth of fine roots and inhibits the uptake of water and cations [55–57]. Al stress on the growth of fine roots is governed by the Ca/Al molar ratio [58]. According to Cronan and Grigal [56], the Ca/Al molar ratio in soil solution provided an indicator for identification of thresholds beyond which the risk of fine root damage from Al stress and nutrient imbalances increased. They estimated that there was a 50% risk of Al toxicity when the Ca/Al molar ratio was as low as 1.0, a 75% risk when the ratio was as low as 0.5, and nearly a 100% risk when the ratio was as low as 0.2.

This study showed that in 0–20 cm and 20–40 cm the Ca/Al molar ratios increased with increasing dose (Fig. 1) and the length densities of fine roots of Masson pine linearly increased with the ratios (Table 1). Al stress on the growth of fine roots of Masson pine was probably excluded in both layers in the 3 and 4 t ha^{-1} lime treatments, where the Ca/Al molar ratios were >1.0, and the best growth of fine roots was found in the 4 t ha^{-1} lime treatment. However, risks seemed to exist in both layers in the 0, 1, and 2 t ha^{-1} lime treatments, where the Ca/Al molar ratios were <1.0. The stress indication was very noticeable in both layers in the 0 and 1 t ha^{-1} lime treatments, where the Ca/Al molar ratios were ≤0.22, and the worst growth of fine roots was observed in the 0 t ha^{-1} lime treatment (Figs. 1 and 2). The application of limestone powder ameliorated Al toxicity on fine roots of Masson pine; this effect increased with increasing dose.

Liming ameliorates upper soil acidity in a relatively short term, but is generally slow in reducing deeper soil acidity [29,47]. The movement of lime to greater depths depends on site, soil type, form of lime, timing and rate of liming, and weather conditions [20,59–62]. Matzner et al. [63] showed a clear increase in exchangeable Ca and Mg in the 0–50 cm mineral soil depth in an experiment in *Fagus sylvatica* and *Picea abies* stands about 10 years after fertilization of NH_4NO_3 and KCl followed by Ca and Ma fertilization by 2 liming applications. Moore et al. [64] reported that in an uneven-aged *Acer saccharum* forest 10 years after liming, exchangeable Ca and Mg contents down to 20 cm depth increased obviously with the amount of added lime (powdered and pelletized dolomitic lime) while Ca contents in 20–40 cm depth were

unaffected. Guckland et al. [47] found a significant increase in exchangeable Mg in the forest floor and exchangeable Ca in the 0–20 cm mineral soil depth in stands of *P. abies*, *F. sylvatica*, *Quercus petraea/Q. robur*, and *Pinus sylvestris* 2–9 years after one-time liming with 3–5 t ha^{-1} lime (mainly calcite and dolomite), and they suggested that the element input, CEC, C stock of the forest floor, and groundwater recharge were the most important parameters connected with the change of Ca and Mg stocks after liming in 0–10 cm, but in 10–20 and 20–40 cm, the C concentration of this depth and the year after first liming gained more influence. In our study conducted at TSP in the humid subtropical zone, in the 0–40 cm mineral soil depth of the limed Masson pine stand 8 years after one-time liming with 2–4 t ha^{-1} limestone powder, significant effects of liming were observed on the chemical properties and the growth of fine roots (Figs. 1 and 2). The fast downward movement of dissolved Ca cations into deeper mineral soil depth (the 20–40 cm) might be mainly related to the amount of finely ground limestone powder applied, rather thin forest floor layer (approximately 1-cm-thick humus layer), and relatively high annual and seasonal rainfall in Chongqing [33,34]. Undoubtedly, the promotion of the growth of fine roots of Masson pine in the 20–40 cm in the 2–4 t ha^{-1} lime treatments, especially the application of high lime doses, is beneficial to enhancing tree resistance to drought and frost.

The growth of aboveground parts (stem, branches, twigs, and leaves) of plants is coordinated with the growth of belowground part (roots) [20,65–67]. This lime-induced stimulation of the growth of fine roots of Masson pine favored Ca uptake by trees [48], resulting in the improvement of aboveground health and growth of trees (Fig. 3). The results in this study showed that the mean tree crown defoliation linearly decreased with the length densities of living fine roots in both soil layers, whereas the values of all aboveground growth parameters linearly increased with them (Tables 2 and 3). Pine trees with more fine roots grew more healthily and faster than those with less fine roots.

In summary, in the acidified and declining Masson pine stand of Chongqing, southwest China, application of limestone powder proved to be effective for ameliorating the soil acidity, improving the crown health condition, and stimulating the tree growth; these effects increased as the amount of applied limestone powder increased.

Acknowledgments

We are grateful to Forestry Administration of Chongqing Municipality and Tieshanping Forest Farm of Chongqing for their help in field works. We acknowledge two anonymous referees for their valuable comments on our manuscript. We used the Masson pine (*Pinus massoniana*) stand in this study with the permission of Forestry Administration of Chongqing Municipality and Tieshanping Forest Farm of Chongqing and the field studies did not involve endangered or protected species.

Author Contributions

Conceived and designed the experiments: ZL YW. Performed the experiments: ZL YL HG TL. Analyzed the data: ZL YW ZHL GS. Wrote the paper: ZL YW.

References

1. Zhao DW, Seip HM (1991) Assessing effects of acid deposition in southwestern China using the magic model. Water, Air and Soil Pollution 60: 83–97.

2. Hruška J, Oulehle F, Šamonil P, Šebesta J, Tahovská K, et al. (2012) Long-term forest soil acidification, nutrient leaching and vegetation development: linking modelling and surveys of a primeval spruce forest in the Ukrainian Transcarpathian Mts. Ecological Modelling 244: 28–37.

3. Liang GH, Liu XZ, Chen XM, Qiu QG, Zhang DQ, et al. (2013) Response of soil respiration to acid rain in forests of different maturity in southern China. PLoS One 8: e62207.

4. Liu HT, Zhang WP, Shen YW, Du XM, Zou XY, et al. (1988) Relationship between acid rain and the decline of a Masson pine forest in Nanshan,

Chongqing. Acta Scientiae Circumstantiae 8: 331–339. (In Chinese with English abstract).

5. Ulrich B (1990) Waldsterben: forest decline in West Germany. Environmental Science & Technology 24: 436–441.

6. Matzner E, Murach D (1995) Soil changes induced by air pollutant deposition and their implication for forests in central Europe. Water, Air and Soil Pollution 85: 63–76.

7. Harrison AF, Carreira J, Poskitt JM, Robertson SMC, Smith R, et al. (1999) Impacts of pollutant inputs on forest canopy condition in the UK: possible role of P limitations. Forestry 72: 367–377.

8. Tkacz B, Moody B, Castillo JV, Fenn ME (2008) Forest health conditions in North America. Environmental Pollution 155: 409–425.

9. Elias PE, Burger JA, Adams MB (2009) Acid deposition effects on forest composition and growth on the Monongahela National Forest, West Virginia. Forest Ecology and Management 258: 2175–2182.

10. Ito K, Uchiyama Y, Kurokami N, Sugano K, Nakanishi Y (2011) Soil acidification and decline of trees in forests within the precincts of shrines in Kyoto (Japan). Water, Air and Soil Pollution 214: 197–204.

11. Durka W, Schulze E-D, Gebauer G, Voerkeliust S (1994) Effects of forest decline on uptake and leaching of deposited nitrate determined from ^{15}N and ^{18}O measurements. Nature 372: 765–767.

12. Burke MK, Raynal DJ (1998) Liming influences growth and nutrient balances in sugar maple (Acer saccharum) seedlings on an acidic forest soil. Environmental and Experimental Botany 39: 105–116.

13. Meiwes KJ, Mindrup M, Khanna PK (2002) Retention of Ca and Mg in the forest floor of a Spruce stand after application of various liming materials. Forest Ecology and Management 159: 27–36.

14. Prietzel J, Rehfuess KE, Stetter U, Pretzsch H (2008) Changes of soil chemistry, stand nutrition, and stand growth at two Scots pine (Pinus sylvestris L.) sites in Central Europe during 40 years after fertilization, liming, and lupine introduction. European Journal of Forest Research 127: 43–61.

15. Rizvi SH, Gauquelin T, Gers C, Guérold F, Pagnout C, et al. (2012) Calcium–magnesium liming of acidified forested catchments: effects on humus morphology and functioning. Applied Soil Ecology 62: 81–87.

16. Matzner E, Murach D, Fortmann H (1986) Soil acidity and its relationship to root growth in declining forest stands in Germany. Water, Air and Soil Pollution 31: 273–282.

17. Hahn G, Marschner H (1998) Effect of acid irrigation and liming on root growth of Norway spruce. Plant and Soil 199: 11–22.

18. Bakker MR, Garbaye J, Nys C (2000) Effect of liming on the ectomycorrhizal status of oak. Forest Ecology and Management 126: 121–131.

19. Persson H, Ahlström K (1990/91) The effects of forest liming and fertilization on fine-root growth. Water, Air and Soil Pollution 54: 365–375.

20. Bakker MR, Kerisit R, Verbist R, Nys C (1999) Effects of liming on rhizosphere chemistry and growth of fine roots and of shoots of sessile oak (Quercus petraea). Plant and Soil 217: 243–255.

21. Helmisaari HS, Hallbäcken L (1999) Fine-root biomass and necromass in limed and fertilized Norway spruce (Picea abies (L.) Karst.) stands. Forest Ecology and Management 119: 99–110.

22. Huettl RF, Zoettl HW (1993) Liming as a mitigation tool in Germany's declining forests—reviewing results from former and recent trials. Forest Ecology and Management 61: 325–338.

23. Ljungström M, Nihlgård B (1995) Effects of lime and phosphate additions on nutrient status and growth of beech (Fagus sylvatica L.) seedlings. Forest Ecology and Management 74: 133–148.

24. Sikström U (1997) Effects of low-dose liming and nitrogen fertilization on stemwood growth and needle properties of Picea abies and Pinus sylvestris. Forest Ecology and Management 95: 261–274.

25. Lundström US, Bain DC, Taylor AFS, van Hees PAW (2003) Effects of acidification and its mitigation with lime and wood ash on forest soil processes: a review. Water, Air and Soil Pollution: Focus 3: 5–28.

26. Børja I, Nilsen P (2009) Long term effect of liming and fertilization on ectomycorrhizal colonization and tree growth in old Scots pine (Pinus sylvestris L.) stands. Plant and Soil 314: 109–119.

27. Tveite B, Abrahamsen G, Stuanes AO (1990/91) Liming and wet acid deposition effects on tree growth and nutrition: experimental results. Water, Air and Soil Pollution 54: 409–422.

28. Kreutzer K, Weiss T (1998) The Höglwald field experiments—aims, concept and basic data. Plant and Soil 199: 1–10.

29. Jonard M, André F, Giot P, Weissen F, Van der Perre R, et al. (2010) Thirteen-year monitoring of liming and PK fertilization effects on tree vitality in Norway spruce and European beech stands. European Journal of Forest Research 129: 1203–1211.

30. Kakei M, Clifford PE (2002) Short-term of lime application on soil properties and fine-root characteristics for a 9-year-old Sitka spruce plantation growing on a deep peat soil. Forestry 75: 37–50.

31. Kuang YW, Wen DZ, Li J, Sun FF, Hou EQ, et al. (2010) Homogeneity of δ^{15}N in needles of Masson pine (Pinus massoniana L.) was altered by air pollution. Environmental Pollution 158: 1963–1967.

32. Teng XR (2005) Status quo of forest resources and management strategy in Chongqing city. Forest Inventory and Planning 30: 73–76. (In Chinese with English abstract).

33. Li ZY, Wang YH, Yu PT, Zhang ZJ (2007) A comparative study of resistance to soil acidification and growth of fine roots between pure stands of Pinus massoniana

and Cinnamomum camphora. Acta Ecologica Sinica 27: 5245–5253. (In Chinese with English abstract).

34. Wang YH, Solberg S, Yu PT, Myking T, Vogt RD, et al. (2007) Assessments of tree crown condition of two Masson pine forests in the acid rain region in south China. Forest Ecology and Management 242: 530–540.

35. Li ZY, Wang YH (2009) Forest Health under Acidification Stress—A Case Study of Tieshanping in Chongqing. China Agriculture Press, Beijing. (In Chinese).

36. Rodhe H, Herrera R (1988) Acidification in Tropical Countries. John Wiley & Sons Ltd., New York.

37. Zhao DW, Xiong JL, Xu Y, Chan WH (1988) Acid rain in southwestern China. Atmospheric Environment 22: 349–358.

38. Jin L, Shao M, Zeng LM, Zhao DW, Tang DG (2006) Estimation of dry deposition fluxes of major inorganic species by canopy throughfall approach. Chinese Science Bulletin 51: 1818–1823.

39. Hao JM, Wang SX, Liu BJ, He KB (2000) Control strategy for sulfur dioxide and acid rain pollution in China. Journal of Environmental Sciences 12: 385–393.

40. Jiang WH, Zhang S, Chen GC, Xiong HQ, Ding Y, et al. (2002) Effect of acid deposition on soil and vegetation of forest ecosystem in Nanshan of Chongqing. Research of Environmental Sciences 15: 8–11. (In Chinese with English abstract).

41. Guo JH, Zhang XS, Vogt RD, Xiao JS, Zhao DW, et al. (2007) Evaluating main factors controlling aluminum solubility in acid forest soils, southern and southwestern China. Applied Geochemistry 22: 388–396.

42. Larssen T, Carmichael GR (2000) Acid rain and acidification in China: the importance of base cation deposition. Environmental Pollution 110: 89–102.

43. Chinese Society of Forestry (1989) Acid Rain and Agriculture. China Forestry Press, Beijing. (In Chinese).

44. Du XM, Tian RS (1996) The relation between aluminium poisoning and decline of Masson pine forest in Nanshan Mountain, Chongqing. Research of Environmental Sciences 9: 21–25. (In Chinese with English abstract).

45. Bäckman JSK, Hermansson A, Tebbe CC, Lindgren PE (2003) Liming induces growth of a diverse flora of ammonia-oxidising bacteria in acid spruce forest soil as determined by SSCP and DGGE. Soil Biology & Biochemistry 35: 1337–1347.

46. Rosenberg W, Nierop KGJ, Knicker H, de Jager PA, Kreutzer K, et al. (2003) Liming effects on the chemical composition of the organic surface layer of a mature Norway spruce stand (Picea abies [L.] Karst.). Soil Biology & Biochemistry 35: 155–165.

47. Guckland A, Ahrends B, Paar U, Dammann I, Evers J, et al. (2012) Predicting depth translocation of base cations after forest liming: results from long-term experiments. European Journal of Forest Research 131: 1869–1887.

48. Huang YM, Duan L, Jin T, Yang YS, Hao JM (2006). Effect of limestone and magnesite applications on Masson pine (Pinus massoniana) forest growing on acidified soil. Acta Ecologica Sinica 26: 786–792. (In Chinese with English abstract).

49. Zhao DW, Larssen T, Zhang DB, Gao SD, Vogt RD, et al. (2001) Acid deposition and acidification of soil and water in the Tie Shan Ping Area, Chongqing, China. Water, Air and Soil Pollution 130: 1733–1738.

50. Vogt RD, Guo JH, Luo JH, Peng XY, Xiang RJ, et al. (2007) Water chemistry in forested acid sensitive sites in sub-tropical Asia receiving acid rain and alkaline dust. Applied Geochemistry 22: 1140–1148.

51. ICP Forests website. Available: http://www.icp-forests.org/Manual.htm. Accessed 2013 July 31.

52. Fischer R, Mues V, Ulrich E, Becher G, Lorenz M (2007) Monitoring of atmospheric deposition in European forests and an overview on its implication on forest condition. Applied Geochemistry 22: 1129–1139.

53. Wu ZY (1980) Plants of China. Science Press, Beijing. (In Chinese).

54. Seftigen K, Moldan F, Linderholm HW (2013) Radial growth of Norway spruce and Scots pine: effects of nitrogen deposition experiments. European Journal of Forest Research 132: 83–92.

55. Hutchinson TC, Bozic L, Munoz-Vega G (1986) Response of five species of conifer seedlings to aluminium stress. Water, Air and Soil Pollution 31: 283–294.

56. Cronan CS, Grigal DF (1995) Use of calcium/aluminum ratios as indicators of stress in forest ecosystems. Journal of Environmental Quality 24: 209–226.

57. Staszewski T, Kubiesa P, Łukasik W (2012) Response of spruce stands in national parks of southern Poland to air pollution in 1998–2005. European Journal of Forest Research 131: 1163–1173.

58. Gao JX, Cao HF (1991) Effects of ionic strength, pH and Ca/Al ratio on aluminum toxicity of Masson pine seedlings. Acta Scientiae Circumstantiae 11: 194–198. (In Chinese with English abstract).

59. Ponette Q, Dufey JE, Weissen F (1997) Downward movement of dolomite, kieserite or a mixture of $CaCO_3$ and kieserite through the upper layers of an acid forest soil. Water, Air and Soil Pollution 95: 353–379.

60. Frank J, Stuanes AO (2003) Short-term effects of liming and vitality fertilization on forest soil and nutrient leaching in a Scots pine ecosystem in Norway. Forest Ecology and Management 176: 371–386.

61. Caires EF, Garbuio FJ, Churka S, Barth G, Corrêa JCL (2008) Effects of soil acidity amelioration by surface liming on no-till corn, soybean, and wheat root growth and yield. European Journal of Agronomy 28: 57–64.

62. Löfgren S, Cory N, Zetterberg T, Larsson PE, Kronnäs V (2009) The long-term effects of catchment liming and reduced sulphur deposition on forest soils and

runoff chemistry in southwest Sweden. Forest Ecology and Management 258: 567–578.

63. Matzner E, Khanna PK, Meiwes KJ, Ulrich B (1985) Effects of fertilization and liming on the chemical soil conditions and element distribution in forest soils. Plant and Soil 87: 405–415.

64. Moore JD, Duchesne L, Ouimet R (2008) Soil properties and maple–beech regeneration a decade after liming in a northern hardwood stand. Forest Ecology and Management 255: 3460–3468.

65. Misra RK, Turnbull CRA, Cromer RN, Gibbons AK, LaSala AV (1998) Below- and above-ground growth of *Eucalyptus nitens* in a young plantation I. Biomass. Forest Ecology and Management 106: 283–293.

66. Štofko P (2010) Relationships between the parameters of aboveground parts and the parameters of root plates in Norway spruce with respect to soil drainage. Journal of Forest Science 56: 353–360.

67. Dhief A, Abdellaoui R, Tarhouni M, Belgacem AO, Smiti SA, et al. (2011) Root and aboveground growth of rhizotron-grown seedlings of three Tunisian desert *Calligonum* species under water deficit. Canadian Journal of Soil Science 91: 15–27.

19

Gaseous Elemental Mercury (GEM) Emissions from Snow Surfaces in Northern New York

J. Alexander Maxwell[1], Thomas M. Holsen[2]*, Sumona Mondal[3]

1 Institute for a Sustainable Environment, Clarkson University, Potsdam, New York, United States of America, **2** Department of Civil and Environmental Engineering, Clarkson University, Potsdam, New York, United States of America, **3** Department of Mathematics, Clarkson University, Potsdam, New York, United States of America

Abstract

Snow surface-to-air exchange of gaseous elemental mercury (GEM) was measured using a modified Teflon fluorinated ethylene propylene (FEP) dynamic flux chamber (DFC) in a remote, open site in Potsdam, New York. Sampling was conducted during the winter months of 2011. The inlet and outlet of the DFC were coupled with a Tekran Model 2537A mercury (Hg) vapor analyzer using a Tekran Model 1110 two port synchronized sampler. The surface GEM flux ranged from -4.47 ng m^{-2} hr^{-1} to 9.89 ng m^{-2} hr^{-1}. For most sample periods, daytime GEM flux was strongly correlated with solar radiation. The average nighttime GEM flux was slightly negative and was not well correlated with any of the measured meteorological variables. Preliminary, empirical models were developed to estimate GEM emissions from snow surfaces in northern New York. These models suggest that most, if not all, of the Hg deposited with and to snow is reemitted to the atmosphere.

Editor: Stephen J. Johnson, University of Kansas, United States of America

Funding: New York State Energy Research and Development Authority (NYSERDA http://www.nyserda.ny.gov/) financially supported this research (Charles Driscoll, Syracuse University PI). It has not been subject to the NYSERDA's peer and policy review and, therefore, does not necessarily reflect the views of NYSERDA and no official endorsement should be inferred. The funders had no role in study design, data collection and analysis, decision to publish, or preparation of the manuscript.

Competing Interests: The authors have declared that no competing interests exist.

* E-mail: holsen@clarkson.edu

Introduction

Hg is a potent neurotoxin and regulated by the U.S. EPA [1], European Union Restriction of Hazardous Substances Directive (RoHS) [2], and other government agencies worldwide as a hazardous pollutant. In the form of monomethylmercury (MeHg) it can adversely impact the development and health of both humans and wildlife [3]. Gaseous elemental mercury (GEM) is emitted into the atmosphere from both natural and anthropogenic sources, and has an atmospheric residence time of 0.5–2 years, allowing it to be transported over great distances [4–6]. Anthropogenic sources can also emit Hg in the form of gaseous oxidized Hg (GOM) and particulate bound Hg (PBM), which have shorter atmospheric lifetimes on the order of days to weeks [4]. GOM is fairly soluble in water, thus allowing it to be readily deposited to terrestrial surfaces through wet deposition, including snow [4–6]. The Hg deposited with snow is then either quickly revolatilized back into the atmosphere or incorporated into the snowpack. Newly deposited Hg has been shown to preferentially revolatilize, depending on the deposition surface, in a process known as prompt recycling [7].

While the role of snow surfaces in Hg cycling has been widely studied in arctic regions [5,8–11], much less is known about its importance in more temperate climates [12–14]. Hg is deposited to snowpacks through both wet (snow) and dry deposition. Once deposited on the snowpack surface, it has been shown that >50% of the Hg deposited is reemitted within the first 24 hours [8,12]. This process is believed to be governed by photoinduced reduction of GOM to GEM. Hg in the snowpack is mainly found in the form

of GOM dissolved in snow grains, while <1% remains trapped in the interstitial air as GEM [8]. Hg concentrations are known to decrease with depth [12] with the higher concentrations up to 1.5 ng m^{-3} (GEM) remaining on the surface [8].

In the arctic, the snow surface-to-air flux of Hg is mainly the result of a diurnal pattern of GEM production in the interstitial air near the surface of the snowpack during the daytime ($\sim$15–50 ng m^{-2} hr^{-1}), with little contribution from deeper snow layers [15,16]. However, internal production of GEM increases slightly with higher temperatures and snowmelt [8]. Since this process has not been well studied in temperate climates, measurements of snow surface-to-air fluxes were made over the 2011 winter season in Potsdam, NY.

Materials and Methods

Site Description, Methods, and Materials

Flux measurements were conducted at an open field site located at the Potsdam Municipal Airport (Damon Field) in Potsdam, NY (44°40.41N, −74°57.06′W) near the Clarkson University Observatory. This site remains largely undisturbed throughout the year and has served as a background site for the New York State particulate matter (PM) monitoring network. Sampling periods were determined based on access to the site and snow conditions. Special considerations were made to ensure that the chamber was never buried in snow and that all inlets and outlets remained above the snow during sampling. Measurements were conducted on a concrete slab, isolating the snowpack from the soil surface.

Concentrations of GEM were measured using a DFC with a method previously described in Choi & Holsen (2009). Briefly, the ambient sampling line (inlet) and chamber sampling line (outlet) of the DFC (described below) were coupled with a Tekran Model 2537A Hg vapor analyzer operated at room temperature in a field shed (Tekran Corporation, Inc., Toronto, Ontario, Canada) using a Tekran Model 1110 two-port synchronized sampler. The Tekran 1110 unit allowed for alternating five minute sampling pairs to be made between the inlet and outlet sample lines every 20 minutes (trap A inlet, trap B inlet; trap A outlet, trap B outlet). During inlet sampling, outlet air is bypassed at the same 1 L min^{-1} flow rate as the Tekran Model 2537A to maintain a constant turnover time (TOT) of 0.78 minutes and an optimized flushing flow rate (FFR) of 5 L min^{-1}[17] through the flux chamber. The inlet and outlet openings were placed next to each other at the same height, roughly 2 cm above the snow surface. Four, 1 cm diameter holes were evenly distributed around the perimeter of the chamber wall to insure the chamber was well-mixed. Although a standard method for the use of DFCs does not exist and this method has not been used in other snow studies, this sampling approach is similar to methods used in past studies over soil surfaces [17–19]. The 5 L min^{-1} FFR and 0.78 minute TOT are also similar to those used in a study by Eckley, et al. (2010).

Modified Teflon fluorinated ethylene propylene (FEP) chambers were used in the study. The modified Teflon chamber was constructed using a polycarbonate (PC) chamber frame and thin, 25 μm thick Teflon FEP film (CS Hyde Company, Lake Villa, IL) to cover the top and side windows (Figure 1). In previous studies [20,21], Teflon film was shown to allow better UV permeability, up to 85±11% of light for wavelengths between 260 and 970 nm.

Figure 1. Modified Teflon Fluorinated Ethylene Propylene (FEP) Chamber with Polycarbonate (PC) Frame and 25 μm Teflon FEP Film Top and Side Windows.

Each DFC had a chamber volume of 3.9 L with a 18 cm diameter opening covering an area of approximately 254 cm^2 of the snow surface.

Manual spike Hg recovery tests were conducted at the start of each sampling period by injecting 20 μL of Hg at roughly 13.23 pg μL^{-1} (20°C) or approximately 0.26 ng into an operating chamber using a calibrated (ANSI/NCSL Z540-1-1994) Hamilton Digital Syringe (Hamilton Company, Reno, NV) and a Tekran Model 2505 Hg vapor calibration unit (Tekran Corporation, Inc., Toronto, Ontario, Canada). The recorded Hg concentrations after each manual spike test were roughly 9 ng m^{-3}, on the same order as the average daytime Hg concentrations around 2 ng m^{-3}. The recovery was 97.5±3.8%. Flow rates were calibrated using a Bios Definer 220 volumetric flow meter (Bios International Corporation, Butler, NJ) at the beginning of each sampling period.

Prior to all field measurements, the Tekran Model 2537A was calibrated with an internal permeation source to ensure acceptable response factors (>6,000,000) and that the concentration difference between the inlet and outlet samples was less than 5%. In addition, all soda-lime traps and 0.2 μm polytetrafluoroethylene (PTFE) membrane filters were replaced at the start of each sampling period.

Meteorological data was collected using a weather station (Vantage Pro 2 Weather Station, Davis Instruments, Hayward, CA) located 1–2 m away from the chamber. The weather station measured ambient air temperature (°C), relative humidity (%), and solar radiation (W m^{-2}) at a 10 minute time resolution.

Sampling Analysis and Calculations

The GEM flux from the snow under the chamber was calculated using the following mass balance equation:

$$F = (C_{outlet} - C_{inlet}) \times (Q/A) \qquad (1)$$

Where F is flux (ng GEM m^{-2} h^{-1}); C_{outlet} and C_{inlet} are the concentrations of GEM (ng GEM m^{-3}) at the outlet and inlet, respectively; Q is the FFR (m^3 h^{-1}) through the chamber; and A is the surface area (m^2) of the snow exposed in the chamber. When fluxes were negative (-), Hg was being deposited on the snow surface, and when fluxes were positive (+) Hg was being emitted from the snow surface. All flux data was then smoothed using a Savitzky-Golay smoothing filter [22], (Eqn 2), to account for random error/noise while also preserving the quantitative information and trends.

$$F_4^* = [(-2 \times F_1) + (3 \times F_2) + (6 \times F_3) + (7 \times F_4) + (6 \times F_5) + (3 \times F_6) + (-2 \times F_7)]/21 \qquad (2)$$

Where F_4^* is the smoothed flux (ng GEM m^{-2} h^{-1}), F_{1-7} are the range of measured abscissa flux values (ng GEM m^{-2} h^{-1}), and 21 is the normalizing factor.

Histograms of the GEM flux and the three meteorological predictor variables (temperature, solar radiation, and relative humidity) showed that none of the variables were normally distributed. Figure 2 provides histograms and residual plots of daytime GEM flux when compared to solar radiation. Similar plots were constructed for each individual variable, temperature, solar radiation, and relative humidity. Shapiro-Wilk normality tests [23] were then employed to confirm that the data deviated from normality. Non-parametric Pearson product-moment tests

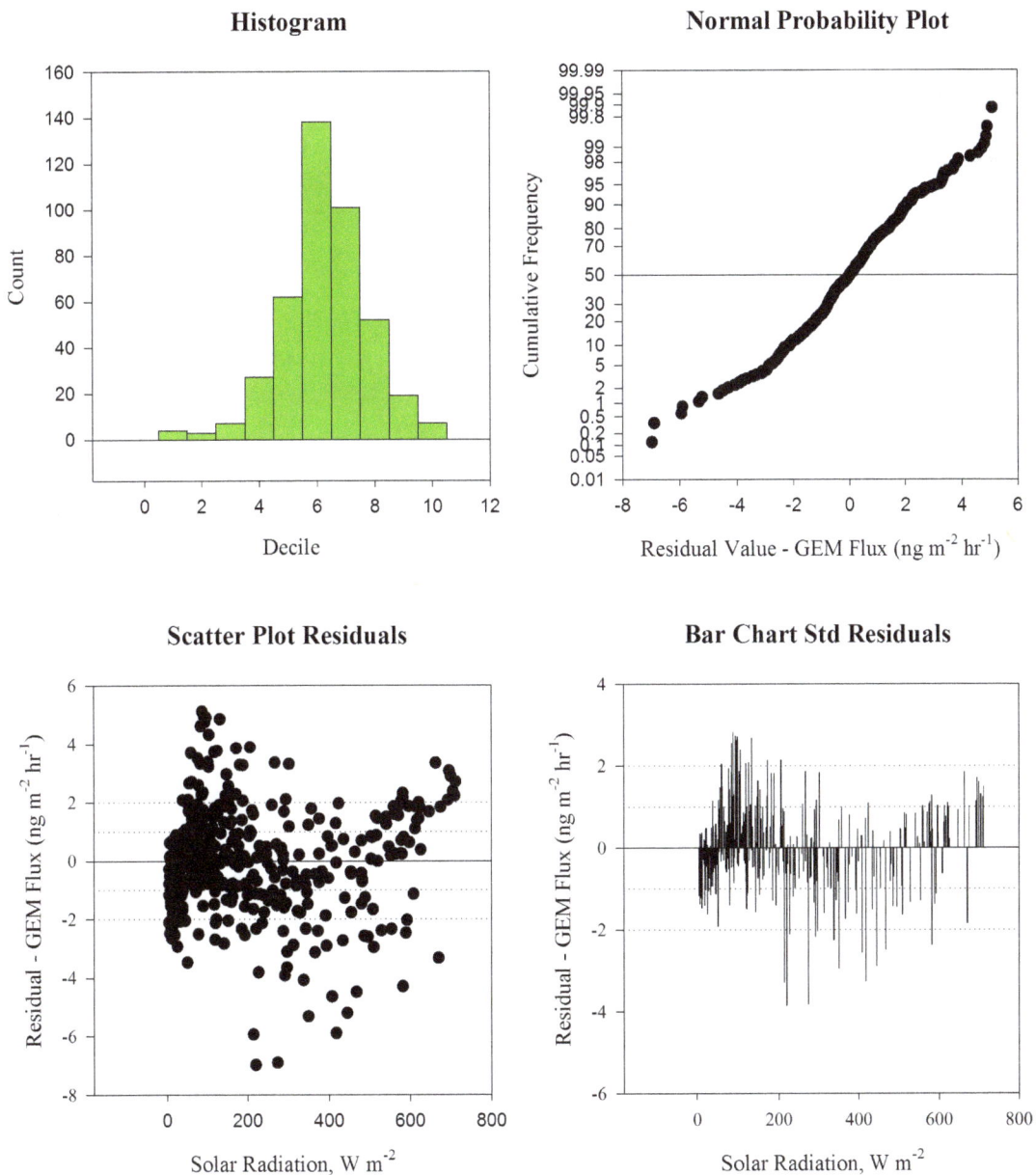

Figure 2. Histograms and Residual Plots of Daytime GEM Flux and Solar Radiation During Winter 2011.

Table 1. Measured Daytime And Nighttime GEM Flux Over 5 Winter 2011 Sampling Periods.

| Date | Diurnal Period | GEM Flux (ng m^{-2} hr^{-1}) | | | | | |
		Mean	Std. Dev.	Range	Max	Min	Median
21–24 Jan	Daytime	1.13	1.37	5.60	4.44	−1.16	0.94
	Nighttime	−0.44	0.29	1.47	0.14	−1.33	−0.43
26–31 Jan	Daytime	2.65	1.92	8.20	6.57	−1.63	2.69
	Nighttime	−0.21	0.42	1.95	0.69	−1.26	−0.22
15–18 Feb	Daytime	1.88	2.96	12.56	8.09	−4.47	1.30
	Nighttime	−0.57	0.45	2.10	0.39	−1.71	−0.50
23–24 Feb	Daytime	3.03	2.60	9.38	8.13	−1.25	2.46
	Nighttime	−0.29	0.23	1.00	0.25	−0.75	−0.30
08–09 Mar	Daytime	3.50	3.08	9.97	9.89	−0.08	2.29
	Nighttime	−0.08	0.22	0.87	0.27	−0.60	−0.10
Overall	Daytime	2.37	2.48	14.36	9.89	−4.47	1.89
	Nighttime	−0.35	0.41	2.41	0.69	−1.71	−0.33

Table 2. Pearson Product-Moment Correlation Coefficients and P-Values For Correlations Between GEM Flux and Temperature, Relative Humidity, and Solar Radiation.

Date	Diurnal Period	Temperature (°C)		Relative Humidity (%)		Solar Radiation (W m^{-2})	
		Coefficient	P Value	Coefficient	P Value	Coefficient	P Value
21–24 Jan	Daytime	0.562	0.000	−0.494	0.000	0.546	0.000
	Nighttime	−0.004	0.959	−0.192	0.026	−	−
26–31 Jan	Daytime	0.189	0.022	−0.046	0.585	0.304	0.000
	Nighttime	−0.183	0.009	−0.319	0.000	−	−
15–18 Feb	Daytime	−0.518	0.000	0.673	0.000	0.820	0.000
	Nighttime	−0.553	0.000	−0.600	0.000	−	−
23–24 Feb	Daytime	0.300	0.027	−0.629	0.000	0.875	0.000
	Nighttime	0.000	0.997	−0.053	0.745	−	−
08–09 Mar	Daytime	0.446	0.001	−0.787	0.000	0.942	0.000
	Nighttime	−0.518	0.000	0.251	0.129	−	−
Overall	Daytime	0.103	0.035	−0.385	0.000	0.684	0.000
	Nighttime	−0.222	0.000	−0.132	0.002	−	−

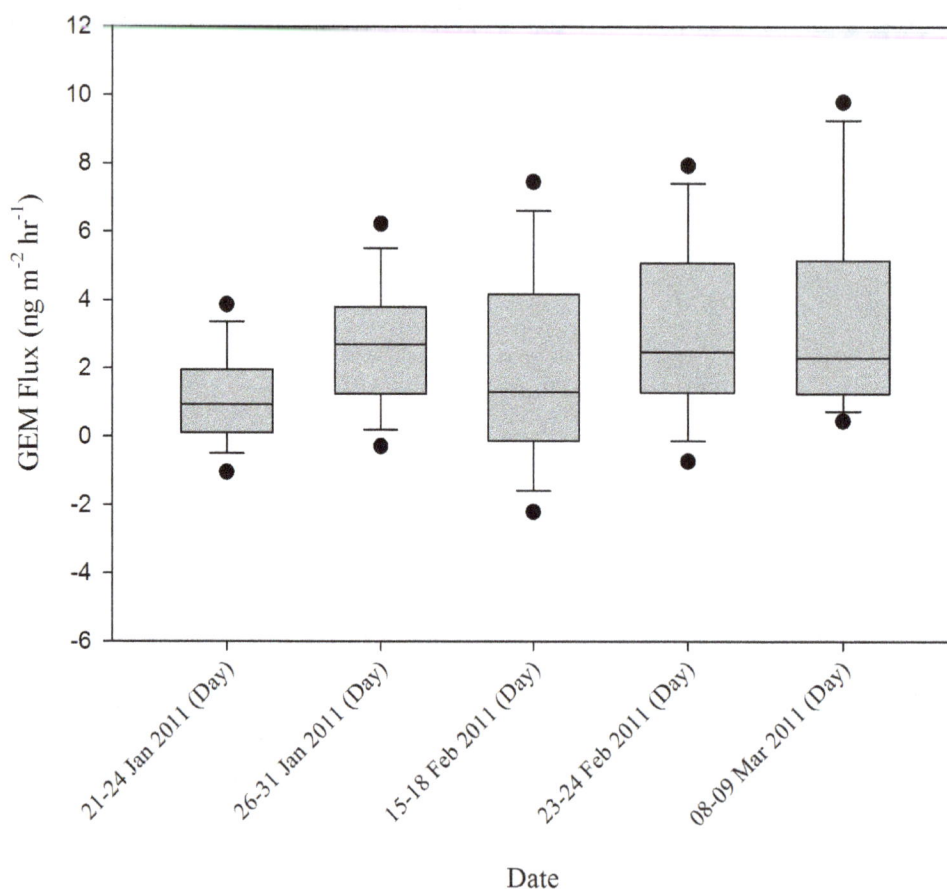

Figure 3. Average Daytime GEM Flux Measurements Made For Each Sampling Conducted During Winter 2011.

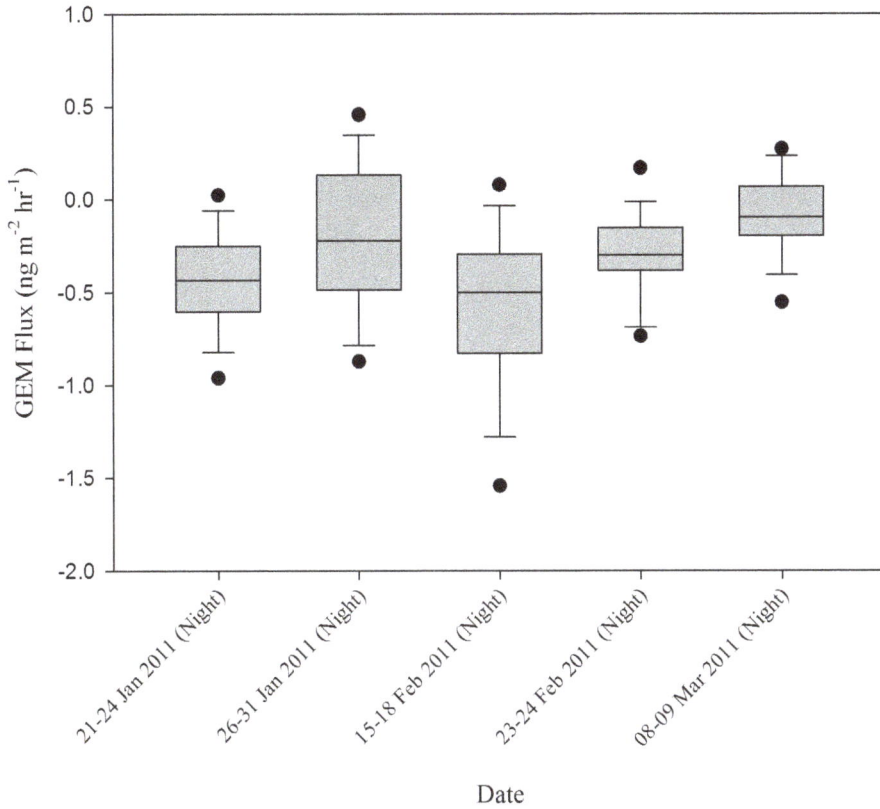

Figure 4. Average Nighttime GEM Flux Measurements Made For Each Sampling Conducted During Winter 2011.

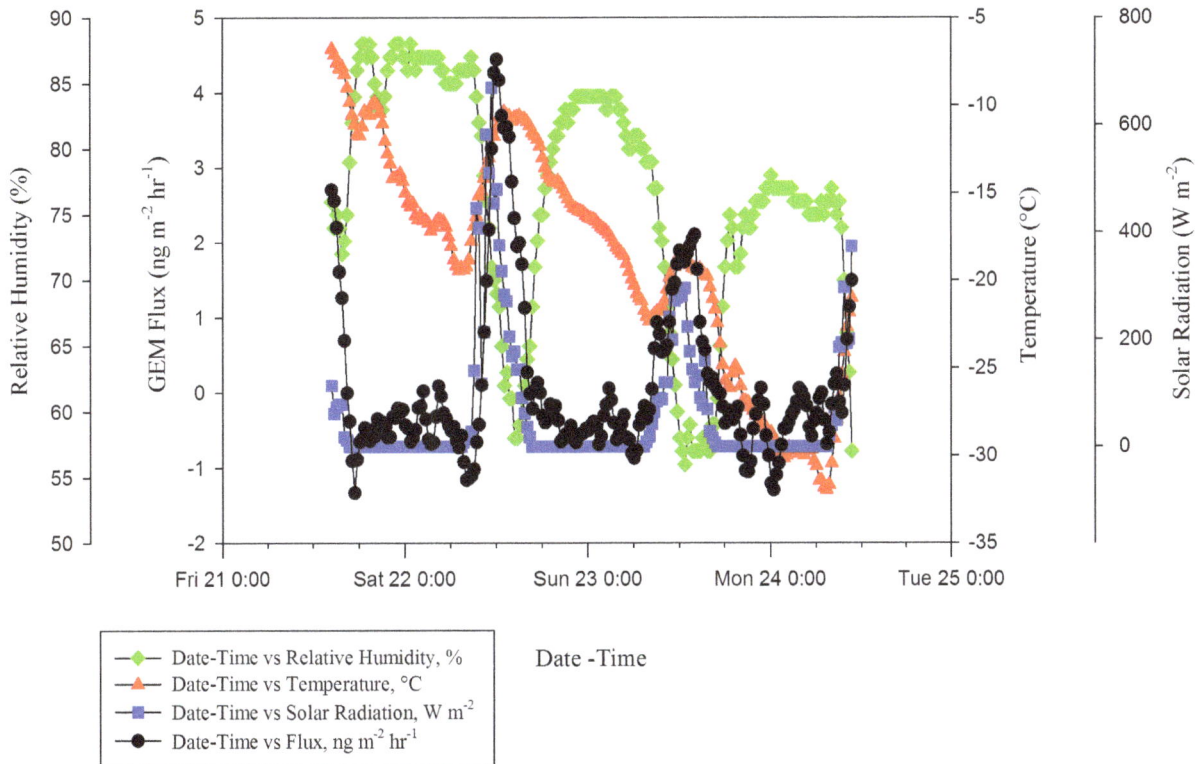

Figure 5. Diurnal Pattern Of GEM Flux For 21–24 January 2011 Sampling With Temperature, Relative Humidity, And Solar Radiation.

22-23 February 2011 (Covered)

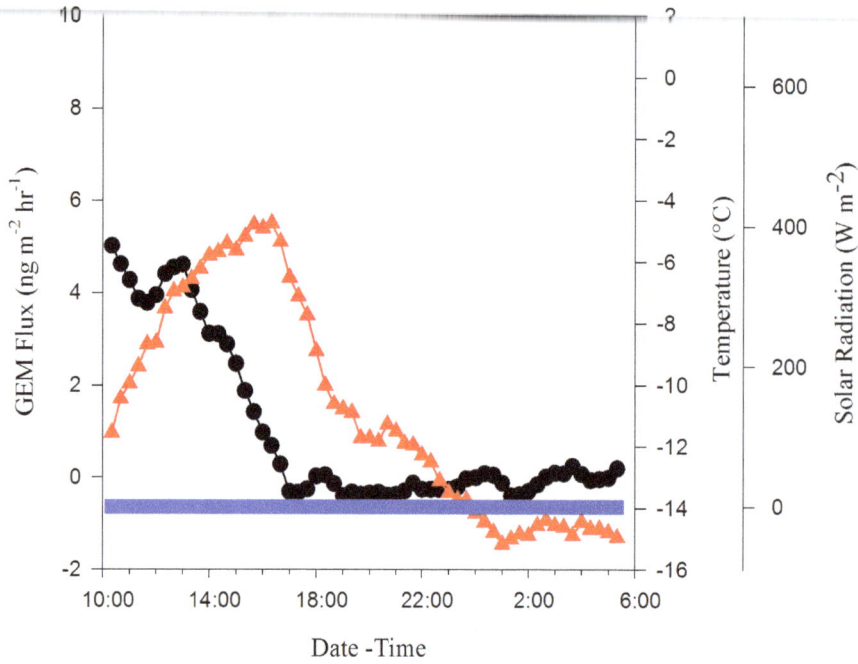

23-24 February 2011 (Uncovered)

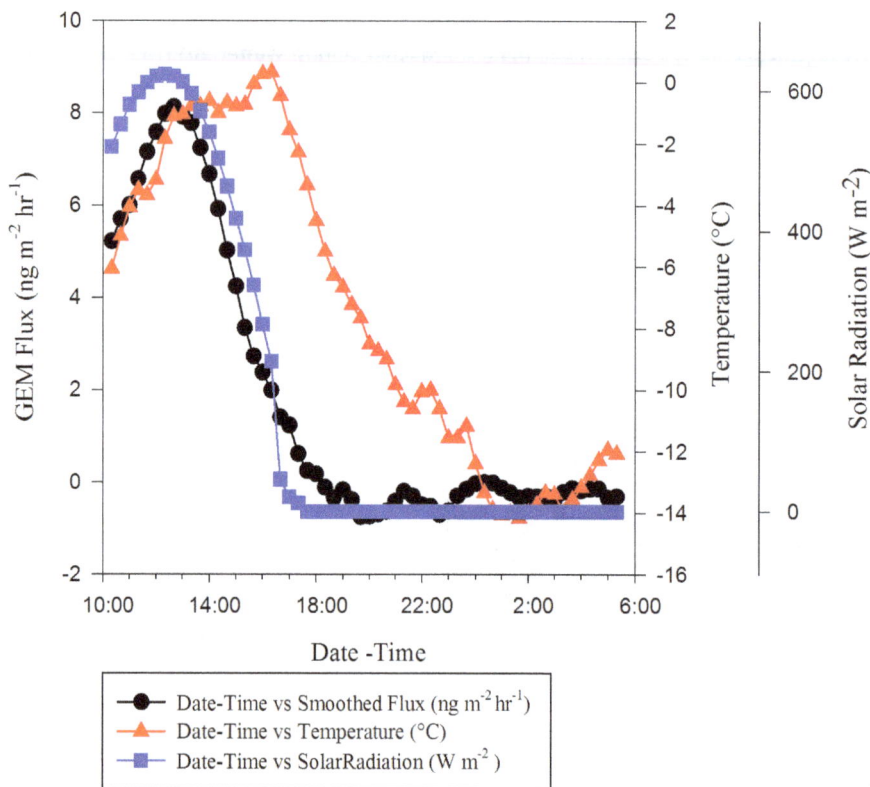

- ●— Date-Time vs Smoothed Flux (ng m^{-2} hr^{-1})
- ▲— Date-Time vs Temperature (°C)
- ■— Date-Time vs SolarRadiation (W m^{-2})

Figure 6. GEM Fluxes Measured Using Covered And Uncovered Chambers To Determine The Impact Of Solar Radiation On GEM Flux.

Table 3. PPMCs For Covered And Uncovered Chamber Tests For Impact Of Solar Radiation On GEM Flux.

Date	Diurnal Period	Temperature (°C)		Relative Humidity (%)		Solar Radiation (W m^{-2})	
		Coefficient	P Value	Coefficient	P Value	Coefficient	P Value
22–23 Feb (Covered)	Daytime	−0.624	0.002	0.489	0.021	–	–
	Nighttime	−0.374	0.025	0.141	0.412	–	–
23–24 Feb (Uncovered)	Daytime	0.300	0.027	−0.629	0.000	0.875	0.000
	Nighttime	0.000	0.997	−0.053	0.745	–	–

were used to determine the correlation coefficients (PPMC) between the variables [24].

Results and Discussion

Flux Measurements

During the 2011 winter sampling season, the flux was measured over five sampling periods, each lasting from one to six days. The measured flux ranged from a minimum −4.47 to a maximum 9.89 ng GEM m^{-2} h^{-1} (Table 1). The average daytime flux was 2.37±2.48 ng GEM m^{-2} h^{-1} and the average nighttime flux was −0.35±0.41 ng GEM m^{-2} h^{-1}. Measured nighttime Hg emission fluxes from other snowpack studies have been ≈0 ng GEM m^{-2} h^{-1} [8], while daytime fluxes have been shown to be much higher, ≈30–50 ng GEM m^{-2} h^{-1} [8]. Daytime fluxes were strongly correlated with solar radiation (PPMC value = 0.684, p-value = 0.000<0.050) and to a lesser extent temperature and relative humidity (PPMC value = 0.103, p-value = 0.035<0.050 and PPMC value = −0.385, p-value = 0.000<0.050 respectively) (Table 2). This strong correlation with solar radiation suggests that the daytime Hg emissions from the snow surface are a result of the photoreduction of GOM associated with the snow to GEM. Similar results have been reported in Ferrari, et al. (2005), where it was also reported that GEM emissions from the snowpack were negligible in comparison to emissions caused by solar irradiation at the surface. Nighttime fluxes were only weakly correlated with both temperature (PPMC value = −0.222, p-value = 0.000<0.050) and relative humidity (PPMC value = −0.132, p-value = 0.002<0.000) (Table 2) and showed a statistically significant difference from zero.

Overall, peak fluxes tended to increase later in the sampling season (Figure 3 and Figure 4). Emissions were highest during the last sampling period, 08–09 March, corresponding with the highest solar radiation peak (Max: 712 W m^{-2}). Fluxes also tended to follow a diurnal pattern (Figure 5) with peaks occurring during the day following increased exposure to solar radiation, and deposition occurring at night, similar to patterns reported in other literature [15,16].

Impact of Solar Radiation

To test the impact of solar radiation on GEM fluxes, the chamber was covered with aluminum foil to simulate zero UV conditions. The uncovered measurements were made on 23–24 February 2011, while the covered measurements were made on 22–23 February 2011 (Figure 6). During the uncovered and covered tests, the average GEM fluxes were 1.76±3.06 and 0.99±1.81 ng GEM m^{-2} h^{-1} respectively. The covered DFC daytime measurements were negatively correlated with temperature (PPMC coefficient = −0.624, p-value = 0.000<0.050) (Table 3). The slow decline in GEM flux after covering the

chamber is likely a result of diffusion of GEM from the interstitial air in the snowpack into the DFC. The uncovered DFC daytime measurements were positively correlated with solar radiation, and to a lesser degree, temperature (PPMC coefficient = 0.875, p-value = 0.000<0.050 and PPMC coefficient = 0.300, p-value = 0.027<0.050) (Table 3), similar to what has been reported in other arctic studies [8]. Overall, solar radiation had the highest positive impact on GEM emissions, and though temperature and relative humidity were correlated to GEM flux, their correlation with solar radiation (PPMC coefficient = 0.711 & −0.686 respectively, p-value = 0.000<0.050) indicate that their influence was likely a result of their codependence on solar radiation.

Modeling

In the past, empirical models have been developed using meteorological data in order to estimate surface GEM flux from soils in temperate regions of eastern North America [17,21,25]. However, no model exists to estimate GEM flux from snow in the temperate climate of northern New York. Previous models for this region [17] excluded winter fluxes from snow surfaces. In order to better model GEM flux throughout the winter season, two multiple linear regression models were developed based on aggregated seasonal flux data:

Winter 2011 (Daytime): ($R^2 = 0.481$)

$$F = 0.722 + 0.0358(T) + 0.00906(SR)$$

Winter 2011 (Nighttime): ($R^2 = 0.0616$)

$$F = -0.167 - 0.00939(T) - 0.00344(RH)$$

where F is GEM flux in ng m^{-2}hr^{-1}, T is ambient temperature in °C, RH is relative humidity in %, and SR is solar radiation in W m^{-2}. Fluxes predicted by this model for the 22–24 January sampling period are shown in Figure 7.

Several nonlinear polynomial and power equation fits and variable transformations were conducted using SigmaPlot, ver. 12 in order to develop a more precise correlative model structure. However, the dynamic fits showed little improvement.

Using the multiple linear regression models in conjunction with 5-year winter (December-March, 2005–2010) EPA Clean Air Status and Trends Network (CASTNET) meteorological data from the National Atmospheric Deposition Program (NADP) site, NY20, it is estimated that the average snow surface emissions from the open Huntington Wildlife Forest (HWF) site range from −0.10±0.07 ng m^{-2} hr^{-1} (nighttime) to 1.53±1.69 ng m^{-2} hr^{-1} (daytime) or ~17.22 ng m^{-2} year^{-1}. During the same time period Mercury Deposition Network (MDN) data from the same site yield

21-24 Jan 2011 Daytime GEM Flux Model Comparison

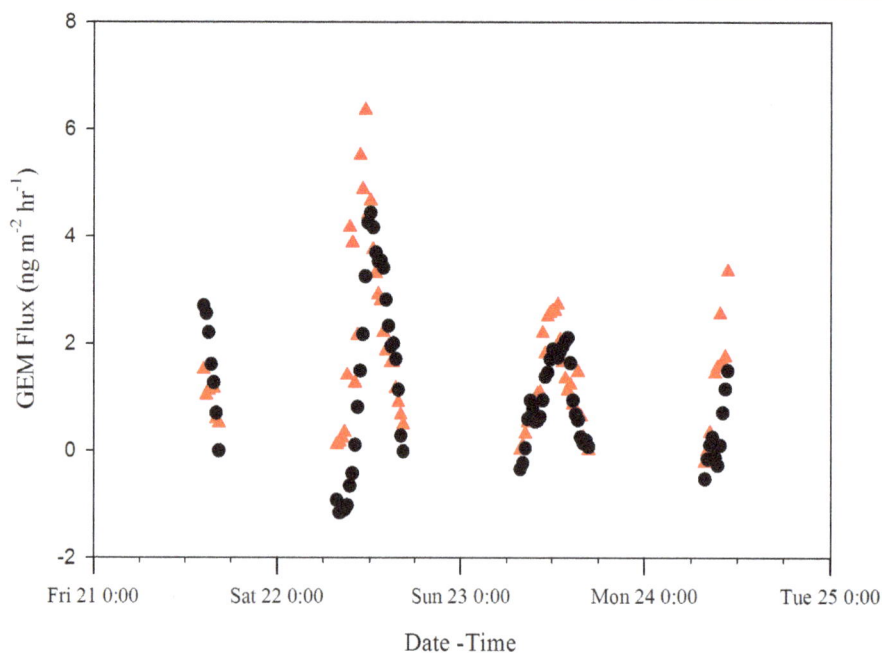

21-24 Jan 2011 Nighttime GEM Flux Model Comparison

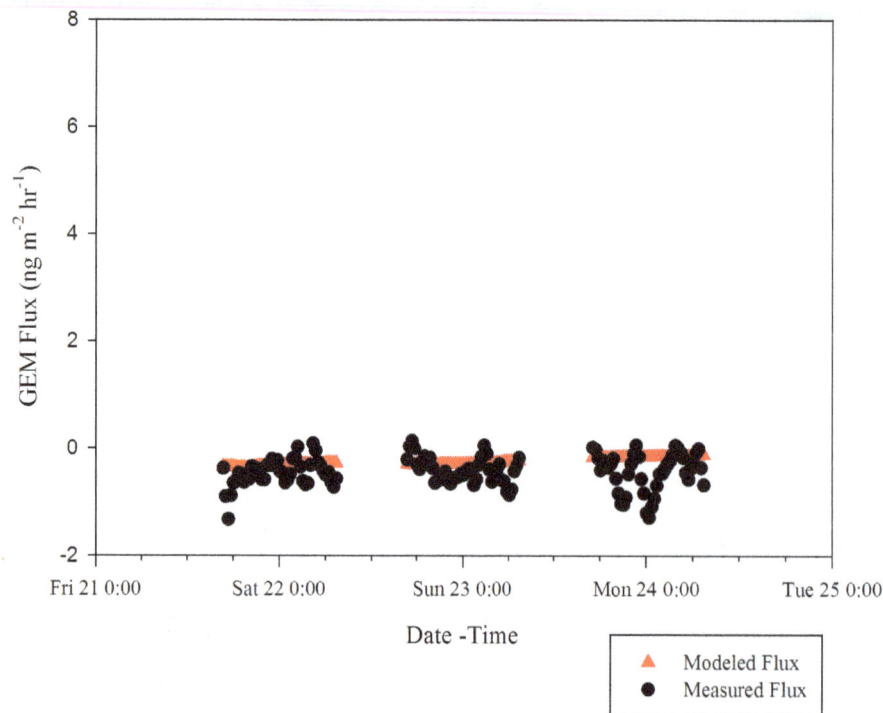

Figure 7. Daytime And Nighttime GEM Flux Model Comparison For 21–24 January Sampling Period.

a similar value with an average deposition flux of 0.48±0.41 ng m^{-2} hr^{-1} or 11.52±9.84 ng m^{-2} $year^{-1}$. The reason for the slightly higher modeled flux compared to the measured flux is likely due to the fact that some of the measurements used to make the empirical model were made after fresh snowfall when GEM fluxes would be at their maximum values.

Overall, these models suggest that most if not all the Hg deposited to snow surfaces is promptly recycled. Similar reemission

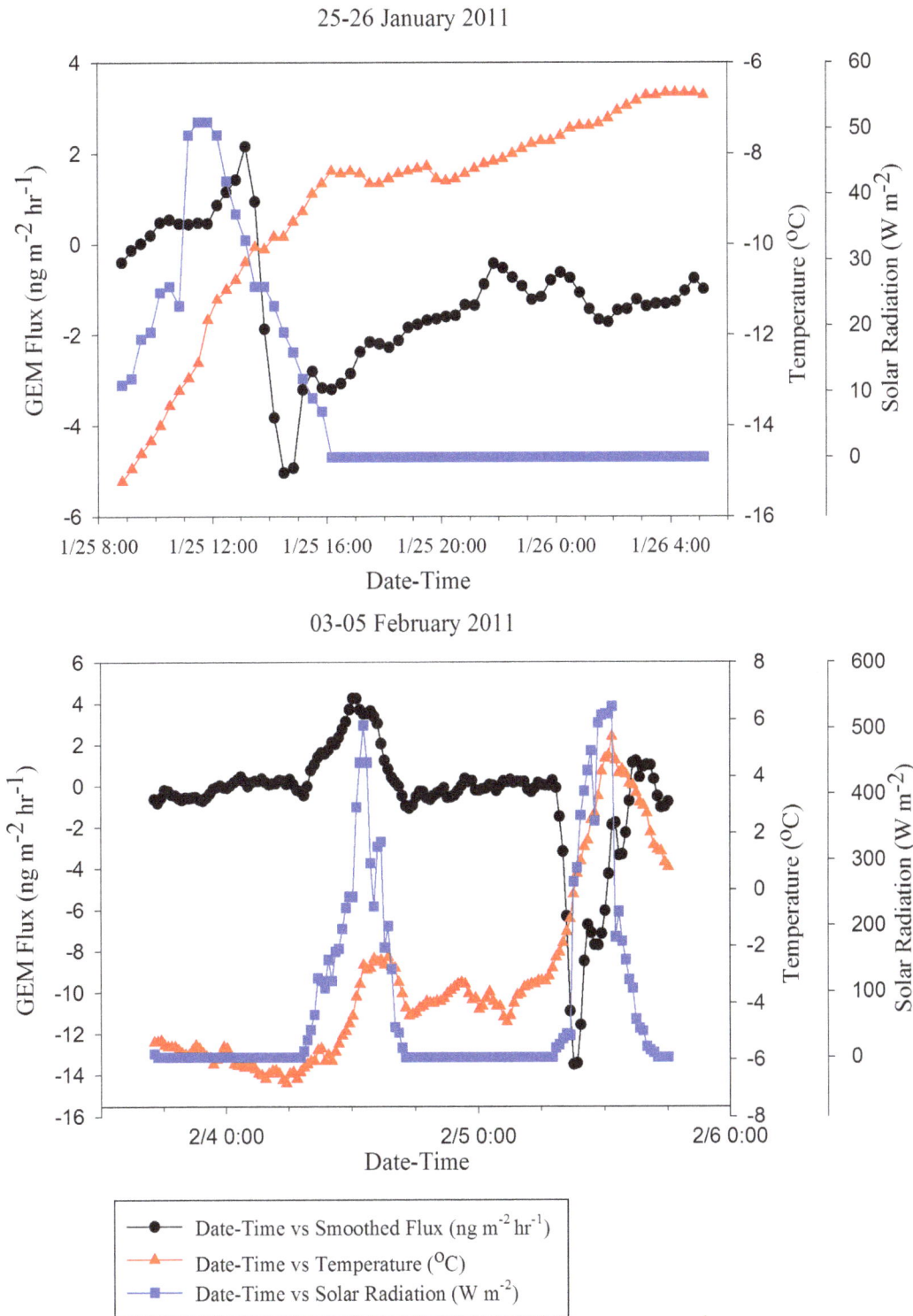

Figure 8. GEM Deposition Events For 25–26 January And 03–05 February 2011 Sampling Period.

phenomena have been reported by other research groups [8,12] with mean emission fluxes of 2–5 ng m^{-2} hr^{-1}, zero change in surface snow Hg concentration after deposition events, and up to 54% GEM reemission during the first 24 hours after a snowfall.

Deposition Events

Two unique deposition events with fluxes as high as -14 ng m^{-2} hr^{-1} occurred during separate sampling periods, one on 25 January and one on 05 February. Both of these event followed snowfalls ≥ 3 cm (Table 4) and melting also occurred during the 03–05 February sampling (Figure 8). During both of these events,

Table 4. Field Observations Made During Various Measurement Periods Throughout the 2011 Winter Sampling Season.

Date	Observation
21–24 Jan.	No recent snow
26–27 Jan.	Fresh snow (dusting, <2.5 cm) prior to sampling
27–30 Jan.	Fresh snow (dusting, <2.5 cm) and intermittent light snowing throughout sampling (no enough to cover inlets of chamber)
30–31 Jan.	No recent snow
15–18 Feb.	No recent snow
23–24 Feb.	No recent snow
08–09 Mar.	Fresh snow (≈7.5 cm); during end of season melt

fluxes were negatively correlated with temperature (PPMC coefficient $= -0.421$ and -0.439 respectively, p-value $= 0.000 < 0.050$) and displayed patterns opposite to the diurnal patterns typically seen. This sudden deposition event is similar to atmospheric Hg depletion events (AMDEs) witnessed in arctic regions during polar sunrise [9–11].

During an AMDE, rapid oxidation of GEM forms GOM that is subsequently deposited to the snow surface. Arctic AMDEs are springtime phenomenon that occur as a result of reactions with ozone and other halogen compounds, especially bromine oxides [26]. Though the cause of the two deposition events seen in Potsdam is unclear, they could coincide with sudden increases in atmospheric oxidant concentrations including free halogens.

Conclusions

Snow surface-to-air exchange of gaseous elemental Hg (GEM) was measured using a modified Teflon fluorinated ethylene propylene (FEP) dynamic flux chamber (DFC) in a remote, open site in Potsdam, New York during the winter months of 2011. The surface GEM flux ranged from -4.47 ng m^{-2} hr^{-1} to 9.89 ng m^{-2} hr^{-1}. For most sample periods, the daytime GEM flux was

strongly correlated with solar radiation. The average nighttime GEM flux was slightly negative and was weakly correlated with all of the measured meteorological variables. Overall, preliminary models indicate that most if not all the Hg being deposited to snow surfaces is being reemitted back into the atmosphere. Two unique deposition events with fluxes as high as -14 ng m^{-2} hr^{-1} occurred during separate sampling periods following snowfalls ≥ 3 cm. During both of these events, fluxes were negatively correlated (PPMC coefficient $= -0.421$ and -0.439 respectively, p-value $= 0.000 < 0.050$) with temperature and displayed patterns opposite to the diurnal patterns typically seen.

Acknowledgments

A special thank you to Dr. Andrea Ferro, Dr. Stephan Grimberg, and Dr. Jiaoyan Huang for their help in reviewing this work, and James Laing for assistance with sampling.

Author Contributions

Conceived and designed the experiments: JAM TMH. Performed the experiments: JAM. Analyzed the data: JAM SM. Contributed reagents/materials/analysis tools: JAM TMH SM. Wrote the paper: JAM.

References

1. EPA (2013) Laws and Regulations | Mercury | US EPA. Available: http://www.epa.gov/hg/regs.htm. Accessed 14 December 2012.

2. EU (2011) DIRECTIVE 2011/65/EU of the European Parliament and of the Council of 8 June 2011 on the restriction of the use of certain hazardous substances in electrical and electronic equipment (recast): 88–110. Available: http://eur-lex.europa.eu/LexUriServ/LexUriServ.do?uri = OJ:L:2011:174:0088:0110:EN:PDF.

3. Mergler D, Anderson HA, Chan LHM, Mahaffey KR, Murray M, et al. (2007) Methylmercury Exposure and Health Effects in Humans: A Worldwide Concern. AMBIO: A Journal of the Human Environment 36: 3–11. Available: http://dx.doi.org/10.1579/0044-7447(2007)36[3:MEAHEI]2.0.CO;2. Accessed 14 December 2012.

4. Lin C-J, Pehkonen SO (1999) The chemistry of atmospheric mercury: a review. Atmospheric Environment 33: 2067–2079. Available: http://dx.doi.org/10.1016/S1352-2310(98)00387-2. Accessed 14 December 2012.

5. Schroeder WH, Munthe J (1998) Atmospheric mercury–An overview. Atmospheric Environment 32: 809–822. Available: http://dx.doi.org/10.1016/S1352-2310(97)00293-8. Accessed 14 December 2012.

6. Lindqvist O, Rodhe H (1985) Atmospheric mercury-a review. Tellus B 37B: 136–159. Available: http://www.tellusb.net/index.php/tellusb/article/view/15010. Accessed 14 December 2012.

7. Selin NE, Jacob DJ, Yantosca RM, Strode S, Jaegle L, et al. (2008) Global 3-D Land-Ocean-Atmosphere Model for Mercury: Present- Day Versus Preindustrial Cycles and Anthropogenic Enrichment Factors for Deposition. Global Biogeochemical Cycles 22: 13.

8. Ferrari C, Gauchard P, Aspmo K, Dommergue A, Magand O, et al. (2005) Snow-to-air exchanges of mercury in an Arctic seasonal snow pack in Ny-Ålesund, Svalbard. Atmospheric Environment 39: 7633–7645. Available: http://dx.doi.org/10.1016/j.atmosenv.2005.06.058. Accessed 14 December 2012.

9. Ariya PA, Dastoor AP, Amyot M, Schroeder WH, Barrie L, et al. (2004) The Arctic: a sink for mercury. Tellus B 56: 397–403. Available: http://www.tellusb.net/index.php/tellusb/article/view/16458. Accessed 14 December 2012.

10. Skov H, Christensen JH, Goodsite ME, Heidam NZ, Jensen B, et al. (2004) Fate of Elemental Mercury in the Arctic during Atmospheric Mercury Depletion Episodes and the Load of Atmospheric Mercury to the Arctic. Environmental Science & Technology 38: 2373–2382. Available: http://dx.doi.org/10.1021/es030080h. Accessed 14 December 2012.

11. Lindberg SE, Brooks S, Lin C-J, Scott KJ, Landis MS, et al. (2002) Dynamic Oxidation of Gaseous Mercury in the Arctic Troposphere at Polar Sunrise. Environmental Science & Technology 36: 1245–1256. Available: http://dx.doi.org/10.1021/es0111941. Accessed 14 December 2012.

12. Lalonde JD, Poulain AJ, Amyot M (2002) The Role of Mercury Redox Reactions in Snow on Snow-to-Air Mercury Transfer. Environmental Science & Technology 36: 174–178. Available: http://dx.doi.org/10.1021/es010786g. Accessed 14 December 2012.

13. Lalonde JD, Amyot M, Doyon M-R, Auclair J-C (2003) Photo-induced Hg(II) reduction in snow from the remote and temperate Experimental Lakes Area (Ontario, Canada). Journal of Geophysical Research: Atmospheres (1984–2012) 108: 4200.

14. Nelson SJ, Fernandez IJ, Kahl JS (2010) A review of mercury concentration and deposition in snow in eastern temperate North America. Hydrological Processes 24: 1971–1980.

15. Faïn X, Grangeon S, Bahlmann E, Fritsche J, Obrist D, et al. (2007) Diurnal production of gaseous mercury in the alpine snowpack before snowmelt. Journal of Geophysical Research 112: D21311. Available: http://www.agu.org/pubs/crossref/2007/2007JD008520.shtml. Accessed 14 December 2012.

16. Dommergue A, Ferrari CP, Poissant L, Gauchard P-A, Boutron CF (2003) Diurnal Cycles of Gaseous Mercury within the Snowpack at Kuujjuarapik/Whapmagoostui, Québec, Canada. Environmental Science & Technology 37:

3289–3297. Available: http://dx.doi.org/10.1021/es026242b. Accessed 14 December 2012.

17. Choi H-D, Holsen TM (2009) Gaseous mercury fluxes from the forest floor of the Adirondacks. Environmental pollution (Barking, Essex□: 1987) 157: 592–600. Available: http://dx.doi.org/10.1016/j.envpol.2008.08.020. Accessed 14 December 2012.

18. Zhang H, Lindberg SE, Barnett MO, Vette AF, Gustin MS (2002) Dynamic flux chamber measurement of gaseous mercury emission fluxes over soils. Part 1: simulation of gaseous mercury emissions from soils using a two-resistance exchange interface model. Atmospheric Environment 36: 835–846. Available: http://dx.doi.org/10.1016/S1352-2310(01)00501-5. Accessed 14 December 2012.

19. Lindberg SE, Zhang H, Vette AF, Gustin MS, Barnett MO, et al. (2002) Dynamic flux chamber measurement of gaseous mercury emission fluxes over soils: Part 2-effect of flushing flow rate and verification of a two-resistance exchange interface simulation model. Atmospheric Environment 36: 847–859. Available: http://dx.doi.org/10.1016/S1352-2310(01)00502-7. Accessed 14 December 2012.

20. Eckley CS, Gustin M, Lin C-J, Li X, Miller MB (2010) The influence of dynamic chamber design and operating parameters on calculated surface-to-air mercury fluxes. Atmospheric Environment 44: 194–203. Available: http://dx.doi.org/10.1016/j.atmosenv.2009.10.013. Accessed 14 December 2012.

21. Carpi A, Frei A, Cocris D, McCloskey R, Contreras E, et al. (2007) Analytical artifacts produced by a polycarbonate chamber compared to a Teflon chamber for measuring surface mercury fluxes. Analytical and bioanalytical chemistry 388: 361–365. Available: http://www.ncbi.nlm.nih.gov/pubmed/17260134. Accessed 14 December 2012.

22. Savitzky A, Golay MJE (1964) Smoothing and Differentiation of Data by Simplified Least Squares Procedures. Analytical Chemistry 36: 1627–1639. Available: http://dx.doi.org/10.1021/ac60214a047. Accessed 31 October 2012.

23. Shapiro SS, Wilk MB (1965) An Analysis of Variance Test for Normality (Complete Samples). Biometrika 52: 591–611. Available: http://www.jstor.org/stable/10.2307/2333709.

24. Pearson K (1895) No Title. Royal Society Proceedings. 241.

25. Gbor P, Wen D, Meng F, Yang F, Zhang B, et al. (2006) Improved model for mercury emission, transport and deposition. Atmospheric Environment 40: 973–983. Available: http://dx.doi.org/10.1016/j.atmosenv.2005.10.040. Accessed 14 December 2012.

26. Steffen A, Douglas T, Amyot M, Ariya P, Aspmo K, et al. (2008) A synthesis of atmospheric mercury depletion event chemistry in the atmosphere and snow. Atmospheric Chemistry and Physics 8.

γδ T Cells Are Required for Pulmonary IL-17A Expression after Ozone Exposure in Mice: Role of TNFα

Joel A. Mathews*, Alison S. Williams, Jeffrey D. Brand, Allison P. Wurmbrand, Lucas Chen, Fernanda M.C. Ninin, Huiqing Si, David I. Kasahara, Stephanie A. Shore

Molecular and Integrative Physiological Sciences Program, Department of Environmental Health, Harvard School of Public Health, Boston, Massachusetts, United States of America

Abstract

Ozone is an air pollutant that causes pulmonary symptoms. In mice, ozone exposure causes pulmonary injury and increases bronchoalveolar lavage macrophages and neutrophils. We have shown that IL-17A is important in the recruitment of neutrophils after subacute ozone exposure (0.3 ppm for 24–72 h). We hypothesized that γδ T cells are the main producers of IL-17A after subacute ozone. To explore this hypothesis we exposed wildtype mice and mice deficient in γδ T cells ($TCR\delta^{-/-}$) to ozone or room air. Ozone-induced increases in BAL macrophages and neutrophils were attenuated in $TCR\delta^{-/-}$ mice. Ozone increased the number of γδ T cells in the lungs and increased pulmonary $Il17a$ mRNA expression and the number of IL-17A$^+$ CD45$^+$ cells in the lungs and these effects were abolished in $TCR\delta^{-/-}$ mice. Ozone-induced increases in factors downstream of IL-17A signaling, including G-CSF, IL-6, IP-10 and KC were also decreased in $TCR\delta^{-/-}$ versus wildtype mice. Neutralization of IL-17A during ozone exposure in wildtype mice mimicked the effects of γδ T cell deficiency. TNFR2 deficiency and etanercept, a TNFα antagonist, also reduced ozone-induced increases in $Il17a$ mRNA, IL-17A$^+$ CD45$^+$ cells and BAL G-CSF as well as BAL neutrophils. TNFR2 deficient mice also had decreased ozone-induced increases in Ccl20, a chemoattractant for IL-17A$^+$ γδ T cells. $Il17a$ mRNA and IL-17A$^+$ γδ T cells were also lower in obese Cpe^{fat} versus lean WT mice exposed to subacute ozone, consistent with the reduced neutrophil recruitment observed in the obese mice. Taken together, our data indicate that pulmonary inflammation induced by subacute ozone requires γδ T cells and TNFα-dependent recruitment of IL-17A$^+$ γδ T cells to the lung.

Editor: Shama Ahmad, University of Colorado, Denver, United States of America

Funding: This work was supported by: F32ES02256, NIH-HL007118, NIEHS: ES-013307 and ES-000002. The funders had no role in study design, data collection and analysis, decision to publish, or preparation of the manuscript.

Competing Interests: The authors have declared that no competing interests exist.

* E-mail: jmathews@hsph.harvard.edu

Introduction

γδ T cells are a key component of the innate immune response, especially at mucosal surfaces. These cells are found throughout the lung, particularly in the subepithelial region, where they may regulate other immune cells including macrophages and dendritic cells [1]. γδ T cells are an important source of IL-17A, a key cytokine involved in neutrophilic inflammation [2]. In mice, the number of pulmonary γδ T cells increases following infection with certain bacteria [3]. Mice deficient in γδ T cells ($TCR\delta^{-/-}$ mice) have attenuated pulmonary clearance of these bacteria, likely as a result of loss of IL-17A production by γδ T cells and consequent reduced neutrophil recruitment [4]. The number of γδ T cells in the lung also increases under conditions associated with oxidative stress, including smoking, bleomycin instillation, and allergen challenge [5–8]. Moreover, the pulmonary inflammation induced by such agents requires γδ T cells.

Inhalation of ozone (O_3), a common air pollutant, has a significant impact on human health. O_3 causes respiratory symptoms and reductions in lung function [9–13]. O_3 also increases the risk of respiratory infections and is a trigger for asthma [14–16]. Exposure to O_3 induces oxidative stress in the lung, damages lung epithelial cells, and causes the release of

numerous cytokines and chemokines that recruit neutrophils and macrophages to the lung [9,17]. We have reported increased $Il17a$ mRNA expression and increased numbers of IL-17A$^+$ γδ T cells in the lungs after subacute O_3 exposure (0.3 ppm O_3 for 24–72 h) [18]. Hence, we tested the hypothesis that γδ T cells, via their ability to produce IL-17A, are involved in orchestrating the inflammatory response to subacute O_3 exposure. We examined IL-17A expression in WT and $TCR\delta^{-/-}$ mice after exposure to air or to O_3 (0.3 ppm for 24–72 h). We also examined the effect of IL-17A neutralizing antibodies on O_3-induced inflammation. Our results indicate an important role for IL-17A$^+$ γδ T cells in the inflammatory cell recruitment induced by subacute O_3 exposure.

TNFα a pleiotropic pro-inflammatory cytokine, enhances the recruitment of neutrophils to the lungs in response to a variety of noxious stimuli, including LPS [19], cigarette smoke [20], and enterobacteria [21]. TNFα is also required for neutrophil recruitment after subacute O_3 exposure [22,23]. However, TNFα does not have direct chemoattractant activity for neutrophils [24]. Instead, TNFα recruits neutrophils in part by inducing expression of other cytokines and chemokines [24,25]. In several pathological states, TNFα induces the expression of IL-17A [26,27]. Hence, we hypothesized that TNFα contributes to neutrophil recruitment following subacute O_3 exposure by promoting recruitment to or activation of IL-17A$^+$ γδ T cells in the lungs. We used two methods to test this hypothesis. First, we assessed the effect of O$_3$ exposure on pulmonary *Il17a* expression and recruitment of IL-17A$^+$ γδ T cells in WT mice and in mice deficient in TNFR2 (TNFR2$^{-/-}$ mice). Others have established that either TNFR1 or TNFR2 deficiency reduces the inflammatory response to subacute O$_3$, and there is no further impact of combined TNFR1/TNFR2 deficiency [22]. Second, we examined the impact of the TNFα antagonist, etanercept, on *Il17a* expression. Our data suggest that TNFα is required for the recruitment of IL-17A$^+$ γδ T cells to the lung after subacute O$_3$ exposure.

Approximately one third of the US population is obese and another third is overweight, but our understanding of how obesity impacts pulmonary responses to O$_3$ is still rudimentary. Such an understanding may have broad reaching implications since oxidative stress also contributes to responses to a variety of other noxious stimuli [5–8], many of which are affected by obesity [28,29]. In mice, the impact of obesity on responses to O$_3$ depends on the nature of the exposure: the pulmonary inflammation induced by acute O$_3$ exposure (2 ppm for 3 h) is augmented in all types of obese mice examined to date [30–33], whereas the pulmonary inflammation induced by subacute O$_3$ exposure (0.3 ppm for 24–72 h) is reduced [34]. Given our findings of the requirement for TNFα-recruitment of IL-17A producing γδ T cells in the induction of pulmonary inflammation after subacute O$_3$, we sought to determine if changes in the activation of γδ T cells might explain the reduced responses to subacute O$_3$ we observed in obese *Cpe*(carboxypeptidase E)^{fat} mice. Data described below indicate that the reduced O$_3$-induced neutrophil recruitment observed in obese mice is likely the result of reduced *Il23* expression leading to reduced IL-17A$^+$ γδ T cells. Given the importance of IL-17$^+$ γδ T cells for responses to viral and bacterial pathogens (see above), these observations might explain the altered response of the obese to bacteria and virus (see review by Peter Mancuso [35]).

Methods

Animals

This study was approved by the Harvard Medical Area Standing Committee on Animals. Male age-matched WT and TCRδ$^{-/-}$ mice were either purchased from The Jackson Laboratory (Bar Harbor, ME) and acclimated for 4 weeks, or bred in house. *Cpe*^{fat} mice are deficient in carboxypeptidase E, an enzyme involved in processing neuropeptides involved in eating behaviors [36]. The breeding strategy used to generate *Cpe*^{fat}/TNFR2$^{-/-}$ mice from *Cpe*^{fat} and TNFR2$^{-/-}$ mice (also originally purchased from The Jackson Laboratory) was previously described [37]. All mice were on a C57BL/6J background, fed a standard mouse chow diet, and were 10–13 weeks old at the time of study.

Protocol

For comparisons of WT and TCRδ$^{-/-}$ mice, mice were exposed to O$_3$ (0.3 ppm) or to air, for 24–72 hours, as previously described [18]. Mice were exposed in normal cages without the microisolator top, but with free access to water and food throughout exposure. Mice were checked daily. At least two mice were placed in each cage to limit stress. After exposure, mice were euthanized with an overdose of sodium pentobarbital. The trachea was cannulated and bronchoalveolar lavage (BAL) was performed. After BAL, the lungs were flushed of blood by injecting 10 ml of cold PBS through the right ventricle, after creating a large excision in the left ventricle. One lung was excised and used for flow cytometry. The other was excised and placed in RNAlater

(Qiagen, Germantown, MD) for preparation of RNA for real time PCR. In another cohort, WT mice were injected i.p. with 100 μg of anti–IL-17A neutralizing monoclonal antibody (Ab) (Rat IgG2A, clone 50104, MAB421; R&D Systems, Minneapolis, MN) or isotype control Ab (clone 54447, MAB006; R&D Systems) in 100 μl of sterile saline 24 hours before O$_3$ exposure. Mice were exposed to O$_3$ for 72 hours, euthanized, and tissues were harvested as described above. In a separate series of experiments, WT, TNFR2$^{-/-}$, *Cpe*^{fat}, and *Cpe*^{fat}/TNFR2$^{-/-}$ mice were exposed to room air or O$_3$ (0.3 ppm) for 48 h followed by BAL and tissue harvest. In other experiments, WT and *Cpe*^{fat} mice were treated twice (48 h and 1 h prior to O$_3$ exposure) with the TNFα blocking drug, etanercept (30 mg/kg s.c.) (Immunex, Thousand Oaks, CA), or vehicle. A similar etanercept dosing regimen has been shown to be effective in inhibiting TNFα in mice over the time course of O$_3$ exposures we used (48 h) [38,39].

Bronchoalveolar Lavage

BAL was performed and cells counted as previously described [18]. BAL supernatant was stored at −80°C until assayed. BAL KC, IL-6, MCP-1, IP-10 and G-CSF were measured by ELISA (R&D Systems). In mice treated with anti-IL-17A, BAL cytokines and chemokines were measured by multiplex assay (Eve Technologies, Calgary, Alberta). Total BAL protein was measured by Bradford assay (Bio-Rad, Hercules, CA).

Flow Cytometry

Left lungs were harvested and placed on ice in RPMI 1640 media containing 2% FBS and HEPES. Lungs were digested and prepared for flow cytometry as previously described [18]. Cells were stained using the following antibodies: Alexa Fluor 647 anti-IL-17A (clone: TC11-18H10.1), PE anti-TCRδ (clone: GL3), PE-cy7 anti-CD45 (clone: 30-F11), and APC-cy7 anti-CD3 (clone: 17A2) (all antibodies from Biolegend). Isotype control antibodies were used to set all gates. Cells were visualized using a Canto II (BD Biosciences) and the data was analyzed using Flowjo (Tree Star; Ashland, OR).

To determine if TNFα impacted IL-12Rβ1 expression on lung γδ T cells, lungs from WT mice were digested as above and then cultured in complete RPMI media (RPMI 1640 (Corning, Tewksbury, MA), 10% FBS (Life Technologies), 2 Mm L-glutamine (Life Technologies), 100 units/ml Pen/Strep (Lonza, Hopkinton, MA) and 20 Mm Hepes (Thermo Scientific, Tewksbury, MA)). Cells were plated at a concentration of 10^6 cells/ml in 24 well plates with or without 100 ng/ml of recombinant murine TNFα (R&D Systems) [40]. Cells were harvested after 24 h, washed with PBS, and stained using the following antibodies: anti-CD16/32 (True Stain biolegend), Strep-APC (Biolegend), PE anti-CD212 (IL-12Rβ1) (BD Biosciences), Biotin anti-TCRδ (clone: GL3, biolegend), PE-cy7 anti-CD45 (clone: 30-F11) and analyzed by flow cytometry as described above.

Real-time PCR

RNA was extracted from lung tissue and prepared for qPCR using the SYBR method as previously described [18]. All expression values were normalized to 36B4 expression using the ΔΔCt method. The primers for *Il17a* and *36B4* were previously described [37]. Primers for *Ccl20*, *Il23* (p19) and *Il12Rβ1* are described in Table 1. For each set of primers, melt curve analysis yielded a single peak. *Il12Rβ1* expression was measured at baseline in order to tease apart the effects of genotype (deficiency of TNFα signaling versus sufficient signaling) versus O$_3$ exposure.

Table 1. Primers used for real time PCR.

Il23p19	F: CCC ATG GAG CAA CTT CAC AC R: GCT GCC ACT GCT GAC TAG AAC
Ccl20	F: AAG ACA GAT GGC CGA TGA AG R: AGG TTC ACA GCC CTT TTC AC
Il12Rb1	F: GTG CTC GCC AAA ACT CGT TT R: GGA TGT CAT GTT GCC TCC CA

Statistical Analysis

Data were analyzed by factorial ANOVA using STATISTICA software (Statistica, StatSoft; Tulsa, OK) with mouse genotype and exposure as main effects. Fisher's least significant difference test was used as a post-hoc test. BAL cells and flow cytometry data were normalized by log transformed prior to analysis. A p value $<$ 0.05 was considered significant.

Results

O_3-induced Inflammation is Reduced in TCR$\delta^{-/-}$ Mice

In WT mice, O_3 exposure caused a time-dependent increase in BAL neutrophils, macrophages, and protein (a measure of O_3-induced lung injury [41]) (Fig. 1A–C), consistent with previous reports by ourselves and others [18,22,23,41,42]. Increases in BAL inflammatory cells were significantly reduced in TCR$\delta^{-/-}$ versus WT mice after 48 (neutrophils) and 72 (neutrophils and macrophages) hours of exposure (Fig. 1A,B). BAL protein was also reduced in TCR$\delta^{-/-}$ versus WT mice after 72 hours exposure, but not at earlier times (Fig. 1C).

Several cytokines, including KC, IL-6, IP-10 (CXCL10), G-CSF, MCP-1 and IL-17A [17,18,22,23,41–44], can contribute to inflammatory cell recruitment to the lungs after O_3 exposure. BAL IL-17A expression was below the limits of detection of ELISA. Consequently, we used q-RT-PCR to measure IL-17A. *Il17a* mRNA abundance increased after 24, 48 and 72 hours of O_3 in WT but not TCR$\delta^{-/-}$ mice (Fig. 1D). O_3-induced increases in BAL concentrations of BAL G-CSF, IL-6, KC and IP-10 were each reduced in TCR$\delta^{-/-}$ versus WT mice at 72 hours of exposure (Fig. 1E–H). For G-CSF and IP-10, there was a similar trend at 24 and 48 hours (Fig. 1E,G). $\gamma\delta$ T cell deficiency had no effect on O_3-induced changes in BAL MCP-1, although MCP-1 trended lower in TCR$\delta^{-/-}$ versus WT mice at 72 hours.

IL-17A$^+$ $\gamma\delta$ T Cells are Increased by O_3 Exposure

Flow cytometry indicated that the number of IL-17A$^+$ CD45$^+$ cells was significantly increased by O_3 in WT mice. This effect was ablated in TCR$\delta^{-/-}$ mice (Fig. 2A). Further analysis indicated that in WT mice, the numbers of IL-17A$^+$ $\gamma\delta$ T cells as well as the total number of $\gamma\delta$ T cells were increased by O_3 (Fig. 2B, C), as reported previously using a similar gating strategy [18].

Effect of Anti-IL-17A Treatment

Compared to isotype control, anti-IL-17A treatment of WT mice caused a significant reduction in BAL neutrophils and macrophages (Fig. 3A). Anti-IL-17A treatment also significantly decreased BAL protein (Fig. 3B) and BAL G-CSF (Fig. 3C). Given this key role for IL-17A, these data indicate that the decreased inflammatory response observed in the TCR$\delta^{-/-}$ mice was likely due to the lack of *Il17a* expression (Fig. 1D) and demonstrate that G-CSF likely contributes to the effect of IL-17A on neutrophil recruitment.

Role of TNFα

BAL neutrophils were significantly lower in TNFR2$^{-/-}$ versus WT mice exposed to O_3 for 48 h (Fig. 4A), consistent with the results of Cho et al [22]. Similar results were obtained in WT mice treated with etanercept versus vehicle (Fig. 4D). O_3 exposure caused a significant increase in pulmonary *Il17a* expression in WT mice (Fig. 4B), consistent with results described above (Fig. 1D). However in TNFR2$^{-/-}$ mice, no such increase in *Il17a* mRNA abundance was observed (Fig. 4B). Similar results were obtained in mice treated with etanercept (Fig. 4E). Flow cytometry also indicated a decrease in IL-17A$^+$CD45$^+$ cells in O_3-exposed TNFR2$^{-/-}$ versus WT mice (Fig. 5A). This change was due to decreased numbers of IL-17A$^+$ $\gamma\delta$ T cells (Fig. 5B). BAL G-CSF was also significantly lower in O_3-exposed TNFR2$^{-/-}$ versus WT mice (Fig. 4C) and in etanercept treated versus vehicle treated WT mice (Fig. 4F).

The requirement of IL-23 and IL-6 for IL-17A expression in $\gamma\delta$ T cells [45,46], suggested that reductions in IL-17A$^+$ $\gamma\delta$ T cells in TNFR2$^{-/-}$ mice might be the result of loss of TNFα-induced expression of IL-23 or IL-6. O_3 increased BAL IL-6 in WT mice (Fig. 1F) and O_3 also increased pulmonary *Il23* (p19) mRNA abundance (Fig. 6B), but neither IL-6 nor IL-23 were affected by TNFR2 deficiency or etanercept treatment (Fig. 6A, C). In contrast, TNFR2$^{-/-}$ mice had reduced expression at baseline of *Il12Rβ1* (Fig. 6H), a component of the IL-23 receptor. A similar trend was observed in etanercept treated mice (data not shown). O_3 exposure had no effect on *Il12Rβ1* (data not shown). Expression of the other component of the IL-23 receptor, *Il23R*, was not affected by TNFR2 deficiency (data not shown). To determine if TNFα was having direct effects on *Il12Rβ1* expression on $\gamma\delta$ T cells, we isolated total lung cells from WT mice, stimulated them overnight with TNFα and examined IL-12Rβ1 expression on $\gamma\delta$ T cells by flow cytometry (Fig. 6I,J). TNFα had no effect on the levels of IL-12Rβ1 on $\gamma\delta$ T cells as measured by MFI and did not affect the percentage of $\gamma\delta$ T cells expressing IL-12Rβ1, suggesting that other cells in the lung accounted for differences in *Il12Rβ1* mRNA expression.

We also considered the possibility that TNFα might impact the recruitment of $\gamma\delta$ T cells to the lung. In WT mice, O_3 exposure caused an increase in pulmonary mRNA expression of *Ccl20* (Fig. 6E), a chemoattractant for IL-17A$^+$ cells [47,48], whereas no such increase was observed in mice treated with etanercept (Fig. 6F), suggesting that the role of TNFα is in the CCL20 dependent recruitment of IL-17$^+$ $\gamma\delta$ T cells to the lungs. Similarly, there was a trend towards reduced *Ccl20* mRNA abundance in O_3-exposed TNFR2$^{-/-}$ versus WT mice (Fig. 6G), although the effect did not reach statistical significance.

Response to O_3 in Obese Mice

Cpe^{fat} mice, regardless of their TNFR2 genotype or exposure, weighed almost twice as much as controls (data not shown). BAL neutrophils were significantly lower in *Cpe^{fat}* versus WT mice exposed to O_3 (Fig. 4A,D), consistent with our previous observations using this exposure regimen [34]. In contrast to the

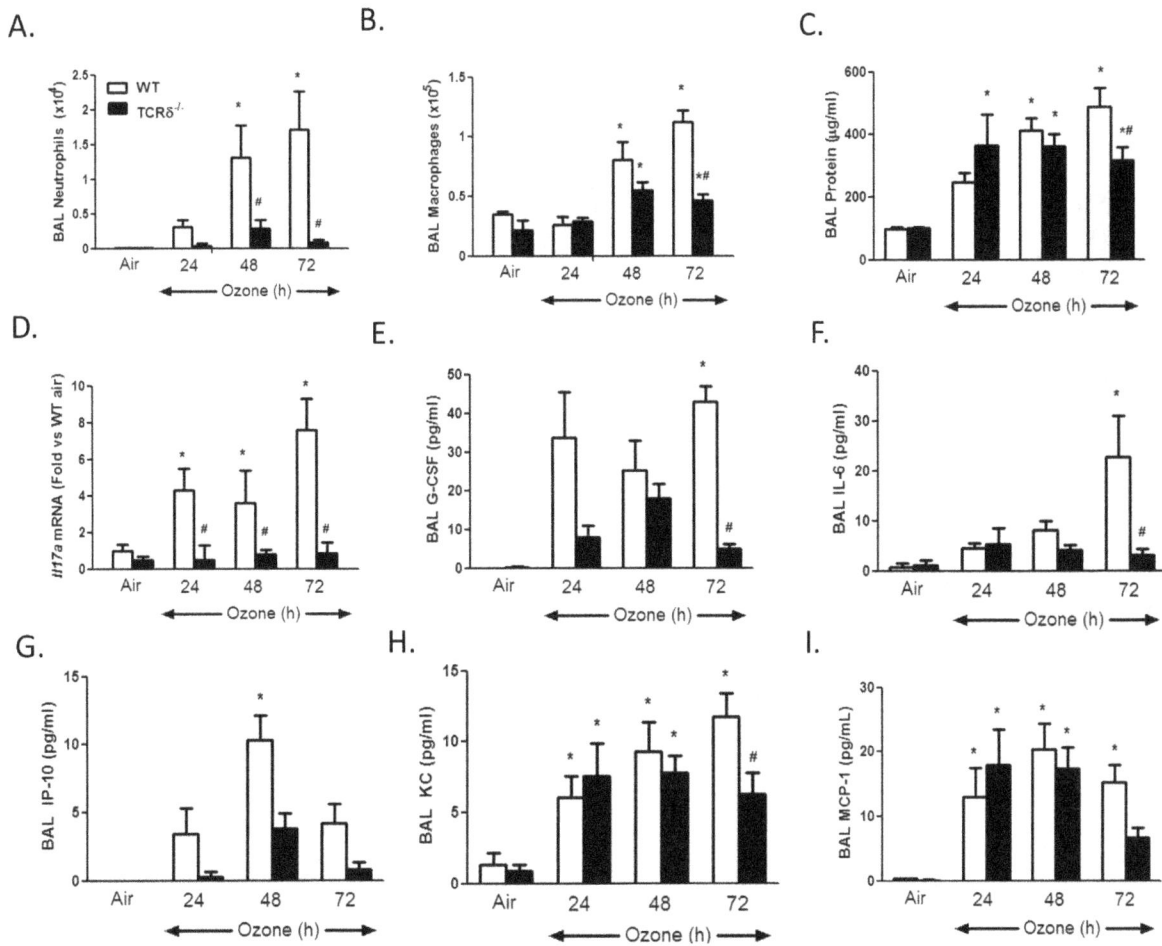

Figure 1. Effect of γδ T cell deficiency on pulmonary inflammation and injury. (A–C) BAL neutrophils, macrophages, and protein; (D) pulmonary *Il17a* mRNA expression; (E–I) BAL G-CSF, IL-6, IP-10, KC, and MCP-1. Results are mean±SEM of 4–11 mice per group. *p<0.05 versus genotype-matched air-exposed mice. #p<0.05 versus WT mice with the same exposure.

substantial reduction in BAL neutrophils observed in TNFR2$^{-/-}$ versus WT mice, TNFR2 deficiency had no significant effect on BAL neutrophils in O$_3$-exposed *Cpe^{fat}* mice (Fig. 4A). Similar results were obtained in etanercept treated WT mice (Fig. 4D). Cpe genotype had no impact on the number of BAL or lung macrophages (data not shown).

Il17a expression was significantly lower in O$_3$ exposed *Cpe^{fat}* versus WT mice (Fig. 4B,E). The number of IL-17A$^+$ CD45$^+$ cells was also significantly lower in O$_3$-exposed *Cpe^{fat}* than WT mice (Fig. 5A). The total number of γδ T cells and the number of IL-17A$^+$ γδ T cells was also reduced in the lungs of *Cpe^{fat}* versus WT mice (Fig. 5B,C). O$_3$-induced increases in BAL G-CSF were also

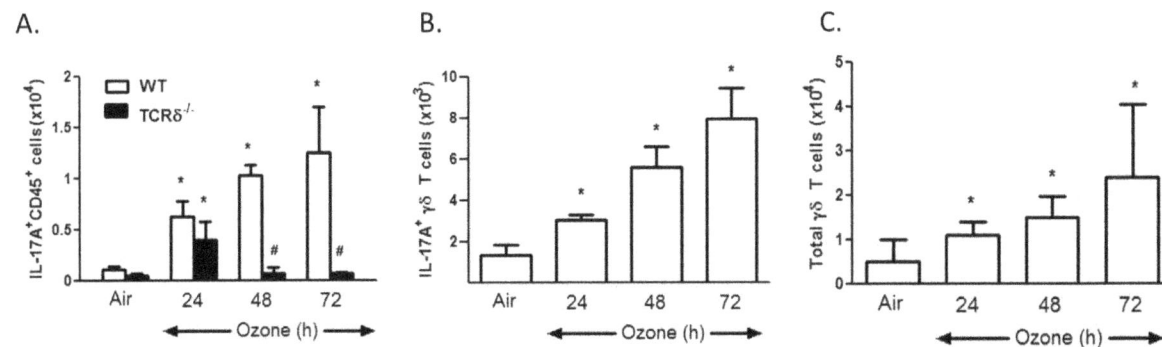

Figure 2. Effect of O$_3$ exposure on IL-17A positive lung cells assessed by flow cytometry. (A) lung IL-17A$^+$CD45$^+$; (B) lung IL-17A$^+$ γδ T cells; (C) total lung γδ T cells. Results are mean±SEM for 3–6 air-exposed and 4–11 O$_3$-exposed mice. *p<0.05 versus genotype-matched air-exposed mice. #p<0.05 versus WT mice with same exposure.

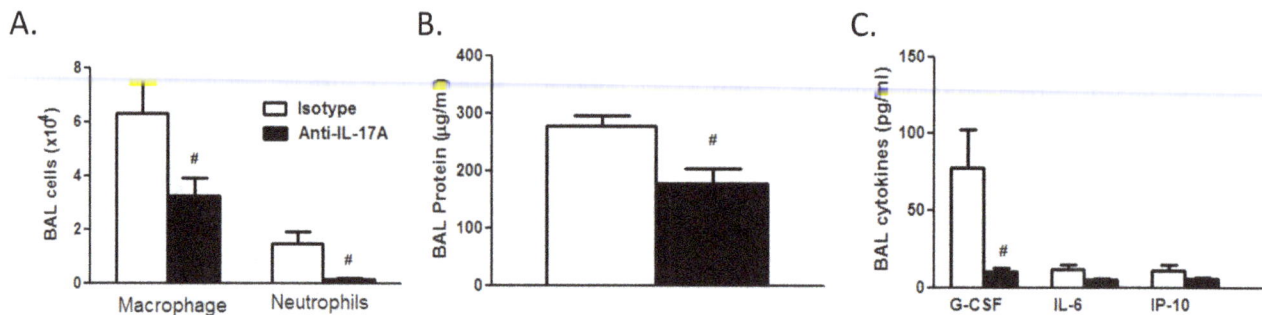

Figure 3. Effect of anti-IL-17A on O$_3$-induced pulmonary inflammation and injury. WT mice were injected with anti-IL-17A or isotype 24 h prior to O$_3$ (0.3 ppm O$_3$ for 72 h). (A) BAL macrophages and neutrophils; (B) BAL protein; (C) BAL cytokines determined by multiplex assay. Results are mean±SEM of 5–7 mice per group. #p<0.05 versus isotype control.

lower in Cpe^{fat} versus WT mice (Fig. 4C, E) consistent with the reductions in IL-17A expression. Both BAL IL-6 and pulmonary $Il23$ mRNA expression were lower in Cpe^{fat} versus WT mice (Fig. 6A, C,D). Reductions in these cytokines would be expected to reduce IL-17A expression, as observed (Fig. 4B, E). Whereas TNFR2 deficiency and etanercept reduced $Il17a$ mRNA, IL-17A$^+$ γδ T cells, and BAL G-CSF in lean WT mice, neither TNFR2 deficiency or etanercept affected these outcomes in obese Cpe^{fat} mice (Fig. 4B,C and 5A–C).

Discussion

Our data indicate a key role for IL-17A$^+$ γδ T cells in the pulmonary inflammation induced by subacute O$_3$. Our data also indicate that TNFα promotes pulmonary inflammation after subacute O$_3$ by inducing recruitment of IL-17A$^+$ γδ T cells, likely via $Ccl20$ expression. Finally, our data suggest that the attenuated

pulmonary inflammation observed in obese mice after subacute O$_3$ is the result of reduced pulmonary IL-17A$^+$ γδ T cells, consequent to reduced IL-23 and IL-6 expression.

Inflammatory cell recruitment to the lungs after subacute O$_3$ exposure required γδ T cells (Fig. 1A,B). γδ T cells have also been shown to be required for the pulmonary inflammation observed 24 but not 8 hours after acute exposure to much higher O$_3$ concentrations (2 ppm) [49,50], consistent with the time needed for recruitment and activation of γδ T cells. However, in those studies, the precise role of these γδ T cells was not assessed. Our data indicate that after exposure to lower concentrations of O$_3$ for much longer periods of time, the role of γδ T cells involved IL-17A expression. Both lung $Il17a$ mRNA and lung IL-17A$^+$ γδ T cells increased after subacute O$_3$ exposure with a time course similar to that of neutrophil recruitment (Figs. 1A, 1D, 2B). Furthermore, O$_3$-induced increases in $Il17a$ mRNA abundance were abolished in TCRδ$^{-/-}$ mice (Fig. 1D). In addition, both BAL neutrophils

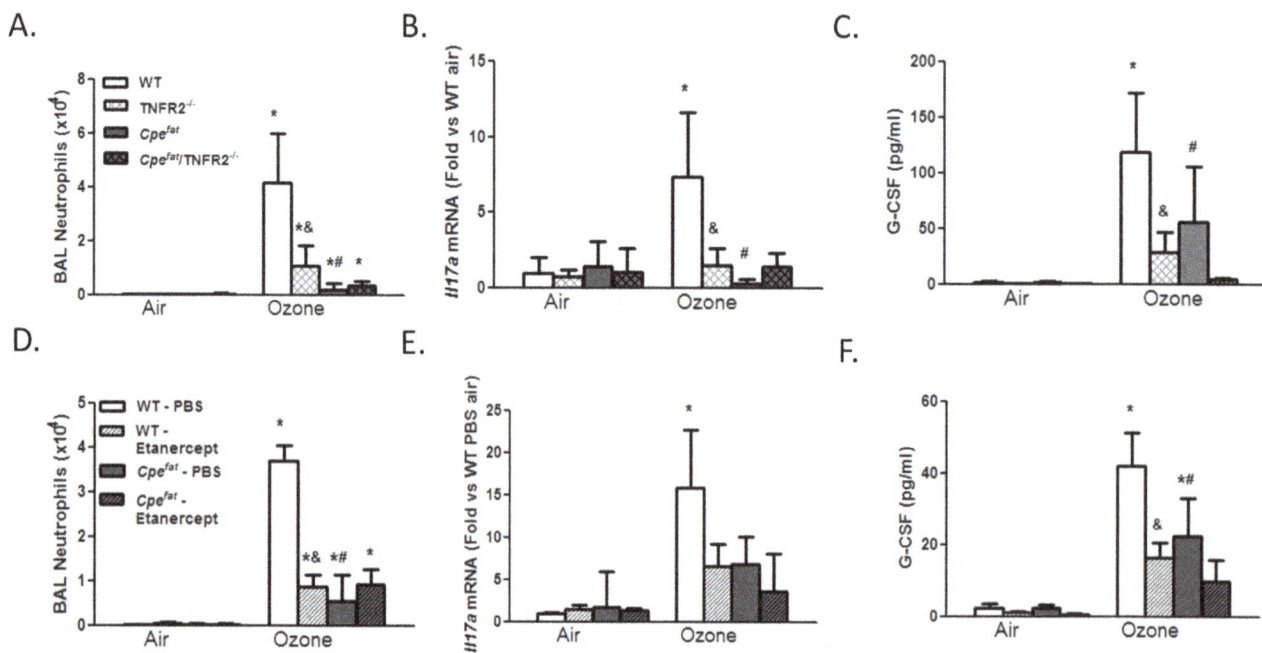

Figure 4. Impact of TNFR2 deficiency (A–C) or etanercept (D–F) on O$_3$-induced inflammation in obese (Cpe^{fat}) and lean (WT) mice. (A, D) BAL neutrophils; (B, E) $Il17a$ mRNA expression; (C, F) BAL G-CSF. Results are mean±SE of data from 3–11 mice in each group.*p<0.05 versus air-exposed mice of same genotype and treatment; #p<0.05 versus exposure matched lean mice with same TNFR2 genotype or treatment; & p<0.05 versus TNFR2 sufficient (A–C) or vehicle treated mice (D–F) with same exposure and Cpe genotype.

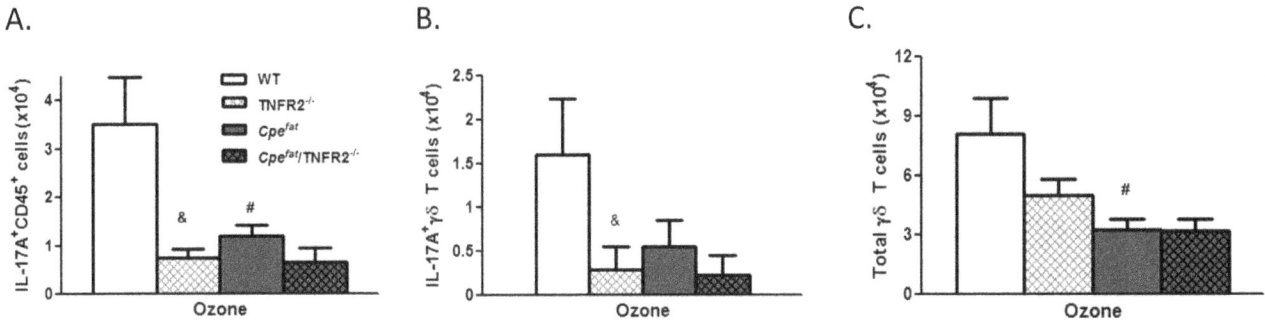

Figure 5. Role of TNFα for IL-17A expression in γδ T cells. Total number of (A) lung IL-17A⁺CD45⁺ cells; (B) lung IL-17A⁺ γδ T cells; and (C) total lung γδ T cells. Results are mean±SE of data from 5–6 mice in each group. #p<0.05 compared to lean mice with same TNFR2 genotype; & p<0.05 compared to TNFR2+/+ Cpe genotype matched mice.

and macrophages were reduced in mice treated with anti-IL-17A versus isotype control antibody (Fig. 3A). This ability of IL-17A⁺ γδ T cells to control the influx of macrophages and neutrophils is consistent with the findings in other models of lung infection and injury [4,51–54]. While our data indicate that IL-17⁺ γδ T cells are *required* for O₃-induced inflammatory cell recruitment, they are not *sufficient*. For example, O₃ is highly reactive and macrophages and epithelial cells are the initial targets of its action. These cells are the likely source of TNFα which is required for neutrophil recruitment (Fig. 4) perhaps via induction of CCL20 and consequent recruitment IL-17A+ γδ T cells (Figs. 5,6). Epithelial cells are also the likely source of CCL20. Furthermore, macro-

phages also produce IL-17A after O₃ exposure [18], and the role of γδ T cells may be to promote these effects. Macrophages and epithelial cells are also the likely source of other chemokines that interact with IL-17A (see below) to promote neutrophil recruitment.

IL-17A has direct chemoattractant effects on macrophages [55], which likely explains the ability of anti-IL-17A to attenuate O₃-induced increases in BAL macrophages (Fig. 3A). In contrast, IL-17A induces neutrophil recruitment to the lungs by inducing expression of other neutrophil chemotactic and survival factors. With subacute O₃ exposure, G-CSF appears to be one of these factors. In WT mice, the time courses of induction of BAL G-CSF

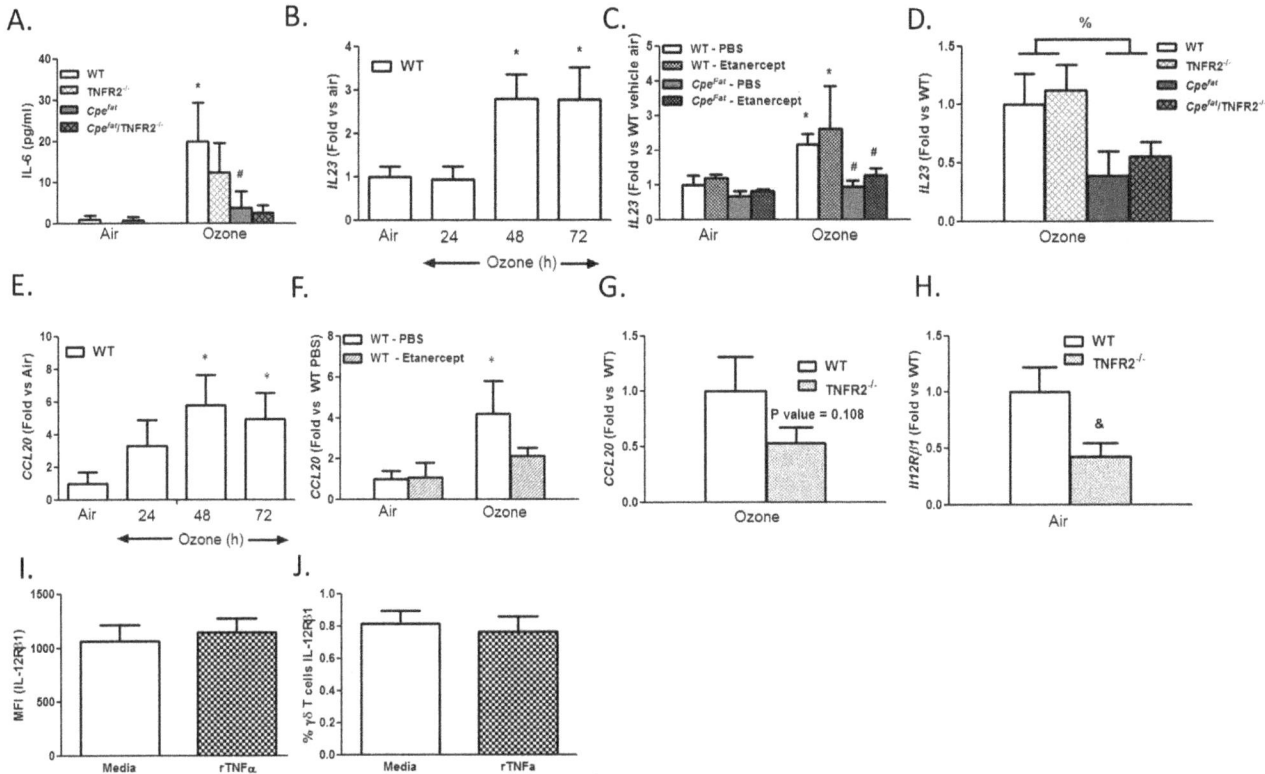

Figure 6. TNFα signaling is required for expression of *Il12Rβ1* and *Ccl20*. (A) BAL IL-6; (B–D)) *Il23 (p19)* mRNA; (E–G) *Ccl20* mRNA; (H) *Il12Rβ1* mRNA; (I) MFI and (J) % of γδ T cells positive for IL-12Rβ1 after stimulation with TNFα Results are mean±SE of data from 3–11 mice in each group. *p<0.05 versus air exposed mice of the same genotype; #p<0.05 versus exposure matched lean mice with the same TNFR2 genotype or treatment; & p<0.05 versus WT; %<0.05 obese versus lean regardless of TNFR2 genotype.

and *Il17a* expression were similar (Fig. 1D,E). Importantly, anti-IL-17A and γδ T cell deficiency each caused a marked and significant reduction in BAL G-CSF in O₃ exposed mice (Fig. 1E, 3C). The data are also consistent with our previous observations showing reductions in BAL G-CSF in O₃-exposed adiponectin-deficient mice treated with anti-IL-17A [18]. The observed role of IL-17A in G-CSF expression is in agreement with previous reports indicating that IL-17A signaling increases the transcription and stability of the *Gcsf* mRNA [56,57], via effects on ERK1/2 activation [58]. G-CSF causes neutrophil release from bone marrow and promotes neutrophil survival [59]. Since serum G-CSF did not increase after subacute O₃ exposure (data not shown), G-CSF is unlikely to act via effects on bone marrow in this model. Instead, G-CSF likely contributes by increasing the survival of neutrophils recruited to the lungs in response to other factors such as IP-10 (Fig. 1G).

TNFα is not directly chemotactic for neutrophils [24]. However, in lean WT mice, TNFR2 deficiency or the TNFα antagonist, etanercept, reduced the O₃-induced increase in BAL neutrophils (Fig. 4A,D) consistent with previous reports [22,23,60] indicating a role for TNFα in neutrophil recruitment induced by subacute O₃. TNFα also contributes to neutrophil recruitment in other conditions (reviewed in [61]), though the mechanism is not well understood. Our data suggest that at least in the setting of O₃ exposure, the ability of TNFα to recruit neutrophils involves IL-17A and that the source of this IL-17A is γδ T cells (Fig. 5). O₃-induced increases in pulmonary *Il17a* expression were attenuated in TNFR2$^{-/-}$ versus WT mice (Fig. 4B) and in etanercept versus vehicle treated WT mice (Fig. 4E). The number of IL-17A⁺ γδ T cells in the lung was also lower in TNFR2$^{-/-}$ versus WT mice exposed to O₃ (Fig. 5A,B). The ability of TNFα to promote pulmonary IL-17A expression after O₃ exposure is consistent with the role of TNFα in other pathogenic states. For example, etanercept reduces the elevated blood and skin Th17 cells observed in patients with psoriasis [26]. Similarly, another anti-TNFα therapy, infliximab, reduces IL-17A in ocular fluid from uveitis patients with Behcet's disease [27].

To better understand the role of TNFα, we examined IL-6 and IL-23 expression. Both these cytokines can contribute to induction of IL-17A in γδ T cells [45,62]. Both IL-6 and IL-23 were induced in the lungs after O₃ exposure, but were not affected by TNFR2 deficiency or by etanercept (Fig. 6A,C,D), indicating that TNFα is not required for their expression. We did observe that mRNA expression of one of the two subunits of the IL-23 receptor, *Il12Rβ1*, was decreased (Fig. 6H) in unexposed lungs from TNFR2$^{-/-}$ mice. Similar trends were observe after etanercept treatment (data not shown). Since others have reported that TNFα can act directly on γδ T cells [40,63], we considered the possibility that TNFα was acting to increase *Il12Rβ1* expression on γδ T cells, thus increasing their ability to respond to IL-23. However, culture of lung cells with TNFα resulted in no change in surface bound IL-12Rβ1 on γδ T cells (Fig. 6I,J). Instead, our data, suggest that effects of TNFα on *Ccl20* expression (Fig. 6F,G) account for the observed effects of TNFα/TNFR blockade on IL-17A⁺ γδ T cells. Ccl20 acts via CCR6, a receptor expressed by IL-17A⁺ γδ T cells that promotes chemotaxis of these cells [64]. TNFα is also required for pulmonary Ccl20 expression after acute O₃ exposure (2 ppm for 3 h) [37]. A role for TNFα in *Ccl20* expression has also been demonstrated in dermal lesions of psoriasis patients based on treatment with the TNFα antagonist infliximab [65].

We observed fewer neutrophils in BAL fluid of obese *Cpe^{fat}* versus lean WT mice after subacute O₃ exposure (Fig. 4A,D), consistent with previous observations [34]. Reduced responses are observed in *Cpe^{fat}* mice not only after 48 h exposure (Fig. 4A,D), but also after 24 or 72 h exposures [34]. Pulmonary *Il17a* expression and IL-17A⁺ γδ T cells were also reduced in the obese mice, as was the total number of γδ T cells (Fig. 4). BAL G-CSF was also lower in *Cpe^{fat}* versus lean WT mice (Fig. 4C,F). Moreover, O₃-induced increases in BAL IL-6 and pulmonary *Il23* expression were also reduced in *Cpe^{fat}* versus WT mice (Fig. 6C,D). TNFR2 deficiency or etanercept treatment in *Cpe^{fat}* mice did not further reduce BAL neutrophils or pulmonary *Il17a* expression, in contrast to what was observed in WT mice (Fig. 4B,E). Given the already reduced numbers of total γδ T cells in *Cpe^{fat}* mice exposed to O₃ (Fig. 5C), and our observations indicating the key role for IL-17A⁺ γδ T cells in the effects of TNFα on neutrophil recruitment, it is not surprising that TNFα had no further effect on the response to O₃ in obese mice. Taken together, the data suggest that obesity-related reductions in neutrophil recruitment induced by subacute O₃ exposure are the result of reduced IL-17A-dependent G-CSF release, consequent to reduced IL-6 and IL-23 expression. However, we cannot rule out the possibility that other factors also contributed. For example, neutrophils from obese mice exhibit reduced chemotactic activity towards CXCR2 ligands [66]. Such defects in neutrophil chemotaxis would also be expected to reduce O₃-induced neutrophil recruitment in *Cpe^{fat}* mice.

In addition to affecting responses to O₃, obesity also impacts responses to bacterial and viral infections [67–71]. As described above, IL-17⁺ γδ T cells contribute to neutrophil recruitment and pathogen clearance after certain bacterial infections [3,4]. IL-17⁺ γδ T cells are also required for clearance of secondary infections after influenza [72]. Hence, obesity-related changes in IL-17⁺ γδ T cells (Figs. 4b, 5a,b) may contribute not only to obesity-related alterations in responses to O₃, but may have broader implications for effects of obesity on host defense. In support of this, obese mice compared to lean mice have fewer skin γδ T cells number and the few γδ T cells they have are dysfunctional [73], which leads to impairment in wound healing. These decreases in γδ T cells numbers and impairment in function of the skin in obese mice are due to altered STAT5 signaling and chronic TNFα signaling [74].

In summary, our data indicate that γδ T cells are required for the pulmonary inflammation that occurs after subacute O₃ exposure in mice via their ability to produce IL-17A. IL-17A then leads to G-CSF expression. Our data also indicate that TNFα is required for recruitment IL-17A⁺ γδ T cells to the lungs likely through its ability to induce *Ccl20*. These results emphasize the importance of γδ T cells not only for pathogen clearance, but also for responses to other insults that induce oxidative stress, and describe a new role for TNFα in these events. Finally, our data indicate that obesity-related reductions in the ability of subacute O₃ to promote neutrophil recruitment to the lungs are the result of reduced IL-17A⁺ γδ T cells. These results suggest that other conditions that impact γδ T cell recruitment or activation will also impact responses to this common pollutant.

Author Contributions

Conceived and designed the experiments: JAM ASW JDB HS DIK SAS. Performed the experiments: JAM ASW JDB APW LC FMCN. Analyzed the data: JAM ASW JDB SAS. Contributed reagents/materials/analysis tools: JAM ASW. Wrote the paper: JAM ASW SAS.

References

1. Wands JM, Roark CL, Aydintug MK, Jin N, Hahn Y-S, et al. (2005) Distribution and leukocyte contacts of γδ T cells in the lung. Journal of Leukocyte Biology 78: 1086–1096.

2. Laan M, Cui Z-H, Hoshino H, Lötvall J, Sjöstrand M, et al. (1999) Neutrophil Recruitment by Human IL-17 Via C-X-C Chemokine Release in the Airways. The Journal of Immunology 162: 2347–2352.

3. Skeen MJ, Ziegler HK (1993) Induction of murine peritoneal gamma/delta T cells and their role in resistance to bacterial infection. The Journal of Experimental Medicine 178: 971–984.

4. Cheng P, Liu T, Zhou W-Y, Zhuang Y, Peng L-s, et al. (2012) Role of gamma-delta T cells in host response against Staphylococcus aureus-induced pneumonia. BMC Immunology 13: 38.

5. Koohsari H, Tamaoka M, Campbell H, Martin J (2007) The role of gammadelta T cells in airway epithelial injury and bronchial responsiveness after chlorine gas exposure in mice. Respiratory Research 8: 21.

6. McMenamin C, Pimm C, McKersey M, Holt PG (1994) Regulation of IgE responses to inhaled antigen in mice by antigen-specific gamma delta T cells. Science 265: 1869–1871.

7. Pociask DA, Chen K, Mi Choi S, Oury TD, Steele C, et al. (2011) γδ T Cells Attenuate Bleomycin-Induced Fibrosis through the Production of CXCL10. The American Journal of Pathology 178: 1167–1176.

8. Pons J, Sauleda J, Ferrer JM, Barceló B, Fuster A, et al. (2005) Blunted γδ T-lymphocyte response in chronic obstructive pulmonary disease. European Respiratory Journal 25: 441–446.

9. Devlin RB, McDonnell WF, Mann R, Becker S, House DE, et al. (1991) Exposure of Humans to Ambient Levels of Ozone for 6.6 Hours Causes Cellular and Biochemical Changes in the Lung. American Journal of Respiratory Cell and Molecular Biology 4: 72–81.

10. Bell ML, Dominici F, Samet JM (2005) A Meta-Analysis of Time-Series Studies of Ozone and Mortality With Comparison to the National Morbidity, Mortality, and Air Pollution Study. Epidemiology 16: 436–445 410.1097/ 1001.ede.0000165817.0000140152.0000165885.

11. Levy JI, Chemerynski SM, Sarnat JA (2005) Ozone Exposure and Mortality: An Empiric Bayes Metaregression Analysis. Epidemiology 16: 458–468 410.1097/ 1001.ede.0000165820.0000108301.b0000165823.

12. Triche EW, Gent JF, Holford TR, Belanger K, Bracken MB, et al. (2006) Low-level ozone exposure and respiratory symptoms in infants. Environ Health Perspect 114: 911–916.

13. Chiu H-F, Cheng M-H, Yang C-Y (2009) Air Pollution and Hospital Admissions for Pneumonia in a Subtropical City: Taipei, Taiwan. Inhalation Toxicology 21: 32–37.

14. Peden DB (1996) Effect of Air Pollution in Asthma and Respiratory Allergy. Otolaryngology – Head and Neck Surgery 114: 242–247.

15. Charpin D, Pascal L, Birnbaum J, Armengaud A, Sambuc R, et al. (1999) Gaseous air pollution and atopy. Clin Exp Allergy 29: 1474–1480.

16. Boutin-Forzano S, Hammou Y, Gouitaa M, Charpin D (2005) Air pollution and atopy. Eur Ann Allergy Clin Immunol 37: 11–16.

17. Zhao Q, Simpson LG, Driscoll KE, Leikauf GD (1998) Chemokine regulation of ozone-induced neutrophil and monocyte inflammation. American Journal of Physiology - Lung Cellular and Molecular Physiology 274: L39–L46.

18. Kasahara DI, Kim HY, Williams AS, Verbout NG, Tran J, et al. (2012) Pulmonary inflammation induced by subacute ozone is augmented in adiponectin-deficient mice: role of IL-17A. J Immunol 188: 4558–4567.

19. Shimizu M, Hasegawa N, Nishimura T, Endo Y, Shiraishi Y, et al. (2009) Effects of TNF-alpha-converting enzyme inhibition on acute lung injury induced by endotoxin in the rat. Shock 32: 535–540.

20. Churg A, Dai J, Tai H, Xie C, Wright JL (2002) Tumor Necrosis Factor-α Is Central to Acute Cigarette Smoke–induced Inflammation and Connective Tissue Breakdown. American Journal of Respiratory and Critical Care Medicine 166: 849–854.

21. Malaviya R, Ikeda T, Ross E, Abraham SN (1996) Mast cell modulation of neutrophil influx and bacterial clearance at sites of infection through TNF-[alpha]. Nature 381: 77–80.

22. Cho H-Y, Zhang L-Y, Kleeberger SR (2001) Ozone-induced lung inflammation and hyperreactivity are mediated via tumor necrosis factor-α receptors. American Journal of Physiology - Lung Cellular and Molecular Physiology 280: L537–L546.

23. Kleeberger SR, Levitt RC, Zhang LY, Longphre M, Harkema J, et al. (1997) Linkage analysis of susceptibility to ozone-induced lung inflammation in inbred mice. Nat Genet 17: 475–478.

24. Yonemaru M, Stephens KE, Ishizaka A, Zheng H, Hogue RS, et al. (1989) Effects of tumor necrosis factor on PMN chemotaxis, chemiluminescence, and elastase activity. J Lab Clin Med 114: 674–681.

25. Pober JS (1987) Effects of tumour necrosis factor and related cytokines on vascular endothelial cells. Ciba Found Symp 131: 170–184.

26. Antiga E, Volpi W, Cardilicchia E, Maggi L, Filì L, et al. (2012) Etanercept Downregulates the Th17 Pathway and Decreases the IL-17+/IL-10+ Cell Ratio in Patients with Psoriasis Vulgaris. Journal of Clinical Immunology 32: 1221–1232.

27. Sugita S, Kawazoe Y, Imai A, Yamada Y, Horie S, et al. (2012) Inhibition of Th17 differentiation by anti-TNF-alpha therapy in uveitis patients with Behcet's disease. Arthritis Research & Therapy 14: R99.

28. Cazzola M, Calzetta L, Lauro D, Bettoncelli G, Cricelli C, et al. (2013) Asthma and COPD in an Italian adult population: role of BMI considering the smoking habit. Respir Med 107: 1417–1422.

29. Ehrlich SF, Quesenberry CP, Van Den Eeden SK, Shan J, Ferrara A (2010) Patients Diagnosed With Diabetes at Increased Risk for Asthma, Chronic Obstructive Pulmonary Disease, Pulmonary Fibrosis, and Pneumonia but Not Lung Cancer. Diabetes Care 33: 55–60.

30. Johnston RA, Theman TA, Lu FL, Terry RD, Williams ES, et al. (2008) Diet-induced obesity causes innate airway hyperresponsiveness to methacholine and enhances ozone-induced pulmonary inflammation. Journal of Applied Physiology 104: 1727–1735.

31. Johnston RA, Theman TA, Shore SA (2006) Augmented responses to ozone in obese carboxypeptidase E-deficient mice. Am J Physiol Regul Integr Comp Physiol 290: R126–133.

32. Lu FL, Johnston RA, Flynt L, Theman TA, Terry RD, et al. (2006) Increased pulmonary responses to acute ozone exposure in obese db/db mice. American Journal of Physiology - Lung Cellular and Molecular Physiology 290: L856–L865.

33. Shore SA, Rivera-Sanchez YM, Schwartzman IN, Johnston RA (2003) Responses to ozone are increased in obese mice. J Appl Physiol 95: 938–945.

34. Shore SA, Lang JE, Kasahara DI, Lu FL, Verbout NG, et al. (2009) Pulmonary responses to subacute ozone exposure in obese vs. lean mice. Journal of Applied Physiology 107: 1445–1452.

35. Mancuso P (2010) Obesity and lung inflammation. Journal of Applied Physiology 108: 722–728.

36. Coleman DL, Eicher EM (1990) Fat (fat) and Tubby (tub): Two Autosomal Recessive Mutations Causing Obesity Syndromes in the Mouse. Journal of Heredity 81: 424–427.

37. Williams AS, Mathews JA, Kasahara DI, Chen L, Wurmbrand AP, et al. (2013) Augmented Pulmonary Responses to Acute Ozone Exposure in Obese Mice: Roles of TNFR2 and IL-13. Environ Health Perspect 121: 551–557.

38. Skerry C, Harper J, Klunk M, Bishai WR, Jain SK (2012) Adjunctive TNF inhibition with standard treatment enhances bacterial clearance in a murine model of necrotic TB granulomas. PLoS ONE 7: e39680.

39. Grounds MD, Davies M, Torrisi J, Shavlakadze T, White J, et al. (2005) Silencing TNFα activity by using Remicade or Enbrel blocks inflammation in whole muscle grafts: an in vivo bioassay to assess the efficacy of anti-cytokine drugs in mice. Cell and Tissue Research 320: 509–515.

40. Lahn M, Kalataradi H, Mittelstadt P, Pflum E, Vollmer M, et al. (1998) Early Preferential Stimulation of γδ T Cells by TNF-α. The Journal of Immunology 160: 5221–5230.

41. Bhalla DK (1999) Ozone-induced lung inflammation and mucosal barrier disruption: toxicology, mechanisms, and implications. J Toxicol Environ Health B Crit Rev 2: 31–86.

42. Backus GS, Howden R, Fostel J, Bauer AK, Cho HY, et al. (2010) Protective role of interleukin-10 in ozone-induced pulmonary inflammation. Environ Health Perspect 118: 1721–1727.

43. Johnston RA, Schwartzman IN, Flynt L, Shore SA (2005) Role of interleukin-6 in murine airway responses to ozone. American Journal of Physiology - Lung Cellular and Molecular Physiology 288: L390–L397.

44. Michalec L, Choudhury BK, Postlethwait E, Wild JS, Alam R, et al. (2002) CCL7 and CXCL1 orchestrate oxidative stress-induced neutrophilic lung inflammation. J Immunol 168: 846–852.

45. Sutton CE, Lalor SJ, Sweeney CM, Brereton CF, Lavelle EC, et al. (2009) Interleukin-1 and IL-23 Induce Innate IL-17 Production from γδ T Cells, Amplifying Th17 Responses and Autoimmunity. Immunity 31: 331–341.

46. Veldhoen M, Hocking RJ, Atkins CJ, Locksley RM, Stockinger B (2006) TGFβ in the Context of an Inflammatory Cytokine Milieu Supports De Novo Differentiation of IL-17-Producing T Cells. Immunity 24: 179–189.

47. Li Z, Burns AR, Byeseda Miller S, Smith CW (2011) CCL20, γδ T cells, and IL-22 in corneal epithelial healing. The FASEB Journal 25: 2659–2668.

48. Mabuchi T, Singh TP, Takekoshi T, Jia G-f, Wu X, et al. (2013) CCR6 Is Required for Epidermal Trafficking of [gamma][delta]-T Cells in an IL-23-Induced Model of Psoriasiform Dermatitis. J Invest Dermatol 133: 164–171.

49. Matsubara S, Takeda K, Jin N, Okamoto M, Matsuda H, et al. (2009) Vgamma1+ T cells and tumor necrosis factor-alpha in ozone-induced airway hyperresponsiveness. Am J Respir Cell Mol Biol 40: 454–463.

50. King DP, Hyde DM, Jackson KA, Novosad DM, Ellis TN, et al. (1999) Cutting Edge: Protective Response to Pulmonary Injury Requires γδ T Lymphocytes. The Journal of Immunology 162: 5033–5036.

51. Umemura M, Yahagi A, Hamada S, Begum MD, Watanabe H, et al. (2007) IL-17-Mediated Regulation of Innate and Acquired Immune Response against Pulmonary Mycobacterium bovis Bacille Calmette-Guérin Infection. The Journal of Immunology 178: 3786–3796.

52. Braun RK, Ferrick C, Neubauer P, Sjoding M, Sterner-Kock A, et al. (2008) IL-17 producing gammadelta T cells are required for a controlled inflammatory response after bleomycin-induced lung injury. Inflammation 31: 167–179.

53. Wozniak K, Kolls J, Wormley F (2012) Depletion of neutrophils in a protective model of pulmonary cryptococcosis results in increased IL-17A production by gamma/delta T cells. BMC Immunology 13: 65.

54. Lo Re S, Dumoutier L, Couillin I, Van Vyve C, Yakoub Y, et al. (2010) IL-17A–Producing γδ T and Th17 Lymphocytes Mediate Lung Inflammation but Not Fibrosis in Experimental Silicosis. The Journal of Immunology 184: 6367–6377.

55. Sergejeva S, Ivanov S, Lotvall J, Linden A (2005) Interleukin-17 as a recruitment and survival factor for airway macrophages in allergic airway inflammation. Am J Respir Cell Mol Biol 33: 248–253.

56. Cai X-Y, Gommoll Jr CP, Justice L, Narula SK, Fine JS (1998) Regulation of granulocyte colony-stimulating factor gene expression by interleukin-17. Immunology Letters 62: 51–58.

57. Jones CE, Chan K (2002) Interleukin-17 stimulates the expression of interleukin-8, growth-related oncogene-alpha, and granulocyte-colony-stimulating factor by human airway epithelial cells. Am J Respir Cell Mol Biol 26: 748–753.

58. Hirai Y, Iyoda M, Shibata T, Kuno Y, Kawaguchi M, et al. (2012) IL-17A stimulates granulocyte colony-stimulating factor production via ERK1/2 but not p38 or JNK in human renal proximal tubular epithelial cells. American Journal of Physiology - Renal Physiology 302: F244–F250.

59. Cox G, Gauldie J, Jordana M (1992) Bronchial epithelial cell-derived cytokines (G-CSF and GM-CSF) promote the survival of peripheral blood neutrophils in vitro. Am J Respir Cell Mol Biol 7: 507–513.

60. Bauer AK, Travis EL, Malhotra SS, Rondini EA, Walker C, et al. (2010) Identification of novel susceptibility genes in ozone-induced inflammation in mice. Eur Respir J 36: 428–437.

61. Vassalli P (1992) The Pathophysiology of Tumor Necrosis Factors. Annual Review of Immunology 10: 411–452.

62. Korn T, Petermann F (2012) Development and function of interleukin 17–producing γδ T cells. Annals of the New York Academy of Sciences 1247: 34–45.

63. Ueta C, Kawasumi H, Fujiwara H, Miyagawa T, Kida H, et al. (1996) Interleukin-12 activates human gamma delta T cells: synergistic effect of tumor necrosis factor-alpha. Eur J Immunol 26: 3066–3073.

64. Kim CH (2009) Migration and function of Th17 cells. Inflamm Allergy Drug Targets 8: 221–228.

65. Brunner PM, Koszik F, Reininger B, Kalb ML, Bauer W, et al. (2013) Infliximab induces downregulation of the IL-12/IL-23 axis in 6-sulfo-LacNac (slan)+ dendritic cells and macrophages. Journal of Allergy and Clinical Immunology 132: 1184–1193.e1188.

66. Kordonowy LL, Burg E, Lenox CC, Gauthier LM, Petty JM, et al. (2012) Obesity Is Associated with Neutrophil Dysfunction and Attenuation of Murine Acute Lung Injury. American Journal of Respiratory Cell and Molecular Biology 47: 120–127.

67. Smith AG, Sheridan PA, Harp JB, Beck MA (2007) Diet-Induced Obese Mice Have Increased Mortality and Altered Immune Responses When Infected with Influenza Virus. The Journal of Nutrition 137: 1236–1243.

68. Mancuso P, Gottschalk A, Phare SM, Peters-Golden M, Lukacs NW, et al. (2002) Leptin-Deficient Mice Exhibit Impaired Host Defense in Gram-Negative Pneumonia. The Journal of Immunology 168: 4018–4024.

69. Wieland CW, Florquin S, Chan ED, Leemans JC, Weijer S, et al. (2005) Pulmonary Mycobacterium tuberculosis infection in leptin-deficient ob/ob mice. International Immunology 17: 1399–1408.

70. Milner JJ, Sheridan PA, Karlsson EA, Schultz-Cherry S, Shi Q, et al. (2013) Diet-Induced Obese Mice Exhibit Altered Heterologous Immunity during a Secondary 2009 Pandemic H1N1 Infection. The Journal of Immunology 191: 2474–2485.

71. Morgan OW, Bramley A, Fowlkes A, Freedman DS, Taylor TH, et al. (2010) Morbid Obesity as a Risk Factor for Hospitalization and Death Due to 2009 Pandemic Influenza A(H1N1) Disease. PLoS ONE 5: e9694.

72. Li W, Moltedo B, Moran TM (2012) Type I interferon induction during influenza virus infection increases susceptibility to secondary Streptococcus pneumoniae infection by negative regulation of gammadelta T cells. J Virol 86: 12304–12312.

73. Taylor KR, Costanzo AE, Jameson JM (2011) Dysfunctional gammadelta T cells contribute to impaired keratinocyte homeostasis in mouse models of obesity. J Invest Dermatol 131: 2409–2418.

74. Taylor KR, Mills RE, Costanzo AE, Jameson JM (2010) Gammadelta T cells are reduced and rendered unresponsive by hyperglycemia and chronic TNFalpha in mouse models of obesity and metabolic disease. PLoS ONE 5: e11422.

21

Ozone-Induced Responses in *Croton floribundus* Spreng. (Euphorbiaceae): Metabolic Cross-Talk between Volatile Organic Compounds and Calcium Oxalate Crystal Formation

Poliana Cardoso-Gustavson[1], Vanessa Palermo Bolsoni[2], Debora Pinheiro de Oliveira[2], Maria Tereza Gromboni Guaratini[2], Marcos Pereira Marinho Aidar[3], Mauro Alexandre Marabesi[3], Edenise Segala Alves[4], Silvia Ribeiro de Souza[2]*

1 Programa de Pós-Graduação em Biodiversidade Vegetal e Meio Ambiente, Instituto de Botânica, São Paulo, São Paulo, Brazil, 2 Núcleo de Pesquisa em Ecologia, Instituto de Botânica, São Paulo, São Paulo, Brazil, 3 Núcleo de Pesquisa em Fisiologia e Bioquímica, Instituto de Botânica, São Paulo, São Paulo, Brazil, 4 Núcleo de Pesquisa em Anatomia, Instituto de Botânica, São Paulo, São Paulo, Brazil

Abstract

Here, we proposed that volatile organic compounds (VOC), specifically methyl salicylate (MeSA), mediate the formation of calcium oxalate crystals (COC) in the defence against ozone (O_3) oxidative damage. We performed experiments using *Croton floribundus*, a pioneer tree species that is tolerant to O_3 and widely distributed in the Brazilian forest. This species constitutively produces COC. We exposed plants to a controlled fumigation experiment and assessed biochemical, physiological, and morphological parameters. O_3 induced a significant increase in the concentrations of constitutive oxygenated compounds, MeSA and terpenoids as well as in COC number. Our analysis supported the hypothesis that ozone-induced VOC (mainly MeSA) regulate ROS formation in a way that promotes the opening of calcium channels and the subsequent formation of COC in a fast and stable manner to stop the consequences of the reactive oxygen species in the tissue, indeed immobilising the excess calcium (caused by acute exposition to O_3) that can be dangerous to the plant. To test this hypothesis, we performed an independent experiment spraying MeSA over *C. floribundus* plants and observed an increase in the number of COC, indicating that this compound has a potential to directly induce their formation. Thus, the tolerance of *C. floribundus* to O_3 oxidative stress could be a consequence of a higher capacity for the production of VOC and COC rather than the modulation of antioxidant balance. We also present some insights into constitutive morphological features that may be related to the tolerance that this species exhibits to O_3.

Editor: Martin Heil, Centro de Investigación y de Estudios Avanzados, Mexico

Funding: The authors acknowledge financial support from Fundação de Amparo à Pesquisa do Estado de São Paulo (FAPESP, 12/11663-8). PCG thanks Brazilian CAPES/CNPq (0469-13-0 and 0588/12-1) for funding. DOP and VPB thank Brazilian CNPq for undergraduate scholarships. SRS and ESA thank Brazilian CNPq for their research grant (307238/2013-2 and 36461/2012-1, respectively). The funders had no role in study design, data collection and analysis, decision to publish, or preparation of manuscript.

Competing Interests: The authors have declared that no competing interests exist.

* Email: souzasrd@pq.cnpq.br

Introduction

Tropospheric ozone (O_3) is considered to be the gaseous pollutant that is most damaging to plants due to its strong oxidation capacity [1]. This gas initially enters plants through the stomata, causing biochemical and physiological alterations [2]; based on the concentration, the duration of the exposure, and the responsiveness of the plant species, O_3 can induce disturbances in organelles and tissues, leading to visible morphological symptoms [3–9].

A biochemical balance is required in the cell to counteract the negative effects of O_3. Enzymatic and non-enzymatic antioxidant compounds, including ascorbate peroxidase (APX), superoxide dismutase (SOD), peroxidase (POD), and ascorbic acid (AA), counteract the increase in reactive oxygen species (ROS) promoted by O_3 [2,10–11]. Another possible metabolic response

against oxidative stress that is actively modulated by plants is the production of terpenoids, which can constitute the emissions of volatile organic compounds (VOC) [12,13]. Being antioxidant compounds, VOC can remove the ROS formed in the intercellular spaces [11–12,14–22]. Mono- and sesquiterpenes are the most dominant VOC emitted by plants in response to O_3 [1,23–25]; they constitute a large family of plant metabolites, with diverse functions in plant growth, development and stress response [26]. These compounds can be produced by various metabolic routes [27], preferentially in phloem and xylem parenchyma or by secretory cells associated with these tissues [28–30].

Intracellular calcium is also modulated during plant defence responses to O_3 [31], which might lead to the formation of calcium salt crystals. Calcium oxalate crystals (COC) have roles in the defence against herbivores and/or the accumulation of excess calcium –see reviews in [32–34]. It is well known that these constitutive inclusions can be quantitatively induced by biotic stressors such as herbivory [35–37]. Although there is only one report of the induction of crystal formation by O_3 in a gymnosperm species, *Picea abies* [38], there seems to be a significant difference in the quantity of COC produced in *Eugenia uniflora* (Myrtaceae), a tropical species, when polluted and non-polluted environments are considered [39].

There are few studies on the responses of tropical Brazilian tree species to O_3 stress [40]. *Croton floribundus* Spreng. is a pioneer tree species widely distributed in the Brazilian forest and recommended for use in ecological restoration [41]. This species seems to be tolerant to O_3 fumigation, not presenting any structural symptoms of oxidative stress even under 80 ppb of O_3 6 h/day during 53 days [42], although visible symptoms such as hypersensitive-like responses (HR-like), peroxide hydrogen accumulation (H_2O_2) and polyphenol compound accumulation occurred after exposure to 200 ppb of O_3 for 3 h/day for three days [8]. In our ongoing research, O_3 has not been able to change the antioxidant levels in this species, reinforcing the hypothesis that *C. floribundus* is tolerant to oxidative stress caused by this pollutant (at least under less-intense acute exposures). Moreover, in our ongoing research, VOC and COC levels appeared to be O_3-dependent. These preliminary findings suggested that the responsiveness of *C. floribundus* to O_3 might be linked with VOC emission and COC formation. Thus, we raised the hypothesis of which there is a metabolic cross-talk between VOC emission and COC formation that confers defences against oxidative stress.

To test this hypothesis, the aims of this study are (1) to assess ozone oxidative stress by measuring biochemical and physiological responses; (2) to verify whether the variation in the emitted VOC and in the quantity of COC co-occur as a response to this pollutant and if so, (3) to evaluate whether there is cross-talk between the emitted VOC and the COC formation, by the direct application of a compound present in the emitted *bouquet* that is probably involved in the induction of the COC formation.

Materials and Methods

Plant material

Six-month old *C. floribundus* plants were purchased from BIOvida Company (São Paulo, Brazil) and immediately planted in 10 L pots filled with a 3:1 mixture of peat and sand, and watered by capillarity. Plants were kept inside the greenhouse for three weeks and then were transferred from the greenhouse to the fumigation chambers, where they were kept for 2 days before the beginning of fumigation experiment (acclimation period).

Using ozone fumigation to assess leaf responses

The ozone fumigation experiments were performed in closed chambers kept inside a laboratory with temperature and humidity controlled with central air conditioner and under artificial illumination supplied by metallic vapour (400 W) and fluorescent (30 W TL05) lamps [43]. The material was divided into two lots, with half exposed to ozone (FA+O3) and the other half receiving filtered air (FA) only. The chambers were composed of a stainless steel structure covered by a film of Teflon, in the dimensions of 85 cm×94 cm×85 cm (W×D×H). The air was filtered by paper filter to remove gross particulate matter (Whatman 40), followed by silica gel (150 g, Merck), active carbon (250 g, Merck), potassium permanganate (500 g, Purafil Select), and paper filter to remove fine particulate (Whatman QMA). The filtration efficiency was assessed by measuring ozone levels in the air passed through the filtering system. The average ozone levels reached a maximum of 5 ppb after filtration, which indicates an efficiency of filtration of 98.5% [44]. After filtration, the air was enriched with 80 ppb of ozone. Ozone was generated under electrical discharge by the dissociation of oxygen contained in filtered air, using an ozone generator (Ozontechenic). The ozone levels were monitored using an Ecotech 9810B photometric monitor. The ozone monitor was calibrated once before each exposure. During the fumigation experiments, the mean temperature, relative humidity, and photon flux density values were 29±1.5°C, 63±17 and 184.2 µmol/m^2.s respectively, simulating appropriate conditions for optimal growth of *C. floribundus*. In this experiment, six plants were exposed to filtered ozone-free air (FA) and another six to filtered air plus 80 ppb of ozone (FA+O3) for 4 hours/day per seven consecutive days. Each experiment was performed in triplicate. In order to reduce chamber effects, plants were switched between two chambers in the end of every day of exposure. Thus, both chambers were used for FA and FA+O3 treatment. The position of the plants was also changed to counteract the positional effect [45]. It is important to highlight that there were no differences in the conditions (illumination, humidity, temperature) in which the experiments were carried out.

Table 1. Values (mean and standard deviation) of biochemical and physiological parameters measured in leaves of *Croton floribundus* exposed to ozone (FA+O$_3$) and control plants under filtered air (FA).

Variable	Treatment	
	FA+O$_3$	**FA**
Biochemical		
AA (mg g^{-1} DW)	7.74±3.39a	7.70±1.85a
POD (10^2 DA min^{-1} DW)	11.13±1.72a	10.55±0.80a
SOD (10^2 U g^{-1} DW)	6.99±1.24a	7.90±0.89a
Physiological		
NPQ	0.657±0.021a	0.077±0.034b
Fv:Fm	0.654±0.017b	0.713±0.004a
ETR	82.7±8.8b	101.9±1.9a

Different letters indicate statistically significant differences among treatments for each parameter analysed ($p<0.05$).

Table 2. Percentages of volatile organic compounds (means $\pm$ S.D) emitted by leaves of *Croton floribundus* exposed to ozone (FA+O_3) and control plants under filtered air (FA), and their linear retention indices in literature (Ri ref) and calculated (Ri cal).

Compounds	FA[a]	FA+O_3[a]	Ri ref	Ri cal[b]
Non-terpenoid Oxygenated				
nonan-2-one	0.67±0.48	0.53±0.52	1091	1091.3
pentadecanoic acid	0.33±0.15	0.50±0.35	1829	1826.3
Nonadecanal	0.33±0.21	0.37±0.24	2105	2105.1
octadecanoic acid	0.34±0.24	0.47±0.17	2200	2200.2
(E)-3-octen-2-one	2.20±1.70	4.07±1.21	1034	1033
Sum	*3.87*	*5.94*		
Aromatic				
(Z)-3-hexenyl benzoate	0.63±0.44	0.48±0.18	1570	1570,27
ethyl benzoate	0.42±0.22	0.62±0.22	1169	1170.68
methyl salicylate	0.30±0.32	0.73±0.13	1191	1191.13
methyl benzyl formate	0.57±0.28	0.57±0.26	1335	1335.56
methyl 3–4 dimethylbenzoate	0.52±0.41	0.34±0.23	1287	1353.73
2,6 dimethylphenol	Nd	0.25±0.20	1510	1504
Sum	*2.44*	*2.99*		
Monoterpene				
Ocimene	0.55±0.48	0.65±0.29	1063	1098
Geraniol	0.39±0.26	0.68±0.22	1276	1235
geranil acetate	0.32±0.12	0.54±0.25	1383	1378
γ terpinene	0.43±0.26	0.50±0.33	1016	1027
Carvone	0.42±0.30	0.45±0.14	1253	1248
Linalool	0.32±0.15	0.26±0.16	1173	1174
α-pinene	0.26±0.20	0.31±0.23	934	--
1,8 cineol	0.70±0.37	1.47±0.18	1027	1030
Myrcene	0.34±0.19	0.40±0.25	988	--
Sum	*3.73*	*5.26*		
Sesquiterpene				
trans-nerolidol	0.53±0.40	0.55±0.18	1566	1564.2
β-gurjunene	0.48±0.37	0.56±0.26	1409	1423
caryophyllene oxide	0.70±0.26	0.84±0.25	1566	1576
aromadendrene	0.48±0.37	0.39±0.26	1636	1638
Copaene	0.55±0.26	0.58±0.29	1369	1372.8
α-caryophyllene	0.23±0.20	Nd	1411	1420
Sum	*2.97*	*2,92*		

Significance of Repeated Measures ANOVA				
Factor[c]	**F**		**P**	
ozone	0.02		0.90	
time	0.54		0.51	
ozone * time	0.94		0.49	

[nd]not detected.
[a]mean percentage and standard deviation of three replicates (including the six young plants after 7 days of exposure).
[b]linear retention index calculated on DB5 capillary column with a homologous series of n-alkanes (C8–C30).
--data insufficient to calculate Ri.
[c]Greenhouse-Geisser values calculated by ANOVA considering all volatiles.

Figure 1. *In situ* **localisation of terpenoids on leaves of** *Croton floribundus.* Control plants under filtered air (A, C) and exposed to ozone for three (D), five (E) or seven (B, F) days. (A, B) Control from histochemical analysis (material without any dye or reagent). (C–F) Positive results for Nadi reagent inside laticifers (C); note a qualitative increase in the terpenoid content inside the parenchyma cells of vascular tissue according to the ozone exposure time (D–F). Arrows indicate laticifers. Bars: 100 μm.

Biochemical and physiological responses to ozone: fluorescence and antioxidant analyses

A Pulse-Amplitude-Modulated Fluorometer (PAM-2100, Heinz Walz GMBH, Effeltrich, Germany) was used to measure leaf steady-state chlorophyll fluorescence parameters after seven days of fumigation and following a 30 min adaptation to the dark. The NPQ was calculated as $NPQ = (Fm–Fm')/Fm'$, where Fm is a maximal fluorescence in the dark and Fm' is a maximal fluorescence in the light. The Fv/Fm values (photosynthetic quantum efficiency) represent averages from 15 measurements taken sequentially on two different fully expand leaves of three plants per experiment. Electron Transport Rate (ETR) is calculated as $ETR = yield \times 0.5 \times 0.82 \times PAR$, where yield is the quantum yield of the PSII (Fq/Fm), 0.5 represents the light absorbed by the PSII, 0.82 represents the absorbance of the leaf and PAR represents the light intensity used (400 µmol photons/$m^2.s^2$ using the halogen light source). Values of ETR represent ten measurements in two different fully expand leaves of three plants per treatment. All measurements were taken in the morning, between 10:00 and 11:00 am.

The antioxidant defences were analysed in six individuals per treatment. Collection of leaves and preparation of extracts for analysis of antioxidants always followed the same sequence in time to avoid diurnal variation. The extraction was carried out with a mix of all expanded leaves. Total ascorbic acid was measured in 0.5 g of fresh leaves and homogenised with 12 mL of EDTA-Na$_2$ (0.07%) and oxalic acid (0.5%). The mixture was centrifuged at 40.000 g for 30 min at 2°C. An aliquot of the supernatant was added to 2.5 mL of DCPIP (0.02%), and absorbance was measured with a spectrophotometer at 520 nm. After the addition of 0.05 mL of ascorbic acid (1%), a second absorbance measurement was taken. Both absorbance measurements were used to estimate the total ascorbic acid content following [46].

Superoxide dismutase activity was measured in 0.35 g of fresh leaves homogenised with 12 mL of potassium phosphate buffer (50 mM pH 7.5), EDTA-Na$_2$ 1 mM, NaCl 50 mM and ascorbic acid 1 mM in the presence of 0.4 g of PVPP 2%. This mixture was centrifuged at 22.000 g for 25 min at 4°C. SOD activity was assayed by measuring the SOD inhibition of the NBT photochemical reduction [47]. Each reaction mixture contained 0.5 mL of EDTA-Na$_2$ 0.54 mM, 0.8 mL of potassium phosphate buffer (0.1 M, pH 7.0), 0.5 mL of methionine 0.13 mM, 0.5 mL of NBT 0.44 mM, 0.2 mL of riboflavin 1 mM, and 0.2 mL of leaf extract.

The samples were incubated for 20 min under a fluorescent lamp (80 W). The absorbance of the reaction mixture was measured at 560 nm. A similar mixture lacking the leaf extract was used as a control, and a dark control mixture served as a blank. The enzymatic activity was expressed as the amount of extract needed to inhibit the reduction of NBT by 50%.

Peroxidase activity was measured in 0.3 g of leaves homogenised with 12 mL potassium phosphate buffer (0.1 M, pH 7.0) in the presence of 0.4 g of PVPP 2%. The homogenate was centrifuged at 40.000 g for 30 min at 2°C. Peroxidase activity was also measured in a reaction mixture of plant extracts using 0.1 M potassium phosphate buffer (pH 5.5) and phenylenediamine (1%) to which an aliquot of H$_2$O$_2$ (0.3%) was added. Unspecific POD activity was measured with a spectrophotometer following the increase in absorbance (DA) at 485 nm due to the formation of an H$_2$O$_2$-POD complex at two different times in the linear reaction curve [48].

Volatile Collection

Volatiles from the headspace of six whole plants from each treatment (FA and FA+O3) were collected into steel tubes containing 150 mg charcoal adsorbent (Supelco, PA, USA) at an airflow rate of approximately 200 ml/min for 60 min. To avoid the VOC from soil, the vessels were covered with aluminium foil. During the collection pure air was continuously supplied into the chambers. One tube was fixed into each chamber (FA and FA+O3) with a line Teflon tubes and connected in the vacuum pumps for volatiles sampling. All collections were performed in the morning after approximately 60 min of the light being switched on and before starting the ozone fumigation. The ozone scrubbers (filters coated with a saturated solution of potassium iodide) were fixed before the adsorbent to avoid any degradation of samples by residual ozone. The collections were made every day after the first day of exposure, comprising a total of six samples per treatment (n = 6). Samples were replicated twice. Blank samples from an empty chamber were also collected twice.

CG-MS analyses

VOC were analysed by gas chromatography-mass spectrometry (GC-MS Hewlett-Packard GC 6890, MSD 5973, Wilmington, DE, USA). Trapped compounds were desorbed chemically with 200 µL of hexane-methylchloride (4:1). The volume was reduced to 50 µL before injection. The separation was performed on a DB-

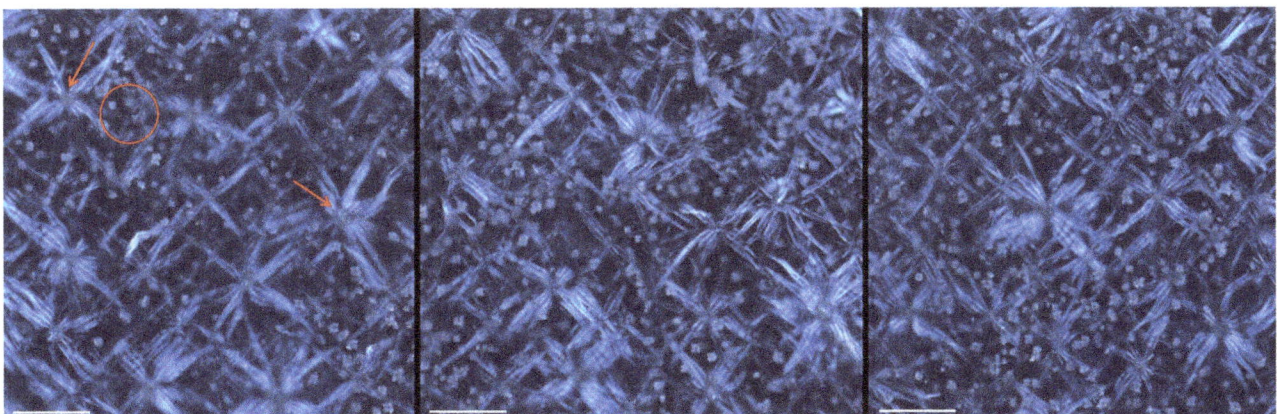

Figure 2. Areas in the adaxial surface of *Croton floribundus* observed under light microscope between crossed polarisers. Leaves before the exposition to O$_3$ (A), and after five (A) and seven (C) days of exposition. Note the abundance of crystals (circle) and non-secretory stellar trichomes (arrows). Bars: 300 µm.

Table 3. Effects of ozone, methyl salicylate and time (days) on the quantity of calcium oxalate crystals from *Croton floribundus* leaves.

Ozone treatment		
	Time (days) of exposure	
	5 (means ± SE)	7 (means ± SE)
FA	565±72	501±177
FA+O3	575±90	787±150
Significance of Repeated Measures ANOVA		
Factor[a]	F	P
ozone	31.42	0.001
time	1.77	0.192
ozone * time	27.12	0.001
MeSA treatment		
	Time (days) of application	
	5 (means ± SE)	7 (means ± SE)
Control	72.2±16.2	71.5±19.9
MeSA	105.6±14.7	111.53±21.9
Significance of Repeated Measures ANOVA		
Factor[a]	F	P
MeSA	165.67	0.001
Time	0.85	3.463
MeSA * time	1.37	0.243

[a]Greenhouse-Geisser values calculated by ANOVA.

5 capillary column (Agilent technologies, USA; 30 m × 0.25 mm ID, 0.25 μm film thickness). Helium was used as a carries gas at a constant flow rate 1.5 ml/min. The inlet temperature was 250°C, splitless injection mode. The oven temperature program was held at 40°C for 1 min and then raised to 210°C at a rate of 5°C/min, and finally to 250°C at a rate of 20°C/min. The mass spectrometer was operated in electron ionization mode at 70 eV, source temperature 230°C and quadrupole temperature 150°C.

Compounds identification and semi-quantification

VOC identification was undertaken by comparing the recorded mass spectra using Wiley library. The peak identification was performed when the similarity of mass spectra was higher than 80%. The linear retention index was used to secure the identification of each molecule. Retention index was calculated by injecting saturated n-alkanes standards solution C7–C30 (Supelco, Belgium) using the definition of Kovats retention [49]. The identification was not confirmed using standards due to limited availability of chemicals. Absolute peak areas were used to calculate the percentage of each compound in the sample. The percentage was performed comparing the sum of peaks areas (hundred percent of compounds, including unidentified compounds) and the individual area of each compound.

Histochemical localisation of terpenoids

For each O_3 exposure time (three, five or seven days), three leaves from the sixth oldest node were used. The medial region of fresh, fully expanded leaf blades was freehand sectioned. Five sections were used for *in situ* localization of terpenoids using Nadi reagent (naphthol+dimethyl-paraphenylenediamine) [50]. Sections were incubated in the dark for 60 min at room temperature in Nadi reagent, prepared immediately prior to staining. After incubation, the sections were rinsed for 2 min in a sodium phosphate buffer (0.1 M, pH 7.2). By oxidation this reagent forms indophenol blue that changes colour with variation in pH and makes it possible to distinguish between essential oils (blue) and resin acids (intense red), in which a purple color is observed when both compounds constitute the secretion [50,51]. Five other sections remained untreated – not submitted to any dye or reagents – for observation of the colour and structure of the cells *in vivo*. Observations and digital images were acquired with an Olympus BX53 compound microscope equipped with an Olympus Q-Color 5 digital camera and Image Pro Express 6.3 software.

Identification and quantification of calcium oxalate crystals

The same leaves used for *in situ* localisation of terpenoids were used for COC quantification purposes. The remainder of each leaf was fixed in FAA$_{50}$ (formalin-acetic acid-50% ethanol, 1:1:18) for

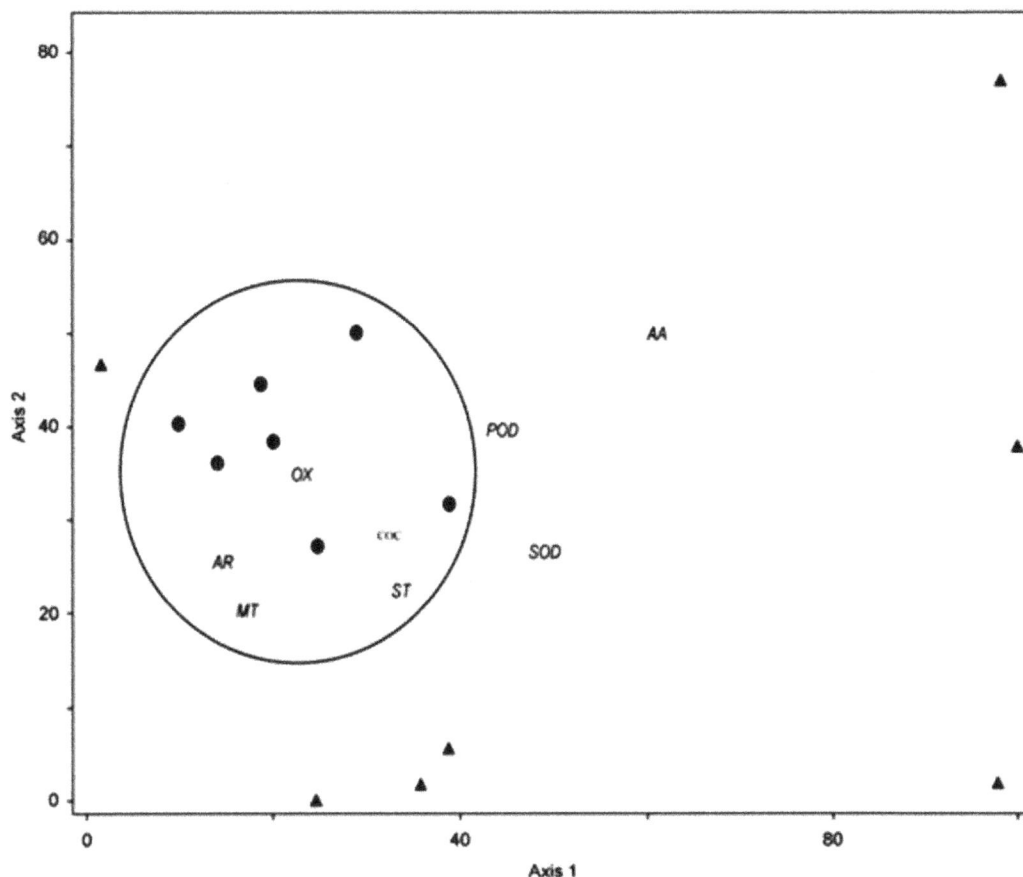

Figure 3. The PCA biplot diagram showing the volatile organic compounds (VOC) and antioxidant defences superimposed on the score of treatments indicated by black circles (ozone exposed, FA +O₃) and black triangles (filtered air, FA). The PC1 axis separates the treatments based on the O_3-induced VOC and calcium oxalate crystals (COC). On the PC2 axis, the treatments were characterised by their antioxidant defences. Abbreviations: VOC as aromatic (AR), monoterpene (MT), oxygenated (OX), sesquiterpene (ST); Ascorbic acid (AA); Peroxidases (POD); Superoxide dismutase (SOD).

24 h and then stored in 70% ethanol [52]. The samples were hydrated and diaphanised using 10% sodium hydroxide and 20% hypochlorite solution (modified from) [52], and the medial region (approximately 3 cm in length) was sectioned and mounted in glycerine. Thirty squares of 5 mm² were observed between crossed polarisers and photographed with an Olympus BX53 microscope. All the crystals present on images were counted using Image J (National Institutes of Health, USA).

For anatomical purposes, we used all the methods above on fully expanded leaves assuming that ozone symptoms initially appear with greater severity in older leaves toward the base of the plant [53] and that total terpenoids significantly increase with plant ontogeny [54].

Is methyl salicylate responsible for the induction of COC?

To verify if methyl salicylate (MeSA) was directly involved in the induction of COC, we performed an independent experiment using young plants acquired from the same company, planted, watered, and two days acclimated in fumigation chambers as aforementioned.

Twelve plants were used for this experiment, separated in two groups of six plants and maintained separately inside two chambers, the same ones used for fumigation experiments, but emitting filtered air only. Plants were kept under the same

conditions as described for filtered air plants from fumigation experiments during seven days.

A solution of 5 mM of MeSA (Sigma-Aldrich, Saint Louis, USA) was sprayed to each plant of the treated group every morning (1 ml/day), while distilled water was applied to each plant of the control group. After five and seven days of application the median region of the expanded leaves from two treated and two control plants were removed and then proceeded the protocols of COC counting as aforementioned. Thirty squares of 1.225 mm² were observed between crossed polarisers, photographed and counted as described above.

Statistical analysis

We used the statistical package SPSS 14.0 for Windows (SPSS; Chicago, IL, USA). Differences between FA and FA+O3 for antioxidant and physiological parameters were determined by the paired t-test with 95% confidence. VOC and COC were analysed using Repeated Measures ANOVA, with time and ozone as factors. Repeated Measures ANOVA was also applied on the independent experiment with MeSA and time. The differences between means were considered to be statistically significant at $P <$ 0.05. As the sphericity assumption was violated in all Repeated Measures ANOVA applied on the VOC data, the Greenhouse-Geisser conservative F-test was used [55].

The correlations among treatments, antioxidants, VOC emission and COC formation were investigated by principal component analysis using the data of seventh day of exposure in order to include all the variables in the statistical analyses (PCA, developed by PCord package 6.0), in which seven principal components explained over 90% of the variation.

Results

Which are the effects of ozone on antioxidant defence and fluorescence?

After seven days of exposure, there was a slight increase in total ascorbic acid and peroxidases in plants exposed to FA+O3 treatment, but no significant differences were found when these data were compared to FA treatment (Table 1). Significant decreases in the values of Fv/Fm and ETR (8.3% and 18.9%, respectively) and increase in the values of NPQ were observed in plants submitted to FA+O_3 treatment.

What are the emitted volatiles?

All the VOC identified were grouped into their respective chemical classes and summarised on Table 2. The relative quantity of oxygenated non-terpenoids and monoterpene tended to be higher in FA+O3 plants, representing 39 and 27% of total relative quantity of VOC emitted, respectively. MeSA, (E)-3-octen-2-one and 1,8 cineole showed higher levels in FA+O3 plants. No significant effects ($P>0.05$) of time and in the interaction between time and O_3 were found.

Where are the terpenes?

Major qualitative results were observed in the midrib of the medial region of the leaf blade. Plants from FA and FA+O3 treatments presented no structural differences in epidermis, ground and vascular tissues (Figures 1A, B). The histochemical tests with Nadi reagent showed a positive reaction inside laticifers and parenchymal cells of xylem and phloem in all plants analysed: there were few cells with terpenoid content in leaves from FA treatment plants (constituent terpenoids, Figure 1C), while a qualitative increase of parenchyma cells involved on terpenoids metabolism was observed with increased exposure time to ozone (induced terpenoids, Figures 1D–F).

Do ozone and MeSA increase the COC abundance?

The abundance of COC and non-secretory stellar trichomes can be visualized in the Figure 2, also representing the areas where the quantification of the crystals was performed. Ozone fumigation resulted in a significant increase in the COC quantity ($P<0.01$) on the fifth and seventh days of exposure (Table 3). No significant effect of time ($P>0.05$) was observed on the COC quantity, whereas the effect of the interaction between O_3 and time was significant ($P>0.05$).

A significant increase was observed in the COC quantity on leaves from plants treated with MeSA ($P<0.01$; Table 3): the increase was not time-dependent and there was no interaction effect between time and MeSA.

Correlations among antioxidant defence, volatiles and crystal formation

Principal components analysis was used to study relationships among the antioxidants, volatiles and crystal formation (Figure 3). The first PC was described by the term 'antioxidant defence' and explained 57% of the total variation in the data, characterised by a high score of ascorbic acid rather than SOD and POD. The second PC, designated as 'induced defence', explained 26% of the total variation and was characterised by a lower score in volatile and crystal levels than ascorbic acid. The treatments were separated into specific groups: FA treatment was more related to antioxidant defences, explained by a high level of AA, while FA+O3 treatment was linked to volatile emissions and crystal formation.

Discussion

Our results indicate that O_3 induces biochemical, physiological and anatomical responses in a native Brazilian species. When these responses are combined, it is possible to understand how the plant can defend itself and how apparently independent features, such as the emission of VOC and production of COC, may play an important role in these responses.

The antioxidant defences act against O_3 to avoid the increase in cellular ROS such that the expression and/or activity levels of enzymes and other compounds change to keep the pro-antioxidant balance [13,56–61]. Changes in the antioxidant levels, such as an increase in the level of ascorbate as a response to O_3 fumigation, have been reported in spinach (*Spinacia oleracea*) and beech (*Fagus sylvatica*) [59]. In contrast, when the O_3-induced ROS does not result in a reducing environment inside the cell, the antioxidant levels might remain constant. In this case, there can be increases in other metabolic compounds that can act as scavengers of ozone before its strong oxidative action occurs in the cell [1,16,62].

In our study, no significant changes were observed in the levels of antioxidants whereas the inducible defences, terpenoid compounds, tended to increase in O_3 exposed leaves. This can be explained by the fact that O_3 uptake does not necessarily cause injury in plants due to the production of metabolic compounds, such as volatiles, that remove ROS before they cause serious damage in the cell [63].

In our experiment, although there was a slight increase in the quantity of volatiles emitted, there was no change in volatile profile and *de novo* volatiles were not induced. Monoterpenes, aromatics and non-terpenoid oxygenated compounds, which have been reported to be protective against oxidative damage due to rapid reaction with ROS [15,18–20] increased with FA+O3 treatment. Thus, the increase in these constitutive metabolites in *C. floribundus* might be a defence response against O_3 damage, removing ROS through gas-phase chemical reactions in the intercellular spaces of the leaves [2] because isoprene synthesis is stimulated by ozone [11,64]. Although terpenoid compound levels might have inhibited the effects of ozone most likely by restricting the damage caused by ROS, the fumigation of *C. floribundus* with 80 ppb of O_3 lead to a significant decrease in Fv:Fm and ETR compared to the FA condition, which might indicate a reduction of photosystem II (PSII) efficiency. A decrease in the ability of PSII to dissipate light energy as non-photochemical quenching (NPQ) can induce the ROS production in chloroplast, signalling the primer mechanism of immune response [65]. In addition, chloroplast-generated ROS are important to up-regulation of defence-relate genes and down-regulation of photosynthesis, such as the regulation of redox state of photosynthetic components, including plastoquinone and glutathione pools [65,66]. Our results suggest no disturbance in enzymes and ascorbic acid, which are compounds of ascorbate-glutathione cycle, indicating that ROS produced by O_3 likely not disturbed the redox state of the chloroplast. Our results also indicate that the ROS produced by ozone exposure in *C. floribundus* might have been sufficient to signal the terpenoid pathway but not to cause a strong oxidative stress.

Secretory glands such as laticifers are involved in plant defences against herbivory [67,68]. Laticifers' metabolite contents are biochemical end products (generally cytotoxic); these cells are involved in sequestering toxic compounds or their precursors independent of vascular tissues [69]. It is assumed that laticifers have no functional plasmodesmatal connections with their neighbour cells [70–71], making it impossible to release their contents without physical injury, and they are not modulated in response to oxidative stress. On the other hand, terpenoids are produced by cells of the mesophyll and parenchyma cells from the vascular system, which are also responsible for translocation of these compounds [28–30]. Indeed, FA treatments presented only laticifer cells with terpenoid content (constituent terpenoids), but a progressive recruitment of vascular parenchyma cells involved with terpenoid production and translocation was observed in the course of the time in which the plants were exposed to O_3. Since the terpenoids are volatile, their diffusion from these cells can be confined over and to the leaf (boundary layer) by means of the non-secretory trichomes, which are abundant in both faces of the epidermis. These stellar non-secretory trichomes can act as a container for volatiles analogous to the corona on flowers of *Passiflora* species [72]. Therefore, the concentration of VOC under the leaf might be increased simply by the presence of these non-secretory trichomes. Because VOC can scavenge O_3, this morphological adaptation can also prevent the entrance of this pollutant into the leaf.

Considering that the production of volatiles is systemic [73–74], the O_3-induced VOC in *C. floribundus* might be transported by phloem throughout the plant and act as signalling defences. Methyl salicylate, an oxygenated volatile that increased when *C. floribundus* plants were exposed to O_3, is one of the key messenger molecules synthesised by plants in response to stress [75]. This compound may act as a mobile signal throughout the plant, triggering the systemic acquired resistance by means of its precursor, salicylic acid, which is able to enhance chemical defences, such as antioxidants [75–76]. It is interesting to note that ROS is mediated by salicylic acid, which plays a key role in the changes in the cytosolic concentration of free calcium ions [77]. Indeed, O_3 stress induces the ROS-mediated opening of calcium channels and increases the intracellular calcium concentration [78], a mechanism involved in COC formation [79]. In our work, *C. floribundus* exposed to O_3 showed slightly increase in MeSA emission and in the number of COC, but there is no evidence of alteration in pro-antioxidant balances. Therefore, we hypothesized that the ozone-induced VOC (such as MeSA) regulate ROS formation in a way that promotes the opening of calcium channels and subsequently, the formation of crystals. To test this hypothesis we performed an independent experiment spraying MeSA on *C. floribundus* leaves and verified that, indeed, MeSA was able to induce an increase on COC number. Thus, we suggest that the increase in the quantity of crystals could be a fast and stable mechanism to stop the consequences of ROS in the tissue, immobilising the calcium excess that can be dangerous to the plant.

Our findings indicate that the tolerance of *C. floribundus* to oxidative stress caused by ozone may be the consequence of a higher capacity to produce volatiles and oxalate crystals rather than the modulation of the antioxidant balance. These metabolic features could be used as biomarkers for ozone tolerance and may also be useful for choosing the functional groups resistant to air pollution and, consequently, for use in ecological restoration plans in urban areas highly impacted by air pollution.

Author Contributions

Conceived and designed the experiments: PCG SRS. Performed the experiments: PCG SRS VPB DPO. Analyzed the data: PCG SRS MTGG MPMA MAM ESA. Contributed reagents/materials/analysis tools: SRS MTGG MPMA ESA. Wrote the paper: PCG SRS. Provided helpful comments on the manuscript: VPB DPO MTGG MPMA MAM ESA.

References

1. Souza SR, Blande J, Holopainen J (2013) Pre-exposure to nitric oxide modulates the effect of ozone on oxidative defenses and volatile emissions in lima bean. Environmental Pollution 179: 111–129.

2. Fares S, McKay M, Holzinger R, Golstein AH (2010) Ozone fluxes in a *Pinus ponderosa* ecosystem are dominated by non-stomatal processes: evidence from long-term continuous measurements. Agricultural and Forest Meteorology 150: 420–431.

3. Günthardt-Goerg MS, Vollenweider P, McQuattie CJ (2003) Differentiation of ozone, heavy metal or biotic stress in leaves and needles. Ekologia Bratislavia 22: 110–113.

4. Vollenweider P, Günthardt-Goerg MS (2006) Diagnosis of abiotic and biotic stress factors using the visible symptoms in the foliage. Environmental Pollution 140: 562–571.

5. Günthardt-Goerg MS, Vollenweider P (2007) Linking stress with macroscopic and microscopic leaf response in trees: New diagnostic perspectives. Environmental Pollution 147: 467–488.

6. Guerrero CC, Günthardt-Goerg MS, Vollenweider P (2013) Foliar Symptoms triggered by ozone stress in irrigated Holm Oaks from the city of Madrid, Spain. PLOS one, DOI 10.1371/journal.pone.0069171.

7. Moura BB, Souza SR, Alves ES (2011) Structural responses of *Ipomoea nil* (L.) Scarlet O'Hara (Convolvulaceae) exposed to ozone. Acta Botanica Brasilica 25: 122–129.

8. Moura BB, Souza SR, Alves ES (2013) Response of Brazilian native trees to acute ozone dose. Environmental Science and Pollution Research 21: 4220–4227.

9. Vollenweider P, Fenn ME, Menard T, Günthardt-Goerg MS, Bytnerowicz A (2013) Structural injury underlying mottling in ponderosa pine needles exposed to ambient ozone concentrations in the San Bernardino Mountains near Los Angeles, California. Trees 27: 895–911.

10. Bagard M, Le Thiec D, Delacote E, Hasenfratz-Sauder MP, Banvoy J, et al. (2008) Ozone-induced changes in photosynthesis and photorespiration of hybrid poplar in relation to the developmental stage of the leaves. Physiologia Plantarum 134: 559–574.

11. Fares S, Park JH, Ormeno E, Gentner DR, McKay M, et al. (2010) Ozone uptake by citrus trees exposed to a range of ozone concentrations. Atmospheric Environment 44: 3404–3412.

12. Loreto F, Pinelli P, Manes F, Kollist H (2004) Impact of ozone on monoterpene emission and evidence for an isoprene-like antioxidant action of monoterpenes emitted by *Quercus ilex* leaves. Tree Physiology 24: 361–367.

13. Vuorinen T, Nerg AM, Holopainen JK (2004) Ozone exposure triggers the emission of herbivore-induced plant volatiles, but does not disturb tritrophic signaling. Environmental Pollution 131: 305–331.

14. Fares S, Loreto F, Kleist E, Wildt J (2008) Stomatal uptake and stomatal deposition of ozone in isoprene and monoterpene emitting plants. Plant Biology 10: 44–54.

15. Logan BA, Monson RK, Potosnak MJ (2000) Biochemistry and physiology of foliar isoprene production. Trends in Plant Science 5: 477–481.

16. Loreto F, Mannozi M, Maris C, Nascetti P, Ferranti F, et al. (2001) Ozone quenching properties of isoprene and its antioxidant role in leaves. Plant Physiology 126: 993–1000.

17. Loreto F, Velikova V (2001) Isoprene produced by leaves protects the photoshynthetic apparatus agains ozone damage, quanches ozone products, and reduces lipid peroxidation of cellular membranes. Plant Physiology 127: 1781–1787.

18. Peñuelas J, Llusià J (2002) Linking photorespiration, monoterpenes and thermotolerance in *Quercus*. New Phytologist 155: 227–237.

19. Sharkey TD, Chen X, Yeh S (2001) Isoprene increases thermotolerance of fosmidomycin-fed leaves. Plant Physiology 125: 2001–2006.

20. Sharkey TD, Wiberley AE, Donohue AR (2008) Isoprene emission from plants: why and how. Annals of Botany 101: 5–18.

21. Velikova V, Fares S, Loreto F (2008) Isoprene and nitric oxide reduce damages in leaves exposed to oxidative stress. Plant Cell and Environment 31: 1882–1894.

22. Vickers CE, Gershenzon J, Lerday MT, Loreto F (2009) A unified mechanism of action for volatile isoprenoids in plant abiotic stress. Nature Chemical Biology 5: 283–291.

23. Blande JD, Tiiva P, Oksanen E, Holopainen JK (2007) Emission of herbivore-induced volatile terpenoids from two hybrid aspen (*Populus tremula* x *tremuloides*) clone under ambient and elevated ozone concentration in the field. Global Change Biology 12: 0520-0550.

24. Holopainen JK, Gerhenzon J (2010) Multiple stresses factors and the emission of plant VOC. Trends in Plant Science 15: 1360–1385.

25. Pinto DM, Blande JD, Nykanen R, Dong WX, Nerg AM, et al. (2007) Ozone degrades common herbivore-induced plant volatiles: does this affect herbivore prey location by predators and parasitoids? Journal of Chemical Ecology 33: 683–694.

26. Shah J (2009) Plants under attack: systemic signals in defence. Current Opinion in Plant Biology 12: 459–464.

27. Hampel D, Mosandl A, Wüst M (2005) Induction of de novo volatile terpene biosynthesis via cytosolic and plastidial pathways by methyl jasmonate in foliage of *Vitis vinifera* L. Journal of Agricultural and Food Chemistry 53: 2652–2657.

28. Hudgins JW, Christiansen E, Franceschi VR (2004) Induction of anatomically based defense responses in stems of diverse conifers by methyl jasmonate: a phylogenetic perspective. Tree Physiology 24: 251–264.

29. Köllner TB, Lenk C, Schnee C, Kopke S, Lindemann P, et al. (2013) Localization of sesquiterpene formation and emission in maize leaves after herbivore damage. BMC Plant Biology 13: 15.

30. Martin D, Tholl D, Gershenzon J, Bohlmann J (2002) Methyl jasmonate induces traumatic resin ducts, terpenoids resin biosynthesis and terpenoids accumulation in developing xylem of Norway spruce stems. Plant Physiology 129: 1003–1018.

31. Bowler C, Fluhr R (2000) The role of calcium and activated oxygen as signals for controlling cross-tolerance. Trends in Plant Science 5: 241–246.

32. Franceschi VR, Horner HT (1980) Calcium oxalate crystals in plants. The Botanical Review 46: 361–427.

33. Franceschi VR, Nakata PA (2005) Calcium oxalate in plants: formation and function. Annual Review of Plant Biology 56: 41–71.

34. Nakata PA (2012) Engineering calcium oxalate crystal formation in *Arabidopsis*. Plant Cell Physiology 53: 1275–1282.

35. Molano-Flores B (2001) Herbivory and calcium concentration affect calcium oxalate crystal formation in leaves of *Sida* (Malvaceae). Annals of Botany 88: 387–391.

36. Ruiz N, Ward D, Saltz D (2002) Calcium oxalate crystals in leaves of *Pancratium sickenbergeri*: constitutive or induced defence? Functional Ecology 16: 99–105.

37. Xiang H, Chen J (2004) Interspecific variation of plant traits associated with resistance to herbivory among four species of *Ficus* (Moraceae). Annals of Botany 94: 377–384.

38. Fink S (1991) Unusual patterns in the distribution of calcium oxalate in spruce needles and their possible relationships to the impact of pollutants. New Phytologist 119: 41–51.

39. Alves ES, Tresmondi F, Longui EL (2008) Analise estrutural de folhas de *Eugenia uniflora* L. (Myrtaceae) coletadas em ambiente rural e urbano, SP, Brasil. Acta Botanica Brasilica 22: 241–248.

40. Moraes RM, Bulbovas P, Furlan C M, Domingos M, Meirelles ST, et al. (2006) Physiological responses of saplings of *Caesalpinia echinata* Lam., a Brazilian tree species, under ozone fumigation. Ecotoxicology and Environmental Safety 63: 306–312.

41. Fragoso FP, Varanda EM (2011) Flower-visiting insects of five tree species in a restored area of semideciduous seasonal forest. Neotropical Entomology 40: 431–435.

42. Moura BB (2013) Análises estruturais e ultraestruturais em folhas de espécies nativas sob influência de poluentes aéreos. PhD Thesis, Instituto de Botânica, São Paulo, Brazil.

43. Souza SR, Santana SRM, Rinaldi M, Domingos M (2009) Short-term leaf responses of *Nicotiana tabacum* 'Bel-W3' to ozone under the environmental conditions of Sao Paulo, SE, Brazil. Brazilian Archives of Biology and Technology 52: 251–258.

44. Souza E, Pagliuso JD (2009) Evaluation of the urban air quality in Brazilian Midwest Region. International Journal of Environmental Protection 2: 14–19.

45. Potvin C, Tardif S (1988) Source variability and experiential designs in growth chambers. Functional Ecology 2: 123–130.

46. Keller T, Schwager H (1977) Air pollution and ascorbate. European Journal of Forest Pathology 7: 338–350.

47. Osswald WF, Kraus R, Hipelli S, Bens B, Volpert R, et al. (1992) Comparison of the enzymatic activities of dehydroascorbic acid redutase, glutathione redutase, Catalase, peroxidase and superoxide dismutase of healthy and damaged spruce needles *Picea abies* (L.) Karst. Plant physiology 139: 742–748.

48. Klumpp G, Guderian R, Küppers K (1989) Peroxidase- und Superoxiddismu-tase-Aktivität sowie Prolingehalte von Fichtennadeln nach Belastung mit O₃, SO₂ und NO₂. European Journal of Forest Pathology 19: 84–97.

49. Lucero M, Estell R, Tellez M, Fredrickson E (2009) A retention index calculator simplifies identification of plant volatile organic compounds. Phytochemical Analysis 20: 378–384.

50. David R, Carde JP (1964) Coloration différentielle dês inclusions lipidique et terpeniques dês pseudophylles du pin maritime au moyen du reactif Nadi. Comptes Rendus de l'Academie des Sciences Paris 258: 1338–1340.

51. Machado SR, Gregorio EA, Guimaraes E (2006) Ovary peltate trichomes of *Zeyheria montana* (Bignoniaceae): developmental ultrastructure and secretion in relation to function. Annals of Botany 97: 357–369.

52. Johansen DA (1940) Plant microtechnique. New York: McGraw-Hill Books.

53. Novak K, Skelly JM, Schaub M, Krauchi N, Hug C, et al. (2003) Ozone air pollution and foliar injury development on native plants of Switzerland. Environmental Pollution 125: 41–52.

54. Goodger JQD, Hesjes AM, Woodrow IE (2013) Contrasting ontogenetic trajectories for phenolic and terpenoids defences in *Eucalyptus froggattii*. Annals of Botany 112: 651–659.

55. Barcikowski RS, Robey RR (1984) Decisions in single group repeated measures analysis: statistical test and three computer packages. The American Statistician 35: 148–150.

56. Hartikainen K, Nerg A-M, Kivimäenpää M, Kontunen-Soppela S, Mäenpää M, et al. (2009) Emissions of volatile organic compounds and leaf structural characteristics of European aspen (*Populus tremula*) grown under elevated ozone and temperature. Tree Physiology 29: 1163–1173.

57. Heiden AC, Hoffman T, Kahl J, Kley D, Klockow D, et al. (1999) Emission of volatile organic compounds from ozone-exposed plants. Ecological Applications 9: 1160–1167.

58. Llusià J, Peñuelas J, Gimeno BS (2002) Seasonal and species-specific response of VOC emissions by Mediterranean woody plants to elevated ozone concentrations. Atmospheric Environment 36: 3931–3938.

59. Luwe M, Heber U (1995) Ozone detoxification in the apoplasm and symplasm of Spinach, broad bean and beech leaves at ambient and elevated concentrations of ozone in air. Planta 197: 448–455.

60. Mehlhorn H, Cottam DA, Lucas PW, Wellburn AR (1987) Indication of ascorbate peroxidase and glutathione reductase activities by interactions of mixtures of air pollutants. Free Radical Research Communications 3: 1–5.

61. Peñuelas J, Llusià J, Gimeno BS (1999) Effects of ozone concentrations on biogenic volatile organic compounds emission in the Mediterranean region. Environmental Pollution 105: 17–23.

62. Loreto F, Schnitzler JP (2010) Abiotic stresses and induced BVOC. Trends in Plant Science 15: 1350–1385.

63. Loreto F, Fares S (2007) Is ozone flux inside leaves only a damage indicator? Clues from volatile isoprenoid studies. Plant Physiology 143: 1096–1100.

64. Fares S, Barta C, Brilli F, Centritto M, Ederli L, et al. (2006) Impact of high ozone on isoprene emission, photosynthesis and histology of developing *Populus alba* leaves directly or indirectly exposed to the pollutant. Physiologia Plantarum 128: 456–465.

65. Shapiguzov A, Vainonen JP, Wrzaczek M, Kangasjarvi J (2012) ROS-talk – how the apoplast, the chloroplast and the nucleus get the message through. Frontiers in Plant Science 3: 1–9.

66. Pfannschmidt T (2003) Chloroplast redox signals: how photosynthesis controls its own genes. Trends in Plant Science 8: 33–41.

67. Hunter RJ (1994) Reconsidering the functions of latex. Trees 9: 1–5.

68. Optiz S, Kunert G, Gershenzon J (2008) Increased terpenoid accumulation in cotton (*Gossypium hirsutum*) foliage is a general wound response. Journal of Chemical Ecology 34: 508–522.

69. Hagel JM, Yeung EC, Facchini PJ (2008) Got milk? The secret life of laticifers. Trends in Plant Science 13: 631–639.

70. Fat E, Sanier C, Hebant C (1989) The distribution of plasmodesmata in the phloem of *Hevea brasiliensis* in relation to laticifer loading. Protoplasma 149: 155–162.

71. Pickard WF (2008) Laticifers and secretory ducts: two other tube systems in plants. New Phytologist 177: 877–888.

72. Garcia MTA, Galati BG, Hoc OS (2007) Ultraastructure of the corona of the scented and scentless flowers of *Passiflora* spp (Passifloraceae). Flora 202: 302–315.

73. Karban R (2007) Plant behaviour and communication. Ecology Letters 11: 727–729.

74. Pateraki I, Kanellis KA (2010) Stress and developmental responses of terpenoid biosynthetic genes in Cistus creticus subsp. Creticus. Plant Cell Reports 29: 629–641.

75. Baldwin IT, Halitschke R, Paschold A, von Dah CC, Preston C (2006) Volatile signaling in plant-plant interactions: "Talking Trees" in the genomics era. Science 311: 811–813.

76. Blande JD, Korjus M, Holopainen JK (2010) Foliar methyl salicylate emissions indicate prolonged aphid infestation on silver birch and black alder. Tree Physiology 30: 404–416.

77. Kawano T, Boteau F (2013) Crosstalk between intracellular and extracellular salicylic acid signaling events leading to long-distance spread of signals. Plant Cell Reports 32: 1125–1138.

78. McAinsh MR, Evans NH, Montgomery LT, North KA (2002) Calcium signalling in stomatal responses. New Phytologist 153: 441–447.

79. Volk GM, Gossi GM, Franceschi VR (2004) Calcium channels are involved in calcium oxalate crystal formation specialized cells of *Pistia stratiotes* L. Annals of Botany 93: 741–753.

22

Effects of Controlled-Release Fertiliser on Nitrogen Use Efficiency in Summer Maize

Bin Zhao, Shuting Dong*, Jiwang Zhang, Peng Liu

State Key Laboratory of Crop Biology, College of Agronomy, Shandong Agricultural University, Tai'an, China

Abstract

Nitrogen (N) is a nutrient element necessary for plant growth and development. However, excessive inputs of N will lead to inefficient use and large N losses to the environment, which can adversely affect air and water quality, biodiversity and human health. To examine the effects of controlled-release fertilisers (CRF) on yield, we measured ammonia volatilisation, N use efficiency (NUE) and photosynthetic rate after anthesis in summer maize hybrid cultivar Zhengdan958. Maize was grown using common compound fertiliser (CCF), the same amount of resin-coated controlled release fertiliser (CRFIII), the same amount of sulphur-coated controlled release fertiliser (SCFIII) as CCF, 75% CRF (CRFII) and SCF (SCFII), 50% CRF (CRFI) and SCF (SCFI), and no fertiliser. We found that treatments CRFIII, SCFIII, CRFII and SCFII produced grain yields that were 13.15%, 14.15%, 9.69% and 10.04% higher than CCF. There were no significant differences in grain yield among CRFI, SCFI and CCF. We also found that the ammonia volatilisation rates of CRF were significantly lower than those of CCF. The CRF treatments reduced the emission of ammonia by 51.34% to 91.34% compared to CCF. In addition, after treatment with CRF, maize exhibited a higher net photosynthetic rate than CCF after anthesis. Agronomic NUE and apparent N recovery were higher in the CRF treatment than in the CCF treatment. The N uptake and physiological NUE of the four yield-enhanced CRF treatments were higher than those of CCF. These results suggest that the increase in NUE in the CRF treatments was generally attributable to the higher photosynthetic rate and lower ammonia volatilisation compared to CCF-treated maize.

Editor: Randall P. Niedz, United States Department of Agriculture, United States of America

Funding: This research was supported by the National Natural Science Foundation of China (31171497, 31071358), System of Maize Modern Industrial Technologies (nyhyzx07-003), Ministry of Agriculture System of Maize Industrial Technologies (CARS-02), The Major State Basic Research Development Program of China (973 Program) (2011CB100105), Special Fund for Agro-scientific Research in the Public Interest (20120306, HY1203096), "Five-twelfth" National Science and Technology Support Program (2011BAD16B14, 2011BAD11B01, 2011BAD11B02). The funders had no role in study design, data collection and analysis, decision to publish, or preparation of the manuscript.

Competing Interests: The authors have declared that no competing interests exist.

* E-mail: stdong@sdau.edu.cn

Introduction

Nitrogen (N) is a critical element for plant growth, and adding N to crops is a valuable agronomic practice. During the past decade, China has made considerable progress in terms of grain yield (GY) and feeding its growing population; however, this increase in agricultural yield has partly resulted from excessive application of N fertilisers [1]. Excessive application can result in inefficiencies and large losses of excess N to the environment, which can impact air and water quality, biodiversity and human health [2]. The overuse of fertilisers contributes to NO_3-N contamination of both surface water and soil water, and high profile NO_3-N accumulation can reduce N use efficiency (NUE) [1,3]. Releases of nitrous oxide (mainly via the application of N fertiliser) can degrade stratospheric ozone and contribute to global warming [4]. Ammonia (NH_3) volatilisation from soil and plants can also aggravate environmental contamination and contribute to acid deposition [5]. Therefore, interventions to increase NUE and reduce N inputs are important not only for reducing environmental risk but also for lowering agricultural production costs [6].

Controlled-release fertiliser (CRF) is a possible alternative to common compound fertiliser (CCF) to increase N uptake efficiency and minimise N losses to the environment. However, current grower acceptance is limited due to a lack of experience with CRF performance and its high relative cost [7]. As one kind of enhanced-efficiency fertiliser, CRF has several advantages compared to CCF. Some of the advantages and disadvantages are listed in Table 1. The greatest benefits of switching from CCF to CRF include increased profitability and reductions in the environmental impact of crop production.

In sandy nursery soils, CRF was shown to be effective for seedling production, due to the increased residence time of CRF in the soil relative to conventional fertilisation [8,9]. Oliet et al. [10] found that CRF promoted suitable morphological values and nutritional status in *Pinus halepensis* planting stock, suggesting that the CRF types used in their study were suitable for the nursery production of *P. halepensis*. Tang et al. [11] reported major increases in rice yield following a single basal application of CRF, that was attributed to increased soil availability of N, superior development of the root systems, better nutrient absorption capacity, delayed senescence and enhanced lodging resistance.

To improve CRF, studies need to describe nutrient release characteristics. Du et al. [12] revealed that the thickness of the coating membrane was the most important parameter for controlling nitrate release, followed by temperature, granule radius and the saturated concentration of nitrate. However, the global use of CRF has so far been limited due to the higher cost (at least 2 or more times the price) compared to CCF.

Table 1. Advantages and disadvantages of CRF over CCF.

Advantages	Disadvantages
Slower release rate – plants are able to take up most of the fertilisers	Very high costs
Reduced fertiliser loss – slower leaching and run-off	
Reduced labour capital – less frequent application is required	Lower consumption
Lower salt index – reduced plant damage from high concentrations of salts	
Fertiliser burn is not a problem with CRF even at high rates of application	Limited to nursery stock

Note(s): CRF, controlled-release fertiliser; CCF, common compound fertiliser.

To date, few investigations have been carried out in the field on the performance of crops grown with CRF. Even if CRF use becomes economical, the widespread acceptance by growers will likely be limited as a result of grower concern about field performance [7]. It has mainly been applied to nursery stock in foreign countries. Until now, reports on CRF in crops have focused mainly on domestic rice and little information is available about the effects of CRF on maize. Consequently, it is valuable to clarify the mechanisms of the impacts of CRF on maize. Therefore, in the present study, we investigated the following: how to select the right application rate of CRF in maize, the photosynthetic traits and physiological mechanisms associated with yields and the NUE of CRF in maize. Our results will assist in the successful application of CRF to maize fields.

Materials and Methods

Experimental Design

Summer maize hybrid Zhengdan958 (released in 2000) was planted on a farm at Shandong Agricultural University, Shandong Province, China (36°10′19′′N, 117°9′03′′E). The soil, classified as a silt loam, is considered to be highly suitable for crop production. The soil pH was 6.1. The average organic matter content in the tillage layer was 19.7 g kg^{-1} and the available N, phosphorus (P) and potassium (K) were 124.38 mg kg^{-1}, 45.23 mg kg^{-1} and 81.78 mg kg^{-1}, respectively. Methods of soil analysis referenced from "Agricultural Soil Analysis" written by Bao S D [13]. Two kinds of CRF, a resin-coated CRF (hereafter, CRF) and a sulphur-coated CRF (hereafter, SCF) offered by Shandong Kingenta Ecological Engineering Co., Ltd. located in No. 19 Xingda West Street, Linshu, Shandong Province were used in our experiment and common compound fertiliser (CCF) was used as a control. The content of N, P_2O_5 and K_2O in each CCF, CRF and SCF was 24%, 8% and 16%; 21%, 7% and 14%; and 18%, 6% and 12%, respectively. The experiment was conducted as a random complete block design with three replications. There were eight treatments: CCF applied at 1250 kg ha^{-1} (the local average commercial fertiliser N application rate; hereafter, CCF); CRF applied at 714.29 kg ha^{-1} (CRFI, 50% CCF), 1071.43 kg ha^{-1} (CRFII, 75% CCF) and 1428.57 kg ha^{-1} (CRFIII, 100% CCF); SCF applied at 833.33 kg ha^{-1} (SCFI, 50% CCF), 1250 kg ha^{-1} (SCFII, 75% CCF) and 1666.67 kg ha^{-1} (SCFIII, 100% CCF); and control plots without fertiliser application (CK). All fertilisers were applied at a basal dose.

Measurement of Net Photosynthetic Rate

Net photosynthetic rate (P_N) was measured with a portable, open-flow portable photosynthetic system (LI-COR, LI-6400 System, UK) in 2006. The photosynthetic photon flux density (PPFD) was 1400 μmolm^{-2}s^{-1} provided by the internal light source of the leaf chamber. The measurements were taken at approximately 10-day intervals following pollination on cloudless days. All of the leaves (22 or 23) were fully expanded at the time of measurement. Only ear leaves were used for P_N measurements, because these structures are metabolically active for the longest period of time and their relative contribution to total photosynthetic assimilates is high [14]. The leaves for experiments were all fully exposed and oriented to normal irradiation during measurements. Five plants per treatment were randomly selected for measurements.

Plant Sampling and N Content Determination

To measure aboveground N uptake, five plants were collected per treatment in about 10-day intervals 30 days after sowing. At the mature stage, five plants were manually harvested per treatment. Rows per ear (RE), kernels per row (KR) and kernels per ear (KE) were counted. Aboveground dry matter (DM) was determined by oven-drying the samples at 80°C until a constant weight was achieved. Subsequently, samples were manually separated into the vegetative and grain portions. Then the GY and thousand-kernel weight (TKW) were determined.

The grains and straw were ground using a cyclone sample mill with a mesh size of 0.5 mm. Then the grain N concentration (GNCT) and the straw N concentration (SNCT) were measured using the micro-Kjeldahl method (CN61M/KDY-9820, Beijing).

The following parameters were calculated:

$Plant\ N\ uptake = GNCT(the\ grain\ N\ concentration)$

$$\times GW(the\ grains\ weight)$$

$$+ SNCT(the\ straw\ N\ concentration)$$

$$\times DMSW(the\ dry\ matter\ of\ straw\ weight)$$

$Agronomic\ NUE(ANUE)$

$= (grain\ weight[fertiliser] - grain\ weight[no\ fertiliser])/N$

$fertiliser\ applied$

$Physiological\ NUE\ (PNUE)$

$= (grain\ weight[fertiliser] - grain\ weight[no\ fertiliser])$

$/[plant\ N(fertiliser) - plant\ N(nofertiliser)]$

$Apparent\ N\ recovery(AR)$

$= (plant\ N[fertiliser] - plant\ N[no\ fertiliser])/N$

$fertiliser\ applied$

The plant N included N in dry matter of straw and grain.

Ammonia (NH₃) Volatilisation Measuring Device and Procedure

Devices. The vented chamber (Figure 1) was made of grey round polyvinyl chloride (PVC) tubing (15 cm internal diameter and 10 cm high), as described by Liao [15]. Two pieces of round sponge (16 cm in diameter and 2 cm in thickness) were moistened with a 15 mL phosphate-glycerol solution (50 mL analytical phosphate and 40 mL glycerol diluted to 1000 mL with pure water) and then inserted into each chamber. Because the volume of the solution only accounted for 3.7% of the sponge's volume, the sponge was still ventilative after being moistened. The sponge inside the chamber absorbed NH_3 volatilised from the soil, and the top sponge absorbed NH_3 from the ambient air. Glycerol in the sponges absorbed moisture from the air inside or outside the chamber and prevented the sponges from drying [16].

Detection of Ammonia Volatilization in Field

On June 12, 2006, when N fertiliser was applied to summer maize at a basal dose, five sets of both vented chambers were evenly placed at different locations in each plot, with the bottom edge pushed 2 cm into the soil. The two pieces of sponge in the vented chamber and the boric acid solution in the closed chamber were replaced every day during the first week and every 2 to 3 days during the second and third week. Simultaneously, each chamber was moved to a new location in the plot. Ammonia in the phosphate solution in each sponge inside the vented chamber was extracted with 300 mL of 1 M KCl after 60 min of oscillation. Ammonium quantities in the KCl extract solution were measured using the micro-Kjeldahl method (CN61M/KDY-9820, Beijing). NH_3 volatilisation from the soil was estimated by the following formula:

$$NH_3 - N\left(kgNha^{-1}d^{-1}\right) = M/(A \times D) \times 10^{-2}$$

where M is the NH_3 (mg N) captured by the vented chamber during each sampling, A is the cross-section area (m²) of the round chamber, D (days) is the duration of each sampling, and 10^{-2} is 10,000 m² ha⁻¹×10⁻⁶ kg mg⁻¹.

In the field, five sets of vented-chamber devices were evenly placed at different locations in each plot in the manner described above. During the summer maize-growing season, NH_3 emission was measured from June 12 to August 3, 2006, after basal N fertilisation. Measurements continued for 53 days.

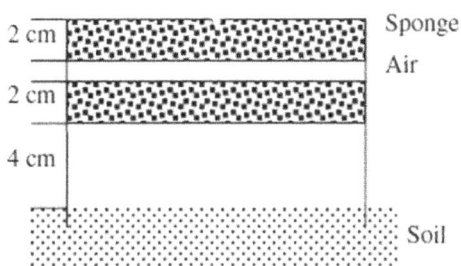

Figure 1. Vented-chamber methods used in field experiments to capture NH₃ emitted from the soil.

Statistical Analysis

Statistical analyses were performed using the analysis of variance (ANOVA) in the General Linear Model procedure of SPSS (Ver. 11, SPSS, Chicago, IL, USA). Results are presented as means of the 2 years of experimentation, because the trends of these parameters were consistent between years. The least significant differences (LSDs) between the means were estimated at the 95% confidence level. Unless indicated otherwise, significant differences among different plants are given at $P<0.05$. LSD was used to compare adjacent means arranged in order of magnitude. Calculations and linear regressions were performed using a *SigmaPlot 10.0* program.

Results

GY and GY Components

The application of fertilisers increased GY significantly compared to that of no fertiliser (Table 2), and the effect of CRF was much more pronounced than that of CCF. Furthermore, CRFIII, SCFIII, CRFII and SCFII were 13.15%, 14.15%, 9.69% and 10.04% higher in GY than CCF. No significant difference in GY was found between CRFI, SCFI and CCF, and there was no significant difference in GY between the two CRFs. The average economic efficiency of CRFIII/SCFIII was 1190.50 yuan hm⁻² more than CCF; CRFII/SCFII was 1753.75 yuan hm⁻² more than CCF; CRFI/SCFI was 758.75 yuan hm⁻² more than CCF.

Net Photosynthetic Rate (Post-anthesis Changes in the Light-saturated Photosynthesis Rate)

There was no significant difference in net P_N among treatments (Figure 2). All P_N of the ear leaves decreased after flowering, and P_N values for CRFIII, CRFII, SCFIII and SCFII decreased more slowly than CCF. The P_{Ns} of CRFIII, CRFII, SCFIII and SCFII on the 10th day after flowering were 24.4%, 21.6%, 23.0% and 24.5%, higher, respectively, than those of CCF ($P<0.05$); on the 30th day after flowering were 13.6%, 14.9%, 15.3% and 17.8%, higher, respectively, than those of CCF ($P<0.05$); on the 50th day after flowering were 27.3%, 20.8%, 23.4% and 26.0%, higher, respectively, than those of CCF ($P<0.05$). No significant difference in P_N was observed among CRFI, SCFI and CCF. These results suggest that the yield increases afforded by the CRF treatments were generally attributable to their higher photosynthetic rates.

NH₃ Volatilisation

The application of N fertilisers in the field increased the volatilisation of NH_3 (Figure 3). The maximum flux of NH_3 increased to 3.36 kg N ha⁻¹ d⁻¹ 2 days after the application of CCF, and then rapidly decreased to approximately 1.18 kg N ha⁻¹ d⁻¹. However, the flux of NH_3 from CRF treatments was significantly lower than that of the CCF treatment. NH_3 volatilisation fluxes from CRF treatments peaked later than those of CCF. The treatments CRFIII and SCFIII reached their peak NH_3 volatilisation fluxes of 1.87 kg N ha⁻¹ d⁻¹ and 1.06 kg N ha⁻¹ d⁻¹, respectively, 9 days following fertiliser application (Figure 3 A B).

Cumulative rates of NH_3 volatilisation generally displayed similar patterns of increase and temporal characteristics up to 53 days following treatment (Figure 3 C D), after which parameters remained relatively constant. Cumulative fluxes of NH_3 emitted from the field were 10.56 kg N ha⁻¹ d⁻¹, 2.23 kg N ha⁻¹ d⁻¹, 3.16 kg N ha⁻¹ d⁻¹, 5.65 kg N ha⁻¹ d⁻¹, 3.79 kg N ha⁻¹ d⁻¹, 4.06 kg N ha⁻¹ d⁻¹ and 5.88 kg N ha⁻¹ d⁻¹ in the CCF, CRFI, CRFII, CRFIII, SCFI, SCFII and SCFIII

Table 2. Effect of controlled-release fertiliser on yield and its component of summer maize.

Treatments	Rows per ear	Kernels per row	Kernels per ear	Wt. per 1000-kernel(g)	Grain yield (kg hm^{-2})		Benefit increase than CK (yuan hm^{-2})	
					2005	2006	2005	2006
CK	14.44	39.02	563.65	281.17e	9383.8d D	8896.1c D	–	–
CCF	14.95	40.74	609.02	293.72d	11380.4c C	11046.1b BC	1056e	1349e
CRFI	14.53	39.93	580.36	298.20cd	11558.6bc BC	10826.8b C	2257c	1790d
CRFII	15.20	40.83	620.67	300.71c	12382.7ab AB	11990.4a A	2882a	3064a
CRFIII	14.60	41.70	608.82	316.29a	12716.9a A	12108.0a A	2571b	2339b
SCFI	14.67	40.57	595.04	296.58cd	11331.8c C	10986.4b BC	1763d	2035c
SCFII	14.80	42.07	622.59	299.77cd	12523.5a AB	11909.5a AB	3060a	2819a
SCFIII	15.07	41.16	620.08	308.53b	12810.5a A	12011.0a A	2629b	2033c

CCF, common compound fertiliser; CRF, a resin-coated CRF; SCF, a sulphur-coated CRF.
CCF, applied at 1250 kg ha^{-1} (the local average commercial fertiliser N application rate); CRFI, CRF applied at 714.29 kg ha^{-1} (50% CCF), CRFII,1071.43 kg ha^{-1} (75% CCF), CRFIII, 1428.57 kg ha^{-1} (100% CCF); SCFI, SCF applied at 833.33 kg ha^{-1} (50% CCF), SCFII, 1250 kg ha^{-1} (75% CCF), SCFIII, 1666.67 kg ha^{-1} (100% CCF); CK, control plots without N application.
Yield component of summer maize includes rows per ear, kernels per row, kernel No. per ear and Wt. per 1000-kernel.
According to the average market price at present, that is 1911 yuan t^{-1} for maize, 2190 yuan t^{-1} for CCF, 2660 yuan t^{-1} for CRF, 2350 yuan t^{-1} for SCF.
All data are means of 3 replications.
Means values marked with different capital letters indicate significant differences at $P = 0.01$ level; different small letters indicate significant differences at $P = 0.05$ level.

treatments, respectively. Volatilisation rates of NH$_3$ were 78.8%, 70.0%, 46.5%, 64.1%, 61.5% and 44.3% lower than those of CCF for CRFI, CRFII, CRFIII, SCFI, SCFII and SCFIII, respectively ($P<0.05$). These results suggest that the application of CRF considerably decreased NH$_3$ volatilisation rates.

N Uptake and NUE

Following the field application of CCF, N uptake increased rapidly during the first phase (i.e., the stage prior to flowering) and then increased slowly after flowering (Figure 4). Phenotypic phenomena such as vigorous growth before flowering and premature senescence after flowering were observed (data not shown). However, N uptake increased relatively constantly with

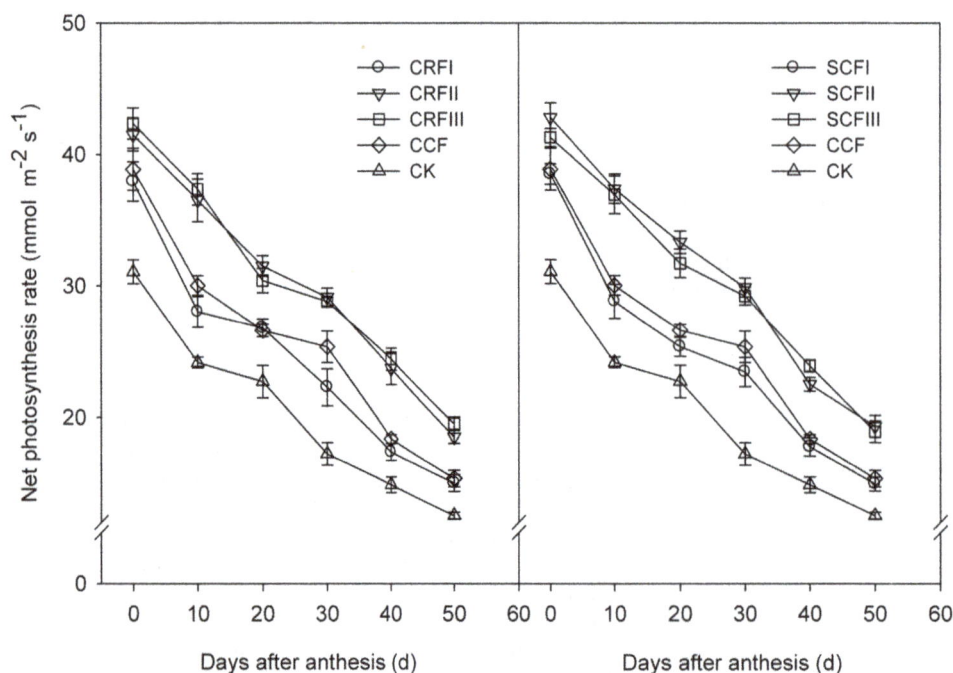

Figure 2. Effects of controlled-release fertiliser on net photosynthetic rate in ear leaves of summer maize. CCF, common compound fertiliser; CRF, a resin-coated CRF; SCF, a sulphur-coated CRF; CCF, applied at 1250 kg ha^{-1} (the local average commercial fertiliser N application rate); CRFI, CRF applied at 714.29 kg ha^{-1} (50% CCF), CRFII,1071.43 kg ha^{-1} (75% CCF), CRFIII, 1428.57 kg ha^{-1} (100% CCF); SCFI, SCF applied at 833.33 kg ha^{-1} (50% CCF), SCFII, 1250 kg ha^{-1} (75% CCF), SCFIII, 1666.67 kg ha^{-1} (100% CCF); CK, control plots without N application. Error bars are SE (n = 5).

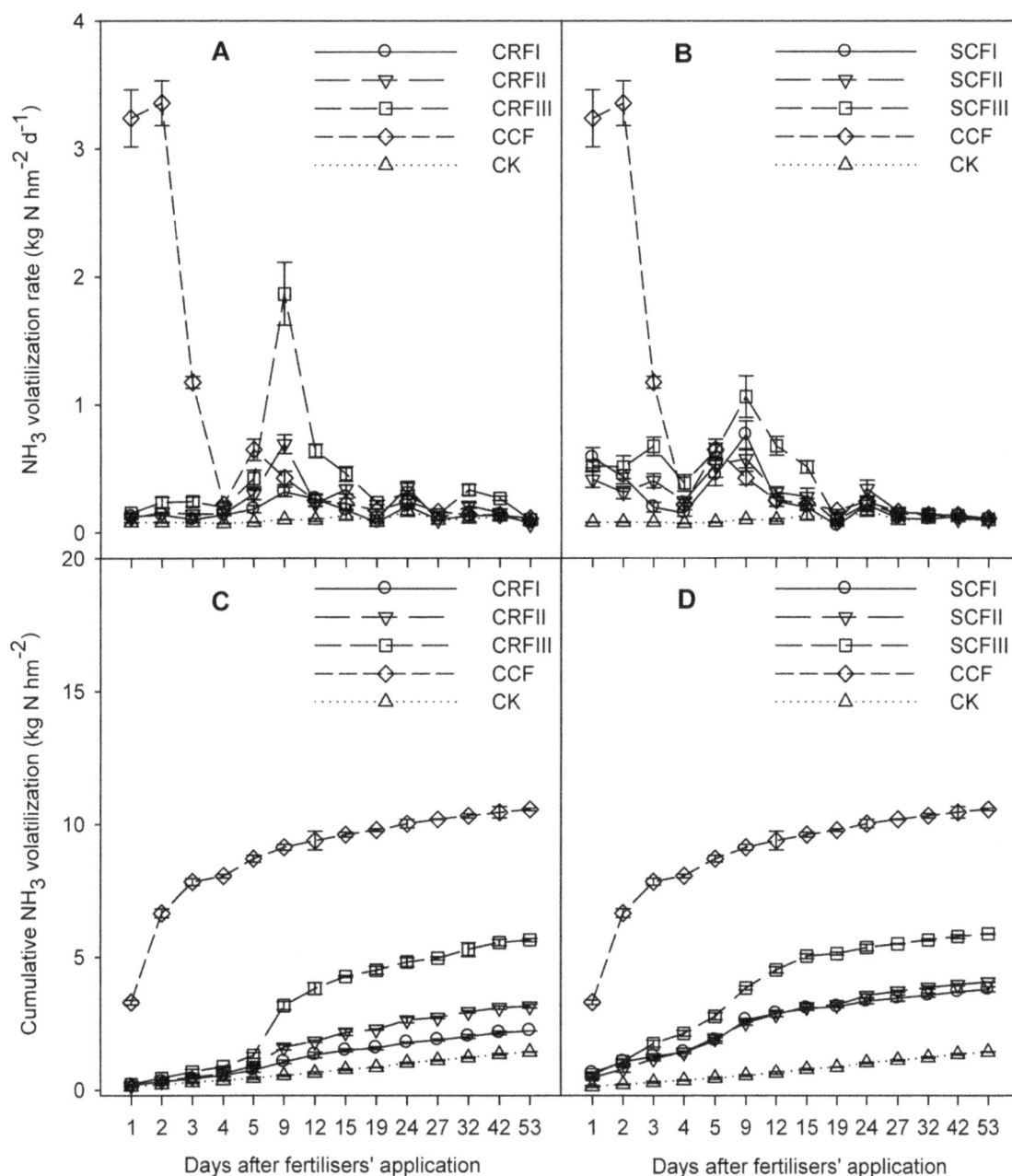

Figure 3. Soil NH₃ volatilisation rates (A and B) and changes in cumulative NH₃ volatilisation (C and D) following basal fertilisation.
CCF, common compound fertiliser; CRF, a resin-coated CRF; SCF, a sulphur-coated CRF; CCF, applied at 1250 kg ha^{-1} (the local average commercial fertiliser N application rate); CRFI, CRF applied at 714.29 kg ha^{-1} (50% CCF), CRFII,1071.43 kg ha^{-1} (75% CCF), CRFIII, 1428.57 kg ha^{-1} (100% CCF); SCFI, SCF applied at 833.33 kg ha^{-1} (50% CCF), SCFII, 1250 kg ha^{-1} (75% CCF), SCFIII, 1666.67 kg ha^{-1} (100% CCF); CK, control plots without N application. Error bars are SE (n = 15; some SE bars are smaller than the symbols).

the increased application of CRF throughout the growth period (Figure 4), and caused obvious delays in leaf senescence. After flowering, the rank of shoot N uptake among all treatments was CRFIII>CRFII>CRFI>CCF>CK and SCFIII>SCFII>CCF >SCFI>CK.

ANUE and NR were significantly higher for CRF than for CCF (91.7% and 89.8%, respectively; $P<0.05$) (Table 3). Although no significant differences in PNUE were found between CRF and CCF, PNUE was slightly higher for CRFIII, CRFII, SCFIII and SCFII than for CCF.

Discussion

Suitable Application Rates for CRF in the Field

Fertiliser use efficiency has become a critical measure of sustainable agriculture. Efforts are underway to improve crop production and enhance N use efficiency in two main ways: by breeding new varieties of maize with high NUE and by improving fertiliser application management [17,18]. Among the improved N management practices, the use of enhanced-efficiency fertilisers such as SRF and CRF, nitrification inhibitors (NI) and urease inhibitors (UI) are being studied extensively under a variety of

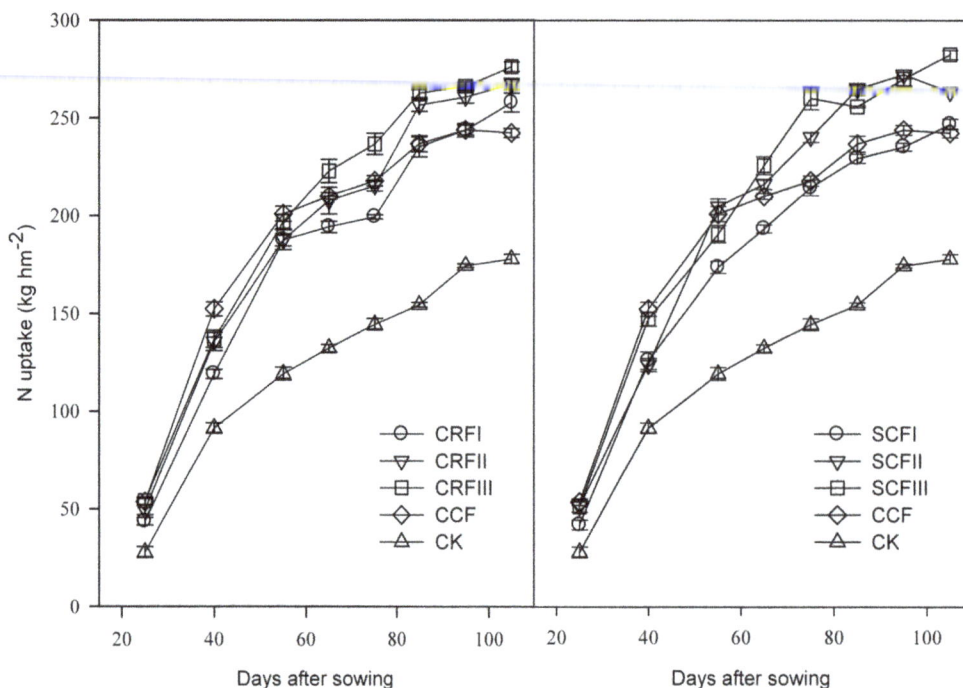

Figure 4. Dynamics of N uptake by aboveground parts after fertilisation. CCF, common compound fertiliser; CRF, a resin-coated CRF; SCF, a sulphur-coated CRF; CCF, applied at 1250 kg ha^{-1} (the local average commercial fertiliser N application rate); CRFI, CRF applied at 714.29 kg ha^{-1} (50% CCF), CRFII,1071.43 kg ha^{-1} (75% CCF), CRFIII, 1428.57 kg ha^{-1} (100% CCF); SCFI, SCF applied at 833.33 kg ha^{-1} (50% CCF), SCFII, 1250 kg ha^{-1} (75% CCF), SCFIII, 1666.67 kg ha^{-1} (100% CCF); CK, control plots without N application. Error bars are SE (n = 3; some SE bars are smaller than the symbols).

environmental conditions and agricultural systems to determine their effectiveness for increasing agricultural production and reducing environmental N losses [19]. CRFs are the fertilisers of the future, especially for open-field crops. Therefore, choosing the appropriate application rate is critical for the successful field application of CRF. In the present study, values of GY were not significantly different between CRFI/SCFI and CCF, CRFIII/SCFIII or CRFII/SCFII. However, values of GY for CRFIII, CRFII, SCFIII and SCFII were significantly higher than for CCF.

Furthermore, agronomic NUE, apparent N recovery and economic efficiency of fertiliser for CRFII/SCFII (equivalent to 75% of CRF/SCF) were higher significantly than CCF. These results suggest that both of the two CRF types used in this study were effective for the agricultural production of maize, which also indicates that the CRFII/SCFII treatments corresponded to the optimum application rate of CRF for the maize fields studied in the North China Plain.

Table 3. Effects of different controlled-release fertiliser treatments on NUE of maize.

Treatments	Grain yield (t hm^{-2})	Total N uptake (kg N hm^{-2})	Agronomic N use efficiency (kg$_{grain}$ kg^{-1}N)	Apparent N recovery (%)	Physiological N use efficiency (kg$_{grain}$ kg^{-1}N)
CK	9.38d	177.83	–	–	–
CCF	11.38c	242.3	6.66c	21.49f	30.97ab
CRFI	11.56bc	257.82	14.50a	53.33a	27.19b
CRFII	12.38ab	267.28	12.56ab	39.76c	31.59ab
CRFIII	12.72a	276.11	11.11b	32.76e	33.92ab
SCFI	11.33c	246.82	12.99ab	45.99b	28.24b
SCFII	12.52a	263.28	13.95ab	37.98cd	36.74a
SCFIII	12.81a	282.48	11.42ab	34.88de	32.74ab

CCF, common compound fertiliser; CRF, a resin-coated CRF; SCF, a sulphur-coated CRF; CCF, applied at 1250 kg ha^{-1} (the local average commercial fertiliser N application rate); CRFI, CRF applied at 714.29 kg ha^{-1} (50% CCF), CRFII,1071.43 kg ha^{-1} (75% CCF), CRFIII, 1428.57 kg ha^{-1} (100% CCF); SCFI, SCF applied at 833.33 kg ha^{-1} (50% CCF), SCFII, 1250 kg ha^{-1} (75% CCF), SCFIII, 1666.67 kg ha^{-1} (100% CCF); CK, control plots without N application.
All data are means of 3 replications.
Means values marked with different letters indicate significant differences at $P = 0.05$ level.

Photosynthesis Rates and Physiological Mechanisms of NUE

Active photosynthesis has always been considered a desirable characteristic during the growing season [20]. Active photosynthesis in maize is primarily associated with the plants' ability to produce grain [20,21]. It had been suggested that yield increases in maize may be at least partly accounted for by increases in net leaf photosynthesis [22]. Leaf photosynthesis has been studied extensively as a plant trait in relation to NUE [23].

In the present study, after flowering, P_N in the yield-enhanced treatments (CRFIII, SCFIII, CRFII and SCFII) was significantly higher than that in the CCF treatments. No significant difference was found in P_N and yield among CRFI, SCFI and CCF. This suggests that the yield increase in the four CRF treatments may be attributable to the higher net photosynthetic rate. Wang et al. [14] also suggested that a yield increase in two cross-pollination treatments was generally due to a higher photosynthetic rate and related photosynthetic traits. Delaying or slowing down senescence may improve yield by increasing photosynthetic leaf area, which increases total photosynthate transported to sink tissue [24]. Indeed, in the present study, phenotypic delayed leaf senescence was obvious in the fourth yield-enhanced treatments (CRFIII, SCFIII, CRFII and SCFII), in accordance with Iain et al. [24].

Reduced NH3 Volatilisation

Ammonium ions (NH_4^+) in the soil exist in equilibrium with NH_3. If this conversion occurs at the soil surface and is accompanied by warm sunny days, NH_3 is subject to gaseous losses to the atmosphere. NH_3 is the most prolific atmospheric reactive N species emitted [25], and agricultural NH_3 emissions are predicted to increase significantly in Asia from 13.8 Tg N $year^{-1}$ in 2000 to 18.8 Tg N $year^{-1}$ in 2030 [26]. NH_3 emissions may result in N deposition to neighbouring ecosystems, which can damage vegetation [27]. In addition, some of the NH_3 may be oxidised and converted into nitric acid, which together with sulphuric acid, make up acid rain. This acidic deposition also damages vegetation, and can acidify both soil and surface water, inducing aluminium toxicity in terrestrial and aquatic organisms [28]. NUE are GY are complex traits that depend on interactions among several component traits [18]. Both appear to be most affected by the NH_3 volatilisation losses of N fertilisers [19].

NH_3 volatilisation in the present study was highest for the CCF treatment, and the majority of N losses occurred within the first 2–

12 days after CCF application. However, the majority of N losses occurred within the first 9–20 days after CRF application. The measured N uptake rates in the CRFIII, SCFIII, CRFII and SCFII treatments were higher than those of the CCF treatment, and ANUE and NR were significantly higher for CRF than for CCF. This suggests that ANUE and NR are significantly and positively correlated with N uptake, and negatively correlated with NH_3 volatilisation. Improvements in NUE and environmental protection are increasingly important issues. The use of CRF can partially resolve both of these issues. The reduced NH_3 volatilisation of CRF will help improve NUE while also decreasing the environmental contamination associated with excess N leaching. In addition, we found that the residual N of CRF in 0–100 cm soil profile was significantly higher than that of CCF (data not shown). Therefore, we conclude that CRF can be used to help conserve both air and water quality by maximising NUE and reducing N losses to the environment.

Conclusion

We found that GY was significantly higher for CRFIII, CRFII, SCFIII and SCFII treatments than for CCF treatment, while no significant difference in GY was found between CRFIII/SCFIII or CRFII/SCFII. These results indicate that 75% CRF/SCF was the optimum application rate for CRF in maize fields of the North China Plain. Further research is necessary to determine the effects, if any, of the type, frequency and timing of CRF applications on maize. In addition, the yield increases afforded by CRF were partly due to higher rates of net photosynthesis and lower rates of NH_3 volatilisation. Finally, we need to develop integrative approaches for enhancing societal *acceptance* of CRF, and for promoting the global application of this technology in economically sound and environmentally friendly agricultural systems.

Acknowledgments

Thanks are due to Dr Ruifang Wang, Qingen Xie, MS, Congfeng Li, MS, Hao Dong, MS, Dejun Gao, MS, for their help in the field and laboratory. We thank the anonymous reviewers for constructive suggestions.

Author Contributions

Conceived and designed the experiments: STD BZ. Performed the experiments: JWZ BZ. Analyzed the data: PL BZ. Contributed reagents/materials/analysis tools: STD BZ. Wrote the paper: BZ.

References

1. Zhu ZL, Chen DL (2002) Nitrogen fertilizer use in China contributions to food production, impacts on the environment and best management strategies. Nutrient Cycling in Agroecosystems 63: 117–127.
2. Goulding K, Jarvis S, Whitmore A (2008) Optimizing nutrient management for farm systems. Philosophical Transactions of the Royal Society B: Biological Sciences 363: 667–680.
3. Ju X, Liu X, Zhang F, Roelcke M (2004) Nitrogen fertilization, soil nitrate accumulation, and policy recommendations in several agricultural regions of China. AMBIO: A Journal of the Human Environment 33: 300–305.
4. Milich L (1999) The role of methane in global warming: where might mitigation strategies be focused? Global Environmental Change, Part A: Human and Policy Dimensions 9: 179–201.
5. Harper L, Sharpe R (1995) Nitrogen dynamics in irrigated corn: soil-plant nitrogen and atmospheric ammonia transport. Agronomy Journal 87: 669–675.
6. Wang RF, An DG, Hu CS, Li LH, Zhang YM, et al. (2011) Relationship between nitrogen uptake and use efficiency of winter wheat grown in the North China Plain. Crop & Pasture Science 62: 1–11.
7. Medina CL, Obreza TA, Sartain JB, Rouse RE (2008) Nitrogen release patterns of a mixed controlled-release fertilizer and its components. Hort Technology 18: 475–480.
8. Dobrahner J, Lowery B, Iyer JG (2007) Slow-release fertilization reduces nitrate leaching in bareroot production of Pinus strobes seedlings. Soil Science 172: 242–255.
9. Zotarelli L, Scholberg JM, Dukes MD, Munoz-Carpena R (2008) Fertilizer residence time affects nitrogen uptake efficiency and growth of sweet corn. Journal of Environmental Quality 37: 1271–1278.
10. Oliet J, Planelles R, Segura ML, Artero F, Jacobs DF (2004) Mineral nutrition and growth of containerized Pinus halepensis seedlings under controlled-release fertilization. Sci Hortic (Amsterdam) 103: 113–129.
11. Tang SH, Yang SH, Chen JS, Xu PZ, Zhang FB, et al. (2007) Studies on the mechanism of single basal application of controlled-release fertilizers for increasing yield of rice (Oryza sativa L.). Agricultural Sciences in China 6: 586–596.
12. Du C, Tang D, Zhou J, Wang H, Shaviv A (2008) Prediction of nitrate release from polymer-coated fertilizers using an artificial neural network model. Biosyst. Eng. 99: 478–486.
13. Bao SD (2005) Agricultural Soil Analysis [M]. Beijing:China Agriculture Press.
14. Wang RF, An DG, Xie QE, Wang KJ, Jiang GM (2009) Leaf photosynthesis is enhanced in normal oil maize pollinated by high oil maize hybrids. Industrial Crops and Products 29: 182–188.
15. Liao XL (1983) The methods of research of gaseous loss of nitrogen fertilizer. Progress Soil Sci. 11: 49–55.
16. Wang CH, Liu XJ, Ju XT, Zhang FS, Malhi SS (2004) Ammonia Volatilization Loss from Surface-Broadcast Urea: Comparison of Vented- and Closed-Chamber Methods and Loss in Winter Wheat–Summer Maize Rotation in North China Plain. Communications in Soil Science and Plant Analysis 35: 2917–2939.

17. Baligar V, Fageria N, He Z (2001) Nutrient use efficiency in plants. Communications in Soil Science and Plant Analysis 32: 921–950.

18. Dawson J, Huggins D, Jones S (2008) Characterizing nitrogen use efficiency in natural and agricultural ecosystems to improve the performance of cereal crops in low-input and organic agricultural systems. Field Crops Research 107: 89–101.

19. Motavalli PP, Goyne KW, Udawatta R (2008) Environmental impacts of enhanced efficiency nitrogen fertilizers. Online. Crop Management. doi:10.1094/CM-2008-0730-02-RV.

20. Ding L, Wang KJ, Jiang GM, Liu MZ, Niu SL, et al. (2005) Post-anthesis changes in photosynthetic traits of maize hybrids released in different years. Field Crop Research 93: 108–115.

21. Araya T, Noguchi K, Terashima I (2006) Effects of carbohydrate accumulation on photosynthesis differ between sink and source leaves of Phaseolus vulgaris L. Plant & Cell Physiology 47: 644–652.

22. Dwyer LM, Tollenaar M (1989) Genetic improvement in photosynthetic response of hybrid maize cultivars 1959 to 1988. Can.J. Plant Sci. 69: 81–91.

23. Foulkes M, Hawkesford M, Barraclough P, Holdsworth M, Kerr S, et al.(2009) Identifying traits to improve the nitrogen economy of wheat: Recent advances and future prospects. Field Crops Research 114: 329–342.

24. Iain SD, Alan PG, Howard T, Keith JF, David E, et al. (2007) Modification of nitrogen remobilization, grain fill and leaf senescence in maize (Zea mays) by transposon insertional mutagenesis in a protease gene. New Phytology 173: 481–494.

25. Galloway JN, Cowling EB (2002) Reactive nitrogen and the world: 200 years of change. Ambio 31: 64–71.

26. Zheng XH, Fu CB, Xu XK, Yan XD, Huang Y, et al. (2002) The Asian nitrogen cycle case study. Ambio 31: 79–87.

27. Newbould P (1989) The use of fertiliser in agriculture. Where do we go practically and ecologically? Plant Soil 115: 297–311.

28. Reuss JO, Johnson DW(1986) Acid deposition and acidification of soil and waters. Ecol Studies No. 59. Springer-Verlag, NY.

PERMISSIONS

LIST OF CONTRIBUTORS

Matthew J. Kesic
Center for Environmental Medicine, Asthma, and Lung Biology, University of North Carolina Chapel Hill, North Carolina, United States of America

Megan Meyer
Center for Environmental Medicine, Asthma, and Lung Biology, University of North Carolina Chapel Hill, North Carolina, United States of America
Department of Microbiology and Immunology University of North Carolina Chapel Hill, North Carolina, United States of America

Rebecca Bauer
Center for Environmental Medicine, Asthma, and Lung Biology, University of North Carolina Chapel Hill, North Carolina, United States of America

Curriculum in Toxicology, University of North Carolina Chapel Hill, North Carolina, United States of America

Ilona Jaspers
Center for Environmental Medicine, Asthma, and Lung Biology, University of North Carolina Chapel Hill, North Carolina, United States of America Curriculum in Toxicology, University of North Carolina Chapel Hill, North Carolina, United States of America Department of Pediatrics University of North Carolina Chapel Hill, North Carolina, United States of America Department of Microbiology and Immunology University of North Carolina Chapel Hill, North Carolina, United States of America

Raul R. Cordero, Alessandro Damiani, Jorge Ferrer and Mario Tobar
Departamento de Física, Universidad de Santiago de Chile, Santiago, Chile

Jose Jorquera
Departamento de Física, Universidad de Santiago de Chile, Santiago, Chile

Escuela Superior Politécnica del Litoral, Guayaquil, Ecuador

Fernando Labbe
Departamento de Ingeniería Mecánica, Universidad Técnica Federico Santa María, Valparaíso, Chile

Jorge Carrasco
Dirección Meteoroló gica de Chile, Santiago, Chile

David Laroze
Instituto de Alta Investigación, Universidad de Tarapacá , Arica, Chile

James K. Chambers, Kazuyuki Uchida, Tomoyuki Harada, Masaya Tsuboi and Hiroyuki Nakayama
Department of Veterinary Pathology, Graduate School of Agricultural and Life Sciences, the University of Tokyo, Tokyo, Japan

Masumi Sato
National Institute of Animal Health, Ibaraki, Japan

Masahito Kubo
Laboratory of Veterinary Pathology, Joint Faculty of Veterinary Medicine, Yamaguchi University, Yamaguchi, Japan

Hiroaki Kawaguchi and Noriaki Miyoshi
Laboratory of Veterinary Histopathology, Joint Faculty of Veterinary Medicine, Kagoshima University, Kagoshima, Japan

Hajime Tsujimoto
Department of Veterinary Internal Medicine, Graduate School of Agricultural and Life Sciences, the University of Tokyo, Tokyo, Japan

Scott L. Graham
School of Biological Sciences, University of Canterbury, Christchurch, New Zealand
Landcare Research, Lincoln, New Zealand

Jason M. Tylianakis
School of Biological Sciences, University of Canterbury, Christchurch, New Zealand
Department of Life Sciences, Imperial College London, Silwood Park Campus, Ascot, Berkshire, United Kingdom

John E. Hunt, Peter Millard, Tony McSeveny and David Whitehead
Landcare Research, Lincoln, New Zealand

Pei-Fen Liao
Department of Pediatrics, Chung Shan Medical University Hospital, Taichung City, Taiwan R.O.C

Ko-Hsiu Lu
School of Medicine, Chung Shan Medical University, Taichung City, Taiwan R.O.C

Hui-Hsien Pan, Hai-Lun Sun, Min-Sho Ku, Ji-Nan Sheu and Ko-Huang Lue
Department of Pediatrics, Chung Shan Medical University Hospital, Taichung City, Taiwan R.O.C
School of Medicine, Chung Shan Medical University, Taichung City Taiwan R.O.C

Chun-Tzu Chen
Department of Pediatrics, Chung Shan Medical University Hospital, Taichung City, Taiwan R.O.C
Department of Health Policy and Management, Chung Shan Medical University, Taichung City, Taiwan R.O

Jar-Yuan Pai
Department of Health Policy and Management, Chung Shan Medical University, Taichung City, Taiwan R.O.C

Jing-Yang Huang
Institute of Public Health, Department of Public Health, Chung Shan Medical University, Taichung City, Taiwan R.O.C

Madeleine S. Günthardt-Goerg and Pierre Vollenweider
Forest Dynamics. Swiss Federal Research Institute WSL, Birmensdorf, Switzerland

Carlos Calderón Guerrero
Department of Silvopasture, Faculty of Forest Engineering (EUIT Forestal), Universidad Politécnica de Madrid, Madrid, Spain

Jing Peng and Li Dan
START Temperate East Asia Regional Center and Key Laboratory of Regional Climate-Environment for Temperate East Asia, Institute of Atmospheric Physics, Chinese Academy of Sciences, Beijing, China

Mei Huang
Key Laboratory of Ecosystem Network Observation and Modeling, Institute of Geographical Sciences and Natural Resources Research, Chinese Academy of Sciences, Beijing, China

Jennifer M. Albertine
Harvard Forest, Harvard University, Petersham, MA 01366, United States of America

William J. Manning and Michelle DaCosta
Stockbridge School of Agriculture, University of Massachusetts, Amherst, MA 01003, United States of America

Kristina A. Stinson
Department of Environmental Conservation, University of Massachusetts, Amherst, MA 01003, United States of America

Michael L. Muilenberg and Christine A. Rogers
Environmental Health Sciences, School of Public Health and Health Sciences, University of Massachusetts, Amherst, MA 01003, United States of America

Atte Penttilá and Tomas Roslin
Department of Agricultural Sciences, University of Helsinki, Helsinki, Finland

Eleanor M. Slade
Department of Agricultural Sciences, University of Helsinki, Helsinki, Finland
Department of Zoology, University of Oxford, Oxford, United Kingdom

Asko Simojoki
Department of Food and Environmental Sciences, University of Helsinki, Helsinki, Finland

Terhi Riutta
Environmental Change Institute, University of Oxford, Oxford, United Kingdom

Kari Minkkinen
Department of Forest Sciences, University of Helsinki, Helsinki, Finland

Navya Mastanaiah and Subrata Roy
Applied Physics Research Group (APRG), Department of Mechanical and Aerospace Engineering, University of Florida, Gainesville, Florida, United States of America

Judith A. Johnson
Department of Pathology, Immunology and Laboratory Medicine, College of Medicine and Emerging Pathogens Institute, University of Florida, Gainesville, Florida, United States of America

Rafaela T. P. Sant'Anna and Roberto B. Faria
Instituto de Química, Universidade Federal do Rio de Janeiro, Rio de Janeiro, RJ, Brazil

Elisa Vanzo1, Andrea Ghirardo and Jörg-Peter Schnitzler
Research Unit Environmental Simulation, Institute for Biochemical Plant Pathology, Helmholtz Zentrum München, Neuherberg, Germany

Juliane Merl-Pham and Stefanie M. Hauck
Research Unit Protein Science, Helmholtz Zentrum München, Neuherberg, Germany

Christian Lindermayr, Werner Heller and Jörg Durner
Institute for Biochemical Plant Pathology, Helmholtz Zentrum München, Neuherberg, Germany

Hayley Saul and Oliver E. Craig
BioArCh, University of York, York, United Kingdom

Marco Madella
Institució Catalana de Recerca i Estudis Avanc ats, Institución Milái Fontanals, Spanish National Research Council, Barcelona, España,

Anders Fischer
Danish Agency for Culture, Copenhagen, Denmark

Aikaterini Glykou
Institute of Prehistoric and Protohistoric Archaeology, University of Kiel, Kiel, Germany

Sönke Hartz
Stiftung Schleswig-Holsteinische Landesmuseen, Schlob Gottorf, Schleswig, Germany

Alejandro Tabas-Diaz and Marcin Budka
Faculty of Science and Technology, Bournemouth University, United Kingdom

Emili Balaguer-Ballester
Faculty of Science and Technology, Bournemouth University, United Kingdom
Bernstein Center for Computational Neuroscience, Medical Faculty Mannheim and Heidelberg University, Mannheim, Germany

David I. Kasahara, Alison S. Williams, Leandro A. Benedito, Lester Kobzik and Stephanie A. Shore
Department of Environmental Health, Harvard School of Public Health (HSPH), Boston, Massachusetts, United States of America

Barbara Ranscht
Department of Neurosciences, University of California San Diego, San Diego, California, United States of America

Christopher Hug
Division of Pulmonary Medicine, Children's Hospital Boston, Harvard Medical School (HMS), Boston, Massachusetts, United States of America

Takahiro Nakajima, Taisuke Kudo, Ryo Ohmura and Yasuhiko H. Mori
Department of Mechanical Engineering, Keio University, Yokohama, Japan

Satoshi Takeya
Research Institute of Instrumentation Frontier, National Institute of Advanced Industrial Science and Technology (AIST), Tsukuba, Japan

Guoan Shi
College of Agriculture, Henan University of Science and Technology, Luoyang, China

Zhiyong Li
College of Agriculture, Henan University of Science and Technology, Luoyang, China
Institute of Forest Ecology, Environment and Protection, Chinese Academy of Forestry, Beijing, China

Yanhui Wang and Zhen-Hua Li
Institute of Forest Ecology, Environment and Protection, Chinese Academy of Forestry, Beijing, China

Yuan Liu
University of Chinese Academy of Sciences, Beijing, China

Hao Guo
Institute of Desertification Studies, Chinese Academy of Forestry, Beijing, China

Tao Li
Hebei University, Baoding, China

J. Alexander Maxwell
Institute for a Sustainable Environment, Clarkson University, Potsdam, New York, United States of America

Thomas M. Holsen
Department of Civil and Environmental Engineering, Clarkson University, Potsdam, New York, United States of America

Sumona Mondal
Department of Mathematics, Clarkson University, Potsdam, New York, United States of America

Joel A. Mathews, Alison S. Williams, Jeffrey D. Brand and Allison P. Wurmbrand, Lucas Chen Fernanda M.C. Ninin, Huiqing Si, David I. Kasahara, Stephanie A. Shore
Molecular and Integrative Physiological Sciences Program, Department of Environmental Health, Harvard School of Public Health, Boston, Massachusetts, United States of America

Poliana Cardoso-Gustavson
Programa de Pós-Graduaçãoem Biodiversidade Vegetal e Meio Ambiente, Instituto de Botânica, São Paulo, São Paulo, Brazil

Vanessa Palermo Bolsoni, Debora Pinheiro de Oliveira, Maria Tereza
Gromboni Guaratini and Silvia Ribeiro de Souza
Núcleo de Pesquisa em Ecologia, Instituto de Botânica, São Paulo, São Paulo, Brazil

Marcos Pereira Marinho Aidar and Mauro Alexandre Marabesi
Núcleo de Pesquisa em Fisiologia e Bioquímica, Instituto de Botânica, São Paulo, São Paulo, Brazil

Edenise Segala Alves
Núcleo de Pesquisaem Anatomia, Instituto de Botânica, São Paulo, São Paulo, Brazil

Bin Zhao, Shuting Dong, Jiwang Zhang and Peng Liu
State Key Laboratory of Crop Biology, College of Agronomy, Shandong Agricultural University, Taían, China

Index